First South Carolina Cavalry

Robert J. Driver, Jr.

HERITAGE BOOKS
2017

HERITAGE BOOKS
AN IMPRINT OF HERITAGE BOOKS, INC.

Books, CDs, and more—Worldwide

For our listing of thousands of titles see our website
at
www.HeritageBooks.com

Published 2017 by
HERITAGE BOOKS, INC.
Publishing Division
5810 Ruatan Street
Berwyn Heights, Md. 20740

Copyright © 2017 Robert J. Driver, Jr.

Heritage Books by the author:
Confederate Sailors, Marines and Signalmen from Virginia and Maryland
First and Second Maryland Cavalry, C. S. A.
First South Carolina Cavalry
The First and Second Maryland Infantry, C.S.A.

All rights reserved. No part of this book may be reproduced or transmitted in any form or by any means, electronic or mechanical, including photocopying, recording or by any information storage and retrieval system without written permission from the author, except for the inclusion of brief quotations in a review.

International Standard Book Number
Paperbound: 978-0-7884-5782-1

First South Carolina Cavalry

This book is dedicated to the memory of those gallant Confederate soldiers who fought in the 1st South Carolina Cavalry. They served bravely for four years, fighting for the cause they believed in. Their sacrifice and courage should never be forgotten.

1st SOUTH CAROLINA CAVALRY

1st PALMETTO CALVARY REGIMENT

With the secession of South Carolina, Governor Francis W. Pickens authorized the enlistment of four companies of cavalry. These four troops became part of the 1st Battalion of South Carolina Cavalry and the nucleus of the 1st South Carolina Cavalry regiment. These companies served in the coastal areas around Charleston.

The 1st South Carolina Battalion of Cavalry was authorized by the Confederate Secretary of War on 31 October, 1861. Lieutenant Colonel John Logan Black, who had attended West Point, was appointed commander. The companies were Captain's James Davis Trezevant's "Fort Motte Rangers", from Colleton and Orangeburg counties, Moses T. Owens' "Abbeville Troop," from Abbeville County, Niles G. Nesbitt's "Ferguson Rangers, from Spartanburg and Laurens counties, John David Twiggs '"Edgefield Rangers," from Edgefield and Barnwell counties, and William Alexander Walker's "Chester Troop" from Chester County. Owens' Troop arrived in Charleston on September 22nd, and served on Sullivan's Island. Nesbitt's company was posted on Edisto Island on October 19th. Trezevant's, Twiggs' and Walker's companies were stationed at Adams Run, near Charleston.

The Abbeville Troop became Company A, Nesbitt's Company B, Twiggs 'Company C, Walker's Company D, and Trezevant's Company E of the Battalion.

During early November, 1861 The "Fort Motte Rangers" moved to Coosawatchie on the 6th, Bee's Creek on the 9th and then to Fort Moultrie on the 14th. They were later ordered back to Adams Run.

Walker's Troop was at Mars Bluff on November 2nd and marched to Florence on the 4th, then to Camp Marion and later to Cut Island on the 19th.

The first mention of the 1st Battalion was by General Robert S. Ripley, commanding the Provisional Forces of South Carolina at Coosawhatchie. was on November 18, 1861. He reported Captain Trezevant's company of cavalry was in advance of Colonel Thomas L.

Clingman's North Carolina Regiment in observation of Boyd's Point and Tenny's Landings.

Colonel Arthur M. Manigault, commanding the 1st Military District of South Carolina at Georgetown, reported on 27 December, 1861, that "Captain Walker's company of horse, Black's battalion, under orders from General [Robert E.] Lee's headquarters, has been ordered to report to Colonel Black, and is now on the road to join him." Lee had been appointed to the command the Department of South Carolina, Georgia and East Florida on November 5, 1861. He had taken command at Coosawhatchie on November 8, 1861.

Lee wrote General Nathaniel G. Evans, commanding Third Military District of South Carolina, at Adams Run, on 24 December, 1861, noting that the 1st South Carolina Battalion of Cavalry, under Lieutenant Colonel Black, was part of Evans command. Black had the battalion headquarters at Adams Run, with detachments scouting and patrolling the coast and outer islands near Charleston.

A correspondent for the 'Yorkville Enquirer,' wrote in "News from the Coast" on January 2nd from Wadmalaw Island, "About 10 o'clock, Capt. Twigg and his men, of the Edgfield Mounted Company, acting as scouts, came hurriedly from Rockville, to our camp, and reported to Lt. Colonel Gaillard that one of their men, who had been on lookout, reported the landing of a small body of the enemy numbering sixty or eighty, at Rockville.

Watching his opportunity, he had suddenly faced the enemy, fired and killed his man. Turning his horse somewhat suddenly, in order to elude his pursuers, he was then, when his horse ran in the direction of Rockville. On recovering, he found his shoulder was dislocated, but managed to secrete himself in a thick wood, and has thus been unable to communicate with his captain. He also reported the enemy advancing." The trooper was not identified.

The same issue of the newspaper published a CARD reading "I want four more Companies of Cavalry to complete my REGIMENT in Confederate Service for the war. These in addition SIX under my command, will constitute the 1st PALMETTO CAVALRY REGIMENT in Confederate Service. From 50 to 72 rank and file, is the desired number. $12 per month is allowed for the hire of a HORSE, for each man, making the pay for each private $24. Corporals, $25. Sergeants, $27. 1st Sergeant, $32. It is desired that each man furnish a DOUBLE BARRELED GUN, for which he will be paid on entering the service. I hope my native District can afford [sic] me ONE company at least. Companies applying will brought into instant service, well armed, splendidly equipped, furnished with wagons, etc. Persons engaged in raising companies will address

JNO. L. BLACK

Lieut. Col. Commanding 1st Bat. S. C. V.

Adams Run, So. Ca."

Additional companies were assigned to Black's Battalion. Captain Elam Sharpe's Company, the "Allen Hussars," was organized for the war at Picken's Court House on December 4th and were ordered to join the battalion at Adams Run later that month. They were assigned as Company F. Captain Leroy J. "Old Hickory" Johnson's Troop was sworn in at Abbeville Court House on 8 January, 1862, and named Company G. They joined the battalion later in the month. Captain Robin Ap. C. Jones unit was organized at Rock Hill the next day, and joined Black soon after. Most of the company were from York County. The 'Yorkville Enquirer' reported on January 16th

"FOR THE WAR"

We learn that Capt. Robert (sic) Jones Company of Cavalry, from this District, was mustered into service at Indian Land on Thursday last, having completed its organization. The Captain is a gallant gentleman, and we doubt not he will be followed by his command through many brilliant successes, shall opportunity offer."

On February 13th the newspaper reported "The Cavalry Company that was to have left last Saturday, did not get off until yesterday (Wednesday) morning, when they left Yorkville and Rock Hill simultaneously, concentrating at Chester. They numbered 72, and have as fine animals and as clever officers and men as the greatest discipline would require. The following are the officers – the roll will be furnished shortly.

R. Ap. P. Jones Captain

J. H. Barry 1st Lieutenant

James White (&) C. A. Abell 2nd Lieutenants

Their destination is   Adams Run, S. Carolina.

The "Round O Troop," led by Captain John Richard Perry Fox, enlisted at Parker's Ferry in Colleton County on April 3, 1862. The officers and men were from Colleton and York counties. Fox and many of his officers and men had served in Captain Sheridan's company of the 11th South Carolina Infantry, which had been disbanded. They were designated as Company I.

With the increased size of the battalion, Twiggs was appointed Major, and Thomas W. Whaley was elected Captain of Company C. Henry L. Mayson became the Quartermaster, and Henry M. Faust Assistant Surgeon of the battalion.

Black's Battalion remained headquartered at "Camp Abbeville," Adams Run from January through April, 1862. Companies A and C were on duty at Wadmalaw Island, Company B on John's Island, while the remaining companies remained at Adam's Run. These companies were not idle, as many of the officers and men were detailed as couriers, scouts and other duties in the Charleston area.

The 'Yorkville Enquirer' reported "Col. John Logan Black, who commands the first Regiment of South Carolina Cavalry, for the war, organized mainly through his influence and exertion, has designed a Skeleton Saddle, which requires very little leather."

General Ripley reported to General Lee on February 16, 1862 from Charleston "Two companies of Black's cavalry have reported lately, which will join the headquarters of the regiment as soon as they can be equipped. Capt. [Robin] Ap C. Jones, of Black's regiment, has found about 20 army revolvers and some 10 Enfield rifles, which have been in the hands of merchants and gun dealers for some time. He has purchased them, trusting to have his purchase authorized." Black had 9 companies present.

Company E, was operating on John's Island by May 6th. They were scouting the island and standing picket to observe any Federal advances. On May 19th they were assigned as General Evans escort.

Evans reported that two companies of cavalry were part of his expedition to drive the enemy off John's and adjoining islands. They forced the Union troops to their gunboats and brought in 200 negroes from the plantations to Church Flats. Evans left one company of infantry and one of cavalry to continue removing the slaves and to destroy the remaining cotton on the island. "When this work is finished I will withdraw all the force except four companies of cavalry." These would be from the 1st South Carolina Battalion of cavalry.

Companies A and D remained on Wadmalaw Island during May and June, while the rest of the battalion were on or around John's Island.

Lieutenant Robert W. Crawford of Company D, reported a skirmish with the Federals on John's Island on June 7th "I was placed in command of the picket post at the fork of the roads leading to Legareville and Haulover Bridge, which picket numbered 26 men. Not expecting to be placed on picket when we left camp, we had not provided ourselves with forage and provisions, and consequently were reduced to the necessity of sending men back to each company to procure them, which, however, was not done without the consent of the captain commanding the regiment, and which reduced us to 18 men. My orders when stationed there were to remain at the forks of the roads with the main body of the picket, and to place two vedettes down each road 2 miles distant, with orders to report to me on the first appearance of the enemy, which was done immediately when the enemy appeared and fired on the vedettes on the Legareville road. I sent couriers forthwith to report to Colonel [John H.] Means [17th South Carolina Infantry] (who with his regiment was only 2 miles in the rear) and to the other commands on the island, and also a courier down the Haulover road for the two vedettes who were 2 miles distant, while I with the remainder of the pickets advanced down the Legareville road and met the enemy, whose force I discovered to consist of a company of cavalry and detachment of infantry (or men on foot). By concealing my force in the woods I managed to engage and detain him until my vedettes on the other road arrived, after which we retreated in good order, firing as we went. The enemy's cavalry, perceiving that we were retreating, charged us, and three or four of my men left and fled. Fortunately for them I do not know their names and am unable to report them, as they justly deserve. The men were all strangers to me except three or four, as our companies had not been together but a few days, and there were only two men from my own company. I learned through Colonel Means that the three men that left me did not stop when they met him with his regiment, but ran through and reported that the enemy, were just behind, which contributed to his mistaking us for the enemy as he did, for he opened fire on us as soon as we came in range, and we were then exposed to his fire and that of the enemy too." Crawford reported 1 man killed and 8 wounded, along with 7 horses killed. The Union losses were 2 killed and 5 wounded, all of the 46th New York Infantry.

The next day Evans advanced part of his command to capture a Federal cavalry picket left at the bridge over Aberpoolie Creek, but found the enemy had retreated to Legardeville. On the 10th he withdrew his infantry, leaving the six companies of the 1st Battalion and the Stono Scouts to guard and scout the islands.

The 1st Battalion remained in the 3rd Military District of South Carolina during the rest of May. The companies were still on John's and Wadmalaw Islands.

The 10th company was not assigned until 25 June, 1862, when the battalion was increased to a regiment. Company K was led by Captain Angus P. Brown, originally a Lieutenant in Company C. The Troop was made up of excess men from companies C, D, and F, plus new recruits. They were stationed on John's Island. Black had been promoted to Colonel in January, although still listed as a Lieutenant Colonel in some reports. Captain Twiggs, promoted to Major on May 26th, was promoted to Lieutenant Colonel on 25 June. Captain Walker was appointed Major, and John Simonton Wilson was elected Captain of Company D. Charles Henry Ragsdale, a graduate of the Citadel, was appointed Adjutant. Captain Mayson remained as Quartermaster. Charles Pinckney was named Surgeon, assisted by Henry Medicus Faust and William Moscow Innabnett.

The 1st Regiment remained under General Evans command, which was re-designated the 2nd Military District of South Carolina, in July. Evans and most of the infantry regiments were organized into a brigade, and ordered to Virginia. Colonel Johnson Hagood took command of the 2nd District.

Company F marched from Adams Run to Logansville on August 5th. They remained there until returning to Adams Run on the 23rd. Company I was already posted at Logansville.

On August 28th companies' B and E were ordered to Edisto Island to bring off all the cattle, horses and mules. They continued to scout the island until September 5th, when they returned, driving 65 cattle, 36 horses and mules across the bridge at Walts Cut. The two companies remained on picket there until ordered to Adams Run.

The Confederate War Department ordered the 1st Cavalry to Virginia. General John C. Pemberton, commanding the Department of South Carolina from Charleston, replied to General Samuel Cooper on September 20th "The removal of Colonel Black's cavalry regiment from this department deprives Charleston of two-thirds of the cavalry for its defense, and leaves the entire coast in the Second Military District without a single cavalryman." The War Department responded by ordering the 6th Battalion of South Carolina Cavalry to the coast.

Colonel Black sent his companies to Virginia in staggered order and by different routes. Most of the horses were driven overland to Staunton, Virginia, while the men traveled by train to Richmond and Staunton. Company B left Summersville, South Carolina on October 1st and arrived at Camp Brown, Staunton on the 23rd. Company C departed on the 23rd of September, but arrived the same day as Company B. Company D left on the 1st of October, arrived Richmond on the 6th, and reached Staunton on the 22nd and 23rd. Their horses left Summersville on September 29th, marched via Richmond, but didn't arrive in Staunton until October 30th. Company E's horses arrived on the 29th. Company F reached Staunton on the 22nd. Company G left Legares Point on September 23rd, reached Richmond on October 7th and then on to Staunton. Company H departed Summersville on September 24th and the men reached Staunton on the 19th of October, but their horses didn't arrive until November 1st. Company I departed the same day, reached Richmond on the 1st of October, but didn't reach Staunton until the 22nd and 23rd. Company K followed the same route and arrived on the 22nd.

The regiment remained at Camp Brown, near Staunton, until being ordered to join the Army of Northern Virginia near Culpeper Court House. They arrived at Camp Whatley in staggered order, over the period of 19-26 November. On November 10th the regiment had been assigned to

General Wade Hampton's Brigade, along with the 2nd South Carolina, 1st North Carolina and Cobb's and Phillip's cavalry legions from Georgia.

General Robert E. Lee wrote General Thomas J. Jackson on the 18th of November that he had ordered the 1st South Carolina Cavalry and a Virginia battery to Gordonsville to protect the town and railroad against any Union raid.

The 1st moved to a camp near Raccoon Ford by early December. On December 10th Hampton made a raid on the Union supply line between Dumfries and Occoquan. Lieutenant Colonel Twiggs led the detachment of the 1st Cavalry on this foray. Arriving near the Telegraph Road between the two towns on the 12th, Hampton divided his force of 520 men. Twiggs and his men were placed under the command of Lieutenant Colonel William T. Martin of the Jeff Davis Legion. After a forced march of 16 miles, Hampton led an attack on Dumfries. "There was only a small force there, and so complete was their surprise that every man was captured after a few shots were fired," he reported. Martin's detachment, which had made a detour around the town, was not engaged. Hampton continued "There was no loss on our side. We captured 50-odd prisoners, with 1 lieutenant and 24 sutler's wagons. "After destroying 2 wagons for lack of horses, the rest were brought off, but 5 broke down on the road and were abandoned. Hampton returned with 17 wagons and the telegraph operator, with his battery. He planned to capture the pickets posted between the two towns, but found a Union corps on the road, and made a 40-mile march to Morrisville that night."

Hampton concluded his report "I can again speak in the highest terms of the conduct of my officers and men. They bore the privations and fatigue of the march – three nights in the snow – without complaint, and were always prompt and ready to carry out my orders. "

Colonel Josiah H. Kellogg, of the 17th Pennsylvania Cavalry, reported that a lieutenant and 30 men of the 10th New York Cavalry were captured at the crossing of Neabsco Creek, and 5 men of the 2nd Pennsylvania Cavalry were taken at the ferry.

On December 17th Hampton made another raid on the Telegraph Road. Colonel Black led the 100 men from the 1st South Carolina Cavalry. Hampton led his command across the railroad crossing on the Rappahannock via Cole's Store. Black led his men and those of the

Phillips and Cobb Legions to attack and capture the Federal pickets at Kanky's Store on Neabsco Creek. Surrounding the post, the entire Union detachment was taken along with 8 wagons, although only 2 had stores in them. Splitting his command into 3 detachments, Hampton moved his men towards Occoquan. Colonel Black led his men and the troopers of Cobb's Legion by the Bacon Race road, so as to get above the town. One detachment, under Major William G. Delony of the Phillips Legion, swept down the Telegraph Road capturing about 20 men and 2 wagons. Learning that a large Federal cavalry force was approaching, Hampton had the wagons and prisoners ferried across the river, while a small force of sharpshooters held back the enemy advance. The Confederate marksmen soon were able to cross the river and held the Federals in check for several miles after they crossed. Hampton marched via Greenwood Church and Cole's Store and camped at Tacket's Fork, on Cedar Run.

Hampton continued "The next day's march brought me safely home without the loss of a man. I brought back about 150 prisoners, besides 7 paroled, 20 wagons with valuable stores in them, 30 stands of infantry arms, and 1 stand of colors. I take great pleasure in commending the conduct of my officers and men. I could desire nothing better on their part, and I feel justified in expressing my pride and confidence in them."

General James E. B. Stuart, commanding the Confederate cavalry, decided to make a much larger raid on the Federal supply lines on the Telegraph Road near Dumfries and Occoquan. Hampton's Brigade furnished 870 officers and men, with 150 of the 1st South Carolina under Captain William A. Walker. The brigade departed on December 27th, taking the route via Cole's Store. Hampton's advance found the area picketed by 15 men. Four were captured by Hampton's advance and the rest picked up by a Virginia regiment. Colonel Matthew C. Butler of the 2nd South Carolina Cavalry, with detachments from the 1st South Carolina and Phillip's Legion, pushed on to Occoquan, where they captured 12 to 19 prisoners from the 17th Pennsylvania Cavalry, 5 wagons loaded with forage and camp equipment, and 2 sutler's wagons, losing only one man wounded. The other brigades not joining him, Hampton withdrew to Cole's Store for the night. On December 28th, Hampton again advance toward Occoquan, sending Butler's command toward Bacon Race Church. Butler ran into a large Federal force, including artillery, which he skirmished with. The rest of the brigade not arriving, he pulled his men out of the fight and retired up the Brentsville road. Butler was later able to rejoin Hampton, with the loss of two men wounded and 8 horses which had to be abandoned. He complimented his command "I cannot speak in terms too high of the conduct of the officers and men. Their behavior could not have been better. They moved forward with unhesitating gallantry and spirit, and rallied at the command without confusion, and promptly faced the enemy under a rapid and terrible fire." The command returned to camp on 1 January, 1863.

While Hampton had been on the raid with part of his brigade, the officers and men left behind had been picketing at Ellis' and Richards' fords on the Rappahannock. Colonel James Barnes of the 18th Massachusetts Infantry reported on the Federal advance to the fords on December 30-31st. Marching towards Ellis's Ford "Mounted pickets belonging to the First South Carolina Cavalry were stationed along the road, who retired as the column advanced, occasionally turning and firing, but the prompt charges of the cavalry in advance of the column dispersed them without difficulty, although their numbers were three or four times our own. In this way the column proceeded to Ellis' Ford without any other conflict than that which was constantly occupying the advanced cavalry. In these skirmishes we succeeded in capturing two of the enemy's pickets, with their horses; but one of the horses, being wounded, was left behind. ... As in the case at Richards' Ford, this was found to consist of a small detachment of the 1st South Carolina Cavalry only."

Captain Johnson of Company G reported "During the month of December the Company as a part of Genl. Hamptons Brigade has been actively engaged in several raids within the enemy's lines which has been attended with a gratifying success." Captain Whatley of Company C reported the same results.

The 1st remained camped near Hamilton's Crossing, close to the railroad, in order to obtain supplies by rail. They continued to picket along the Rappahannock for the first two months of 1863. Their horses suffered from a lack of forage. March and early April were spend recruiting their horses at Camp Hepburn in Lunenburg County. Company E, for example, reported 16 horses disabled and 4 had died. However, only four men were dismounted and not available for duty.

1st Sergeant A. Martin N. Kee, of Company H, wrote from "Camp Oakland," Amherst County on March 30th "We have been constantly on the march for about 5 weeks. We are lying by for a day of rest and we expect to go even farther South. We are merely moving to recruit our horses for they are entirely broken down. We have had a hard time of it on the Rappahannock. We were on several raids and about half our men were constantly out on picket duty which was very hard for we have had the hardest winter I ever saw. Snow often 4 feet deep where it had drifted."

During the period of 4-6 April Company F moved back to near Stevensburg in Culpeper County. Company K didn't depart Lunenburg County until 15 May and rejoined the regiment the next day.

Seeing the plight of the citizens of Fredericksburg who were forced to evacuate their homes, the 1st donated $27.00 to the Fredericksburg Relief Fund near the end of April. The refugees were camped in the woods without shelter, proper clothing and provisions.

The 1st was assigned picket duty at Rogers Ford on May 20th. The orders included instructions that one company would be stationed at the ford for several days at a time.

General Stuart reviewed Hampton's, Fitzhugh Lee's and William H. F. Lee's brigades on May 22nd, near Culpeper Court House. Frank S. Robertson, of Stuart's staff wrote "The grand Cavalry Review took place this morning and was one of the most imposing scenes I have ever seen." Lieutenant William R. Carter of the 3rd Virginia Cavalry agreed "The most magnificent sight I have ever witnessed. A beautiful day & quite a turnout of the ladies, considering the times." Another Virginia officer noted "A fine & very imposing affair – 8,000 men mounted." Major Henry B. McClellan of Stuart's staff wrote "Eight thousand cavalry passed under the eye of their commander, in column of squadrons, first at a walk, and then at the charge, with the guns of the artillery battalion, on the hill opposite the stand, gave forth fire and smoke, and seemed almost to convert the pageant into real warfare. It was a brilliant day, and the thirst for the 'pomp and circumstance' of war was fully satisfied."

Black reported 37 officers and 375 men present for duty on May 25th. 2 officers and 144 men were listed as non –effective and 9 officers and 129 men were absent, many on horse detail.

On May 29 Hampton wrote "In pursuance of orders from Div Hd Qt for a detail of Dismounted Men Col Black Comdg 1st So. Ca. Regt will furnish a detail of Ten (10) dismounted men to report at once as Provost Guard to Lt H. T. Moyler Provost Marshall Culpeper."

General Robert E. Lee wanted to review the cavalry, as he was always concerned about the condition of the horses, for which he had exerted great effort to bring into trim for a planned advance into Maryland. Learning that Lee had ordered the review, the grumblers, who considered it a waste of time and horse flesh, ceased their comments. Stuart chose the John Minor Botts farm, "Auburn," for the review. Botts, a former Congressman, was a staunch Unionist. The review took place on June 8th. Lee, with Generals James Longstreet, Richard S. Ewell and John Bell Hood, were among the large number of reviewers. Hood had arrived with his entire division. William W. Blackford of Stuart's staff noted "Many men from Hood's Division were present and enjoyed it immensely. During the charges past the reviewing stand the hats and caps of the charging column would sometimes blow off, and then, just the charging column passed and before the owners could come back, Hood's men would have a race for them and bear them off in triumph."

J. M. Monie, a North Carolina trooper, wrote after the war "Marshaled on the plain were one hundred and five squadrons [22 regiments and battalions] of cavalry, six light field batteries, ambulances with their trained corps of field attendants, etc. At no time of the war was this branch of service in such fine condition. In members, discipline, and equipment were good enough to create within the mind of our great commander the utmost satisfaction." The line of units was 3 miles long.

Lee congratulated Stuart, and discussed the arms and equipment of this mounted force. Stuart complained of the poor quality of the saddles manufactured in Richmond, which ruined the backs of the horses, and the defective carbines being sent to them. Lee personally wrote Colonel Josiah Gorgas, the Confederacy's Chief of Ordnance, suggesting improvements.

Meanwhile, across the Rappahannock, the Federal cavalry, under General Alfred Pleasonton, had been scouting the movements of Stuart's cavalry. His force of nine thousand infantry and three thousand cavalries, were poised to cross the river and attack the Confederates.

Following the review, Stuart had posted Fitzhugh Lee's Brigade to cover Stark's Ford, William H. F. "Rooney" Lee's Brigade near Welford's Ford, William E. "Grumble" Jones' Brigades held Beverly Ford and Beverly Robertson's Brigade at Kelly's Ford. Hampton's Brigade, with the 1st South Carolina, was posted near Brandy Station. The regiment was the "Grand Guard" for the brigade.

General John Buford 's cavalry led the Union advance across Beverly Ford in the early dawn of June 9th. The Confederate pickets were alert, and fired on the Federals as they splashed across the river. They had erected log and brush barricades across the road to slow any Union advance. After a brief skirmish the troopers of the 6th Virginia Cavalry fell back down the road, firing as they retreated. Brethed's battery of horse artillery was camped near the road and were alerted by the firing, and had their guns ready for action. Part of the 6th Virginia counterattacked down the road, further delaying the Federal advance, and giving the artillery time to withdraw to a position near St. James Church. Couriers arrived at Stuart's Headquarters on Fleetwood Hill, and he and his staff rode toward the scene of the action, passing retreating wagons trying to get away from the fight. Stuart helped rally Jones's unit and was soon joined by the two Lee's and portions of their commands.

Black recalled "I had ordered my Regt. to be mounted at dawn as we were relieved without further orders at sunrise. In these grand guards the men except the deploy[ed] horse guards & out pickets were allowed to sleep. The horses were kept saddled, girths loosened so as to rest the animals. About dawn I was awakened & told by Chaplain [Richard] Johnson that ... some firing could be heard in our front. I was up and in my saddle in a moment & by this time could hear a few guns over and beyond Gen'l Stuart's Headquarters, which were on a hill, one fourth of a mile in my front. Ordering the Adjt. to form the Regiment and follow me with it, I galloped over to Stuart's Hdquar.... I found Gen'l. Stuart standing in the road in front of the house he was quartered in, & I think Hart's [South Carolina] Battery was parked in the yard."

Stuart ordered Black to move forward down the road leading to Rappahannock Station. "I here found the first squadron of my regiment Captains Owens and Jones dismounted and skirmishing with the enemy on my left. This squadron had been ordered to this point on the evening of the 8th, and remained there through the night. I immediately dismounted a party of sharpshooters from the fifth squadron, and moved them forward to support this squadron, at the request of Major [C. E.] Flournoy, commanding the [6th] Virginia Regiment, drawn up on the right of the line; but about the time the second line came up, a report reached me that the enemy was advancing on the road from Kelly's Ford and Rappahannock Station. Communicating with Major Flournoy, I at once withdrew my second line, and moved to the right, crossing the railroad, and selecting a position at the junction of the roads leading to Kelly's and Rappahannock Station Fords, not knowing at the time that General Robertson's brigade was in front of me.

Soon after I had changed position, Captain Owen, commanding my first squadron, retired from Major Flournoy's left. This was done by a misconstrued order, delivered by a courier. As the ammunition of this squadron was exhausted, I at once replaced it with my fifth squadron (Captains Nesbitt and Fox), which retired from this position, as I am informed, by the direction of the officer in charge of this part of the line."

Black continued "At this time I was ordered to join General Hampton's brigade, on the north side of the railroad. Here, by the direction of General Hampton, I dismounted my fifth squadron, and deployed them as sharpshooters, under Captain J. R. P. Fox; afterward the fourth squadron, under Captains Johnson and Wilson, in command of their respective detachments of sharpshooters; one company of the second squadron, under Lieutenant [F. A.] Sitgreaves, the other company of the squadron (Captain Sharpe) having been deployed as flankers on the extreme right. These companies deployed and moved forward steadily, and although they, with the sharpshooters from the other regiments of the brigade, were charged by the enemy's cavalry, they held their position and charged on foot in return, and held their position until ordered by General Hampton to retire, which they did in proper order, coming out with very few rounds of ammunition in their boxes. "

Captain Fox was wounded in the arm and crawled into a ditch to hide from the enemy, but his big white horse, faithfully followed and stood over him, which gave away his position, causing him to be captured.

"Before my sharpshooters could remount." Black stated, "I moved rapidly to the left, toward Brandy, as ordered, following Colonel [P. M. B.] Young, of the Cobb Legion, to support him. This march was made in a column of squadrons. As the head of the Cobb Legion was near General Stuart's headquarters, the enemy was seen approaching on my then left.

Colonel Young immediately changed the head of the column to the left and charged. A portion of the enemy's force turned to the right, along the railroad, to avoid Colonel Young's column. I immediately changed the head of my column to the half left, and ordered my first squadron to charge, and immediately after ordered the second squadron to charge, changing its direction at right angles to the direction of the first, to intercept the enemy escaping in that direction. Both squadrons charged in gallant order, as well as the second, which was in the rear.

The companies in the charge were Captains Owen, Jones, Trezevant, Whatley, Sharpe and Fox (Lieutenant [Frederick] Horsey commanding the latter, Captain Fox having been previously wounded).

In this charge, the first squadron was separated from the others entirely. The second and third were checked by a cut in the railroad, but the men delivered the fire of their rifles on the retreating enemy with effect. The squadrons were soon joined by the first on the hill, as a support on the left of the batteries. "

The First struck the 2nd New York Cavalry in flank. The Yankees "scattered and headed for the protection of the railroad embankment, pursued by Black's screeching South Carolinians, who, "were cutting down the fugitives without mercy," wrote Frederick Whitaker of the 2nd. He continued "Charge after charge, retreat and advance, rally and scatter, firing, clubbing, cutting, with pistol, carbine and sabre. Batteries of flying artillery were taken and re-taken on either side, till in amidst the surging masses, the ground was strewn for acres around with dead and wounded, horses and riders, blankets, baggage, broken arms and equipment." Captain William W. Blackford of Stuart's staff added "It was a thrilling sight to see these dashing horsemen draw their sabres and start for the hill at a gallop. "The Cobb Legion and the 1st South Carolina Cavalry struck the flanks of the 10th and 2nd New York regiments, causing those Yankees to "float of like feathers in the wind."

"From this position I was ordered to rejoin General Hampton south of the railroad, and, by order, changed position several times," wrote Black. "The fourth change of position brought my regiment into line in the ravine between General Stuart's headquarters and Brandy, on the east side of the run, my right resting on the road. Here I was directly in the rear of our battery, on which the enemy was firing rapidly, and a storm of shells passed over the regiment, one exploding in the column as the regiment was coming into line, but, fortunately, inflicting little damage, though many exploded nearby.

I was here ordered to move to the left, to support Brigadier-General [W. H. F.] Lee, and moved up the ravine for that purpose, and reported to Colonel [John R.] Chambliss, commanding Lee's brigade, before coming into action. Here I was ordered to return to General Hampton, near Brandy, and from thence was ordered to hold the road leading from Brandy to Madden's, where I remained until ordered into camp. "

The Federals finally retreated back across the Rappahannock, having sorely tested the Confederate cavalry and proven to themselves that they could hold their own against the Southern horsemen.

Black continued his report "There were 14 or more prisoners and as many horses captured by the regiment under my command. Some of the latter were turned over on the field, besides arms and equipment."

The loss of the regiment in action was 3 killed, 9 wounded (1 since dead), and five missing. Company A reported Corporal John I. Crawford, Privates James A. Richey and Thomas P. Martin were wounded and Private George W. Cox was slightly wounded and captured.

"I regret to report the fall of Captain Robin Ap. C. Jones, who fell, gallantly leading his company in the charge, near division headquarters. A gallant and accomplished officer, his loss cannot be easily repaired." Captain J. R. P. Jones was severely wounded early in the action while in charge of the sharpshooters from his squadron, which he fought and managed well. I fear he fell into the enemy's hands in leaving the field. "Jones had charged past a Federal cavalryman, who he thought would be taken prisoner, but the man shot him in the back."

Corporal Robert A. Steele recalled "Captain Jones, mounted on his beautiful horse, rode among his men cheering and talking to them, while they were preparing to charge. When the time came to make the attack, Captain Jones, followed by his gallant men, rode forward. Presently, his company came to a meadow and Captain Jones and a few men dashed into it. The others of the Company, seeing the horses getting through the mud with such difficulty, made their way around the marsh to rejoin the others. Just as Captain Jones got out of the marsh, he encountered a dismounted Union Cavalryman, and the Captain struck at him with his sword, cutting of his thumb, then dashed past him in the charge. Immediately, the Union man fired at him and Captain Jones fell from his horse. (I) rushed to the rescue, but too late to save [the] Captain; in time, however, to take the Union man prisoner, and deliver him to General Stuart's headquarters. Upon returning with a detachment and ambulance, (I) found him (my) friend and Captain dead. J. W. Marshall of Rock Hill was present and was later made Captain of the company. The power and fleetness of this animal 'Hercules' accounts for the fact Captain Jones got in advance of him men, after making his way through the marsh."

Black added in the post war that Captain Jones was killed in the charge up Fleetwood Hill. "Chaplain [Richard] Johnson was at the head of the squadron with Jones and myself and his horse was shot but not killed, tho I believe he fell."

"Fully satisfied that every man under my command did his duty," Black concluded his report, "and his whole duty, and at the same time proud that not a man who left camp with the regiment at sunrise left it during the day, without first being wounded or ordered off on duty, until marched back at night."

He recalled "I marched two miles to one of Colonel Slaughter's clover fields. I had rested and opening my ranks unbitted the horses and let them graze till 2 a. m. The men generally tied their halters to one foot, laid down & went to sleep. This grazing saved the horses in my Regiment. A few days before the Battle I got 250 Enfield Rifles and opened them. The men cursed awfully at the idea of carrying Rifles but after the Battle of Brandy they made friends with rifles. In fact, all of our Cavalry should have been armed with rifles as we had no long range carbines."

General Hampton wrote on the charge on Fleetwood Hill "The leading regiments (Cobb's Legion and First South Carolina Cavalry) charged gallantly up the steep hill upon which the enemy were strongly posted, and swept them off in a perfect rout without pause or check. Their guns were abandoned and many of their men killed and wounded."

Stuart praised his command in part "... Your Sabre blows inflicted on that glorious day have taught them again the weight of Southern Vengence. You confronted with Cavalry and Horse Artillery this force –

held the Infantry in check, routed the Cavalry and Artillery and captured three pieces of the latter without loss of a gun, and added six flags to the trophies of the Nation besides inflicting the loss in killed and wounded and missing at least double our own, Causing the entire force to retire beyond the Rappahannock. Nothing but the Enemy's Infantry strongly posted in the woods saved his Cavalry from Capture.... With an abiding faith in the God of Battles and a firm reliance on the sabre, your successes will continue = Let the example and heroism for our lamented fallen Comrades prompt us to renewed vigilance and inspire us with devotion to duty."

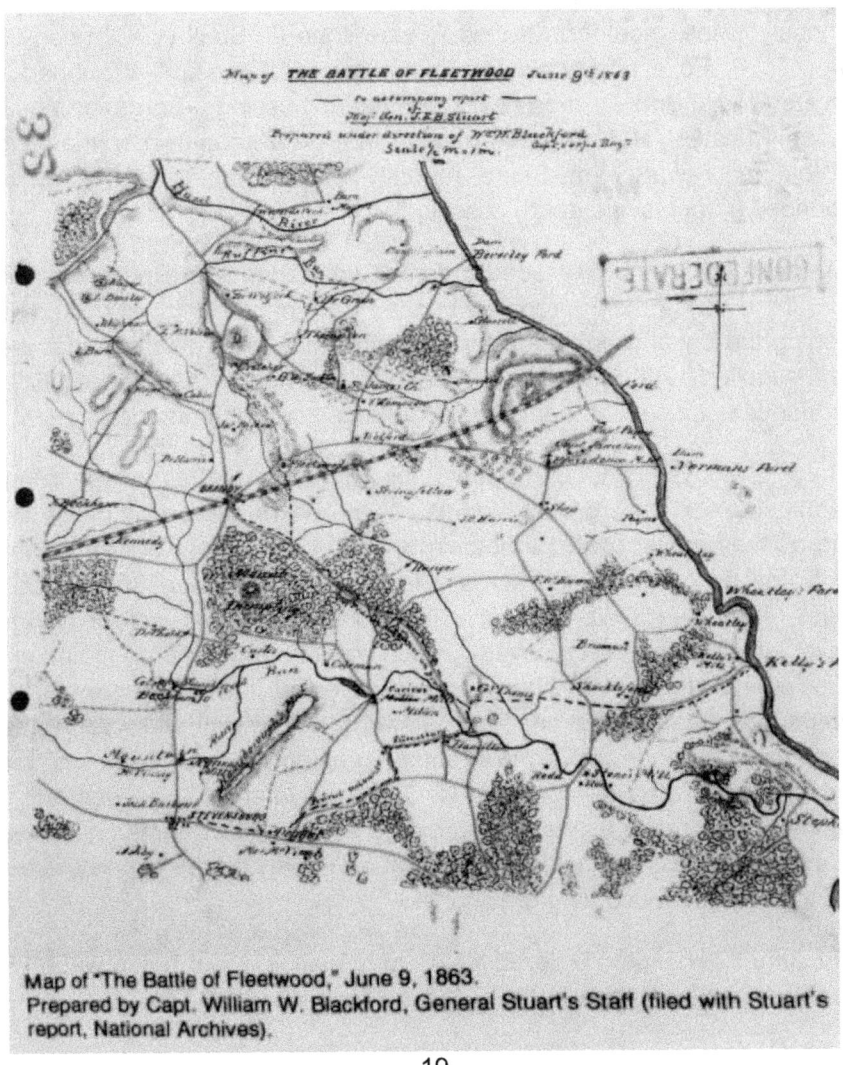

Map of "The Battle of Fleetwood," June 9, 1863.
Prepared by Capt. William W. Blackford, General Stuart's Staff (filed with Stuart's report, National Archives).

Assist Surgeon Joseph Yates sent a casualty list to the "Abbeville Press". "Company A. Captain M. T. Owen, Wounded - Corpl. J. I. Crawford, slightly in the thigh, Missing – Private G. W. Cox. Company B, Capt. N. Nesbitt. Wounded- Private A. P. Willis, in hip seriously. Missing – Private J. W. Keller. Company C, Capt. T. W. Whatley – Wounded – John L. Carey [sic], slightly in the arm. Company D, Capt. Wilson – Killed Corpl. W. T. Wilkes. Wounded – Private A. W. Wade, severely in the hand. Company E, Capt. Trezevant – No casualties. Company F, Capt. Elam Sharpe – Killed Private [Hezekiah] H. Butler. Company G, Capt. L. J. Johnson – Wounded - Private J. S. Williams, arm fractured. Company H. Capt. R. Ap C. Jones killed. Wounded Private H. M. Jackson, seriously in the shoulder. Missing Private John G. Steel(e). Company I, Capt. R. P. Foxe (sic) wounded in the arm and missing. Wounded – Private H. M. Braker, mortally, since dead. Missing –Sergeant W. J. Rhode, Private J. H. Cone. Company K, Lt. F. A. Sitgraves, commanding – Wounded Private Wm. Stone, mortally and missing. Total: Killed 3, Wounded 9, (one since dead) missing 5 – 17."

The next day Hampton issued orders "Commanders of Regiments will have three days rations prepared by their Commands and will make this evening the usual preparation for marching. They will also see that the men supplied with ammunition Forty (40) rounds of Carbine or Rifle Cartridges to be carried in the Cartridge Boxes or their persons."

The 1st enjoyed a brief reprieve following the battle, reorganizing, reequipping and resting their mounts. The break in the action didn't last long as they moved with Hampton's Brigade on the right of Lee's army as it moved northward and across the mountains into the Shenandoah Valley. Stuart's cavalry moved in echelon by brigades, screening the movement from the Union cavalry, which was probing the area to find out if Lee was on the move. Hampton's Brigade was the last to move from camps near Brandy Station. Black stated the regiment was ordered to cross the Rappahannock to Warrenton, keeping the prying eyes of the Federals scouts from locating the Southern infantry. He reported "I marched and grazed as we had no forage supplies. Our horses gathered food from the luxuriant fields of Clover and Timothy on the wayside & we had to consume much time grazing."

On June 12 Hampton ordered the 1st to furnish "(1) four horse wagon & team" to supplement the "Artillery of the Division."

Sergeant Turner W. Holley wrote "Eliza" from "Camp at Stevensburg June 13th 1863. ... this leaves me well. Yesterday morning we were informed that the Yankeys were crossing the river. We were soon in readiness to meet them. We went up to Brandy Station and there we learned that they had not crossed. The gallant Stone [William F.] has not been heard from. Wm. [A.] Bolick and Sergt [Charner T.] Kitchens says that he was shot in the thy [thigh] near the hip that his thigh was very much shattered. There was three horses shot in our company two of them is doing very well. I think that the other will die. We live pretty hard at this time. We have had nothing to eat but pickled Beef and bread and that none of the best. This morning we got some fresh beef and some bacon. We got about half enough corn for our horses. The pare (sic) turn in fine, one of my horses is in very good order the other is improving some. I received two letters from Wm Manning day before yesterday ... I have made arrangements to get him in our company by swapping another man for him. [Didn't happen]. He is near Fredericksburg. Tell Harberson, that if he wishes to come to our company he can come on with Lt Lewis. ... I have just heard we are going to move this morning some mile or so from here. ... Kiss all the children for me. Nothing more from your devoted husband. T. M. Holley."

Hampton directed Colonel Black on June 16 "... tomorrow morning at daylight you will relieve the Pickets at Rogers, Kelly's [,] Rappahannock Bridge and Beverly Fords, with a force from your command (a Sergeant & 8 men will be enough for each Post) and a reserve at Humphreys communicating with your Head Quarters. Communications will also be arranged between the different posts. Your Regiment will take position in the neighborhood of Brandy Station, so as to cover Culpepper, from which place you will draw your supplies = and await orders."

On June 18th Hampton met the Union advance near Warrenton "and drove him back without difficulty, a heavy storm and night intervening to aid the enemy's retreat," wrote Stuart. Black recalled "we had a light skirmish, after which I was ordered to leave Warrenton at 10 P. M. in rain and more than Egyptian darkness to get and hold a certain pass on the pike before me. I reached the pass and occupied it about 1 A. M. wet, wet, & cold, cold. Then we bivouacked in silence without fires (I had orders to have none) until morning. I had my men under arms before dawn but soon learned I had no enemy to meet & here awaited orders.

At 10 A. M. I was ordered to move and move slowly. As I was flanking a train, I did so and kept a bright look out."

Black reached Salem about noon and halted for a rest. The 1st soon resumed the march, and the train of wagons passing, followed the rear guard. "About 10 P. M. cold and raining, I was ordered to have 4 cos. [companies] 2 squadrons under comd. Of a field officer & to move to No. 6, a station on the Upperville Pike & report to Gen'l. Hampton. I did so, passing by and around the wagons and getting near No. 6 at 1 A. M., reported by courier to Gen'l. Hampton who ordered me to go into camp – the first time I had done so for several days. I did so in the dark & rain. We all dismounted & rolled on the ground.... Wm. Carey [Cary], my clerk, found a fly tent and bro't it to me. I told him to spread it over me and Sgt. Major Ed. Yarboro[ugh]. He did so. Wet, but then warm, I went to sleep and was awakened next morning at 7 A. M. by Major [William A.] Walker, riding into camp with my two squadrons and asking for orders. I raised up, looking at Walker, who was in terrible trim, replied: 'Orders, Why, go into camp, of course.'"

The 1st had camped near Ashby's Gap on the night of the 19th. Black continued "I spent the next day as ordered, looking around & inspecting the pickets on Upper Wolff [Goose] Creek that [John S.] Mosby was supposed to have there. I here saw Mosby's command of Cavalry & am free to say they were in very loose order and a loose set – a kind of set of free lancers. I formed a poor opinion of Mosby from his camp discipline. I rode to Mosby's Headquarters but he was away. Finding no picket in his front & mine, I put out one and reported to Gen'l. Hampton what I had done and he approved."

Black spent the rest of the day scouting and mapping the area west of the pike and submitted a sketch to Hampton. A citizen helped him complete his drawing, which was later used by Hampton to allow Hart's battery to escape capture.

Stuart placed his regiments in front of Upperville and picketed the roads to keep watch on the Federal cavalry. General Pleasanton led a reconnaissance in force towards Upperville on June 21st. Black recalled "The enemy struck our outposts at dawn and speedily pushed them in. ... Stuart, who was in command, in person resisted stubbornly & soon showed them that we was not to be put out of the way by a light line of skirmishers. The attack was made 8 to 10 miles to my left & front & we heard little of it. About an hour by sun I got a written order from Gen'l. Hampton to move out to the main pike at No. 6 (3/4 of a mile from my headquarters or place we had bivouacked) and to march down the main pike 'at a walk.' I started and moved at a walk until I passed No. 6 & the firing was so brisk in my front that I did go faster then ordered a trot – for this I was not censored but I paid dearly in person. I moved out, crossed the branch of Wolff [Goose] Creek and up the hill, along [a] slope to the other side. As soon as I rose this, I saw the fight in front & as I rose the crest of the hill I heard Stuart ... call to Gen'l. Hampton & say: 'Here is Black with a fresh regiment. Get him into position. I am going to retreat across Wolff [Goose] Creek.'"

I knew instantly what this meant, "Black continued. "He [Stuart] was badly thumped and I was to take the brunt and cover the retreat. I saw at a glance the heavy long line of Yankee Infantry skirmishers ¾ mile off advancing and our line rapidly retreating but in good order, firing all the time. Hampton, I think, in person ordered me to prepare. I, at once, ordered 200 men dismounted under Captain Angus [P.] Brown. (I selected Brown on account of his stubborn coolness because I saw the matter would be serious). I ordered Brown to advance 200 yards and take possession of a cross [roads] and a few small buildings which he did. I then drew up balance of my rgt. with drawn sabres – right resting on the road & sustain Brown.

By this time, which did not comsume 10 minutes, 17 of our Regiments had passed to the rear and 17 regimental flags – Brown's line was sustaining a heavy attack. I could see Brown handled his command well. I rode up to him and explained my plan. It was to protect him with the sabre & I said: 'Brown, I give you the Post of Honour, hold fast.' Brown was a man of few words. His eye glistened & he stood firm with his crooked legs. I then rode back to the front and right of my Regiment. I had ordered Major Walker, my second in command, to the left of the Regiment & explained to him if I attempted to retreat in presence of the enemy, we could only retreat in a column of twos or fours. I would break to the left & rear.

By the time I got to the front of my Regt. I got an order to withdraw my regt. or the mounted part of it & at once ordered it to 'Break by twos from left to rear at a walk.'

The words underscored, words of command – Walker managed the breaking and it in as quiet a manner as if he had been at drill & not in battle. I sat on my horse & saw my front melting away & my men moving in order. I never felt prouder of my of my regt. but once ... Now and then I could see a nervous man's horse or a nervous horse jogging into a trot. I called out 'At a Walk.' It 'was' a walk.

In the meantime, Brown was pressed sorely & fought gallantly at the last instance my last files were breaking (Walker had by order led the column to the rear). I heard Stuart call to Major McClellan, his adj. general, & say: 'Tell Col. Black to Gallop. Ask him why he doesn't gallop."

From the heavy odds that were on us, it seemed to me a very sensible way of doing and especially as we had to cross Wolff [Goose} Creek Bridge by twos but I quietly waited until Major McClellan came up to me and delivered Gen'l. Stuart's order. I then gave the command: 'Gallop. March.'

These commands given amid the roar of battle were passed from man to man, from front to rear & rear to front. I then ordered Brown to fall back with his line & save his men as best he could & went to the front of my Regt., then in Rear. As soon as the front commenced to get over Wolff [Goose] Creek I passed the order forward: 'Form fours March.'

Our Regts. were occupying the hill and preparing for a stubborn resistance. Hart's battery was on the right of the road (facing the enemy on the right) and by this time the enemies' guns had come up and occupied the cross road where I had stationed Brown half an hour before and opened on us; almost the first shell blew up one of Hart's caissons near us & one took out nine horses out of my front squadron. I saw at once, orders or no orders, I must reverse my column & change from Columns of fours to line of battle facing the enemy. I rode forward to see a piece of ground I could affirm on –following me was a boy as orderly, Harvey Hood, of Chester – about 16 years of age- a nice boy and good soldier... I selected a point. I wished to break thro a stone fence & as I raised my hand to point to the spot & to order Hood to get down and break it, I was struck in the back & top of my head by a fragment of shell. The blow very much stunned and practically paralyzed me. My head fell forward & mouth open. Hood moved by my side & caught me by the arm

& at the same time Carter Singleton, one of Hampton's orderlies, rode up.

I ordered Hood to go and tell Major Walker to change the order of the Regt. from columns into line of battle, faced to the front as soon as possible, so as to protect it from shells. 'You tell Walker to be d[ame]d quick about & move promptly.'"

Walker questioned Hood about the order, as he had seen Black wounded. Hood confirmed that although seriously wounded, Black had called him back and made him repeat the order.

"Walker said but for the repetition of the order he would not have obeyed it. ... Well knowing we were soon to lose the position taken, and be pushed back for I had seen the enemies' strength ... & as I was too far disabled to sit my horse without being held, I moved to the rear." Aided by Hood and John Cary, Black made it several hundred yards to were the surgeons were attending to the wounded. He was first attended to by Assistant Surgeon William D. Jones of the 1st North Carolina Cavalry and then by Surgeon Walter Taylor, Hampton's Brigade Surgeon. Taylor examined his head wound and stated 'but the thickness of a piece of paper more would have done you in 'and bandaged his head. His couriers helped him ride to Upperville where found a place in the shade and Black fell asleep.

Brown recalled "General Stuart ordered a squadron of the 1st South Carolina Cavalry, under my command, to dismount and hold a bridge just in our front. We took our position behind a rock fence running parallel with the creek, but very near the bridge. The Yankee cavalry soon approached in large force and attempted to cross the bridge, but the fire from our carbines was too much for them and they retreated in disorder, but reformed and made a stubborn attack and were again repulsed. They soon found fords above and below the bridge and began to cross.

"I then ordered a retreat, and in falling back several of my men, like myself, were severely wounded and captured. The Yankees carried us some distance to the rear, where our wounds were dressed by an old surgeon who was very kind and attentive. We were soon sent to the railroad and in due time were quartered in the old capitol [prison], Washington."

Brown, who was wounded in the head and leg and unable to move, ordered his men to leave him and retreat. Two other men were wounded and twenty-four were captured. Another report showed only 9 men taken prisoner. Perhaps some troopers escaped during the night.

Stuart fell back fighting to the village of Paris. Here he was able to stop the Federals from getting to Ashby's and Snicker's gaps in the mountains. The 1st South Carolina was not mentioned as being engaged in this fighting, and probably remained in reserve. Longstreet's Corps was thus able to cross through the gaps and into the Shenandoah Valley.

Pleasonton, finding he could not break through Stuart's lines, retreated on the 22nd. Having completed his mission of covering Lee's right flank, Stuart left Jones' and Robertson's brigades to guard the gaps and moved Hampton's, and the two Lee's brigades between Rector's Cross Roads and Ashby's Gap were battling.

Ordered by Lee to march into Maryland covering the right flank of the army, Stuart departed on June 24th, and led the three brigades through Glascock's Gap and moved via Haymarket to near Wolf Run Shoals on the Occoquan, before halting to graze the horses on the 26th. Hampton's brigade was engaged near Fairfax Court House the next day, driving the Federal cavalry from the area. The 1st South Carolina was engaged, but in a supporting role.

Hampton's Brigade led the way to Rowser's Ford on the Potomac, crossing early the night of the 27th. The brigade advanced through Darnestown to Rockville, where they captured some wagons enroute to the enemy. W. H. F. Lee's Brigade came up and captured a larger wagon train, chasing them to within 3 or 4 miles of Washington. One hundred and twenty-five wagons were secured, along with the mules pulling them. Some 400 prisoners were taken, who were paroled the next day.

Fitzhugh Lee's Brigade now led the advance through Brookeville and Cooksville. The troopers reached the railroad the next morning, the 29th, after marching all night. The railroad bridge at Sykesville was burned and the track destroyed as far as Hood's Mills. Learning that Federal cavalry was on the prow looking for them, the Confederates reached Westminister about 5 p.m. and had an engagement with the 1st Delaware

Cavalry, driving them from the field. The 1st South Carolina took part in the action but suffered no casualties. Stuart noted that a "full supply of forage "was obtained for the first time on the raid. Fitz Lee's cavalrymen camped near Union Mills between Westminster and Littlestown on the Gettysburg road.

"Early the next morning (June 30)," wrote Stuart, "We resumed the march direct by a cross route for Hanover, Pa., W. H. F. Lee's brigade in advance, Hampton in the rear with the wagon train, and Fitz Lee's brigade moving on the left flank, between Littlestown and our road." W. H. F. Lee's troopers ran into the Federal cavalry at Hanover, bringing on an all day fight. While the two Lee's men battling with their Union counterparts, Hampton was finally able to get to the scene of the action. Deploying his sharpshooters, Hampton's men advanced on the right of Stuart's line, driving the enemy from the town, and the Federals withdrew. The 1st reported 1 man killed and three wounded, including one officer.

Stuart, needing to locate Lee's army, marched his troopers through the night enroute through Jefferson to York, Pa. "The night's march over a very dark road was one of peculiar hardship, owing to the loss of rest to both man and horse. ... Whole regiments slept in the saddle, their faithful animals keeping the road unguided. In some instances, they fell from their horses, overcome with physical fatigue and sleepiness. Reaching Dover, Pa., on the morning of July 1, I was unable to find our forces. ... After as little rest as was compatible with the exhausted condition of the command, we pushed on for Carlisle. ... I arrived at that village, via Dillsburg, in the afternoon. Our rations were entirely out." The Federal commander refused to surrender Carlisle and Stuart was forced to invest the town and burned the Carlisle Army barracks. The Confederate troopers gathered what supplies they needed from the town and countryside.

"The whereabouts of our army was still a mystery; "Stuart continued," but, during the night I received a dispatch from General Lee that the army was at Gettysburg, and had been engaged on this day (July 1) with the enemy's advance. I instantly dispatched to Hampton to move 10 miles that night on the road to Gettysburg, and gave orders to the other brigades, with a view to reach Gettysburg early the next day, and started myself that night.

My advance reached Gettysburg July 2, just in time, to thwart a move of the enemy's cavalry upon our rear by way of Huntersville, after a fierce engagement in which Hampton's brigade performed gallant service, a series of charges compelling the enemy to leave the field and abandon his purpose. I took my position that day on the York and Heidlersburg roads, on the left wing of the Army of Northern Virginia.

Hampton reported "The brigade was stationed on July 2, at Hunterstown, 5 miles to the east of Gettysburg, when orders came from General Stuart that it should move up, and take position on the left of the infantry. Before this could be accomplished, I was notified that a heavy force of cavalry was advancing on Hunterstown, with the view to get in the rear of our army. Communicating this information to General Stuart, I was ordered by him to return, and hold the enemy in check. . .. I moved back, and met the enemy between Hunterstown and Gettysburg. After skirmishing a short time, he attempted a charge in front by the Cobb Legion, while I threw the Phillips Legion and Second South Carolina Cavalry as support forces on each flank of the enemy. The charge was most gallantly made, and the enemy was driven back in confusion to the support of his sharpshooters and artillery. ... Night coming on, I held the ground until morning, when I found that the enemy had retreated from Hunterstown, leaving some of his wounded officers and men in the village.

On the morning of July 3, I was ordered to move through Hunterstown, and endeavor to get on the right flank of the enemy. In accordance with these orders, the brigade passed through the village, ... across the railroad, and thence south till we discovered the enemy. I took position on the left of Colonel Chambliss [W. H. F. Lee's Brigade], and threw out sharpshooters to check an advance the enemy was attempting on my left. Soon after, General Fitz. Lee came up, and took position on my left. The sharpshooters soon became actively engaged, and succeeded perfectly in keeping the enemy back, while the three brigades were held ready to meet any charge made by the enemy.

In the afternoon (about 4.30 o'clock, I should think) an order came from General Stuart for General Fitz. Lee and myself to report to him, leaving our brigades where they were. Thinking that it would not be proper for both of us to leave the ground at the same time, I told General Lee that I would go to General Stuart first, and on my return, he could go. Leaving General Lee, I rode off to see General Stuart, but could not find him. On my return to the field, I saw my brigade in motion, having been ordered to charge by General Lee. This order I countermanded, as I did not think it a judicious one, and the brigade resumed its former position; not, however, without loss, as the movement had disclosed its position to the enemy.

A short time after this, an officer from Colonel Chambliss reported to me that he had been sent to ask support from General Lee, but that he had replied my brigade was nearest and should support Chambliss' brigade. Seeing that support was essential, I sent to Colonel [Laurence S.] Baker, ordering him to send two regiments to protect Chambliss, who had made a charge (I know not by whose orders), and who was falling back before a large force of the enemy. The First North Carolina and the Jeff. Davis Legion were sent by Colonel Baker, and these two regiments drove back the enemy; but in their eagerness they followed him too far, and encountered his reserve in heavy force.

Seeing the state of affairs at this juncture, I rode rapidly to the front, to take charge of these two regiments, and while doing so, to my surprise I saw the rest of my brigade (excepting the Cobb Legion) and Fitz. Lee's brigade charging. In the hand-to-hand fight which ensued, as I was endeavoring to extricate the First North Carolina and the Jeff. Davis Legion, I was wounded, and had to leave the field, after turning over the command to Colonel Baker. The charge of my brigade has been recently explained to me by Captain [Theodore G.] Barker, assistant adjutant-general, who supposed that it was intended to take the whole brigade to the support of Colonel Chambliss – a mistake which was very natural brought about by the appearance of affairs on the field." Barker had led the 1st, under Lieutenant Colonel John D. Twiggs, the 2nd South Carolina and the Phillips Legion into the melee.

Peter J. Malone of Company E. wrote Colonel Black after the war "About 2 o'clock in the afternoon of July 3rd ... our brigade moved to its position on the left of the army. There was one incessant roar of artillery and the ground was shaken.... Soon a battery opened upon us from the enemy's line ... and shells exploded in our very ranks, little damage was done. ... I was of the color-guard on the right of J. H. Koger, the bearer of the standard, whose heroism in keeping it proudly in the face of the enemy, and afterwards in bearing it in triumph from the field, where he had narrowly escaped death and capture, became so well known. On my right was Sergeant T[homas] P. Brandenburg, whom you will remember as a peerless soldier and truly imperial spirit.... Gen. Fitz Lee encountered the enemy on our right, and being overwhelmed by numbers it became necessary for us to attack them in front, to divert their attention from his brigade. General Hampton proposed to lead our regiment. We started out in fine style, and one continued shout arose from the charging column. The enemy now appeared in a black compact line.... The intervening ground over which we were passing was so crossed and seamed with fences and ditches as to greatly impede our progress, and the sharpshooters, concealed wherever concealment was possible, found in the moving mass of beings an excellent mark for their rifles. It was, no doubt, by one of these chance balls that I was wounded. We had not advanced beyond two hundred yards from the cluster of trees where we had taken shelter, when I was struck, the ball entering my right side, ... Believing it to be no more than a fragment of a shell ... I kept on with the regiment. We were soon at the sabre point and fighting desperately. The color-guard, ... became precipitated from its posture to the head of the column and met the enemy at a small opening in the fence, which soon became so by blockaded by the regiment as to prevent those in the rear coming to the assistance of the few who had first entered the enclosure, or any of us who might be wounded to secure our escape to the hospital. General Hampton ... here engaged a number of the enemy, and cut his way through them ... bearing upon his noble form the marks of cruel wounds."

Malone's right side and arm became paralyzed and Surgeon Joseph Yates came to his assistance and helped him over the fence. He was able to remove his blankets from his horse before the animal, "infuriated by the crash of cannon, the explosion of shells, the sight of blood, rushed desperately to the rear."

Malone was unsure how he made it to the hospital, but recalled passing Private William D. Shirer, also of Company E, whose right arm was broken. "When aroused to consciousness Corporal H[enry] L. Culler, of Company E, Private Ch[arles E.] Franklin, of Company B,".. and others were around, with hundreds of others, friends and foes, receiving medical attention." Surgeon Yates was attending to the men, having charged at the head of the 1st, and later arrested by the Federals, but later released. Malone was informed by the Surgeon that his "chances [of recovery] were against me." The Confederates retreated the next day, but Yates remained to tend to the wounded. "When taken, we were sent to Gettysburg Hospital where our treatment, though kind, was rendered repugnant by the flippancy of some of the United States surgeons. They examined Malone and Culler and stated that they "must die in any event." .... Culler died a few days later. "We were then removed to New York [David's Island], where we received considerate attention." Malone recovered from his wounds and was later exchanged, but was transferred to the Invalid Corps.

Captain John Esten Cooke of Stuart's staff wrote "The struggle was bitter and determined, but brief. For a moment the was full of flashing sabres and pistol smoke, and a wild uproar deafened the ears; the Federal horse gave back, pursued by their opponents. We lost many good men, however; among the rest General Hampton was shot in the side, and nearly cut out the saddle by a sabre stroke. Ten minutes before I had conversed with the noble South Carolinian, and he was full of life, strength, and animation. Now he was slowly being borne to the rear in an ambulance, bleeding from his dangerous wounds. General Stuart had a narrow escape in the charge, his pistol hung in his holster, and as he was trying to draw it, he received the fire of barrel after barrel from a Federal cavalryman within ten paces of him, but fortunately sustained no injury."

"Having failed in this charge the enemy did not attempt another; the lines remained facing each other, and skirmishing, while the long thunder of artillery beyond, indicated the hotter struggle of Cemetery Hill. "

The 1st Cavalry, who charged as part of W. H. F. Lee's brigade, lost 1 killed, 9 wounded and 4 captured in the action. Another reported showed 2 killed and 17 wounded. Some of the slightly wounded men may not have been included in the first report. Major Walker was among the wounded. Sergeant William G. Jarrell of Company D was the only fatality in the regiment. Sergeant Malcolm L. Cox and Private Henry D. Russell were wounded and captured, Corporal William T. McClinton received a sabre cut to the head, and Private Samuel Lee Russell was captured, all members of Company A.

Meanwhile, Colonel Black had recovered enough to rejoin the regiment. He crossed the Potomac at Williamsport and rode toward Hagerstown. "I found, marching on, nearly an entire co[mpany] of my own Regt. and one of 2nd S. C. Cavalry that had been cut off from Stuart in crossing the mountains as well as Lt. Maxwell's 1st N. C. Cavalry. I took command of these detachments and moved on. "Black continued picking up stragglers from cavalry commands until he was leading 200-300 men. He deployed these men on several occasions to protect Confederate ordnance trains enroute to Lee's army. Upon reaching Gettysburg, Black was ordered to report to General Lee's headquarters. He was sent with his group of cavalrymen and Hart's battery to the village of Fairfield as a rear guard for the infantry. 1st Lieutenant Frederick K. Horsey and 2nd Lieutenant John Wilson Marshall were used as scouts to check the ground that Longstreet's Corps would move over on July 2nd. The next day Black was directing Hart's battery as it fired on Federal cavalrymen.

General Elon Farnsworth, commanding a brigade of Union cavalry, attempted to flank the Confederates near Fairfield. General Evander Law's Brigade of Alabamians, with support of Texas and Georgia regiments opposed the Federal advance. Hart's battery and Black's make-shift unit supported the infantry. General Judson Kilpatrick, Farnsworth's division commander, became frustrated at the feeble efforts of his cavalrymen to overcome the Confederate resistance, and ordered him to make a mounted charge. The attack was a valiant effort, but the Southern infantrymen, armed with Enfields, shot many of the Federals from their horses. One shot brought down Farnsworth, and the Union troops retreated.

Black described the event "Gen'l. Farnsworth had perhaps spotted a small body of mounted men, 150 about I had moved in the course of the day on [Captain Mathis W.] Henry's right, * concluded to make a dash at them ... He came down the road at a gallop and ran square into Henry's Battery before he saw it, who saluted him with a sweeping discharge of grape & canister with literally annihilated the head of his column. A second discharge did its work & the Alabama Regiments springing to their feet, pound in a murderous volley ... Farnsworth, ... had his bowels shot out. After the firing ceased, a Confederate ordered him to surrender. He had a pistol, 'No, damn you' & turning the pistol, blew out his own brains.

In the charge in the wheat field two of my men were shot thro the thigh and ... were gathered up and reported to me as bleeding to death. We had no surgeon. An inquiry was made if there was a doctor in the ranks. A recruit of my own Regiment said he was a doctor. I ordered him to improvise some means and stop the hemorrhage. He did and saved the men and attended to others. I said to him 'I will make you and assistant

Surgeon' & did so, recommending him as having won his spurs on the battlefield. It was Dr. Horace Drennan of Abbeville."

The 'Yorkville Enquirer' reported "1st S. C. Cavalry. Among the casualties in this Regiment as reported by Col. J. L. Black, we found the following in Capt. J. H. Barry's company from this District, ... viz.: Wounded: Private A. Ross, back, severely, in hands of the enemy; J. S. Williams, hand and head, sabre[cuts]; J. C. Reed, hip; J. Atkinson, elbow, sabre cut; A. M. Biggers, wrist, sabre cut. Missing: Pvt. S. B. Hall."

The next day Black continued, "I had made application the night before to have my regiment sent to me and about 2 p. m. an order came for me to march my command across, or rather, in the rear of the line of Battle and to report to Hampton's Brigade Hdquar... and reached my Brigade Headquarters about night fall. ... Soon after dark, in gloomy silence, we commenced our retreat. I think my Regt. led off. At all events, Col. Baker, Lt. Col. [Joseph] Waring, Jeff Davis Legion, and I were at the head of the column.

We compared notes & I heard of our losses. Lt. Col. Walker had been wounded, captured, and soon after released by our men's running over the Yankees again. He had been sent off and was safe. Old Hill (Lt. [John R.] Hill, Co. E) had been badly hacked over the head by a Yankee sabre but Hill's head was hard, very hard – tho a good head and a brave one. I believe he was on duty with a bandana bound around it.

We moved out in silence until 2 or 3 A. M. when we halted and bivouacked in the mountains. I sat by a fence all night as it rained incessantly and I got no rest but at the head of my Regt. marched on the next day. ... we came to Fincastle [Greencastle, Pa.]. Col. Baker ordered me to ride ahead for some purpose & look out for something. I did so and down in the suburbs of Fincastle [Greencastle]. I had only my orderly with me. ... a detachment of men had come up to report to me to do look out for our right flank. Ordered them thro the town and told the officer if he was whipped back to burn the town to create a diversion." This was stated in hearing of a citizen of the town who had boasted that our trains would never be able to pass through Greencastle.

"The brigade came up and we passed on," Black recalled. "Of the next 24 hours I have little recollection save marching on slowly ... The next day I was sent by Col. Baker to ride on to Williamsport to try and recruit

& rest. But as I rode into Williamsport where our entire wagon train was parked, unable to cross the swollen Potomac, crowded with thousands of our wounded. All we could bring off, all capable of being moved and many whose wounds were of such character they ought not to have been removed. ... It was reported the Yankee Calvary was approaching. In the town was Gen'l. Imboden's Calvary Brigade, only two regiments strong. I rode at once to Gen'l. Imboden's Hdqar, and explaining my presence, offered my services, which were accepted.

The General asked me to go on the pike north of the town and to stop and organize all the slightly wounded & stragglers that came in. I did so, & soon got a number of lame horse cavalry men, some of them good soldiers... Our train was pouring into the town and we soon picked up a number of men. We also had a number of guns & small arms in abundance which had been brought off from Gettysburg. Our supply for the guns of shells was less than 3 to a gun but a plenty of grape and canister. We got our forces marshalled and had many a wagoner under quartermasters, who volunteered for the occasion. At Gen'l. Imboden's request, I took command of the left of his line and here ... was my commissary Sergeant ... in command of 10 to 15 of my own dismounted men.

Late in the day the enemy began to advance, brought up a battery and shelled us for a while but the shells flew mostly harmlessly over us & over our wagon train which was parked on the alluvial land on the margin of the River. We also had all streets leading out of town barricaded with wagons and Gen'l. Imboden had been very active and made admirable arrangements for defense. The most of the fight took place on my right, yet my part of the line sustained a light attack or was begging to be attacked as all at once the Yankees drew off to their left. The reason why we could not at once understand. They suddenly learned that Gen'l. Stuart was camped up on the Hagerstown Pike directly in their rear, so the Yankee officer adroitly slided off to his left & cleared our front – Stuart marched up as well ... about dark & it was soon too dark to follow them. Our trains were saved & the 'Wagoners Fight,' ... was over.

My Brigade came up that night and, as the enemy was about, I took command of my Regiment tho entirely unfit to do so. We marched out toward Hagerstown and had some skirmishes at a place called Funkstown. "Records show the 1st was on the Antietam on the 7th & 8th of July and were engaged at Funkstown on July 10.

Black continued "In one of these skirmishes Capt. M[oses] T. Owen, Co. A, was wounded in the heel. He lived to get to his home in Abbeville C. H., S. C. & died of tetanus (lockjaw) ... Owen was a gallant man. In one of the same skirmishes a beautiful beardless boy named [Jacob J.] Foreman of Barnwell was killed – shot in the skull."

Suffering from his wound, Black now came down with a fever. "My Regimental Surgeon [Calhoun] Sams [who] ... said I had typhoid fever. I was sent to the tent of Surgeon [Harvey] Black [4th Virginia Infantry] ... [who] said I had Typhoid fever & sent me over the River. "There I met Dr. [James J. H.] Brewton, a private in my own Regt." who agreed with the diagnosis. Black was placed in an ambulance and sent to a hospital in Winchester. Because of the nature of his illness Black was sent to a private home to recover. He was attended by Dr. Andrew Paul, a Private in Company A, who was working in the hospital. Black soon had Paul attending to the "sick & wounded of my own Regt." who included Major Walker and Private Washington Wilkes of Company E. Wilkes died later in a Manchester, Virginia hospital. Black was later evacuated to the hospital in Mt. Jackson.

Meanwhile, the 1st continued to confront the Union cavalry near Funkstown and acted as part of the rear guard until re-crossing the Potomac on July 14th. There was no rest for the weary, as Baker now had to defend the Potomac crossings against Federal incursions. Stuart assigned him to picket the river from Falling Waters to Hedgesville. General George Meade, pushed by President Lincoln, had his cavalry across the river near Shepherdstown and skirmishes took place daily from 16 to 22 July. Baker's brigade not only had to defend the river crossings but the infantry who were busy tearing up the Baltimore & Ohio Railroad near Martinsburg. 2nd Lt. Samuel T. Anderson & Privates Thomas L. McFadden and William R. Wix, all of Company D, were captured near Martinsburg on the 19th. One report shows the regiment losing 10 wounded during this period, including one officer.

Lee moved his infantry back near Culpeper Court House and the cavalry were again assigned to protect the gaps in the Blue Ridge. Lee deployed his foot soldiers on a line behind the Rappahannock River. Baker's Brigade took position on the Beverly's Ford – Kelly's Ford line, guarding the fords with pickets. The regiment was at Gordonsville on July 24th and at Culpeper on the 29th.

Meanwhile Meade had advanced his army to the Warrenton area north of the river and decided to send his cavalry, under General John Buford, across the stream. Baker, warned by his scouts, marched his troopers to Brandy Station on July 31st. His pickets tangled with the Union advance at Beverly's Ford. Baker's outposts were driven back about a mile from the Rappahannock Station area. Meade also sent infantry across the river at Kelly's Ford, keeping other Confederate cavalry from assisting Baker.

Black was able to return to the regiment on the 31st and found them camped "just a short distance north of Stevensburg on the grounds of a Mr. Beckman," known as 'Camp Jackson. We were summoned to march again to Brandy Station on the 13th [1st]. day of August & arrived at 9 or 10 o'clock. My recollection is there were 935 Rank & file all told in the Brigade present and in the saddle... Lt. Colonel Twiggs was present with the 1st." The Brigade was rough in appearance, dirty, ragged and worn from the effects of long Pennsylvania Campaign. They had been back so short a time that the men and horses were not rested & washed up.

The Brigade was assembled ¾ of a mile East of Brandy Depot ... & here we lay listlessly about one o'clock p. m. – when the enemy advanced and deployed, commencing the attack. After an hour's skirmishing we were ordered by Stuart to fall back on Brandy Depot. This we did & here took up the fight but as the attacking party outnumbered us two to one we were soon flanked out of the position and fell back across a run or branch and to the house of a Mrs. Wise. My Regt. halted on the left of this house and Gen'l. Stuart ordered some guns to be unlimbered near this house. In fact, he gave the order to me to have them put there. I objected to this as useless and likely to draw the enemies' fire on Mrs. Wise's house. Stuart insisted & I replied, 'I'll be d----d if I will do it. Send the guns on the hill to our rear.'

In his good-natured way he asked me why I was so emphatic. I replied.

'I know that lady & by G—I will not draw fire of the enemies' batteries on her house.'

'But,' says Stuart, 'Suppose they are not checked?'

'We must then use the sabre,' I replied.

He ordered the battery to the rear. But he had better foresight than I, as the enemy did soon make a dash at us. At that time, I, with my Regt. & the Jeff Davis & Phillip's Legion, were on the right of the R[ail] Road, 1st N. C. C., 2nd S. C. C. & Cobb Legion on the left when the enemy pressed us. We met and repulsed them tho it cost us dear, as Col. Baker was wounded seriously in the arm, Col. [Pierce B. M.] Young shot in ribs, & Lt. Col. [Gilbert J.] Wright also wounded. The enemies' attack was principally on the left of our line, hence our loss was worse at that point. After we had repulsed the enemy, we made a stand at the Kennedy House. As I was rallying & re-forming my Regiment there, Stuart rode up to me & raising in his stirrups in front of my Regt. said."

'Well done, South Carolina, nobly done, South Carolina.'

While drawn up here & while Stuart was with me in front of my Regt., Lt. Donald McLaughlin, who had his back to the enemy & face to the ranks of his co. [G], directing the formation, was struck on the spine with carbine ball, doubtless partly spent. The ball went thru the folds of his overcoat strapped on the cantle of an English saddle & in its course made 27 holes thro his coat, pants, vest & stopping at his shirt made a blue spot on his spine. McLaughlin with an oath called to Lt. [John F.] Livingston to look and see where the d----d rascals had shot him in the back, remarking he would have faced to the front if they had waited a moment. Livingston looked and found no [wound]. Gen'l. Stuart was by me and inquired McLaughlin his name, saying he took gunshot wounds coolly.

Capt. L[eroy] J. Johnson was about the same time hit by a spent ball on the shin bone tho making no wound. McLaughlin always swore that the ball split in two pieces on Old Hickory's shins with excited great wrath in Old Hickory. About the same time Captain [illegible] was struck with a spent ball on the underside of his arm that him much pain but neither of the three quit the field.

We were nowhere in a good position & the wily Yankees, not caring to try and force our front, went to work to flank us out. They got a few sharpshooters in some woods on my right flank & opened fire on me. Suspecting their design, I rode obliquely to the right, followed by Private Wallace [W.] Miller, Co. C, to see & reconnoiter for myself, going over a small eminence 200 paces to my right & front I became a target for this little batch of sharpshooters who let fly above & killed Miller's horse instantly. I rode on until I saw where they were located & then back to the Regt. & ordered Capt. [Elam] Sharpe to deploy his Co. (F) and drive them off, which he did promptly, and I was not again annoyed on my right flank.

The comd. of the Brigade now devolved on me. Tho we had been in a rough & terrible fight for two hours and had lost very little over two miles of ground, Gen'l. Stuart then ordered a further fall back of a hundred yards, where my flank rested on the right flank of a Va. Infantry Regt. Posting my my own Regt. & the 2 Ga. Regts. [Phillips & Cobb Legion] with me, I passed along the line of the Va. Regt. to see who was in command of it.

I soon saw Col. [Thomas J.] Lipscombe [sic] of the 2nd. S. C. C. was engaged with a superior force on the left of their Regt. & sent back for my own Regt. to come to me in a hurry. As soon as I (illegible) to sustain (illegible) for Lipscombe [sic], who made a very gallant attempt to charge and drive back the enemy. In this, tho, he ran into a trap the Yankees had set for him and was only saved by his promptly realizing the whole and ordering a retreat. In falling back, his Regt. mingled with my own which was advancing and created some confusion. Lipscombe [sic] promptly called on his Regt. to rally & reform on line with the infantry. This he did & we were forming our Regiments under a heavy fire of the enemy in front who were attempting to break up our formation when I received a very painful wound in my right hand .. . As soon as the formation was made, as the field was now safe, I left the field in much pain, rode back to Culpepper C. H. I called here at a house where Lawrence [sic] Baker was supposed to be dying & saw him. He was insensible & very weak from loss of blood & I thought from all appearances, dead.

Private Bart Dill was wounded, and Privates William T. Penny and John W. Robinson were captured in the action. All were members of Company A.

"I laid myself down in the yard. Gen'l. Stuart came to see me and on enquiring as to my wound - was sent from Gen'l. R. E. Lee & to know if he (Gen'l. Lee) could do anything for me. I replied to this message that both Col. Baker & myself had lost our servants & that he could favor us by allowing my orderly, John Carley [sic], to accompany us to Richmond. A pass was immediately sent to me to pass my orderly to Richmond to remain there four days to wait on Baker & me." Baker recovered and Black was furloughed to his home in South Carolina for 60 days, which was later extended for an additional 30 days.

On the battlefield, the Confederate cavalry, supported by the infantry drove the Federals back almost to the Rappahannock. Both sides lost heavily. A Captain in the 1st North Carolina Cavalry wrote home that "our loss was quite heavy." About 500 men in the brigade had been killed, wounded or put on foot by the loss of their horses either by enemy fire or because of the heat. Stuart's troopers continued to engage the Federal cavalry south of the river, in an attempt to drive them to the north bank. Finally, on August 9, Meade ordered his troops back across the river. The 1st Cavalry was camped at Gordonsville on August 19th.

The 'Richmond Enquirer' reported on August 2nd: "There was a cavalry fight yesterday near Brandy Station between Hampton's Brigade and three brigades of the enemy, lasting several hours. The Confederates fell back upon the infantry supports and the enemy was then repulsed. Col. Baker, commanding the Brigade, was seriously wounded in the right arm, and Col. Black, of the first S. C. Cavalry, was wounded in the right hand. Both arrived here this afternoon."

Both sides resumed their positions along the river, guarding the fords as before. The stalemate lasted until Meade learned of the departure of Longstreet's Corps to Tennessee. On September 13 he had his cavalry cross at three different locations. The 1st had been camped with the brigade near Stevensburg, and rode forward to engage the advancing Federals near Pony Mountain. The brigade fought stubbornly but was forced to fall back in the direction of Raccoon Ford" wrote an officer in Company C. "On the morning of the 14th we crossed the river. Mounted the heights on the south bank and there made a stand. The enemy seeing our advantage in position desisted from his pursuit but mounting his artillery upon a prominent hill within ¾ mile of us. Contented himself with pouring down on us for the remainder of the day a destructive shower of shell. In the early part of this engagement near Pony Mountain and on the 13th we lost three horses and 1 man wounded in the arm."

Black, who had just rejoined the regiment, stated they were supporting Hart's Battery on Stevensburg Hill and forced to retreat. He reported Lt. Colonel Twiggs' horse lost a leg and Twiggs was injured when his mount fell. Major Walker became second in command. Black recalled "the enemy were unlimbering a battery on the very ground we had been halted on in column & exactly at the spot on which Twiggs' Horse had been shot." Black quickly concealed the regiment in the woods while the Federals shelled the area his troops were last seen. "The skirmish had ended and we halted as it was late in the evening," Black continued. "That night we retreated out of Culpepper Co. – crossing at Raccoon Ford. "

Company H went into the fight with 55 men and lost only 1 horse wounded. Company B was not so fortunate, losing Private David P. R. Layton killed, and Sergeant Johnson C. Woods and Private William Davis lost their horses to enemy fire. In Company I, Private John Keller was severely wounded in the thigh, and died later in a Richmond hospital. Private Charles Gall was slightly wounded.

Stuart aligned his cavalry brigades to cover the crossing at Liberty Mills, 6 miles west of Orange and on the Madison Court House and Gordonsville roads. His patrols and pickets also watched the Peyton's, Barnett's and Cave fords. Fitz Lee's division covered the Sommersville, Raccoon, Morton's, Stringfellow's and Germanna fords. The loss of Culpeper Court House and the railroad running through that town forced the Confederates to draw supplies from Orange Court House. Meade has his cavalry keep the pressure on Stuart's troopers, but found them backed up by Confederate infantry. Stuart, now personally leading Hampton's Division, had his headquarters only a quarter of a mile from Peyton's Ford, near Rapidan Station. The 1st was now assigned to W. H. F. Lee's Brigade, with the 9th, 10th and 12th Virginia regiments.

The 'Abbeville Press' reported on 4 September, that Wm. C. Moore was at Abbeville Court House and those desiring to send clothing to Company A, 1st S. C. Cavalry, should have it to him by October 1st.

Meade maneuvered his infantry into the area near the Rapidan and Robertson's rivers, while Buford and the cavalry kept pressure on the Confederates. On September 21st, Meade ordered Buford to cross the Robertson and advance on Madison Court House. The Federals crossed Russell's and Barnett's fords and captured Madison Court House and advanced beyond the town before halting for the night.

The next day Buford split his force, sending Kilpatrick's division towards Burtonsville, while Buford, with Colonel George Chapman's Brigade, took the road leading to Jack's Shop, about 5 miles from the Court House. The overpowering Union force drove the three small Confederate brigades back. Stuart was on the field and realized he and his troopers were caught in a trap. Stuart rode along the lines informing them that they were surrounded "Boys, it is a fight to captivity, death or victory." While part of each brigade fought the enemy in their respective fronts, Stuart was able to blast through the Union troopers defending the road running from Jack's Shop to Liberty Mills with Baker's and Young's brigades. Lee's Brigade was able to get out of the trap also. Buford and the Federals pursued south of Liberty Mills but soon halted. The next day he retired across the Rapidan.

Company C of the 1st was on picket at Great Run on the Robertson, picketing along the banks of the river, when the Federal cavalry advanced. Privates D. O. Bateman and John A. Howard were captured.

Stuart gathered more of his cavalry and pursued the enemy the next day. Buford fell back across Robertson River, ending the fighting. The 1st resumed picketing along the river but, in doing so, engaged the Federals, losing Corporal John F. James and Privates Miles A. Eads and Nathanial M. Madden, all of Company F, captured.

The two foes continued to face each other across the rivers. Company I, on picket at the Robertson River crossing, had a skirmish with the enemy of September 28, but suffered no loss. The two sides had sent off reinforcements to the west. Lee sent Longstreet's Corps to Tennessee in time for the battle of Chickamauga, while Meade sent several Union Corps westward.

Lee was determined to drive the Union army out of Northern Virginia. The Federals had turned most of the occupied territory into a virtual desert. Stuart, camped in Madison County, led the way northward, crossing at Peyton's and Barnett's fords on the Rapidan on October 8th and occupied Madison Court House. Pushing forward, the 1st South Carolina engaged the enemy at Robertson River "Where the Regt. made a gallant charge upon the infantry capturing a great number of them," according to one officer. Another reported the 1st "charged a regiment [124th New York] of infantry capturing over one hundred men being at that time more men than we had for duty." General Young reported "I was ordered by General Stuart to move forward and attack the enemy's right flank while General Gordon pushed him in the front. I moved my brigade around through the woods, and upon coming in sight of the enemy, drawn up in line of battle, I ordered the leading regiment to charge, which was responded to in a most cheerful and gallant manner by the 1st South Carolina, led by its brave lieutenant-colonel (J. D. Twiggs), who then commanded he regiment. The enemy broke and fled in all directions, utterly routed. The number of prisoners captured was about 87, General Gordon capturing many others." Stuart reported that "Between 75 and 100 excellent arms were captured."

The 1st continued through James City, skirmishing with the enemy cavalry. General Young continued "About 4 p.m. the enemy made a dash upon our battery [Captain William H. Griffin's Baltimore Artillery], charging up to within 200 yards of the pieces; but about 50 sharpshooters, masked behind a stone wall, under command of Captain [Samuel Henry] Jones, of the First South Carolina Cavalry, delivered a volley into them, killing several and wounding others, which caused their speedy retreat."

On October 10th the Federals fell back. An officer reported "We pursued them to Culpeper CH where we had a skirmish with the enemy, the enemy being completely bluffed by the determined stand of the little Brigade. We pursued them." Young added that the brigade found itself defending the town against a Federal advance. "the commissary and quartermaster's trains of the army were at that time loading with supplies at the depot, and would have been an inestimable loss.

My battle line was about 1 mile in extent, Colonel Lipscomb on the right, Colonel Waring on the left, and Colonel Twiggs commanded the center. The enemy came rapidly on... Colonel [Thomas L.] Rosser [5th Virginia Cavalry] fell back skirmishing until reaching my line, when a heavy volley was poured into the enemy simultaneously from my artillery and dismounted troops. Both sides remained stationary until night, keeping up a heavy fire of artillery and sharpshooting. I encamped upon my line of battle, causing a great number of fires to be built and kept up all night along my entire line."

Stuart's troopers continued pushing the Federals back through Culpeper Court House to Stone House Mountain and to Stevensburg on the 11th. He had tried several times to capture the Federal cavalry near Brandy Station, but the Union troopers, who greatly outnumbered the Confederates, escaped the trap. Unfortunately, neither Black nor his others officers gave an account of the roll the 1st played in these heavy skirmishes. Both sides lost heavily, about 400 men each. The Union cavalry fell back behind the Rappahannock. Meanwhile the Confederate infantry moved northward from the west of Culpeper Court House and crossed the upper fords of the river.

The Confederate pursuit continued on the 12th and 13th, Meade pulling his infantry back, while his cavalry retreated slowly, engaging Stuart's Brigades as they advanced. Stuart advanced with two brigades in pursuit of the large Federal wagon trains, to near Auburn on the 13th, finding himself penned in by Federal infantry marching by on the adjacent roads and camping for the night. Black recalled "Stuart was at the head of the Column and my Regt. in front. He ... soon halted, passing back an order for all to keep quiet. Even the very horses seemed to realize the situation and stand still. Gen'l. Stuart asked me for a good man, a scout that he could send thro the enemies' camp to reach Gen'l. Lee's lines. I ordered the adjut. To detail a scout, Sergeant [Manson S. J.] Jolly, of Co. F., asked Capt. [Alfred T.] Clayton to let him go. Clayton directed him to come to me. He did so & I referred him to Gen'l. Stuart. Poor Jolly was a bad talker but a better man or soldier never lived. He did not impress Stuart, who, however, told him to go back and get an overcoat (Yankee) to cover and conceal his uniform. This Jolly did. While away, Stuart asked me about him. I told him there was no better man in the corps and Jolly soon came back and departed on his mission & went thru the lines. We reversed our column and backed out, made a detour, and before daybreak had taken up our position on the flank of Gen'l. Lee' Army."

A total of six of Stuart's scouts reached Lee's headquarters, and he ordered infantry units to Auburn and the cavalry's relief. Black continued on the 14th "Here we lay all day in line of battle, cold, bitter cold, & suffering. ... [Later] "I was sent by Gen'l. Hampton with my own and another Regt. around on the flank of some Yankee Cavalry that had deployed a line and attack their flank. I moved around and did as ordered. My line of skirmishers soon lapped the enemy perpendicular to their line & we rolled them up steadily ... we all at once came to an open field with a farm house in the centre & directly on the Yankee line. At first they seemed to be determined to take possesion [sic] of this house & enclosures and to give us a warm reception. I called on my men to move forward & we made a dash at them but they gave way and fled at our approach. My line was here halted by Gen'l. Hampton's order & he joined me with one or two Regts. of the brigade. The sun was near setting and the enemy were with drawn into woods beyond the farm enclosure and all firing ceased.

While we were halted, two or three hay stacks were seen in the farm enclosures but 300 yards to the right of the house. One or two men made an application to be allowed to go and get some hay. With Gen'l. Hampton's assent, who was present, I gave permission to half a dozen to go for that purpose. They rode off to the haystack and, alighting, commenced to prepare bundles of hay to carry off. But no sooner did the Yankees see this than they opened fire on them. My men laid down their hay and, getting their guns, returned the fire and, reloading, fired again and again. Seeing this, several more applied to go for hay & were allowed to do so. They met the same reception & used their rifles for a while and they took up a bundle of hay and came away. More men applied to go and were allowed, so we had a continual stream of men going and returning & not perhaps over 6 or 8 men at the hay stacks at any one time. A Lieutenant applied to go & carry a detachment. This, by Gen'l. Hampton's direction, I refused & none but a few at a time were allowed to go and on their own hook. Every man fired a few rounds, got his load of hay & came away & the thing was kept up until not a wisp of hay was left, myself and other officers looking on from the farm house. This was always alluded to as the 'Battle of the Hay Stacks' and was decidedly a rich affair & the casualties were none killed – none wounded."

Stuart, with Hampton's division, continued the pursuit of the Federals. An officer in the 1st reported they reached Fairfax Court House on the 15th and Haymarket the next day. Young reported that his brigade, apparently minus the 1st South Carolina, had remained behind to guard the fords, and moved up to Bealton Station, on the right of Lee's Army. The brigade didn't rejoin Stuart until the night of 15 October, near Manassas Junction. "On the evening of the 16th, General Hampton's division, under the personal command of General Stuart, moved around to ascertain the position of the enemy's right flank," wrote Young.

About 4 p. m. on the 17th, my brigade, moving in front, came up with the enemy near Frying Pan. The front squadron charged and captured a number of the enemy's picket. The line of battle was immediately formed and we advanced upon the enemy. After about two hours' skirmishing the position of the enemy was ascertained by General Stuart, and other information gained. The object of the expedition having been accomplished, the troops were withdrawn and I took up the line of march for Hay Market. On the 18th we encamped at Hay Market."

Lee had determined that he could not bring Meade to battle, nor could he subsist his troops in the barren waste land the area had become. He determined to fall back. Meade immediately sent his cavalry in pursuit. By the 19th Lee's infantry was re-crossing the Rappahannock.

The Union cavalry believed they had Stuart on the run and became very aggressive in the chase. Stuart set up a defensive line along Broad Run. The Union troopers attacked the Confederate troopers behind the stream but were turned back by accurate fire from the horse artillery and sharpshooters. In the afternoon George Armstrong Custer's Michigan Brigade found a crossing of the stream and Stuart fell back. Custer pursued aggressively, being lured into a trap. As the Confederates fell back, Stuart sent Fitz Lee's division to get on the flank of the Federals. Custer stopped his command to eat and feed their horses, while Kilpatrick led Henry E. Davies' Brigade after Stuart. Fitz Lee's division struck the Federals in the flank and routed them in confusion. It became a running fight all the way back, some five miles, to Broad Run where Custer was. It was a disaster for the Federals, but the Confederates gathered in some 250 prisoners, horses and wagons. An officer in the 1st recorded "retreating to Buckland where we again met and fought them and completely routed them." The regiment reported no casualties during the "Buckland Races," as the fight became known as. General Custer's headquarters wagons and papers were captured, "report Captain Cooke, "... and the pursuing force, under Kilpatrick, gave Stuart no more trouble as he fell back."

The regiment recrossed the river on the 20th and resumed their camp near Stevensburg, with pickets along the stream. An officer in Company K reported the regiment was without tents or tent flys, and the horses were "in very bad condition." The common problem was lack of long forage. Company K was on picket on October 31st, and could not get a complete muster of the men. During the month of October, the regiment lost only one man killed in action.

Stuart decided to have another review of his command. On November 5th the event was held on the John Minor Botts farm near Brandy Station. General Robert E. Lee and Governor John Letcher of Virginia were the guests of honor and reviewed the troopers. The divisions of Wade Hampton and Fitzhugh Lee rode proudly by. The massed gray-clad horsemen fell into column, making an awe inspiring sight. Bugles sounded, sabres were drawn, and on command, the horsemen turned and thundered past the reviewers. They yelled like demons, reported one participant. Captain Cooke reported it as "a fine spectacle." He noted that General Lee was "sitting his horse by the tall flagstaff in his old gray coat and whitening beard." Stuart was proud of his command. General P. M. B. Young sat erect on his horse as he led his Georgians, Mississippians, and South Carolinians past the reviewers. Black did not comment on the review.

Stuart's reviews seemed to always draw the enemy into battle. The 1st confronted the Union advance at Stevensburg on November 7th and 8th losing only one man wounded. Wade Hampton rode along his division's battle lines for the first time since being wounded at Gettysburg, to wild cheers from his men. The Confederate cavalrymen successfully fended off the Federals. The regiment marched to near Culpeper Court House, where it again skirmished with the Federal advance, who they turned back.

Black led his troopers into Louisa County in search of fodder for his depleted horses. By the 16th they were again on picket along the south bank of the Rapidan. They continued to skirmish with the enemy along the river, driving them back across the stream. Once relieved of picket duty the regiment moved into Caroline County, still seeking food for their mounts, which were in bad condition. The regiment had suffered one man wounded on the 13th.

When Meade advanced across the Rapidan on November 26th, Hampton brought his men up to face the Union cavalry which had crossed at Ely's Ford. An officer in the 1st reported the regiment was engaged with the enemy along the river on the 27th, 29th and 30th of November, without loss.

General Young reported the role of his brigade at Parker's Store on November 29th. Finding Rosser's and Gordon's brigades falling back. "The skirmishing was brisk. I immediately dismounted two regiments (the Cobb and Phillips Legions), bringing them up on the enemy's right flank, charging on foot. The First South Carolina I carried up the plank road at a trot, hoping to have the opportunity to charge. We moved on briskly under a pretty warm fire, but which did me little damage. I soon found myself in possession of the field. I was still pushing the enemy when I was ordered to withdraw my troops and cover the rear of the column. We encamped that night at Antioch Church."

Black reported that when the Federals retreated "our brigade followed them down to Germanna Ford." The enemy had occupied a Confederate fortification on the south bank of the river. "This work was occupied by 3 or 400 Yankee Infantry, a rear guard," he continued. "When we came in 400 yards of it to the edge of the open field in which it was situated, Hampton ordered a halt & soon after ordered me to dismount 150 men. I did this under Capt. Johnson (Old Hickory). Johnson was ordered by Hampton to deploy his men on the left of the road and make a feint on the work but Capt. Johnson either did not hear or did not understand the order & by deploying his line, he called on his men to move forward and capture the work, or would he listen to Lt. McLaughlin when he told him he had mistaken the orders. He was ordered to take the work & he would do it but the order to halt was given as he was putting his line in motion. The enemy about this time were seen to scamper from the work & as the opposite heights were crowned with the enemies' guns in point blank range, our advance into open ground was useless and inadvisable." On December 1st Black led his regiment to Hamilton's Crossing, on the railroad, below Fredericksburg.

Young's Brigade picketed from Fredericksburg to the lower Rappahannock the rest of the winter. Black complained "We were poorly foraged and badly supplied. Our corn issues for near two months not exceed 3 lbs. to the horse... We procured thro our Regimental quartermasters & weak trains a little additional drawn from the lower banks of the Rappahannock. But with all this, it was little better than actual starvation to our poor horses. I had often had my servant, Howard, to lead my grey horse, Roderick, to a fresh white oak sapling that the poor creature might eat the bark from it."

On December 13 Black was directed to select 5 of his men most deserving a new pair of cavalry boots donated by a lady in Richmond.

"Our men were poorly clad & in the coldest part of the winter I had nearly 300 men without blankets." Black continued, I ... wrote ... to Professor Maxmillian La Borde [1804-1873, President of South Carolina College and Head of Association of Relief of South Carolina soldiers. In the winter of 1864 he provided $1.1 million in material and supplies.]... to help supply my men with blankets. Dr. La Borde responded promptly and sent me about two-thirds of a supply at once & more soon after.

While we were without tents, our men improvised shanties and kept warm and under cover when not out on picket. The rations for the men were generally good and my Regt. enjoyed good health... The best winter huts were partly under and partly above ground. The dirt dug from the excavation often forming the roof or covering and chimneys were easily constructed as the clay cut away left a fine place with slight chimneys built above. Confederates were experts in burrowing in the ground."

The regiment remained camped near Hamilton's Crossing until called out due to a Federal advance in early February. They reached Milford Station on the 8th, where they learned they were not needed, and returned to their old camp the next day. They continued to picket along the Rappahannock.

On 13 February General P. B. M. Young present a new battle flag to the Regiment.

"Orders

Men of the 1st S. C. Cavalry. The Brigadier-General commanding, presents to you a new battle-flag, which he hopes will be appreciated by you as a token of the appreciation and admiration which you deserve at the hands of your countrymen; and let it ever remind you that your deeds of heroism at Brandy Station, at Upperville, and at Gettysburg will never be forgotten. But above all, the Brigadier-General commanding desires to express to you his unqualified admiration of your conduct at Bethel City Church on the 18th of October 1863, and upon his memory is engraved in indelible characters the maddening shout that arose when the 1st S. C. Cavalry was called upon to charge, and the flash of sabres as hand-to-hand they rushed upon the dense lines of the enemy's infantry, led by the gallant Lieut. Col. J. D. Twiggs.

The Brigadier-General commanding feels assured that this flag will never trail in the dust, but that the gallant men of the 1st S. C. Cavalry will protect it now and forever, even if this war should last till the crack of doom."

The 1st responded:

"Resolution of the 1st S. C. Cavalry. Whereas the various commands throughout the Army of Northern Virginia and other armies of the Confederate States, are reenlisting for the war and adopting spirited resolutions expressive of their determination to continue in service until the foot of the invader shall be driven from the soil of the Confederacy; and whereas this Regiment, having enlisted 'for the war' from the beginning, has no opportunity of expressing its determination by re-enlisting.

In order, therefore, that the sentiments of the 1st S. C. Cavalry on the subject they be made known both in the army and elsewhere, be it

'Resolved,' That we reiterate our former determination to continue in the service until the last foot of the enemy is driven from our territory, and peace upon honorable terms restored to our beloved countrymen.

'Resolved,' That the commanding officer of this Regiment be requested to forward a copy of the above resolutions to Gen. Young.

        T. W. Whatley, Capt. Co. C,

        J. S. Wilson, Capt. Co. D,

        S. H. Jones, Capt., Co. A.

Committee Appointed to Draft Resolutions."

1st Sergeant Kee wrote on the 16th "we have had beautiful weather for the last month." He noted that Lt. James W. White was commanding Company H. White reported the regiment was camped near Fredericksburg on the 22nd.

"Towards spring our Regiments [1st and 2nd South Carolina], much reduced, were ordered back towards Hanover Junction & only lay here a few days when ... were ordered back to South Carolina and our places to be supplied by the 4th, 5th & 6th S. C. Regiments of Cavalry, prior to that time doing duty on the coast of South Carolina. I had at that time only 27 first class serviceable horses. I had tho a detachment of 65 Remount horses marching from South Carolina to join me and I had 250 or more in a recruiting camp. Even after this ---- winter I thought I could go into the approaching campaign with 550 to 600 fully mounted men." The regiment had 300 men present.

General Samuel Cooper issued the order on March 17th, having Beauregard sent the 3 South Carolina regiments, plus the 7th Georgia Cavalry and other units to Virginia.

"I resented being ordered back to South Carolina from an active to an inactive field," wrote Black. "I ran down to Richmond and got the order so modified as to order my Regt. to Columbia to halt there for Remount purposes and, detailing merely men enough to drive my horses thro, I put the others on the cars via Richmond to Columbia."

"I marched my Regiment to the Exchange Hotel [in Richmond] and there procured quarters. This Hotel had been rented as a soldiers' house, either by the State of S. C. or by the Ladies Home Association and was kept by that most excellent gentleman, Geo. H. McMaster of Winnsboro and was supplied by charitable contributions from our people at home. Officers paid small rates for the times & fare was free for enlisted man. He seemed to think so many would create a disturbance but I assured him he could be easy. I was in command & not an act of an improper kind was committed or a boisterous word heard while we were in the House."

"At one o'clock we marched away to the Petersburg train & embarked. Before marching I had procured from the Adjt. & Insp. Gen'l. Office an order vesting me with some discretionary powers as to furloughing men to procure horses. My men on the train would reach Columbia in three days. It would take the horses two weeks and over to march overland. Before getting this order from Gen'l. [Samuel] Cooper in person I had pledged my word that I would send on from Columbia 250 horses & men ... the day after we arrived there. To do this I had the 27 serviceable horses I had in camp and the 70 Remount horses that had joined me the day before I marched. The 150 others I had to collect in about 10 or 14 days. In order to do this, I let men who had broken down horses go home for 14 days with the promise that they would bring in a fresh horse and some means to send their old broken down horses back home."

"In my camp in Va., the night before I left, I shut myself up with two or three clerks and wrote 300 Furlough orders in Blank and Adj. Ragsdale to carry them in his satchel on the march. Of these the men knew nothing. I also furloughed Lt. Col. Twiggs for 14 days to report to me in Columbia at the end of that time & so let him go home…"

"At Petersburg we had an hour's delay & some trouble to get cars. A. R.[ail] Road man wanted me to put some of my men on top of the cars. This I refused to do and gave him a round rating for proposing such a way of transferring soldiers. "

Company records indicate they departed Petersburg in detachments between the 20th and 23rd of March and arrived in Columbia between the 25th and the 1st of April. The horses arrived in Columbia between the 14th and 20th of April. The regiment numbered 730 men on the 4th.

Black continued "At Raleigh we lay over six or 8 hours waiting for cars & here some of my boys (never know exactly who) fixed up to go and hang Holden. I got a hint that some mischief was brewing and confined my men inside of lines. ... From Raleigh we went ahead & thru Charlotte. At Charlotte I commenced handing out 14 day furloughs and dropping men along the road. At Rock Hill a crowd of relatives & friends of Co. H were waiting to see them." Black allowed Private E. Preston Williams, who was under sentence of a general court martial, to see his wife and child. Black later furloughed him for 24 hours to report to him in Columbia, and Williams kept his word.

Black spend one day visiting his wife but was able to come home occasionally while the regiment was posted in Columbia. ".. I was very busy arranging my Regt. I had in camp a mere skeleton Regiment organization, not over 80 men, and at first no horses at all. In a few days we got in some horses and new men. I camped near Barhamville one mile from the city limits."

"The Conscript laws were vigorously executed and every day a set of the most unsuitable skunks, lame, and blind & halt were being turned out and all such infernal cusses wanted to 'jine' (join) the cavalry. Major David Melton was the Conscript Officer for the State & Melton was disposed to humor me... I carried my point & took none but boys. These I got rapidly and of a good select class, mostly the sons of farmers & many of them fresh from the plow. They were docile & tractable & readily adapted in full into the habits of soldiers' life as shown them by the veteran soldiers – I followed this up afterwards until I had 200 boys & Napoleon was never prouder of his guard ... than I was of my boys. "

"In about 14 days my horses came thro & I ordered Lt. Col. Twiggs to Charleston with about 280 men Rank and file... On my arrival in Columbia I had gone to Charleston & reported in person to Gen'l. [Pierre G. T.] Beauregard. I carried with me my Regt. Morning Report, explained the situation of my Regt. and prospect of recruiting, and was directed by him to carry out my plans as begun."

Black met General Hampton at Branchville and they both exchanged their feelings of the 1st being transferred. Hampton assured Black he had nothing to do with the exchange of regiments. Black asked the general to inspect his regiment and Hampton did so."

"Our old regiments [1st and 2nd], ragged and battle scarred, met the three new and fresh regiments in Columbia on their way to Va. They were dressed in new clothes & much of them. The men of my own and Lipscomb[e]'s Regt. 2nd S. C. C. dubbed themselves 'Old Issue' and the new Regiment[s] 'New Issue' and both took to the name kindly."

Starting April 16th detachments of the regiment were stationed at Green Pond. This enabled them to draw forage and rations from the area.

"I continued recruiting & with success. Gen'l. Beauregard was relieved of command by Major General Samuel Jones of Va... I was ordered to Charleston in person by Gen'l. Jones. He did this to order me to come and make an examination of the mountain passes from Walhalla to opposite Greenville to see to the roads & points of defensive lines... I examined the Rabun Gap as far as Steep House Mountain & Jones Gap east of Greenville." Black and Captain Sharpe conducted the survey and made their reports to Jones on May 6th and 10th.

"Soon after [14 May]," Black related, "I was ordered by Gen'l. Jones... to put my Regt. on the first train and report in Charleston. I got the order about sunset and, striking camp, marched at once to the S. C. depot where I found I could get no train until morning."

"Arriving in Charleston, my own & 2nd. S. C. Cavalry, dismounted, as our horses were being driven thro were ordered at once to James Island and landed in the ---- and were ordered to march three miles up the Island. This we did, wading in the sand."

The company commanders reported arriving on James Island between May 16-24. Black had the stronger of the two regiments and remained on

James Island. "My horses soon arrived & my Detachment under Lt. Col. Twiggs [from Green Pond] was ordered to me and so I at once had a Regiment in good trim of about 600 men Rank and file," Black concluded.

General Jones ordered Brigadier General William B. Taliaferro, commanding the 7th Military District of South Carolina on 9 June "You will proceed with as little delay as possible to equip the companies of Black's regiment, South Carolina Cavalry (now with you), as to render them serviceable for active duty, with their horses, at any point..." He further directed General Beverly H. Robertson to "Send all the companies of the First South Carolina Cavalry at once to Brigadier-General Taliaferro. The detachment of the 2nd South Carolina Cavalry, now in this city, will be sent to you as soon as relieved by a company from James Island."

"Our defensive lines commenced at Fort Johnson & ran to Secessionville, thence by an oblique line of earthern works strengthened by a series of Redouts & batteries to the River on west of Island, ending in a strong battery... mounting heavy guns," noted Black. "The Island was ... divided into the east and west lines – East lines Fort Johnson, Legare's Point and Secessionville, 3 Post & connecting but not continuous lines, the West line all beyond Secessionville. I was assigned to command of the East line... My Regiment was camped in rear of the centre of the line & served as a moveable force of mounted Rifles with which we could, in an emergency, reinforce any part of our long and attenuated lines."

"It took about 4500 men to man the lines and even this force would not allow a moveable reserve of 1000 men to concentrate on any point of attack. In fact, the lines were scarcely defensible against a vigorous attack made at one point by that force &, had the enemy attacked two or more points simultaneously, they would have certainly penetrated the line."

".. the innumerable night pickets required our men on duty, at least every third and often every other night. The preservation of this Island, was much of it due to the ability, skill & untiring energy of Major [Edward] Mangault [sic], an officer specially entrusted with the management of the pickets on the West lines, the assailable point. He slept with an eye open, if indeed he slept at all."

"Port [sic] Johnson was ably commanded by Lt. Col. Joseph [A.] Yates of the Regular artillery. Yates's Garrison consisted of several regular companies of the 1st [S. C.] Artillery (Col. [Alfred] Rhett's Regiment). These were magnificent companies & well officered – and kept in the highest state of discipline by Yates, who was a vigilant Post commander and, at the same time, a skillful artillery officer... As an auxiliary garrison he had two companies of Col. --- Volunteer Regt. of Artillery & these were by Yates kept in good order as it was possible to keep volunteers. "

"Legare's Point was Commanded by Lt. Col. [William H.] Campbell [Palmetto Battalion South Carolina Artillery] ... This was a post of not much importance such as a centre of the East line. The garrison was not very strong and the discipline rather lax."

"Secessionville, which was a salient angle of the general line, was under command of Lt. Col. [J. Welsman] Brown of [2nd South Carolina Artillery] ... The works were strong & the same well supplied with smooth bore guns & some old Howitzers. There was tho fronting Folly Island a battery of two rifles 42's, old U. S. ordnance rifles. These guns were sufficient to often silence a Parrott battery the enemy had on Folly Island, tho our supply of rifle shells was so limited we could not often use them, having always to keep two or 300 rounds on hand for an emergency."

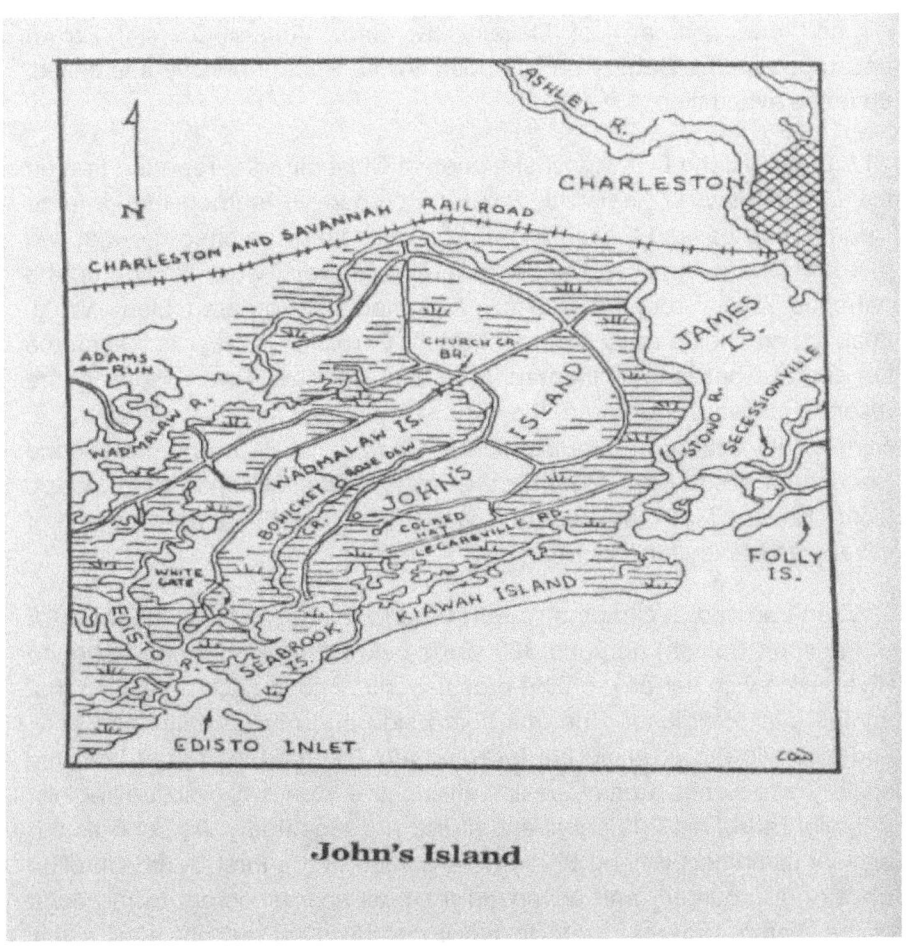

John's Island

"The armament of Legare's Point were mostly old smooth bores & seacoast Howitzers & many of them quaint out --- guns... The armament of the main and outer works of Fort Johnson were supplied with a good class of guns bearing on the channel."

"At the Point was a heavy [battery] of 4 ten inch columbiads. This battery was mounted by a company of Rhett's Regt. commanded by Captain [Christian] Gaillaird [Gaillard], a most excellent officer, and accomplished gentleman."

The other flanking lines were well armed with good guns and s perfect was Yates' management that they were always ready for service."

"I had one company of Cavalry (K, Capt. Angus O. [sic] Brown encamped on the Battery on city point, White Point, I believe it is called, up to the evacuation."

Lt. Colonel John D. Twiggs, stationed at Chisholmville, reported that on the 24th of May, Lt. Isaiah J. Fox, of Company I, learned that a large steamer had passed Chapman's Fort. "Later he dispatched me that two more steamers, supposed to be Federal gun-boats, were coming up the Ashepoo River," reported Black. "I immediately ordered Lieut. W. J. Leak, commanding detachment 1st South Carolina Cavalry, to re-enforce the pickets between Chapman Fort and Chehaw River. Holding the reserve midway, I ordered the first section of Capt. W. [William] E. Earle's light battery to take position at Chapman's Fort, and the second [section] at Mr. William Means' causeway, on the Ashepoo road. Capt. [John Raven] Mathews' [battery] was supported by a detachment of cavalry and ready for action."

"When I arrived at Chapman's Fort I found a large Yankee steamer [U. S. transport Boston] aground 300 yards below the piling in the Ashepoo River, lying with her bow on the breast-works, and in such a position that my left gun commanded her starboard side and my right her port bow, and also allowed a raking fire fore and aft. The third shot fired from the artillery struck her steam-chest, causing the steam to escape rapidly. Captain Earle fired 200 shots and struck her repeatedly. As soon as the artillery commenced firing I ordered a portion of the First South Carolina Cavalry dismounted, and advanced them as sharpshooters to the edge of the marsh, ordering them to fire upon the steamer; but after a few rounds, finding that it caused no diminution of the numbers on her decks, I caused the fire to cease."

"The enemy did not reply to Captain Earle's fire until he had fired 20 shots. His fire was not returned by the vessel aground, but by another Yankee vessel 2 miles off. The fire was kept up by first section of Earle's light battery and the farthest Yankee steamer until 8:30 a. m., when two more steamers made their appearance, and after a very short period two more came up, making five in all. The four steamers afloat then commenced shelling from Chatman's (sic) Fort to Chehaw River."

"At 10 a. m., I ordered Captain Earle to cease firing, his ammunition becoming short. The boats which came up last had a great many men on their deck, and fearing an infantry flank movement I told them to retain a few shell, grape, and canister. During the cessation of Captain Earle's battery, the enemy shelled furiously, making two trips to the steamer aground; took off her crew and troops. Finding her very disabled, they set her on fire and retired down the river. ... The steamer that was burnt was boarded by a detachment from my command, and a number of burnt horses were found."

Great praise is due Captain Earle, Lt. [James F.] Furman, and the men of Earle's light battery for the admirable manner in which they handled their guns... My thanks are due to: Brevet Second Lieutenant Isaiah J. Fox, Company I, Private J. M. Schnierle, Company I, and Sergeant [Manson S. J.] Jolly, Company F, First South Carolina Cavalry, for their efficient services and gallant manner in which they acted. The officers and men of the First South Carolina Cavalry, who have met the enemy before on so many battle-fields in Virginia, were eager again to show on their native soil the daring spirit of the old Hampton brigade."

The 'Yorkville Enquirer' reported on June 29th that "W[illiam] E[M.] Doby, C, 1st S. C. Cav will take charge of packages, etc., [for the men of his company] at Hamburg on 10 August 1864."

"Service on the Island was monotonous and dull," Black wrote. "It was a daily routine and never varied. The enemy was incessantly shelling the city, day and night, from a battery on the main side of Morris Island known as the 'Swamp Angel.' From the battery they thru shell [that]... flew harmlessly over our head on their mission to batter down & tear to pieces the houses of a deserted city as all the of lower & the finer part of Charleston was deserted by its inhabitants & the military garrison of the city proper was very small."

Captain Trezevant complained that the horses were getting light rations, only 2 bushels of grain per day. The other company commanders had the same complaint.

The Federals seized Seabrook Island and advanced on John's Island, in an attempt to capture and destroy the railroad bridge and trestlework over the South Edisto River and the one over the Ashepoo.

Meanwhile, on the night of July 1st, the Federals attempted an amphibious landing by long boats near Fort Johnson and Battery Simpkins. An alert Confederate sentry fired on the boats as they approached, alerting the garrisons of both positions. Colonel Henry M. Hoyt of the 52nd Pennsylvania Infantry, was the only Federals to land successfully, but only part of his regiment got ashore. Hoyt was able to seize Battery Simpkins, but met strong resistance when he approached the defenses near Fort Johnson. His command was soon surrounded and forced to surrender. His regiment lost 7 men killed and 16 wounded, while he and 5 of his officers and 134 men were taken prisoner. There were also some casualties in other units who tried to get ashore. Following his exchange Hoyt reported "we carried Fort Simpkins, the Brooke battery and pushed to the parapets of Fort Johnson and the garrison had actually begun to leave. The battery (Tynes) was in our possession. All who landed (five boat-loads, 135) were captured." He later added "… the parapet was entered near the main fort with a brisk movement of about 30 of the advance, who exchanged shots within the work, but were compelled to retire. The whole of our force was then conducted along the entire line from the rebel left to right, with repeated efforts to enter it, until at the extreme right another assault was attempted. It was only partially successful and resulted in the capture of most of the troops who joined in the attempt."

"At this time my forces were very largely outnumbered; the controversy was prolonged some little time, but in a feeble and desultory manner, and the undertaking was of necessity abandoned. The entire party was taken prisoner."

Lt. Col. Yates reported "I have the honor to report the captured of 140 prisoners, including 5 commissioned officers, including some wounded; also 5 barges, 114 stand of small-arms with accouterments. Enemy's loss in killed and wounded cannot be estimated, as most of the wounded were taken off. …. Our loss very small." Later he added "The enemy immediately after dark made an attempt to take Battery Simpkins, but we opened immediately upon them, heavily, with artillery and infantry, driving them back…. We are all right and ready for them."

General William B. Taliaferro. Commanding the defenses on James Island, reported another attempt by the Federals to make another barge attack on Fort Johnson and Battery Simpkins on July 10th. They were again repulsed, as Yates had been reinforced by a light battery and a company of C. S. Marines. On July 23rd the general praised Colonel

Black for the "energy and vigilance" displayed while commanding the east lines, Captain Trezevant, and the whole 1st South Carolina Cavalry for their bravely under fire.

Black remarked "This attack was fruitless and only lasted a day. It tho showed to our Engineers the necessity of a flanking battery on the West bank of the Stone [sic] and such a battery was erected, tho I do not know if it ever was fully armed."

The regiment number 600 men on James Island on July 21st.

On July 28th a deserter from the 1st South Carolina Artillery told the enemy "At Fort Johnson five companies of heavy artillery are behind the breast-works every night, one to serve the guns the other four used as infantry; one company of Black's cavalry regiment also reports at Fort Johnson for duty every night."

The return of the Department of South Carolina, Georgia and Florida showed General Taliaferro commanded the 7th Military District and Black commanded a brigade consisting of the 1st Georgia Regulars, 5th Georgia Infantry, 47th Georgia Infantry; 1st South Carolina Cavalry with the Pee Dee Artillery and Company D, South Carolina Siege Train, all under Lt. Colonel Twiggs; 3 companies of the 1st South Carolina Artillery. Company A, 1st South Carolina Light Artillery, 8 companies of the 1st South Carolina Heavy Artillery, Major J. John Lucas' South Carolina Battalion of Artillery, Chatham Georgia Artillery and Company B, South Carolina Siege Train Artillery. Captain Angus P. Brown led his company and one from the 2nd South Carolina Cavalry in Ripley's Brigade in the 1st & 4th Districts.

Black's brigade totaled 181 officers and 2,984 men present for duty, 3,215 effective total, 3,472 aggregate, and 5,133 aggregate present and absent. Eight pieces of light artillery were with the artillery units of the brigade.

Company I, Captain Fox, was ordered to Mount Pleasant on August 24th for several days. Company H was on duty at Coosawatchie at the same time. 1st Sergeant Kee reported from James Island "great many sick with chills & fever" on September 9th. On September 19th Company I was back on James Island.

Part of Company H was ordered to Florence during August and remained on duty there until the end of the year. A prisoner of war camp was built there.

During September Black granted Twiggs a ten-day furlough. Twiggs went to his home and had an altercation with a citizen and was shot and killed. Walker was promoted to Lt. Colonel and Captain Thomas W. Watley, advanced to Major.

On September 9th the enemy was firing on Fort Johnson and Black requested permission for Lt. Colonel Yates to fire on the enemy guns in forts Gregg and Wagner on Morris Island. Yates believed he could fire effectively, without endangering the Confederate prisoners in front of the Union guns. General Jones's reply is not recorded.

The return of the Department of South Carolina, Georgia and Florida for 31 October, now commanded by Lieutenant General William J. Hardee, shows General Taliaferro commanding his brigade in the Third Sub-District of South Carolina. Five companies of the 1st South Carolina Cavalry were included in his command. Captain Brown's company was still in Ripley's Brigade. Lieutenant John Henry Copeland, of Company B, commanded a detachment of the regiment assigned to the post of Florence, South Carolina.

The same day General Robert E. Lee recommended to the Secretary of War that a battalion of cavalry or more be send to aid in the defense of Wilmington, North Carolina. "The First and Second South Carolina regiments which formerly belonged to this army might spare the desired force, as they are veteran troops accustomed to severe picket duty." Other units were sent to Wilmington.

The 1st was reported at Florence on November 29th

1st Sergeant Kee reported part of Company H was on James Island on December 5th.

When General Taliaferro returned from a month's leave of absence, he was assigned as a division commander. General Stephen Elliott was placed in command of the troops on James Island.

General William T. Sherman's captured Savannah, Georgia, on December 21st, and advanced towards South Carolina. General Hardee, with the troops evacuated from Savannah, retreated into South Carolina. Elliott was ordered to bring most of his command to reinforce Hardee's Army. He requested Colonel Black to determine the number of troops that could be spared, and still hold James Island for at least 3 days.

Black believed he could hold the position with 4,200 men. Elliott took the balance of his command and joined Hardee.

Captain Trezevant, with 16 dismounted men of his company, were ordered to the Coosawatchie. By 28 December there were 121 effectives there, 140 present and 145 present and absent. By 28 January his command was reduced to 89 effectives of 91 aggregate.

The mounted portion, under Lt. William D. Conner, remained on James Island. By the 28th of January Trezevant, now joined by the rest of his company, was in Colonel George P. Harrison, Jr.'s Brigade, and posted at Braxton's Bridge. He reported 89 effectives and 71 aggregate effectives. The 1st was listed in Elliott's Brigade in Taliaferro's division at the end of January. Company K, Captain Angus P. Brown was in General Hugh W. Mercer's Brigade in General Ambrose R. Wright's division. Company E, led by Lt. L. W. Lusk, was listed as not being in a brigade, and was probably assigned to courier duty. Company H remained at Florence, guarding the prisoner of war camp.

Sherman's advance into South Carolina had been delayed by bad weather. On February 1st he started his Army forward, with some 60,000 troops, 2,500 wagons and 600 ambulances. Beauregard, in overall command of the Confederate forces, had some 22,500 troops to oppose the Federals.

Trezvant reported only 43 men present for duty of 49 aggregate present at Cooswatchie on February 10th.

On February 12th Black received the following orders from Taliaferro:

"You will march with your command to Holly Hill, a point northwest of Sandy River, taking the road via Summerville and Ridgeville. You will examine the passes of the road through the 'by pass,' and the crossing of the road through Four Hole Swamp near Dr. Murray's. You will make your headquarters near Burch's Ford beyond Holly Hill, and dispose of your command so as to cover any advance of the enemy from the direction of Orangeburg, particularly guarding the passage of the Five Notch road through the Four Hole Swamp. You will establish a courier-line from Burch's Ford to George's Station, and keep me advised of your movements and dispositions and those of the enemy. You will also communicate with General [Lafayette] McLaws at Branchville and with any part of General [Carter L.] Stevenson's force that may fall back toward the Santee, Four Hole Swamp, or Sandy River. You will especially observe and report if the enemy seek to strike at the Santee River toward Keysville or Vance's Ferry, and will resist, as far as you can, any force that you can cope with. Should the telegraph station at George's Station be interrupted by the threatening of the enemy you will establish a line of couriers, via Eutaw Spring and Pineville, with Saint Stephen's Depot on the Northeastern Railroad."

Later the same day Taliaferro directed Captain Peyton N. Page, his Assistant Adjutant General to:

"See Colonel Black and tell him to send thirty of his men, with a good commanding officer, to the Sandy River country as ordered, and with the rest to proceed himself to reconnoiter and hold the country towards Bull's Bay, passing from Summerville early to-morrow morning to the first crossing by the bridge above Strawberry Ferry, and holding the enemy in check as far as he can if he advances toward the Northeastern Railroad, and keeping us well advised of the movements of the enemy and his own, via Monk's Corner."

Captain Page wrote Black again two days later in response to a letter from him to General Taliaferro:

He directs you to "push on to Bull's Bay. Resist any effort of the enemy may make to strike the Northeastern Railroad, and keep him informed on the movements and position of the enemy. He directs that you will guard especially against any attempt on the part of the enemy to advance from or land at McClellansville, and send him as accurate and estimate as you can make of the enemy's force. A telegraph office will be situated at Monk's Corner tomorrow."

Black's efforts were of no avail, as Sherman, who had reached the Augusta-Charleston Railroad on February 7th between Bamberg and Windsor. Hardee was forced to withdraw across the Edisto River to northwest of Charleston and Stevenson to Orangeburg. Sherman sent Oliver O. Howard's Corps through Orangeburg toward Columbia, and Henry W. Slocum's Corps from Windsor toward Lexington.

Beauregard was now in command in South Carolina and Wade Hampton in charge of all the cavalry. With their defensive lines penetrated, Beauregard ordered Hardee to evacuate Charleston and move his forces to Cheraw. He ordered General D. H. Hill to have his arriving divisions from the Army of Tennessee, to Chester.

Colonel Harrison made another report on his brigade on February 16th, which showed Trezevant's Company with only 40 men present for duty and an aggregate of 49 men. Former Governor Millege L. Bonham was reappointed Brigadier General on this date.

Beauregard reported the evacuation of Columbia on February 17th. He was moving his command toward Charlotte, North Carolina. Charleston was evacuated the same day.

The route the 1st took was not reported but Black reached Florence by rail in time to help drive off the Federal cavalry and saved the trains, railroad and supplies there. Black reported "Lieut. John Hankerson and five men of my Regt. had been left by me west of the Pee Dee [River]. He was out as a scout or picket. He had just got into Florence ... when he heard the Yankees advancing. He ... ordering his men to mount, reported to Major [John] Jenkins, & offered to assist him. When the enemy advanced and on firing one or two shots, one of Hankerson's men, a new recruit... made tracks for the rear... As soon as the fight was over Lt. Hankerson" chased down and captured the man. Black wanted to prefer charges against the man for "cowardice in action, deserting his colors in the presence of the enemy & stealing a citizen's horse." Black wanted to have a court martial, but the end of the war saved the man. "In a few hours, finding the enemy gone from Florence, I recrossed the Pee Dee with my command and into camp."

"It was apparent to me that the enemy would cross the Pee Dee at Cheraw &, turning to the right down the River, cut us or my part of the command off from Fayetteville. "Black recommended to his superiors "burning our heavy baggage, cutting out our wagon mules, and mounting every available man, making a rapid march of 40 to 50 miles obliquely up & from the River, boldly throw ourselves in Sherman's front and fall back before him to Fayetteville, impeding his advance as much as possible and annoying him all we could." His plan was turned down by General Felix H. Robertson. "I begged him to order me throw with my own Regiment, this he would not do and, angry & disgusted... I quit the [railroad] car abruptly, swearing like fury."

Soon came an order for me to re-cross to Florence with the forces under my command. This I did and, what was worse, had orders to burn the R. Road bridge. This I did... We were to march to Cheraw following Hardee's trail. I saw at once that demoralization was at hand and coming from all sides. We were having frequent desertions, or rather leave takings, as many of our best men were slipping off home with the avowed purpose of coming back... They were principally men of families from sections that it was known Sherman had over run and plundered and burned – I did all I could to stem the tide. I detailed Sergeant James Atkinson, Co. D, to take charge of a mail of letters, to carry them to Camden, Winnsboro and Chester so as to let the men write home and hear from home. ... This task was carefully executed and Atkinson reported back without unnecessary loss of time."

Black had his men take homespun sacks and loaded them with cooking utensils and axes on mules, and did away with wagons. "These served until the end and were of much use as they were always at hand and could easily keep up with the column of march." He also issued his command all the chewing tobacco they wanted from abandoned stores at Florence.

"Our march from Florence began and we only moved a few miles above Darlington C. H. the first day. The second day a few miles above Society Hill. ... We had moved a few miles above Society Hill and gone into camp. Soon after dark a Captain came and took me aside and told me privately that a number, 40 to 60 men, were banded together to leave that night for home. He also informed me that Lt. R—of Co.—said that the war was over & that we had as well go home." A dumbfounded Black was surprised to hear both of the defection of the men, who knew were good soldier and Lieut. R. who was a good and brave man. Black was gloomy himself and suffered from chills and fever nearly every day.

"I ordered to have my Regiment turned out on foot and formed in close columns of squadrons. The guard formed separately under the command of the officer of the day. The bugle sounded and the Regiment was formed in less than five minutes & was formed in rear of a wagon standing near my headquarters' fire. I mounted in the wagon & gave command 'Attention.' All were silent. I commenced by saying that they all knew I never allowed speech making, they all knew I was no speaker & they certainly knew I was no preacher, yet I had a few words to say and would say it... & commenced by saying I have been told tho I don't believe it, that there is some demoralization in my command --... that one of my officers had said the War is over and it no use to hold out longer- [Black gave several examples such as Gates defeat in the Revolutionary War, and others]. And now, fellow soldiers, I know we are low down and many of our great men have fallen but as for my humble self, I mean to stick to the ship and by as long as I can get a single corn ash cake – I then called on the officer of the day by name and ordered him to relieve all the pickets and out guards so as to let any man that would leave would go without being seen.

'Go,' I said, if you will. I am powerless to prevent you –but I & my orderly & that old flag ... will march on and report for duty with Gen. Hampton as soon as we can come up with him. Even if all leave me, I will go alone & report the tale of sorrow to Hampton that one of his old regiments has gone home. But this will not be the case, as I well know the old Regt. will stick to its colors to the last --- But a word more and I am done. All of you, everyone are either married or single. Married men, beware your wives, all South Carolina Ladies, and if you desert & go home they will not sleep with you and, as to you that are single, you each have a South Carolina Sweetheart. Desert if you dare and they, your sweethearts will desert you---They, Carolina's daughters, will never marry dishonored men. But in conclusion, soldiers, I think better of you. You may do wrong under the impulse of the moment but you are now too much battle scarred to act up in an improper manner."

'Break ranks, march,' "were my concluding words & words of command & such a yell as came up from the whole mass never was heard. They yelled & yelled again & the yell was heard for miles. They broke off to their quarters and I lost few if any more men. When I heard that yell I knew I was in command & that I had touched the right cord & awakened a spirit of pride that would brace them up. I was at the very instant proud of my Regt. tho I always was proud of it. Gen'l. Robertson, who was camped a mile off, sent to know what the matter that my command made so much noise."

"We moved on from our camp above Society Hill. My Regt. was in the centre of the column which was composed of Col. [Charles J.] Colcock's Regt. [4th South Carolina Cavalry], Capt. Kit [Christopher] Gailliard's [South Carolina] Battery & some other Cavalry, perhaps [M. J.] Kirk's [19th South Carolina] Battalion. All at once a halt took place in front and a message came back to me from Gen'l. Robertson that there was a broken down bridge in front & that he would be pleased if I would come ahead & see that it was properly repaired. I rode forward and soon came to a sand hill creek, the bridge over the creek proper or its main run was not over (10) feet long was gone. This could be speedily repaired but the worst part was a wash out in the log causeway that led to the bridge. This wash out was 15 to 18 feet wide and most of the creek water was flowing through it. But there was no bottom as it was a peat bog or swamp... It was totally out of the question to go around or avoid this place."

"I sent back to my train to Capt. [James] Henderson to send me up 500 new sacks ... we had saved from the wreck at Florence. These soon came up & getting a detail from Capt. Gailliard's battery, I speedily had them filling sand bags... and forming a small party of cavalry in single file, made the men carry the bags on horse to the break. Here I stood and directed the filling in by a Sergeant & two men. In an hour or less we had the break well paved with sand bags. The water cleared away rapidly... The bridge soon repaired, the passage of the hoofs commenced. I stood by until my own Regiment had passed and then crossed... I left a man to stay & see [how it would stand the pressure of the wagons] and report to me. He reported all in good order... after 200 animals passed. "

"The middle of the same afternoon we reached Thompson Creek. The bridge... was free from water but the swamp on the North was flooded... A halt was ordered & preparations to go into camp. "Black requested Robertson allow him to move the 1st across, and received permission. "I got all over safely... I marched on 4 miles, turning to the left towards Chesterfield C. H. and camped, procuring some forage and some food for my men."

"The next morning, having no orders, I moved 5 or 6 miles & halted to gather up forage and to await orders & the coming up of our Brigade Commander. He came up later in the day, not in a good humor, as he and his part of the command were half starved. My own Regt. had some supplies to go and were refreshed. We moved on, leaving Chesterfield C. H. to our left & marched towards the town of Wadesboro—halted ... 8 miles short of the place and not much forage."

"As we were now leaving the sand belt, it became a matter of great importance to have our horses shod at once... the day before I had send forwards towards Wadesboro not only a swarm of foragers but a large horseshoeing detail under Lieutenant Frank Blassingame. When he got within 4 miles of Wadesboro he found the wreck of a bayonet factory which had been burnt... in the wreck about 500 lb. of nail Rod and other Iron. "Blassingame sent back for a wagon, loaded it with the iron and moved on to Wadesboro "& put all the blacksmith shops to work for my Regt. I sent on 75 horses to him to shoe & so rapidly did the work go on that I had 150 horses shod in 24 hours."

When the 4th regiment arrived the next day, Colonel Colcock found all the blacksmith shops busy shoeing the 1st's horses. Black furnished him with some iron to have his horse shod.

"All of the commands except my Regt. were coast troops that had never seen service other than on the S. C. coast... were poorly adapted to make a rapid march thro a country, almost a desert by the ravages of war. I was very apprehensive our horses would starve and break down. I kept a cloud of scouts all the time out on flanks and front looking for forage... I kept quiet as to my operations, never acknowledging that I had any food for man and horse. Each of my men had one or two of the homespun sacks ... and every man was instructed to keep always in store in his sack one or two days' horse food. Our old veterans were by this time ... [hungry] and food for themselves was ... doled out with a sparing hand... Our discipline was so good that, with good supplies today and one day's rations in the bags on horseback, we could make a march at 50 to 60 miles in 36 hours and not break down our horses."

"From Wadesboro we moved up the Pee Dee towards Christian's Factory below the Enharie [Uwharrie River]. The next evening, my Regt. in front, we came to a small town called Ansonville... Here my scouts found a large amount of wheat in the hand of the commissary. I piled into it and had the local agent... transfer it to my quartermaster and had it promptly removed. By the time was removed one of my men found 100 sacks of wheat bran. Tho full, I had this at once seized and moved on two miles to a place I intended to camp as my scouts reported I could. I did get an abundance of long forage at this spot. "

"I moved at dawn and, marching rapidly, I came to a small Cross Road Store 5 miles east of Christian's Factory and here finding some more wheat in a commissary store. I halted at 12 M[midnight] and went to camp. At the same time I sent on to examine the ferry on the River and determined to cross in the night as I had just got an order from Gen'l.

Robertson to cut loose and march on Raleigh to report to Gen'l. Hampton... I ordered out 3 companies to the ferry at once and I kept sending detachments so that during the night my entire Regt. was ferried over the river and went into camp as fresh as ever ... I crossed with the extreme rear about sunrise next morning. Robertson's Brigade had been assigned to Hampton's command."

"At Christian's Factory I found another commissary store of wheat. From this I got first a supply of breadstuffs for men and had the balance cracked and, after feeding horses well, required each man to carry forward on his horse 16 lb. of cracked wheat as I was assured here by citizens I could get no corn or grain supply for the next 60 miles or until I got to Asheboro. I, therefore, commenced this march with man and horse well fed and with over one full and two days' scant rations. I moved up the East bank of the Enharie [Uwharrie] River – and direct to Asheboro."

"At Christian's Factory I found I was in need of over 30 horses to keep up my remount and determined to send out a party to press that number of horses... I had an order from Dept. HdQuarters directing me to keep my regiment well mounted and in serviceable trim... I was always careful to select the very best men and officers I had to make the impressment. In this case I selected Capt. A. T. Clayton and Lt. J. Wilson Marshall with 30 to 40 select men. I directed Clayton to divide his men and to move up one bank of the Enharie [Uwharrie]... the other ascending the other bank and join me at Asheboro, the next day..."

"About 4 P. M., having made a considerable march & being within four miles of Asheboro, I saw on the wayside some stacks of hay. As I was then taking a chill, I ordered Lt. Col. Walker to take command of the Regt. and to go into camp and feed upon the hay. I rode on into Asheboro accompanied by a single orderly."

"As I rode into town intending to find Dr. Worth & to be his guest for the night, I was surprised to meet on the wayside in the corner of a fence, Private [John A.] Murrell of Co. A, one of the men Capt. Clayton had on his horse impressing detail. Murrell told me hurriedly that Clayton had succeeded in getting a number of horses, that he had them with his guard in a lot in town, that there was a big militia muster in town that day, and that the people were much excited and the militia were talking freely of taking the horses away. ... I told Murrell to make his way back to

Capt. Clayton & tell him I was in town & the command near at hand & for him to keep quiet & not provoke a collision but to keep himself well fortified and to hold the fort to the last."

"I muffled in my overcoat with the chill just leaving me & I rode on into the public square where there were assembled a crowd of perhaps 150 people, plain country farmers. I rode past the crowd, dismounted, threw my bridle to my orderly & walked into a drug store and asked for Dr. Worth. "The druggist explained that Dr. Worth was out of town visiting the sick. "I laid aside my sabre, threw off my overcoat and, showing my uniform, walked out into the crowd & my appearance seemed to attract some attention. "Black explained that he had only impressed the horses by orders and explained that he would hold a court to investigate any complaints, and return such horses that were wrongly seized. An older gentleman volunteered to sit on the court and selected two other citizens to sit also. The court heard over 30 complaints and returned 4 horses, one to woman whose husband was with Lee's Army, and the only animal she had.

Black continued "My Regt. came up next morning & we marched due north towards Union Factory. I sent Lt. [Thomas S.] Miller ahead with orders to have some supplies ready for the command at Union Factory. But when I came up I found Miller with his finger in his mouth & no forage. He said the old Quaker who owned the factory denied he had any. "A local Confederate soldier, home on leave, advised Black that there were corn and bacon on the premises. Lt Miller acquired the needed supplies. The soldier told Black that if he "would march three miles ahead he and his neighbors could supply me with long forage... I arranged with my soldier acquaintance where I would camp & there he & his neighbors supplied us bountifully with hay."

"We were well fed and much refreshed after our long march from Christian's Factory, over 50 miles in two days & moved on next day. About 11 a. m. I was taken with a chill, sudden and like an epileptic fit. I rolled off my horse all at once. Dr. Yates was nearby and he and some of the men gathered me up and carried me in a house nearby. I gave orders to Lt. Col. Walker to march on so many miles & go into camp. "Black's orderly, John Carey, stayed with him.

"Yates watched me & I slept several hours. Awakening suddenly, I jumped to my feet & called out to know where my Regt. was & where was my horse. Yates came to me and told me I had ordered my Regt. 12 miles on & that I had had a severe chill & fever. I called for my horse. No entreaty would stop me. After a cup of tea ... I mounted and galloped 10 or 12 miles. I had some fever on me and felt terrible but my wish was to be with my command. ... Dr. Yates got some ... quinine for me at $2 per grain... I found my Regt. near the house of a very kind Quaker who gave me a bed in his house & here I had one of those night sweats, wet & clammy."

"The next day we marched on in rain and mist. Towards evening more rain began to fall and, well knowing I would have my usual chill, galloped ahead to select a camp and to try and get shelter in a house for myself. I rode 3 or 4 miles and came up with ... my commissary ... & he at once informed me we would have to camp near there as his chance to get bread stuff for men was from a mill a few miles off on a cross road." Meanwhile, Black stopped the impressing of a mule from a lady where they had stopped. "My regiment soon came up and soon after Mr. [General] Wheeler's Corporal rode up with four men and 8 or 10 horses. I had the whole posse put under guard at once, after having them disarmed. They had several carbines and nine army or navy pistols, four or five horses they had plundered from the people around I restored to their owners."

"Two days later after our getting to Raleigh I reported the arrest of these men to Gen'l. Beauregard and the circumstances." Black had the men turned over to the Provost Marshal. Two weeks later Wheeler demanded the arrest of Colonel Black by General Hampton. "For a wonder, Gen'l. Hampton ordered them to be confined & me to send in charges and I heard no more from Mr. Wheeler."

"The day before we reached Raleigh Dr. Yates found a traveling acquaintance, Rev. Mr. Tripp of the Methodist Church, a low country man. He was introduced to me and, as he was making his way to the main army, I invited him to march with us and he did. In fact, Rev. Bro. Tripp took up with us & tho he was a chaplain of another Regiment, his headquarters from this out was with us as he became ours by adoption."

On March 29th General Bonham was assigned by the War Department to command a brigade consisting of the 1st, 2nd and 3rd South Carolina regiments. However, it doesn't appear he ever assumed that role, as the

2nd was with General Butler in South Carolina, and not with the 1st and 3rd.

The regiment was engaged at Gulley's on March 31st and reported one man killed and 3 captured.

Upon reporting to Hampton, Black explained his experiences under the command of General Robertson and would not serve under him again. "Hampton assured me I would not have to."

"Had orders to report to Army Hd Quarters with my command to Gen'l. Joe Johnson [sic] under which order I moved down the Neuse [River] to Smithfield & leaving my Rgt. reported in person to Gen. Jos. E. Johnston near Smithfield. ... I made my report and handed in my morning report. Gen. Johnson [sic] directed me to send my orderly to order my Regiment to go into camp where it was and that I would remain at his headquarters as his guest for the night. "Johnston had been expecting heavy reinforcements of cavalry from South Carolina and was in ill temper by the few units that had appeared with reduced numbers.

Black defended the 1st regiment and declared them ready for any service. "My ranks are full for the long retreat we have been in. By the hardest exertion, marching thro a devastated country, my quartermaster returns show an average of 11 lb. of corn to each horse, and over issue but I had made a desperate attempt to keep up – That my Regt. was in good trim & not demoralized." Johnston expressed regret for his outburst, and agreed to inspect the troops.

"The nest morning I again asked for orders and got them. It was to march down the Neuse 18 miles and report to Gen'l. Hampton at Moccasin Creek. I did so and found Gen'l. H. at Atkinson's house. The day I reached Hampton's Headquarters he moved to Boone Hill & left only my command on the road leading up and down the Neuse River. He was six or seven miles from me & on the Rail Road. In a few days Kirk's Battalion of So. Car. Cavalry was sent to me & remained under my command some time. It was exchanged for Beard's [6th] Regiment of North Carolina Cavalry. This regiment remained with me until after the fight at this place."

"While laying here, I did all I could to forage so as to feed up my horses. I found a pretty good supply of corn in an Island in the Neuse belonging to one Smith, a Union man. ... Mr. Smith had destroyed all boats but I managed to have some built & ferried it out, taking the last ear he had. I also procured a seine and made a detail to fish for shad in the Neuse to aid the men's rations."

"The enemy was now in Goldsboro and we were picketing within three or four miles of that place. Their Cavalry force was evidently small as they kept quiet. ... I had chills every few days, and indeed had half of the men in the entire command & my health was very poor all the time tho I kept up and on duty & sleeping every night in camp. "Black explained that all the troops who had been stationed around Charleston were subject to chills and fever, which was malaria.

"I saw very little of Gen'l. Hampton, now commanding the Cavalry Corps. He came once probably to my camp and I was at his headquarters perhaps once. The headquarters of the Infantry army was now at Smithfield and the cavalry Hdquar. at Boon Hill, 18 miles below and passing thro Boon Hill, the cavalry formed a line perpendicular to the Neuse River. Our pickets were some miles in advance. In fact, we held the line of a creek in two or three miles of Goldsboro. ... I was not out of my cavalry camp same to visit our outposts which I did as often as I could."

"In my front was Moccasin Creek, 800 yards from my Headquarters was the crossing on the Goldsboro pike. About the same distance above the creek & as far from my hdquar., was Atkinson's Mill and Dam. The creek swamp was boggy and some 300 yards wide or over. Either of these places was defendable to some extent but neither were strong positions."

"The signs of advance from Goldsboro were apparent and we began to make our preparations. Our pickets were doubled and ordered to be on alert & every measure taken to avert surprise. I had with care examined the ground. The day before the advance I went over Moccasin Creek & at a house had a barricade erected so as to protect my pickets in falling back & to give full warning to our main force of an advance. I wished to make sufficient resistance to see that they were approaching in force before I gave [up] the East bank of the creek. The position was such I could not command the crossings with the battery I had (Hart's) or could these guns be brought near either of the crossings with any degree of propriety unless they were covered and supported by a much larger force than I had at my command. I had the main bridge, which was a short one, well prepared so I could at once throw the planks off of it & so destroy it for an hour or so as it was useless to do more to the bridge which the enemy could rebuild in an hour...."

"Over the creek Private [Joseph J.] Lawson, Co. K, of Pickens County, was shot thru the heart and fell by me. This was accidental as he was shot by a rifle, perhaps his own falling from a stack of arms. Lawson spoke and talked to me after he was shot. ... He was 54 years old & had been a good soldier. He was sent across the creek and, dying in a few hours, was buried in front of Atkinson's house near the Public Road."

"The next morning the enemy advanced and, about 2 or 3 P. M., came on driving in our pickets who had orders to fall back on Atkinson's house, east of Moccasin Creek. Here a slight resistance was made and we withdrew."

"I had sent Capt. [Leroy J.] Johnson to the mill with 140 men. The balance I posted at the main bridge, keeping a proper reserve to support Hart's near Atkinson's on the west of the creek. They made repeated attacks on the bridge but the planks had been taken up and they could not force it. I had there one half of my own men & half North Carolinas of Col. Beard's Regiment under a Captain Goff a good man."

Colonel James C. Rogers, 123rd New York Infantry reported "On the 10th day of April, 1865, at dawn of day, the regiment broke its camp near Goldsborough and took up the line of march for Raleigh, via Smithfield, having the advance of the infantry of the Twentieth Army Corps. After marching several miles, the enemy was reported in force too great for the scouting party in advance to drive, and two companies and soon after the entire regiment was deployed as skirmishers. It advanced steadily, driving the enemy (consisting of several hundred mounted men of First South Carolina Cavalry and Sixth North Carolina Regiments, under command of Colonel Black) easily before them and taking possession of an entrenched line intended as a cover to a bridge across Moccasin Creek, a deep and rapid stream flowing in two channels through a wide and difficult morass. The enemy in his flight had displaced the planking and timbers of the bridge, and had cut a mill-dam a short distance above, swelling the current of the stream and flooding the adjacent swamps. The regiment was ordered to cross the first channel on the stringers of the bridge yet remaining, and advance far enough to hold it while being repaired. It crossed under a very brisk fire; plunged into the water to their waists of the men, pressed steadily forward and gained a position commanding the bridge over the second channel."

Black wrote "Finding they could not carry the bridge at a dash they moved some forces to the right and attacked Capt. Johnson but not so vigorously by far, as the attack made on the main pike. But, failing to force a passage at either place, they began to try and put a line of scouts thro the swamp & soon a Yankee was seen to emerge from the swamp half way from the Bridge to the mill. I promptly detailed a small force and ... deployed them at 20 paces to watch this interval. In the source of half an hour this slim line was having a brisk skirmish and the Yankee line in the swamp was being rapidly reinforced by men wading over to them. Seeing I had done all I could & that I was likely to have my extended line cut in two parts which left Capt. Johnson with no line of retreat, I decided to fall back and, preparatory to this, I ordered Capt. Johnson (Old Hickory) to swing back in his left flank and I, at the same time, prepared to let loose the bridge."

Rogers continued his report "The first bridge having been completed and supports brought up, the regiment was ordered to again advance and take the second bridge. This was done with spirit, and the enemy retired leaving the working parties to complete the repairs. As soon as the bridge was sufficiently rebuilt to enable other troops to come up should it be necessary, the regiment again deployed as skirmishers, and advancing rapidly again developed the line of the enemy, which a vigorous charge set in hot flight to the rear."

Black stated "The retreat was made in good order and we fell back on Hart's Battery, which I caused to open on the bridge as soon as was ... by our lines retreating from Atkinson's house to our old camp. I formed a new line and as the enemy begun to pour over the bridge and to form an infantry brigade in full view of us. Finding my position untenable in open ground, I ordered Hart to fall back beyond and, mounting all my command, moved back to the woods, three hundred yards to the rear."

"My loss was not serious, tho had several killed and wounded. We had brushed the enemy and, with a small force of 800 men, held one of the infantry columns in check for over two hours and that, too, in a weak position and one that only served us good purpose until we had to yield the bridge & that was to hide from the enemy the smallness of our force. Of the killed I saw a Private [Charles] Tyler, Co. C, fall near us. He was on the line of skirmishers at the very foot of Lawson's grave ... and when shot, fell forward on his face on the grave full length. Wm. Carey, who was near, turned him over & speaking to his twin brother, Elmon Tyler, of the same company, said to him that his brother was dead. Poor fellow, he was a good soldier & the last I saw killed or the last of my men killed in the war. Capt. Gott, N. C. cavalry, was badly wounded and Sgt. [John W.] Ward of Co. B, of my Regt. shot in the mouth and out of the jaw. He was near me &, as he turned around, I told him to go the rear & said: 'You will yet live to kiss the prettiest of the girls.'"

Rogers concluded his report "Our loss in this skirmish was 1 killed and 3 wounded. The enemy left 2 dead on the field, and a citizen reported that 8 wounded had been taken to his house and afterward removed."

"I saw Mr. Atkinson after the war," Black related, "and he told me that eleven Yankees were killed dead alone."

"As Hart's Battery was now beyond the next small creek," he continued, "I moved on after him & crossed the same, the enemy not following. By this time it was near sunset and at this instant I got an order from Gen'l. Hampton to fall back two miles as he had been forced back that far and the Yankees at that time were actually a mile or more ahead of me on the other road. This road was two miles from my flank and converging towards the one I was on, therefore, having a few scouts and pickets of observation, I move back two miles."

"The night after the fight I was ordered to send Col. Beard's Regt. to the left to report to Col. [Gilbert J.] Wright, which I did. As dusk came on, I ... came to a large farm of cleared land on both sides of the road. I at once had a long line of fires made, reaching over half mile on each side of the road & the men warmed & refreshed themselves by the fires which I kept up until about 9 p.m. when I moved my command away to the rear one fourth of a mile. This ruse de Guerre succeeded as the enemy was certain they had run on an infantry camp ... deployed in line of battle and moved forward in fine style with a heavy line of skirmishers in front. When this line of skirmishers got in the middle of the farm, my small line of 50 dismounted men commenced to fire on them and a few shots were exchanged when an imperative & hurried order came from Gen'l.

Hampton to me to move 3 miles at a gallop – so we mounted and moved away. This was about 8 a. m. The enemy in heavy force had pushed both Hampton and Wheeler back. "

"The enemy did not come up with us again on my road & I fell back as ordered & continued to march to Smithfield. I detached a small party to cross a bridge on our right flank and burn it as ordered. While the enemy were not pressing me, they were piling in to Gen'l. Hampton heavily. As I reached the edge of the town of Smithfield on the south side, they were driving in Gen'l. Wheeler on my left. I here met Gen'l. Hampton who ordered me to [send] down 100 men in a hurry to cover the rear & move my mounted men over the bridge in a hurry. I did this, ordering my Regt. over and staying mounted myself. In five minutes our mounted men were all over the River. Our dismounted men fell back. I crossed the bridge with Gen'l. Hampton, the last on horseback, and our dismounted men followed, firing the bridge as they came... the burning bridge cut off the pursuit of the Enemy that evening. My command was ordered on a few miles and went into camp, getting a good supply of corn for our horses, and thus ended our retreat from Goldsboro to Smithfield & the crossing of the Neuse."

"The next day we commenced our retreat, having camped some 5 miles above Smithfield, and fell back before the enemy who had crossed the Neuse at perhaps Smithfield & perhaps at the Railroad crossing higher up the river. This they were able to do with their Pontoon trains as the River is a small one and of an excellent kind to bridge with Pontoon. "

"When the march began I was in the centre of the column & had only to move on quietly. We moved slowly & could hear some firing in our rear – light skirmishing. About 11 A. M. I got an order from Maj. Gen'l. [Evander M.] Law in person to halt until ... our rear guard came up & support it. My orders were to keep 200 to 300 paces in its rear and to prevent its being 'gobbled up.'" Law was in command of the brigade.

"Directing my Lieut. Colonel Walker to assume Command of 8 companies of the Regt. & to keep within 100 yards of me ready to support me, I took a select squadron under Capt. Henry Jones and with drawn sabers got in the road. But to support the rear guard was no fun as we were one moment [told] ... to keep our distance and when we halted, before we could face about, were compelled to wheel about again & move to the rear. Our rear guard now and then fired spasmodically –

Capt. Jones & myself rode forward to the extreme rear skirmish line and sat on our horses. Our men were firing rapidly. I could see no enemy & Jones said he could see none & Jones also asked me if I could hear the whistle of the enemies' balls but not a 'Whiz' could I hear. Jones was generally very cool tho he now lost his temper & wanted to charge our own men in rear to make them stand. We were so amazed that I asked to let Jones' squadron be the rear guard. By this time the rear guard was to be relieved. ..." This ended Black's account of the actions of the 1st South Carolina Cavalry.

The 1st is shown as part of General Thomas M. Logan's Brigade on April 23, 1865. Most of the regiment took their paroles but refused to surrender their horses and arms. Generals Hampton and Wheeler refused to surrender the cavalry and started to Texas to join the Confederate forces there, but all eventually turned back upon learning of the surrendered of that command.

Lieutenant Isaac I. Fox, of Company I, had received a letter from D. F. Towles, Superintendent of Colored People in St. Paul's Parish, on about April 26th. He replied from St. Bartholomew, S. C. on April 27th:

"Your communication has been received in which your 'order' me to lay down my arms and become peaceful citizens, submit to Union Authority & disband my unlawful organization.

I take much pleasure in informing you, I am here by Authority of Lieut. Gen. Hardee comdg. Dept. My command a force of Confederate Cavalry, Commanded by Confederate States Officers. That I received orders from my Commanding Officer and not from the 'Supt. Of Colored People of St. Paul's Parish.'

I have just received notice that an armistige [sic] between Gens Johnston & Sherman had been agreed upon and have extended orders to my commander to cease hostilities during the truce. Your emphatic assertion that Johnson had surrendered must be false or there could not be an armistice.

I am awaiting orders to know what to do in the case of your Negro outlaws arresting my men & citizens. You will do well to warn U. S. authorities that if those men are harmed I will retaliate.

Very respectfully

I. I. Fox

Lt. Co. I., 1st S. C. Cavalry

Commanding Scouts"

The outcome of this confrontation is unknown.

Black wrote a comment on a large sheet bearing the title 'Roll of Field & Staff, 1st. Regt. S. Car. Cavalry, South Carolina Volunteers' reporting: "When Johnson [sic] surrendered this Regiment fully armed & equipped, marched out of camp and was disbanded by companies in York County.' Thus he ended the saga of his regiment in the War for Southern Independence.

ABBREAVIATIONS

Certain standard abbreviations have been used throughout the book.

Ab.=	Absent
AWOL =	Absent without leave
Acad. =	Academy
Adj. =	Adjutant
Arty.=	Artillery
Asst. =	Assistant
AAG =	Assistant Adjutant General
AIG =	Assistant Inspector General
AQM =	Assistant Quarter Master
Asst. Surgeon =	Assistant Surgeon
Att. =	Attended
Bn. =	Battalion
Bty. =	Battery
b. =	Born
Brig. =	Brigade
Brig. Gen. =	Brigade General
Bur. =	Buried

Capt. =		Captain
Cap. =		Captured
Cav. =		Cavalry
Cem. =		Cemetery
Coll. =		College
Col. =		Colonel
Comm. =		Commissary
Co. =		Company
C.S. =		Confederate States
CSN =		Confederate States Navy
Cpl. =		Corporal
Co. =		County
CH =		Court House
CM =		Court-Martial
Des.=		Deserted
d. =		Died
DIS =		Died in Service
DOW's =		Died of Wounds
Dist. =		District
Div. =		Division
Enl. =		Enlisted

1stLt. =	First Lieutenant
1stSgt. =	First Sergeant
Ft. =	Fort
Gen. =	General
GCM =	General Court-Martial
Gd. =	Graduated
Hosp.= =	Hospital
Inf. =	Infantry
KIA =	Killed in Action
Lt. =	Lieutenant
Lt.Col. =	Lieutenant Colonel
Lt. Gen. =	Lieutenant General
Maj. =	Major
MIA =	Missing in Action
Mt. =	Mount or Mountain
Mt.'d. =	Mounted
NFR =	No Further Record
Obit. =	Obituary
Ord. =	Ordnance
OSgt. =	Orderly Sergeant
Pt. =	Point

PO =	Post Office
Pvt. =	Private
PM =	Provost Marshal
QM =	Quartermaster
Reenl. =	Reenlisted
Res. =	Resident
RTD =	Returned to Duty
2ndLt. =	Second Lieutenant
Sgt. =	Sergeant
Sgt.Maj. =	Sergeant Major
SCV =	Sons of Confederate Veterans
Sqd. =	Squadron
3rdLt.=	Third Lieutenant
Trans. =	Transferred
UCV =	United Confederate Veterans
UDC =	United Daughters of the Confederacy
U. =	University
WIA =	Wounded in Action

PRISONER OF WAR CAMPS

Camp Chase, Ohio ....Camp Chase

Camp Morton, Indiana.. Camp Morton

David's Island, New York..David's Island

Elmira, New York..Elmira

Fort Columbus, New York......................................Ft. Columbus

Fort Delaware, Delaware.......................................Ft. Del.

Fort McHenry, Maryland..Ft. McHenry

Fort Monroe, Virginia...Ft. Monroe

Fort Warren, Mass. ....Ft. Warren

Hart's Island, New York...Hart's Island

Johnson's Island, Ohio..Johnson's Island

Newport News, Virginia...Newport News

Point Lookout, Maryland..Pt. Lookout

Rock Island, Illinois..Rock Island

Old Capitol, Washington, D. C. ....Old Capitol

Wheeling, West Virginia...Wheeling

INTRODUCTION TO THE MUSTER ROLLS

The primary source for the muster rolls of the First South Carolina Cavalry is the Compiled Service Records in the National Archives, which consists of date and place of enlistment, rank, age (not in all cases), and occupation (not in all cases). Other entries include presence or absence for specific periods. Muster rolls were made up in two month increments for pay purposes. Records of leave, hospitalization, documents of capture and exchange, absence on horse details, transfer between units and prisons, clothing and equipment receipts, horse appraisals, pay vouchers, paroles, oaths, discharges, and other miscellaneous matters are included in each man's file. Pay records and other muster rolls or rosters were found and the data included. Muster rolls of all companies are not complete and end in February, 1865.

The records of each soldier has been supplemented by casualty reports, hospital records, pension records, records of the Old Soldier's Homes, county histories, county birth, marriage, death and cemetery records, family histories, genealogies, UDC and UCV records, Bible records, census records, newspaper accounts and obituaries. Spelling is as found in these sources. Each soldier's record of service is arranged chronologically from place and date of birth to date of death and burial site. Prewar and postwar careers have been included when available. Periods of confinement as a prisoner of war are not shown specifically but run from the date of capture until the soldier was exchanged, released, paroled, took the oath of allegiance, or died. The term NFR (no further record) is used to show when a soldier's record ended and no reason is stated. Periods of unauthorized absence are stated as absence without leave (AWOL) or deserted. These entries are based on the soldier's record. A lack of court-martial records may indicate that the absence was excused upon return to duty, or the man received some minor punishment without trial. Descriptive lists have been taken from enlistment, hospital, pay, discharge, leave, prisoner of war, parole, desertion and death records. In a few cases postwar descriptions have been included.

MUSTER ROLLS 1ST SOUTH CAROLINA CAVALRY

ABELL, CHARLES ALEXANDER. 2ndLt., Co. H. b. S. C. 6/15/36. Res. Chester Co. 1860 census. Enl. Rock Hill as 3rdLt. 1/9/61. Present until ab. 3-4/62. Ab. sick in Church Flats hospital 5-6/62. Elected 2ndLt. AWOL 7-8/62. Resigned for ill health 10/15/62. NFR. Laborer, West Chester, Chester Co. 1870 census. Farmer, Fairfield Co. 1880 census. d. 7/31/97. Bur. Sandy Level Bapt. Ch., Richland Co.

ABELL, WILLIAM HENRY. Pvt., Co. D. b. S. C. 1/40. Enl. Co. F, 6th S. C. Inf. Summerville 6/11/61. Present until WIA (shoulder) Dranesville 12/20/61. Ab. wounded in Manassas hospital 12/31/61. Ab. on leave 1/62. Reenlisted 3/29/62. Ab. wounded through 6/62. Company Disbanded. Reenl. Co. D, 1st S. C. Cav. Chester CH 6/3/63. Horse KIA Brandy Station 8/1/63. Paid $450.00 for horse. Ab. on leave to Chester CH for a horse 8/18-31/63. Present 9-10/63. Present dismounted 11-12/63. Present 1-2/64. Ab. on detail Green Pond 4/16-30/64. Present 5-6/64. Present sick in quarters 7-8/64. Present 9-10/64. Ab. detailed Pontoon Bridge 12/8-31/64. NFR. Farmer, Chester Co. 1900 census. d. 4/5/01. Bur. Evergreen Cem., Chester Co.

ABERCROMBIE, ABRAM. Pvt., Co. B. b. S. C. circa 1842. Enl. Camp Tripp 5/1/62 age 20. Present through 6/62. On picket duty Ellis Ford 7-12/62. Ab. on wagon guard 3-4/63. Present through 8/63. Captured near Madison CH 9/20/63. Sent to Pt. Lookout. Died of disease there 1/29/64. Bur. Confederate Cem., Pt. Lookout, Md. Father, John Abercrombie, received his pay on 8/19/64.

ABERCROMBIE, JOHN M. Pvt. Co. B. b. S. C. 1836. Res. Tylersville, S. C. Enl. Clinton 8/22/61 age 24. Ab. sick 10/31/61. Present 11-12/61 and 3-4/62. Ab. at home on sick leave 5-6/62. Present 9-12/62. Ab. on detached duty Gordonsville 3-4/63. Ab. on horse detail to S. C. 8/18-31/63. Ab. detailed as Teamster 9-10/20/63. Ab. detailed as Teamster for Brigade Commissary Dept. 12/22-31/63 & 1-2/64. Ab. detailed Green Pond 4/16-30/64. Present 5-8/64. Ab. on picket 9-10/64. Present sick in quarters 11-12/64. NFR. Farmhand, Dials, Laurens Co. 1880 census. d. Austin, Greenville Co. 1910. Bur. Confederate Cem., Greenville Co.

ABERCROMBIE, JOHN M. N., SR. Pvt., Co. B, b. S. C. circa 1838. Enl. Clinton 8/22/61 age 22. Ab. sick

10/31/61. Shot accidentally with a pistol and in hospital 11-12/61. Present 3-10/62. On picket Ellis Ford 11-12/62. Present 3-4/63. Ab. on horse detail to S. C. 9-10/63. Returned 12/6/63. Present through 12/31/63 and 1-12/64. NFR. d. 3/21/14. Bur. Laurel Creek Meth. Ch., Greenville Co.

ABERCROMBIE, JOHNATHAN. Pvt., Co. B. b. S. C. 11/36. Wife's pension application only record of service. Also listed in Co. D, 4th S. C. Reserves. Res. Laurens Co. 1870 & 1880 censuses. d. 11/7/01. Wife receiving pension Woodruff, Laurens. Co.

ABERCROMBIE, MARTIN VAN BUREN. Pvt., Co. B. b. S. C. 3/15/39. Enl. Clinton 8/22/61. Present through 6/6/2. Ab. on leave 7-8/62. Ab. on march with horses from Summerville, S. C. to Staunton, Va. 9-10/62. Ab. with wagon hauling forage 11-12/62. Ab. on horse detail 3-4/63. Present 5-6/63. Ab. as Teamster with Ordnance Train 8/31/63. Present 9-12/63 & 1-10/64. Ab. on detached duty Pocotaligo 12/2-31/64. NFR. Farmhand, Dials, Laurens Co., 1870-1900 census. d. Laurens Co. 7/28/19. Bur. Dials Meth. Ch.

ACKERMAN, JOHN G. Pvt., Co. I. b. S. C. 3/36. Res. Colleton Dist.1860 census. Enl. Co. C, 5th S. C. Cav. Cheraw, S. C. 6/9/62. Transferred Co. I, 1st S. C. Cav. in exchange for Thomas C. Moore 10/12/63. Not yet reported on rolls 10/31-12/31/63. Reported 1/17/64. Present through 2/64. Ab. detailed Green Pond 4/16-30/64, dismounted. Ab. on sick leave 5-6/64. Present 7-10/64, however, in Summerville, S. C. hospital 10/30-31/64. AWOL 12/11-31/64. Transferred 2nd S. C. Arty. 2/4/65, but not on muster rolls of that unit. d. after 1900. Bur. Ackerman Cem., Colleton Co.

ACKERMAN, JOSEPH O. Pvt., Co. I. b. S. C. 1844. Res. Colleton Dist. 1860 census. Enl. Columbia 4/12/64. Ab. on detail at Green Pond, S. C. 4/14-30/64. Ab. on sick leave 5-6/64. Ab. in Mt. Pleasant hospital 8/28-31/64. d. 8/31/64. Bur. Ackerman Cem., Colleton Co.

ACKERMAN, R. NALLY. Pvt., Co. I. b. S. C. circa 1840. Enl. 11th S. C. Inf. Fenwicks Island 10/15/61. Present through 12/31/61. NFR. Enl. Co. I, 1st S. C. Cav. Parker's Ferry 4/3/62. Present through 6/62. Ab. on leave 8/20-31/62. Present 9-10/62. Ab. on scout 11-12/62. Present 3-8/63. Ab. on horse detail 9-10/63. AWOL 12/30-31/63. Present 1-2/64. Ab. detailed Green Pond 4/16-30/64. Present 5-12/64. NFR. Farmer, Burns, Colleton Co. 1880 census. Bur. Pine Grove Bapt. Ch. Cem., Dorchester Co., no dates.

ACKERMAN, STEPHEN WILSON. Sgt., Co. I. b. Round O, Colleton Dist. 4/28/36. Laborer, Walterboro, S. C. 1860 census. Enl. Parker's Ferry 4/3/62 as 1st Cpl. Present through 6/62. Present sick in quarters 7-8/62. Promoted 3rdSgt. Present, sick in quarters 11-12/62. Promoted 2ndSgt. Present 3-6/63. Ab. on horse detail 7-8/63. AWOL 12/30-31/63. Ab. on detail Green Pond 1/1-2/29/64. Remounted 3/1/64. Present 4-10/64. Ab. on detail Pocotaligo Bridge, James Island 12/29-31/64. NFR. Served to May, 1865 on pension application, from Round O, Colleton Co. 4/29/19. Minister & Farmer, Sheridan, Colleton Co. d. 6/5/19. Bur. Ackerman Cem., Colleton Co.

ADAMS, JAMES D. Pvt., Co. B. b. Lownes Dist., S. C. circa 1840. Enl. 6th S. C. Bn. Inf. 12/1/61. Discharged 6/1/62. 6', fair complexion, blue eyes, light hair, Farmer. Enl. Co. B, 1st S. C. Cav. Camp Means 6/1/62. Ab. sick in hospital Church Flats through 6/30/62. Ab. sick at home & on horse detail 7/27-30/62. Present 9-10/62. Ab. on picket Ellis Ford 11-12/62. Ab. as wagon guard 3-4/63. Present 5-6/63. Ab. on horse detail to S. C. 8/18-31/63. Present 9-10/63. Ab. on picket at Chancellorsville 11-12/63. Present 1-2/64. Ab. on horse detail to S. C. 3/31-4/30/64. Present 5-10/64. Present sick in quarters 11-12/64. NFR. Res., Williamston, Anderson Co. 1910 census.

ADAMS, JAMES HENRY. Pvt., Co. D. b. S. C. circa 1838. Blacksmith, Rich Hill Cross Roads, Chester Dist. 1860 census. Enl. Chester CH 7/14/62. Ab. sick Chester 8/4/-31/62. Present 9-10/62. Ab. sick in Lynchburg hospital 11-12/62. d. of pneumonia in Lynchburg hospital 3/23/63. Bur. Lynchburg City Cem. Removed to Chesterville, S. C. on hospital records.

ADAMS, JOSHUA M. Pvt., Co. H. b. Marlboro, S. C. 12/2/43. Farmer, Bennettsville, S. C. 1860 census. Enl. Rock Hill 1/9/62. Ab. on picket Jacksonborough 3-4/62. Ab. on duty as Courier, at McCord's Battery 5-6/62. Ab. sick in quarters 7-8/62. Present 9-10/62. Ab. on raid 11-12/62. Transferred Co. I, 6th S. C. Inf.

2/15/63. Present through 8/64. WIA (flesh - left thigh) 9/30/64. Ab. wounded in Richmond hospital through 10/31/64. Ab. on leave to Rock Hill 11-12/64. Present 1-2/65. Captured High Bridge 4/6/65. Sent to Newport News. Released 6/27/65. 5'11", dark complexion, dark hair, grey eyes, res. Yorkville, S. C. Farmer, Brightsville, S. C. 1880 & 1900 censuses. d. Marlboro, S. C. 1/5/05.

ADAMS, L.A. Pvt., Co. A. On postwar roster. Reenlisted Co. A, Rutledge Mounted Rifles 2/25/62. Ab. detailed to A. A. S. Dept. in S. C. 5-6/63. NFR. Transferred Co. B, 7th S. C. Cav. Paroled Tallahassee, Fla. 5/12/65.

ADAMS, SAMUEL ALEXANDER. 1stSgt., Co. C. b. Edgefield Dist. 2/4/39. Clerk, Hamburg, S. C. 1860 census. Enl. Hamburg 8/27/61 as 2ndSgt. Promoted 1stSgt. & present 3-4/62. Present 5-6/62. Ab. detailed Adjutant General's Dept. 7-8/62. Transferred Gen. Pemberton's office 8/5/62. NFR until ab. on detail 3-4/64 & reduced to Pvt. Ab. on detail 5-8/64. Ab. on detail Gen. Jones Dept. as 1st Class Clerk 9-12/64. Paroled 4/65. Bookkeeper, Augusta, Ga. 1880 census. d. Augusta, Ga. 11/11/86.

ADDISON, JAMES D. Pvt., Co. I. on postwar roster. Reenlisted Co. C, 5th S. C. Cav. at Cheraw 4/14/63. Present until Ab. sick 9/5-10/31/63. Present until ab. on horse detail 8/14/64 for 40 days. Present in dismounted Bn. 9-10-/64. Ab. sick with rubeola in Raleigh, N. C. hospital 4/3-/9/65. NFR.

ADKINS, JACKSON A. Pvt., Co. H. b. S. C. 1/6/38. Enl. Rock Hill 3/19/62. On picket Simons Bluff 3-4/62. Ab. as baggage guard Church Flats 5-6/62. Ab. on picket 7-8/62, no horse. Ab. dismounted guarding baggage Staunton, Va. 9-10/62. Ab. on picket 11-12/62. Present 3-4/63. WIA (left arm) Boonsboro, Md.(saber cut) & in Charlottesville hospital 7/17/63-8/30/63. Transferred to another hospital 9/21/63. Ab. on leave 9-10/63. Present 11-12/63 & 1-2/64. Discharged for disability (wound left elbow) 3/14/64. Fair complexion, blue eyes, dark hair, 5' 3", Farmer. Farmer, York Co. 1870 census. Receiving pension York Co. 1901. d. 5/7/08. Bur. Adnah Meth. Ch, York Co.

ADKINS, JOHN JACKSON. Pvt., Co. H. b. S. C. 6/10/37. Enl. Rock Hill 3/19/62. Present through 4/62. Ab. on duty as Courier at McCord's Battery 5-6/62. Ab. on sick leave 7-8/62. Present 9-10/62. Ab. on picket 11-12/62. Present 3-6/63. Ab. on horse detail 8/18-31/63. Present 9-12/63 & 1-2/64. Ab. on leave 4/25-5/1/64. Present 5-10/64. Ab. on detail Pocotaligo 10-12/64. Served to April, 1865 on pension application from Tirzah, York Co. 1902. d. 5/15/10. Bur. Bethel Presb. Ch. Cem., York Co.

ADKINS, JOSEPH. Pvt., Co. F. b. S. C. circa 1830. Factory Laborer, Spartansburg Dist. 1860 census. Enl. Pickens CH 12/14/61. Present through 4/62. Ab. sick in hospital 5-6/62. Ab. on sick leave 7-8/62. d. Raburn Co., Ga. 9/1/62. Left widow.

ADKINS, WILLIAM DANIAL. Pvt., Co. H. b. York Co. 10/27. Enl. Rock Hill 1/9/62 age 33. Present through 6/62. Ab. on picket 7-8/62. Present 9-12/62 & 3-4/63. Captured Gettysburg 7/4/63. In Letterman hospital, Gettysburg as nurse 8/10/63. Sent to Ft. McHenry. Exchanged 8/20/63. Ab. on parole 9-10/63. Present 1/15-2/29/64. Ab. on leave 3-4/64. Present 5-12/64. NFR. d. Pulaski Co., Ark. 12/5/16. Bur. Sumner Cem., Jacksonville, Ark.

ADKINS, WILLIAM H. Pvt., Co. H. b. S. C. 1843. Farmhand, age 17, Ft. Mill, York Co. 1860 census. Enl. Rock Hill 1/9/62. Present until ab. on picket McCord's Battery 3-4/62. Ab. on duty with wagons 5-6/62. Ab. on sick leave 7-8/62. Present 9-10/62. Ab. sick in hospital 11-12/62. Present 3-10/63. Ab. at Horse Recruiting Station 11/15/63-2/64. Ab. on detail Green Pond 4/16-31/64. Present 5-10/64. Ab. on duty Pocotaligo 1-12/64. NFR.

AIKEN, JOSEPH. Pvt., Co. F. Res. Pickens Co. Died during the war on postwar roster.

ALBRIGHT, WILLIAM C. Pvt., Co. D. b. S. C. circa 1829. Enl. Chester CH 9/10/61 age 32. Present 11-12/61, 3-12/62 & 3-6/63. Ab. on horse detail to Chester, S. C. 12/23-31/63. & 1-2/64. Present 3-4/64. Ab. sick Brigade Hospital, James Island 5/27-6/26/64. Ab. on sick leave Chester Dist. 6/27-8/31/64. d. in Brigade Hospital, James Island 10/11/64.

ALEXANDER, ELIAS FRANKLIN. Pvt., Co. F. b. S. C. 7/29/39. Enl. Pickens CH 12/4/61. Present through 12/31/61 and 3-4/62. Ab. on picket Camp Means 5-6/62. Present 7-10/62. Under arrest 11-12/62. Present 3-6/63. Ab. on detached service 8/18-31/63. AWOL 9-12/63 & 1-2/64. Dropped as a deserter 4/1/64. NFR. Farmer, Keowee, Oconee Co. 1900 census. d. 8/31/02. Bur. Pleasant Ridge Cem., Oconee Co.

ALEXANDER, FREDERICK T. Pvt., Co. H. b. S. C. circa 1831. Enl. Rock Hill 6/6/61 age 32. Ab. on picket Jacksonborough 3-4/62. Ab. as Courier, McCord's Battery 5-6/62. Ab. on sick leave 7-8/62. Ab. to drive horses from Secessionville, S. C. to Staunton, Va. 9-10/62. Ab. on picket 11-12/62. Captured Falmouth 1/2/63. Paroled POW 3-4/63. Present 8/31/63. Ab. on horse detail 9-10/63. Present 11-12/63 & 1-2/64. Present 5-6/64. Ab. on leave 7/27-8/30/64. Ab. on detail 9-10/64. Present 11-12/64. Transferred Capt. Erwin's Co., 5th N. C. Cav. 1/10/65. Not on muster rolls of that unit.

ALEXANDER, J. MATHEW. Pvt., Co. B. b. S. C. circa 1821. Enl. Spartanburg Co. 1861, age 40. Present until transferred 2nd S. C. Cav. 9/15/64. Ab. on sick leave at home for 60 days 9-10/64. NFR. d. by 1898.

ALEXANDER, JAMES MADISON. Pvt., Co.I & B. b. S. C. 1843. Enl. Co. I Columbia 2/15/64. Ab. on detail Green Pond, S. C. 4/26-30/64. Transferred to Co. B 5/31/64. Present 5-8/64. Transferred Co. E, 2nd S. C. Cav. 9/15/64. Ab. on sick leave for 60 days 9-10/64. NFR. Livery Stable Keeper, Charleston 1870 census.

ALEXANDER, W. JORDON. Pvt., Co. F. b. S. C. 6/24/37. Enl. Pickens CH 4/1/63. Present through 9/63. Ab. driving ambulance 10/1/63-2/64, dismounted. Transferred Co. A, 20th S. C. Inf. 5/13/64. Joined 6/16/64. Present until ab. sick with phthisis pulmonalis in Richmond hospital 8/25-10/9/64. Paroled Greensboro, N. C. 4/26/65. Retired, Cypress, Lee Co., Fla. 1910 census. d. 6/17/17. Bur. Alexander Cem., Oconee Co.

ALLSTON, ROBERT AUGUSTUS. 2nd Lt., Co. A. b. Macon, Ga. 12/31/32. Married Charleston, S. C. 1857. Enl. 11/15/61. Served on Gen. Morgan's staff in Ky. by 7/17/62. Promoted 2ndLt., 1st Ga. Regulars 10/4/62. Promoted Captain and Aide-de-campe Gen. Morgan by 12/30/62. Captured Bardstown, Ky. 7/5/63. Sent to Camp Chase & Ft. McHenry. Paroled 10/8/63. Exchanged 10/13/63. Promoted Lt. Colonel, 1st Ky. Bn. Cav. by 8/1/64. Transferred 9th Tenn. Cav. 10/22/64. NFR. d. Atlanta, Ga. 3/11/79. Bur. Old Decatur Cem., Decatur, Ga.

ALTMAN, A. E. Pvt., Co. I. b. S. C. circa 1812. Enl. Parker's Ferry 4/3/62 age 50, 5'11", light complexion, blue eyes, dark hair, Farmer. Present through 4/30/62. Ab. on sick leave 5-6/62. Ab. on leave from hospital 8/14-31/62. Ab. on sick leave 9-10/62. Ab. on leave 11-12/62. Ab. on detached service 3-4/63. AWOL 5-10/63, however, ab. sick in hospital 2/21-12/31/63. Discharged for phthisis pulmonalis Columbia, S. C. 1/1/64. Paid 1/5/64. NFR.

ANDERSON, DAVID E. Pvt., Co. C. b. S. C. circa 1841. Laborer, Edgefield Dist. 1860 census. Enl. Camp Butler 9/18/61 age 21. Present through 12/31/61. Presence or absence not stated 3-4/62. Present 5-6/62. Ab. on sick leave 7/31-8/30/62. Present 9-10/62. Present as Teamster 11-12/62. Present 3-12/63. Ab. on leave 2/12-28/64. Present 3-8/64. Ab. on duty Florence, S. C. guarding POW's 9/17-10/31/64. Present 11-12/64. Present as Teamster 1-2/65. NFR. Alive 1898.

ANDERSON, F. J. Pvt., Co. K. b. S. C. circa 1844. Enl. date unknown. Captured Bennettsville, S. C. 3/6/65. Sent to New Bern, N. C. 4/3/65. Sent to Pt. Lookout. Released 6/6/65. Farmer, age 26, Cokesbury, Abbeville Dist. 1870 census.

ANDERSON, HIRAM J. Pvt., Co. D. b. S. C. circa 1842. Age 18, Chester, 1860 census. Enl. Chester CH 9/10/61 age 18. Present 11-12/61 & 3-4/62. Ab. sick in Richmond hospital 10/13/62. d. of acute dysentery in South Carolina hospital, Manchester, Va. 10/18/62. Claim filed by his mother 2/13/63.

ANDERSON, JACOB B. Pvt., Co. C. b. S. C. circa 1830. Overseer, age 30, Leesville, Lexington Dist. 1860 census. Enl. Camp Butler 9/17/61 age 30, light complexion, light hair, blue eyes, Farmer. Ab. on sick leave 11/5-12/31/61. Presence or absence not stated 3-4/62. Present 5-6/62. Ab. sick leave 7/31-8/30/62. Present 9-10/62. Ab. on scout 11-12/62. Ab. sick 3-4/63. Ab. sick at home 6/11-8/31/63. Present 9-10/63. Ab. sick in hospital 11/28/63-2/28/64. Ab. on horse detail 1/4-5/15/64. Present 6/64. Transferred Co. H, 2nd S. C. Arty. 7/1/64. Present through 12/64. Paroled Augusta, Ga. 5/29/65.

ANDERSON, JAMES L. Pvt., Co. G. b. S. C. 3/4/37. Age 22, Talladega, Ala. 1860 census. Enl. Abbeville CH 1/8/62. Sick in quarters 3-4/62. Discharged 5/11/62. NFR. Farmer, Talladega, Ala. 1880 census. d. 4/20/03. Bur. Childsburg Cem., Childsburg, Ala.

ANDERSON, SAMUEL THOMAS. 1stLt., Co. D. b. Chester Dist., S. C. 6/29/38. Gd. Reform Medical College (now Mercer U.) 1860. Res. Cedar Shoals, Chester Dist. 1860 census. Enl. Co. G, 6th S. C. Inf. Chester CH 4/11/61. Present through 8/61. Present sick in quarters 9-10/61. Ab. sick in hospital 11/1-1serving 2/31/61. Discharged 1/22/62. Enl. Co. D, 1st S. C. Cav. Chester CH 6/18/62 as Pvt. Present 5-6/62. Captured and on parole 7-8/62. Present 9-10/62. Elected 1stLt. and absent on scout 12/25-31/62. Ab. on horse detail 4/1/63. Captured Martinsburg, W. Va. 7/19/63. 6' 2 1/2", ruddy complexion, sandy hair, blue eyes. Sent to Camp Chase, Ft. Pulaski, Pt. Lookout & Ft. Delaware. One of the "Immoral 600" who were held under fire at Battery Wagner, S. C. Released 6/12/65. M. D., Rossville, S. C. 1870 census. M. D., Anderson Community, Chester Co. d. 11/10/94. Bur. Heath's Chapel Cem., Chester Co.

ANDERSON, WILLIAM B. Pvt., Co. A. b. Abbeville Co. 1842. Enl. age 19 by 8/61 on roster. Not on muster roll 10/11/61. Living Abbeville Co. 1902.

ANTHONY, SARGENT DANIEL. Pvt., Co. F. b. Pickens Dist. S. C. 5/46. Serving in Pickens Dist. Militia 1858, 6'1", dark complexion, hazel eyes, dark hair, Farmer. Enl. Columbia 4/20/64. Ab. on duty Green Pond 4/25-30/64. Present 5-12/64. d. of congestive chills and remittent fever in Fork of Yadkin River, N. C. 4/18/65. Age 18, 11 months and 9 days.

ARCH, THOMAS. Black Servant, Co. B. b. S. C. circa 1843. Enl. 6/61 and served to April, 1865 on pension application from Spartanburg Co. 1919. Receiving pension Woodruff, Spartanburg Co. 4/9/23.

ARCHER, JOHN E. Pvt., Co. F. b. Ga. 1834. Farmer, age 26, Pickens Dist. 1860 census. Enl. Pickens CH 12/4/61 age 30. Present through 12/31/61. Ab. on detail Adams Run 3-4/62. Present 5-12/62 & 3-8/63. On detached service 9-10/63. Ab. at horse recruiting station 10/26/63-2/64. AWOL 4/20-30/64. Present 5-6/64. Ab. driving cattle 8/29-31/64. Present 9-12/64. NFR. d. by 1900 on postwar roster.

ARIAL, JOHN A. Pvt., Co. F. Res. Pickens Co. On postwar roster.

ARIAL, JOHN HARVEY. 4thSgt., Co. F. b. Pickens Dist. 8/30/33. Farmer, age 26, Pickens Dist. 1860 census. Enl. Pickens CH 12/4/61 age 30 as 1stCpl. Promoted 4thSgt. 12/31/61. d. Adams Run 4/6/62. Bur. family cem., Easley, S. C. Left widow.

ARMSTRONG, A. Pvt., Co. B Res. Laurel Dist. Captured and died of disease in prison.

ARMSTRONG, A. C. Pvt., Co. B. d. 9/9/62.

ARMSTRONG, JOHN BRYSON. Pvt., Co. G. b. Anderson Dist. 1837. Enl. Co. B, 7th S. C. Inf., Abbeville CH 1/9/61. Transferred Co. G, 2nd S. C. Cav. 4/10/62. Discharged 1/64. Enl. Co. G, 1st S. C. Cav. Chesterfield Co., Va. 3/4/64. Ab. on horse detail 4/12-30/64. Present 5-6/64. Ab. sick in James Island hospital 8/13-31/64. Present sick in quarters 9-10/64. Present 11-12/64. Paroled Greensboro, N. C. 4/26/65. Applied for pardon 9/15/65. Farmer, Oaklawn, Greenville Co. 1880 census. d. Williston, S. C. 12/17/87. Bur. Little River Bapt. Ch., Abbeville Co.

ARMSTRONG, WILLIAM H. (1). Pvt., Co. B. b. S. C. Enl. age 40 on postwar roster. d. by 1898.

ARMSTRONG, WILLIAM H. (2). Pvt., Co. B. b. S. C. circa 1846. Student, age 14, Charleston, 1860 census. Enl. Columbia 3/25/64. Present through 8/64. Ab. on picket 10/30-31/64. Present 11-12/64. NFR. Engineer, age 36, Charleston 1880 census.

ARNOLD, J. A. MORRIN. Pvt., Co. A. b. S. C. 3/28/43. Enl. Abbeville CH 9/12/61. Present through 10/61. Ab. on duty Bears Island 11-12/61. Present sick 3-4/62. Present 5-10/62. Ab. on scout 12/62. Present 3-8/63. Present detailed as Blacksmith 9-10/63. Ab. on horse detail to S. C. 12/23-31/63. Present as Company Blacksmith 1-2/64. Present 3-6/64. Ab. sick 9-10/31/64. Ab. sick in hospital 11-12/31/64. NFR. Living Abbeville Co. 1902. d. 2/12/24. Bur. Rosemont Cem., Union, S. C.

ARTHUR, JOHN. Free Negro. Enl. Columbia 4/18/64. On duty with Infirmary Corps, 1st S. C. Cav., Lancaster 7/1/64. NFR.

ASHER, ABRAM A. Pvt., Co. C. b. Prussia circa 1842. Clerk, age 18, Hamburg 1860 census. Enl. Hamburg 8/27/61 age 18. Ab. detached duty on Wadmalaw Island 11-12/61. Presence or absence not stated 3-4/62. Present 5-6/62. Ab. on sick leave 7/28-31/62. Present 9-10/62. Ab. on picket 11-12/62. Ab. on detached duty 3-4/63 and 8/1-10/31/63. Ab. on horse detail to S. C. 12/22/63-2/64. Ab. detached at Green Pond 4/16-30/64. Ab. sick in hospital 6/25-30/64. Present 7-12/64. NFR. Alive 1898.

ASHWORTH, JOHN H. Pvt., Co. F. b. N. C. circa 1836. Farmer, age 24, Casears Head, Greenville Co. 1860 census. Enl. Pickens CH 12/4/61 age 25. Present through 12/31/61. Ab. on picket Jacksonborough 3-4/62. Present 5-6/62. Present in hospital under sentence of CM for AWOL 10/17-12/31/62. AWOL 3/63-2/64. Dropped as a deserter 3/1/64. NFR. Res. Castle, Okla. postwar.

ATKINSON, ABNER W. Pvt., Co. C. b. S. C. 1839. Enl. Wadmalaw Island 3/17/62. Presence or absence not stated through 4/62. Farrier for Company 5-6/62. Ab. on sick leave at home 7-8/62. Transferred Gen. Hagood's staff 9/1/62. Hired as Wagon Master 4/63. Wagon Master, James Island Lines 10/1-31/63. Messenger, Augusta, Ga. 3-4/64. Detailed in Major N. Smith's Dept. 5-8/64. Messenger, Augusta, Ga. 10-12/64. Paroled Augusta, Ga. 5/20/65. d. 1/29/85. Bur. Hammond Cem., Beech Island, S. C.

ATKINSON, H. W. Pvt., Co. C. Present 2/28/63. Not on muster rolls. NFR.

ATKINSON, J. Pvt., Co. H. WIA (elbow, sabre cut) Gettysburg 7/3/63 on casualty list. NFR.

ATKINSON, JAMES BELDON. 2ndSgt., Co. D. b. S. C. 7/26. Enl. Chester CH 9/10/61 as Pvt., age 35. Detailed as Commissary of Company. On duty transferring baggage from depot 11-12/61. Ab. assisting in QM Dept., Adams Run 3-6/62. Present 7-8/62. Ab. with horses of 4th Squadron 9-10/62 (enroute from Summerville, S. C. to Staunton, Va.). Ab. on scout 12/29-31/62. Promoted 4thCpl. & present 3-4/63. Promoted 3rdCpl. 8/31/63. Ab. on horse detail to Chester, S. C. 10/23-31/63. Promoted 2ndSgt. 11/1/63. Present dismounted 1 month 12/31/63. Ab. on picket 2/28/64. Detailed Green Pond 4/16-30/64. Present 5-12/64. Paroled Hillsboro, N. C. 4/65. Paid taxes Sandersville, S. C. 1865. Magistrate. d. Chester 9/27/09. Bur. Evergreen Cem.

ATKINSON, VALENTINE. Pvt., Co. D. b. S. C. circa 1829. Enl. Chester CH 9/10/61 age 32. Present 11/61-6/62. AWOL 8/19-31/62. Present sick in quarters 9-10/62. Present 11-12/62. Ab. on horse detail 4/7-30/63. Present 8/31/63. Ab. on horse detail 10/23-12/31/63. Ab. with Provost Guard, Fredericksburg 2/25-28/64. Detailed Green Pond 4/26-30/64. Present 5-12/64. Paroled Hillsboro, N. C. 4/65.

AVINGER, JAMES FRANCIS. Pvt., Co. E. b. S. C. 1/16/34. May have served in Co. A, Manigault's Bn. S. C. Arty. Enl. Co. E, 1st S. C. Cav. James Island 6/1/62. No record for the remainder of 1862. Present 3-12/63 & 1-8/64. Present in arrest 10/5-31/64. On picket duty Western Lines, James Island 11-12/64. Paroled Hillsboro, N. C. 4/65. Laborer, Monk's Corner, Berkeley Co., S. C. 1900 census. d. 7/1/19. Bur. Biggin Ch. Cem., Berkeley Co.

BABB, ALBERT. Pvt., Co. B. b. Laurens Co. 9/24/36. Blacksmith, Laurens Co. 1860 census. Enl. Camp Tripp 4/1/62. Present through 6/62. Present sick in quarters 7-8/62. AWOL 9-10/62. Present 11-12/62 &

3-4/63. WIA Brandy Station 8/1/63. Ab. wounded through 10/63. AWOL 12/1-31/63. Present 1-2/64. Ab. on horse detail 3/31-4/30/64. Present 5-8/64. Ab. on picket 10/31/64. Present 11-12/64. Paroled Greensboro, N. C. 5/2/65. Farmer, Pickens Co. 1900 census. d. 7/7/13. Bur. 2nd Bapt. Ch. Cem., Pickens Co., S. C.

BABB, SAMPSON. Pvt., Co. B. b. S. C. circa 1844. Farmhand, age 16, Laurens Co. 1860 census. Enl. Camp Tripp 5/1/62. KIA in skirmish at Walpole Gate, John's Island 6/7/62 age 18.

BABB, WILLIAM MARTIN, JR. Pvt., Co. B. b. Laurens Co. 10/9/33. Farmer, age 26, Laurens Co. 1860 census. Enl. Co. E, Hampton Legion Inf. Columbia 6/19/61 age 21. Present through 8/61. Discharged for phthisis pulmonalies 10/7/61. Enl. Co. B, 1st S. C. Cav. Camp Tripp 4/1/62. Present through 6/62. Ab. on picket Little Brittan 7-8/62. Ab. on march with horses from Summerville, S. C. to Staunton, Va. 9-10/62. Present 11-12/62. Ab. on wagon guard 3-4/63. WIA (lost arm) 6/24/63. Ab. wounded through 10/63. AWOL 12/1/63-2/29/64. Ab. in Columbia, S. C. hospital 3/1-4/30/64. Placed on retired list 5/20/64. Farmer, Woodruff, Spartanburg Co. 1900 census. d. 7/22/20. Bur. Dial Meth. Ch. Cem., Laurens Co.

BAFFETT, WILLIAM. Pvt., Co. F. On postwar roster.

BAILES, ANDREW BAXTER. Pvt., Co. H. b. Ft. Mills, S. C. 5/2/44. Farmer, age 17, York Co. 1860 census. On roster 1stCo. H, 6th S. C. Inf. as Sgt. NFR. Enl. Co. H, 1st S. C. Cav. Rock Hill 5/7/62. Present through 6/62. Ab. sick in hospital 7-8/62. Present 9-10/62. Ab. on raid 11-12/62. Present 3-4/63. Ab. sick in hospital 8/31/63. Present 9-10/63. Ab. on detached service 12/20-31/63. Present 1-8/64. On picket 9-10/64. Present 11-12/64. Paroled Greensboro, N. C. 4/26/65. Farmer, Ft. Mill, York Co. 1870 census. d. 4/18/21. Bur. Flint Hill Bapt. Ch. Cem., York Co.

BAILEY, ROBERT W. Pvt., Co. G. b. S. C. circa 1844. Age 16, Crawfordsville, Spartanburg Co. 1860 census. Enl. Columbia 6/16/64. Present through 8/64. d. of "malignant county fever" in hospital, James Island 10/15/64.

BAILEY, W. C. Pvt., Co. G. d. in hospital James Island 9/15/64.

BAILEY, W. R. Pvt., Co. G. Res. Greenville Dist. age 35. d. James Island 9/64.

BAIRD, SAMUEL H. 1st S. C. Cav. b. 1833. Not on muster rolls. d. 12/11/07. Bur. Confederate Memorial Cem., Davidson Co., Tenn.

BAKER, DAVID D. 2ndCpl., Co. G. b. New Hampshire. circa 1843. Res. of Va. Enl. Abbeville CH 3/15/62 as Pvt. Ab. on picket Jacksonborough through 4/63. Present 5-6/62. Ab. on leave in Charleston 7-8/62. Present 9-10/62. Promoted 3rdCpl. 11/27/62. AWOL 12/31/62. Ab. sick with affection of the lungs in Farmville hospital 4 months through 4/63. Deserted from the hospital 6/9/63. Present until ab. on horse detail 6/15-8/31/63. Promoted 2ndCpl. Ab. sick in Richmond hospital with "Syphilis secondary" 9/63-2/64. Returned to duty 3/28/64. Transferred Co. B, Lucas Bn. S. C. Arty. 5/12/64 as Pvt. Reported 5/20/64. Present through 12/64. Captured and sent North from Charleston 3/18/65. Age 22, 6', red hair, blue eyes, sandy complexion. Took oath and released. d. by 1902.

BAKER, W. A. Pvt., Co. I. Enl. Co. E, 26th S. C. Inf. Chesterfield CH 12/21/61. Appointed 1stLt. 5/19/62. Resigned 7/11/64. Enl. Co. I, 1st S. C. Cav. James Island 5/12/64. Present through 12/64. NFR.

BALL, ELIAS NONUS. 4thSgt., Co. A. b. S. C. 8/8/34. Enl. 11/7/61. Detailed by order of Gen. Ripley 11/22/61. Reenlist. Co. K, 4th S. C. Cav. Green Pond 3/25/62. Detailed with Gen. Pemberton 4/28/62-1863. Resigned as Sgt. 6/7/62. Ab. on leave for 15 days 12/20/63. Ab. on leave for 20 days 2/18/64. AWOL 3-8/64. Dropped as a deserter. However, transferred Co. H, 1st S. C. Arty. 3/24/64. NFR. d. 1872. Bur. St. Mary's Episc. Ch., Burlington, N. J.

BALLENTINE, J. W. Pvt., Co. K. Enl. Pickens Dist. 4/27/62. Teamster. d. Summerville, S. C. by 9/1/62. Left widow.

BALLENTINE, WARREN CHANDLER. Pvt., Co.'s F & K. b. S. C. 12/26/39. Farmer, age 20, Brewerton, Laurens Co. 1860 census. Enl. Co. C, 14th S. C. Inf. 8/12/61. Present through 10/31/61. Ab. on wound furlough 2/28/62. Enl. Co. F, 1st S. C. Cav. without permission Pickens Dist. 4/27/62. Transferred Co. K, 6/25/62. NFR. Returned Co. C, 14th S. C. Inf. by 2/63. WIA 1864 & detailed light duty in Richmond through 10/64. NFR. Farmer, Sullivans, Laurens Co. 1870 & 1880 censuses. d. 10/13/34. Bur. Poplar Springs Bapt. Ch., Laurens Co.

BALLINGER (BELLINGER), JAMES A. Pvt., Co. F. b. Spartanburg Co. 1819. Farmer, Camp Ground, Pickens Co. 1860 census. Enl. Co. G, 12th S. C. Inf. 7/11/61 age 42 as 2ndLt. Promoted 1stLt. 1/4/62. Resigned 3/30/62. Enl. Co. F, 1st S. C. Cav. Columbia 3/7/64. On duty Green Pond 4/20-30/64. Present 5-8/64. Ab. on sick leave 10/31/64. Present 11-12/64. NFR. Farmer, age 50, Center, Oconee Co. 1870 census. d. 5/4/99. Bur. Retreat Presb. Ch. Cem., Oconee Co.

BANKS, SAMUEL M. Pvt., Co. D. b. S. C. 3/10/45. Student, age 14, Fairfield Dist. 1860 census. Enl. Co. F, 9th S. C. Inf. Ridgeville 7/10/61. Present until ab. sick 9-12/61. Discharged for debility 1/24/62. Reenl. Co. D, 1st S. C. Cav. Chester CH 6/18/62. Present through 10/62. Ab. on scout 12/25-31/62. Present 3-4/63. Ab. on detached duty with Brigade Ordnance Dept. 6/1-10/31/63. Ab. on leave 12/28-31/63. Ab. on horse detail 1-2/64. Ab. on detail Chester CH 4/8-30/64. Ab. detailed in Brigade Commissary Dept. 6/10-12/31/64. Paroled Greensboro, N. C. 4/26/65. Receiving pension Chester, S. C. 1919. d. 6/19/23. Bur. Evergreen Cem., Chester.

BARAFER, HARLY. Pvt., Co. E. On rosters but not on muster rolls. NFR.

BARBER, JAMES GEDDINGS. Pvt., Co. D. b. Chester, S. C. 7/14/45. Student, age 15, Chester, 1860 census. Enl. Chester CH 7/1/63. Ab. on guard 8/31/63. Ab. sick with debilitas in Richmond hospital 9/23/63. Furloughed for 30 days 9/30/63. Ab. on sick leave in Charleston, S. C. 10/31/63. AWOL 11/12-12/31/63. Present 1-4/64. Ab. on duty with Brigade Quartermaster 6/10-30/64. Ab. sick in Brigade hospital 8/28-30/64. Ab. detailed guarding POW's Florence 9/18-12/31/64. Also acting as Asst. Commissary of Subsistance. Paroled Greensboro, N. C. 4/26/65. Farmer, Rossville, Chester Co. 1870 census. d. 5/19/97. Wife receiving pension Fort Lawn, Chester Co. 10/1/20.

BARENTINE, PETER. Pvt., Co. E. b. S. C. 12/21. Enl. 2nd S. C. Reserves (90 days) Georgetown, S. C. 2/3/63. NFR. Conscript enlisted Co. E, 1st S. C. Cav. 1/9/64. Present through 2/64. Ab. at home on sick leave 4/11-31/64. Ab. sick in hospital James Island 6/24-30/64. Discharged for disability 7/23/64. Age 40, 5'9", fair complexion, hazel eyes, dark hair, Farmer. NFR. Farmer, age 45, Cuthbert, Ga. 1870 census. Merchant, age 55, Wilmington, N. C. 1880 census. Retired, Baltimore, Md. 1900 census.

BARKER, JAMES HAMILTON. Pvt., Co. E. b. Barnwell, S. C. 10/23/38. Farmer, Barnwell Dist., 1860 census. Conscript assigned 4/12/64. Detailed Green Pond 4/16-30/64. Present 5-12/64. Captured Barnwell 2/5/65. Sent to New Bern 3/30/65. Transferred Pt. Lookout. Released 6/24/65. 5' 8 3/4", light complexion, brown hair, hazel eyes. Farmer, Allendale Co. 1870 census. d. Beaufort, S. C. 11/7/28.

BARKLEY, JOHN W. Pvt., Co. F. b. Anderson Dist., S. C. 11/25/32. Farmer, Pickens Dist. 1860 census. Enl Pickens CH 12/4/61. Present through 6/6/2. Ab. on leave 7-8/62. Ab. on detached service 9-10/62. Ab. sick in private quarters 11/62-4/63. Discharged 5/28 or 6/13/63 for phthisis pulmonalis. Age 27, 5'10", fair complexion, grey eyes, light hair, Famer. Farmer, Pickens Co. 1870 census. Receiving pension 1901. d. Brushy Creek, Anderson Co. 1/25/01.

BARKSDALE, JAMES HENRY. Pvt., Co. G. b. Abbeville, S. C. 7/1/46. Student, age 15, Abbeville Dist.1860 census. Enl. Abbeville 7/4/63. Present through 8/31/63. Ab. on detached service 9-10/63. Present 11-12/63. Ab. on duty with Provost Guard, Hanover Junction 1/12-2/28/64. Transferred 20th S. C. Inf. 5/13/64. NFR. Farmer, Abbeville Co. 1880 census. d. Abbeville 7/10/22.

BARKSDALE, S. E. Pvt., Co. I. May have served in Co. D, 9th S. C. Reserves in 1863. Enl. Co. I, 1st S. C. Cav. Columbia 2/15/64. AWOL through 4/30/64. Present 5-6/64. Ab. sick in hospital 8/8/31/64. Present

9-10/64. AWOL 12/3-31/64. NFR.

BARKSDALE, T. W. Pvt. Co. A. Enl. by 8/61. Not on muster rolls 10/11/61. NFR.

BARNETT, JOSEPH C. Pvt., Co. B. b. S. C. 3/13/45. Res. Woodruff, Spartanburg Dist. Enl. Clinton 5/9/63. Present through 8/31/63. Ab. sick in Charlottesville hospital with erysipelas 10/17-12/15/63. Ab. on horse detail to S. C. 12/22/63-2/64. Ab. on leave 4/26-30/64. Present 5-10/64. Ab. on detached duty Pocotaligo 12/2-31/64. Served to April, 1865 on pension application. Farmer, Woodruff, Spartanburg Co. 1870 census. d. 9/4/02. Bur. Bethel Bapt. Ch. Cem., Spartanburg Co.

BARNETT, WILEY. Pvt. Co. G. On postwar roster.

BARNETT, WILLIAM HENRY. Pvt., Co. B. b. Woodruff, Spartanburg Dist. 3/13/42. Enl. Clinton 8/22/61. Present through 4/62. Ab. sick at Church Flats 5-6/62. Ab. on sick leave at home 7-8/62. Ab. on march with horses from Summerville, S. C. to Staunton, Va. 9-10/62. Present 11/62-4/63. WIA (left arm) Brandy Station 6/3/63. Present through 12/63. Ab. sick in Columbia, S. C. hospital 1/20-2/28/64. Present 3-4/64. Ab. sick in Charleston hospital 6/1-30/64. Ab. sick McLeod House hospital 8/20-31/64. Ab. sick at home 9/3-10/31/64. Present 11-12/64. Present Fayetteville, N. C. 4/65. NFR. Applied for pension from Woodruff, Spartanburg Co. 3/29/19. d. Spartanburg 7/27/27. Bur. Bethel Bapt. Ch. Cem., Spartanburg Co.

BARNWELL, E. M. Pvt., Co. A. b. S. C. circa 1836. Enl. 4/62 age 26. Reenlisted Co. K, 4th S. C. Cav. Transferred on post war roster.

BARRETT, WILLIAM M. Pvt., Co. F. b. S. C. 1831. Laborer, age 19, Pickens Dist. 1850 census. Enl. Pickens CH 12/4/61. Present until ab. on leave 4/25-5/25/62. Ab. on horse detail 5-6/62. Ab. in hospital Summerville 7-8/62. d. there 9/1/62 age 31. d. 8/31/62 on tombstone. Bur. Ponder Cem., Lathem, S. C.

BARRON, DANIEL WATSON. Pvt., Co. H. b. S. C. 4/12/29. Farmer, York Dist. Enl. Columbia 5/28/64. Age 35, 110 lbs., 5' 7", fair complexion, dark hair, hazel eyes. Ab. on leave 20 days 8/18/64. NFR. d. 4/11/14. Bur. Ebenezer Presb. Ch. Cem., York Co.

BARRON, DAVID H. Pvt., Co. H. b. S. C. 3/44. Enl. 1864. Ab. on detail 9-10/64. Ab. on sick leave 11/12-12/31/64. Paroled Charlotte, N. C. 5/11/65. Res. St. Helena Island 1880 census.

BARRON, J. LEROY. Pvt., Co. H. b. S. C. 1/17/26. On postwar roster. Enl. Co. G, 6th S. C. Reserves, York 11/18/62. Ab. on leave 2/5-14/63. Paid taxes York Co. 1866. d. 1/21/76. Bur. Ebenezer Presb. Ch. Cem., York Co.

BARRY, JAMES H. Captain, Co. H. b. S. C. circa 1819. Planter, Rock Hill 1860 census. Enl. Rock Hill 1/9/61 age 43 as 1stLt. Ab. on picket Jacksonborough 3-4/62. Present 5-10/62. Ab. on CM 11-12/62. Sick in quarters 3-4/63. WIA (lost toe from right foot) 6/20/63. Promoted Captain. Ab. wounded 8/17/63-4/64. Present in arrest 5-6/64. Ab. on leave 8/31-9/20/64. Suspended 9/64. Present 10-12/64. Resigned 1/27/65. NFR. Livery Stable Keeper, York 1870 & 1880 censuses. d. 4/14/88.

BARTON, WILLIAM. Pvt., Co. F, b. Tigerville, Greenville Dist. 4/23/33. Farmer, Pickens Dist. 1860 census. Enl. Pickens CH 12/4/61. Present until ab. with Commissary at Adams Run 3-4/62. Ab. on leave 7-8/62. Present 9-12/62 & 3-5/63. Ab. on sick leave 6/23-7/8/63 & dismounted 25 days. Present through 8/31/63. Ab. on detached service 9-10/63. AWOL 10/22-12/31/63. Present as Sutler 2/1/64, dismounted since 9/12/63. Ab. on detached service 4/30/64, without horse since 4/22/64. Present 5-6/64. Present sick in quarters 7-8/64. Ab. on detail guarding POW's Florence 9/17-12/31/64. NFR. d. Greenville, S. C. 10/7/08.

BASS, JOHN BRYANT. Pvt., Co. B. b. S. C. 4/22/40. Enl. Clinton 8/22/61 age 21. Ab. sick 10/21--31/61. Present 11/61-6/63. Ab. on duty as Scout on Edisto Island 7-8/62. Ab. with horses on march from Summerville, S. C. to Staunton, Va. 9-10/62. Present 11/62-6/63. Ab. sick in hospital 8/9-31/63. Present

9/63-2/64. Horse died 1/31/64. Ab. on detail for a horse 4/15-30/64. Present 5-12/64. NFR. Farmer, Youngs, Lauren Dist. 1870 census. Living Beech Springs, Spartanburg Co. 1910 census. d. 12/9/10. Bur. Zoah Meth. Ch. Cem., Spartanburg Co. Wife applied for admission to Old Soldiers' Home, Columbia 1923 & 4/23/1951.

BATEMAN, DYONICIOUS OLIVER. Pvt., Co. C. b. S. C. 1843. Enl. Wadmalaw Island 3/18/62. Sick in camp 5-6/62. Ab. on sick leave 8/11-31/62. Present 9-12/62. Ab. on detached service 3-4/63. Present through 8/31/63. Captured near Robinson River 9/22/63. Sent to Old Capitol. Transferred to Pt. Lookout & Elmira. Sent for exchange 3/10/65. NFR. d. 12/05. Bur. Magnolia Cem., Waynesboro, Ga.

BAXTER, JOSEPH D. Pvt., Co. G. b. S. C. 1841. Enl. Parker's Ferry 4/3/62. AWOL through 4/63. NFR. Enl. Co. F, 1st S. C. Arty. James Island 1/18/64. Present until captured 5/12/64. Exchanged. Ab. on leave 8/21-30/64. On picket 9-12/64. Paroled Greensboro, N. C. 5/1/65. d. 1919. Bur. Denmark Cem., Bamberg Co., S. C.

BEARD, SAMUEL H. Pvt., Co. A. b. S. C. 1847. Enl. Abbeville 3/19/62. Present sick through 4/62. Transferred Co. E, 16th S. C. Inf. 5/22/62. On sick leave Greenville 6/16-30/62. Discharge 7/4/62, probably for underage. Reenlisted Co. E, 16th S. C. 4/1/63. Hired a substitute and discharged 4/27/63. Reenlisted again 3/1/64 at Dalton, Ga. AWOL 8/15-31/64. NFR. Cloth Maker, Augusta, Ga. 1900 census.

BECKHAM, HOYDEN D. Pvt., Co. H. b. S. C. 1844. Age 16, Chester Dist. 1860 census. Enl. Rock Hill 1/20/64. Ab. on horse detail through 2/28/64. Present 3-10/64. Detailed Pocotaligo 12/10-31/64. NFR. Brother of William H. Beckham.

BECKHAM, WILLIAM H. 2ndLt., Co. H. b. Chester Dist. 1833. Farmer, age 26, Chester Dist. 1860 census. Enl. Rock Hill 1/9/61 as Pvt., age 27. Promoted Sgt. Ab. on leave3-4/62. Present 5-6/62. Sick in camp 7-8/62. Present 11-12/62, promoted 3rd Sgt. Ab. on horse detail 3/20-31/63. Elected 3rdLt 6/15/63. Present through 12/63. Present 1-2/64. Commanding Co. F 3-4/64. Promoted 2ndLt. Present 5-12/64. NFR. Farmer, Cedar Creek, Lancaster Co. 1870 & 1880 censuses.

BEE, JAMES LADSON. Pvt., Co. A. b. S. C. 6/19/43. Enl. 11/14/61. Reenlisted Co. K, 4th S. C. Cav. Grahamville 3/25/62. Present until ab. on leave 6/24-12/31/63. Present 1-2/64. WIA (right tibia fractured) and captured Old Church 5/30/64. Sent to Lincoln General Hospital, Washington, D. C. Leg amputated 6/29/64. DOW's 7/8/64. Bur. Magnolia Cem., Charleston, S. C.

BEE, JOHN STOCK. Pvt., Co. A. b. S. C. 2/28/41. On postwar roster. Appointed 2ndLt. of Arty. 6/17/62. Assigned to Co. E, 1st S. C. Arty. Present until transferred Co. I 9-10/62. Present until ab. on recruiting duty 4/63. Present until WIA (face) and captured Morris Island 7/10/63. Sent to Hilton Head hospital. DOW'S 7/18/63. Body sent to Charleston for burial.

BEILLING (BIELLING), E. S. Pvt., Co. C. Enl. Wadmalaw Island 4/2/62. Present 5-6/62. Ab. on sick leave 7/28-8/31/62. Ab. sick at home 9/62-10/63. AWOL 12/26-31/63. Transferred Capt. Hayns' Co. A, 15th S. C. Bn. Arty. 6/16/64. Present through 12/64. NFR. d. by 1898.

BELCHER, HENRY C. Pvt., Co. A. b S. C. circa 1842. Student, age 18, Willington, Abbeville Dist. 1860 census. Enl. Abbeville 9/17/61. Ab. on sick leave 12/4-31/61. Present 3-8/62. Present sick 9-10/62. Ab. sick 11-12/62. Ab. detached 3-4/63, dismounted since 4/8/63. Present through 8/31/63. Ab. on horse detail to S. C. 9-10/63. Present 11-12/63. Ab. sick in hospital with "Primary Syphilis" 2/11-28/64. Remounted 3/28/64. On duty Green Pond 3-4/64. Present 5-6/64. Captured James Island 7/12/64. Exchanged Port Royal Ferry 8/12/64. Detailed Pocotaligo 12/10-31/64. Dismounted and transferred to 7th S. C. Inf. NFR. Farmer, Hall, Anderson Co. 1870 census. Brother of James N. Belcher.

BELCHER, JAMES NORWOOD. Pvt., Co.'s A & G. b. S. C. 6/15/43. Student, age 16, Willington, Abbeville Co. 1860 census. Att. U. of Ga. Enl. Abbeville 3/19/62. Ab. on leave 4/62. Present 5-6/62, dismounted. Present 7-10/62. On picket 11-12/62. Present sick 3-4/63. Present dismounted 5-6/63. Ab. on horse detail 8/13-31/63. Present 9-12/63 & 1-2/64. Ab. detached Green Pond 3-4/64. Remounted 3/20/64. Present 5-8/64. Transferred Co. G. 8/8/64. AWOL 10/26-31/64. Ab. detached Pocotaligo 12/10-31/64. Paroled Greensboro, N. C. 5/1/65. Farmer, Abbeville Co. postwar. d. Abbeville 11/2/16. Brother of Henry C. Belcher.

BELCHER, JOHN HAMPDEN. Pvt., Co. A. b. S. C. 7/31/40. Student, Abbeville 1860 census. Att. U. of Ga. Enl. Abbeville 9/12/61. Ab. as Secretary to Col. Black 3-4/62. Present 7-12/62 & 3-6/63, dismounted one month. Ab. on horse detail 8/18-31/63. Present 9-10/63. Transferred to Capt. Carlton's Troup Ga. Arty. 10/30/63. Present until WIA (fractured left thigh) 5/8/64. Ab. wounded in Richmond hospital until furloughed for 60 days to Abbeville CH. NFR. Farmer, Bossier, La. 1870, 1910 & 1920 censuses. d. 6/13/27. Bur. Rocky Mount Cem., Bossier Parish, La. Brother of William H. Belcher.

BELCHER, WILLIAM HAYNE. Pvt., Co. A. b. S. C. circa 1839. Student, age 21, Abbeville 1860 census. Att. U. of Ga. Enl. Abbeville 9/13/61. Ab. as Courier for Gen. Evans 11-12/61. Ab. with accidental gunshot wound 3-8/62. Discharged 9/28/62. "It will several months before he recovers use of his right arm." Age 23, 5' 11", light complexion, blue eyes, light hair, Student. NFR. d. Benton, La. 1875. Brother of John H. Belcher.

BELL, JAMES. 4thSgt., Co. B. b. S. C. circa 1826. Carpenter, Columbia 1860 census. Enl. Clinton 8/22/61 age 33 as Pvt. Ab. detailed Camp Beauregard through 10/31/61, promoted 4thCpl. Present 11-12/61 & 3-6/62, promoted 3rdCpl. Ab. on detached service 7-8/62. Ab. on march with horses from Summerville, S. C. to Staunton, Va. 9-10/62. On guard duty in camp 11-12/62. Promoted 1stCpl. Present 3-10/63. Ab. at horse recruiting station 12/18-31/63 & 1-2/64. Present 3-8/64. Promoted 4thSgt. Present 9-12/64. NFR. Farmer, age 44, Hibler, Edgefield Co. 1870 census.

BELL, JOHN HENRY. Pvt., Co. B. b. Pickens Dist. 1/6/43. Farmer, age 17, Pickens Dist. 1860 census. May have served in Co. H, 1st S. C. Arty. Enl. Co. B, 1st S. C. Cav. Columbia 3/3/64. Ab. detached Green Pond 4/16-30/64. Present 5-8/64. On picket 9-10/64. Present 11-12/64. NFR. Bank Clerk, Columbia 1880 census. d. Greenville, S. C. 12/13/22.

BELL, WILLIAM, JR. Pvt., Co. A. b. Charleston, S. C. 4/1/37. Enl. 11/7/61. Detailed to Gen. Ripley 1/7/62. Reenlisted Co. K, 4th S. C. Cav. WIA twice and captured Cold Harbor in postwar account. Paroled 5/4/65. d. Mountain Creek, Ala. 12/18/13. Bur. Confederate Cem., No. 2, Chilton Co., Ala.

BELLING, MARTIN. Pvt. Co. C. On post war roster.

BELLINGER, WILLIAM HENRY. Pvt., Co. A. b. S. C. 1844. Age 14, Colleton Co. 1860 census. Enl. Green Pond 7/14/63. Present through 2/64. Ab. in Richmond hospital 5/30-6/26/64. Transferred 4th Va. Cav. Present through 10/64. Paroled Greensboro, N. C. 5/2/65. Lawyer, Walterboro, S. C. 1870 census. Traveling Salesman, Charleston 1880 & 1900 censuses. d. 3/31/97. Bur. Poco Sabo Plantation Cem., Ashepoo, Colleton Co.

BENNETT, THOMAS M. Musician. b. S. C. 1847. Student, age 14, Charleston, 1860 census. Enl. Columbia 8/21/64. Present through 12/64. Paroled Hillsboro, N. C. 4/65. Asst. Miller, age 23, Charleston 1870 census.

BENTON, HENRY RICHARD. Pvt., Co. I. b. Colleton Dist. 7/4/38. Enl. Parker's Ferry 4/3/62. Present through 8/62. Ab. on march with horses from Summerville, S. C. to Staunton, Va. 9/62. Ab. sick in private house near Staunton 10-12/62 & 3-4/63. Ab. sick in Charlottesville hospital with acute diarrohea 7/12-8/1/63. Present through 10/63. Ab. sick in hospital 12/20/63-2/64. Returned 4/1/64. Detailed Green Pond 4/14-30/64. Present 5-10/64. Detached Pocotaligo 12/10-31/64. NFR. Farmer, South River Dist., Augusta Co., Va.1880 census. Farmer, Lynhurst, Va. 1909. d. Stuart's Draft, Va. 9/9/15. Bur. Miller's Ch. of the Brethren, Augusta Co., Va.

BENTON, J. R. Pvt., Co. I. Enl. Co. F, 9th S. C. Inf. (4 month unit) Hilton Head 7/26/61. Present 11/26/61. Disbanded. Enl. Co. E, 24th S. C. Camp Grist 3/31/62. Enl. Co. I, 1st S. C. Cav. Parker's Ferry 4/3/62, without authority. AWOL through 4/30/62. Present in Co. E, 24th S. C. Inf. 4/62-2/64. WIA (hip) 7/21/64.

Ab. wounded through 8/31/64. NFR. May have died of wounds.

BENTON, LUKE L. Pvt., Co. E. Enl. Colleton Dist. 1862, age 35. Transferred 3rd S. C. Cav. 5/62. Not on muster rolls of either unit.

BENTON, SAMUEL S. Pvt., Co. E. Enl. Adams Run 3/18/62. Present through 4/62. Transferred Co. C, 2nd S. C. Cav. 6/121/62. Present through 12/62. AWOL 2/5-28/63. Transferred Co. A, 24th S. C. Inf. 4/1/63. Present through 8/63. Ab. sick leave 9-10/63. AWOL 11-12/63. Ab. in arrest since 2/4/64 on 2/28/64. AWOL 8/18-30/64. Detailed in hospital and issued clothing 7/28/64 & 9/12/64. NFR.

BERRY, BENJAMIN BRUTON. Pvt., Co. E. b. S. C. 2/7/35. Farmer, age 25, Colleton Dist. 1860 census. Enl. Branchville 10/25/61 age 25. On picket White Point 12/31/61. Ab. on sick leave at home 4/15-30/62. Present 5-6/62. On picket Lawachs 7-8/62. Present 9-10/62. Present sick in camp 11-12/62. Ab. sick in hospital 3-4/63. Ab. detailed to rear with disabled horses 8/31-2/64. Detached Green Pond 4/16-30/64. Present 5-6/64. Detailed to drive cattle 5-12/64. Paroled Hillsboro, N. C. 4/65. Living Colleton Co. 1880 census. d. 11/25/92. Bur. Old St. George Bapt. Ch., Dorchester Co.

BERRY, JAMES. Pvt., Co. E. b. S. C. 1843. Enl. Adams Run 1/8/62. Discharged for disability 3/30/62. NFR. Farmer, Colleton Co. 1870 census.

BERRY, JOHN HENRY. Pvt., Co. H. b. 1830. On postwar roster. Reenlisted Co. I, 5th S. C. Cav. NFR. d. 1899. Bur. Bur. Butler Meth. Ch., Saluda Co., S. C.

BERRY, RICHARD G. M. Pvt., Co. E. b. Branchville, S. C. 6/27/29. Farmer, age 29, Branchville, Orangeburg Co. 1860 census. Enl. Branchville or Ft. Motte 10/26/61 age 34. Present through 12/61. Sick in camp 3-4/62. Ab. on sick leave at home 5/12-31/62 & 7-10/62. AWOL 11-12/62 & 3-4/63. Present 5-6/63. Detailed at horse recruiting station 9-10/63, however, ab. on sick leave 9/15/63, now AWOL from hospital through 2/64. Ab. on horse detail 3/31-4/30/64, but detailed to Charleston Gas Light Co. 4/23/64. Present 7-8/64. Present sick in camp 9-10/64. Ab. detailed to Supt., Charleston Gas Light Co. 11/23/64-2/1/65. NFR. d. Bowman, S. C. 9/19/87. Bur. Shiloh Meth. Ch.

BERTIN, J. M. Pvt., Co. C. Captured and died Pt. Lookout, Md. 4/15/65. Bur. Confederate Cem., Point Lookout, Md.

BICKLEY, JOHN C.,2ndSgt., Co. A. b. S. C. 1/16/30. On postwar roster. Reenlisted Co. K, 4th S. C. Cav. Grahamsville 3/25/62 age 32. Ab. sick through 6/30/62. Present 7/62 until promoted 1stSgt. 1/15/63. Present through 8/64. Ab. sick 10/22-31/64. NFR. d. 6/11/01. Bur. Magnolia Cem., Charleston, S. C.

BIGGER, ALEXANDER BARNETT. Pvt., Co. H. b. York Dist. 10/4/17. Planter, Yorkville, 1860 census. Enl. Rock Hill 3/20/62. On picket Ranatole Bridge 4/30/62. Present 5-10/62. On picket 11-12/62. Present 3-4/63. Died of disease in Mt. Jackson, Va. hospital 7/31/63 age 42. Bur. Mt. Jackson Confederate Cem. Marker in Bethel Presb. Ch. Cem., York Co.

BIGGER, MOSES ANDREW. 3rdSgt., Co. H. b. York Dist. 9/26/25. Farmer, Cherry Hill, York Dist. 1860 census. Enl. Rock Hill 3/20/62 as Pvt. On picket Simons Bluff 4/30/62. Present 5-6/62. Ab. on sick leave 7/62-4/63. Ab. on horse detail 8/18-31/63. Present 9-10/63. Promoted 4thSgt. Present 11/63-6/64. Promoted 3rdSgt. Present 7-10/64. Ab. detached 12/26-31/64. NFR. Farmer, Bethel, York Co. 1880 & 1900 censuses. d. 2/18/06. Bur. Bethel Presb. Ch. Cem., York Co.

BILLINGSLEY, ANDREW JACKSON MILTON. Pvt.& Bugler, Co. F. b. S. C. 7/2/43. Enl. Pickens CH 12/4/61 age 20 as Pvt. Present through 4/62. Present as Bugler 5/62-6/63. Ab. detached with Quartermaster's wagon 8/1-10/31/63. Ab. detached as recruiting station for horses 11/63-3/64. Present 3-4/64, dismounted since 4/23/64. Present 5-6/64. Ab. on daily duty at telegraph office 8/18-31/64. Present 9-12/64. Served to April, 1865 on wife's pension application. Farmer, Chatlooga, Oconee Co. 1910 census. d. 5/7/17 at Mt. Rest, Oconee Co. Bur. Double Springs Meth. Ch. Cem., Oconee Co. S. C.

BINGHAM, ROBERT S. Pvt., Co. D. b. New Hampshire circa 1816. Enl. Chester CH 9/10/61 age 25. Ab. on sick leave 10/31/61. Ab. sick leave 125-31/61. Present 3-4/62. Discharge 5/7/62, probably for overage. Reenlisted and paroled Hillsboro, N. C. 4/65. Farmer, Ortonville, Minn. 1880 census.

BIRD, G. Pvt., Co. K. Enl. Pickens Dist. 1862. d. Staunton, Va. 1862.

BIRD, W. JOSEPH. Pvt., Co.'s F & K. b. S. C. circa 1840. Enl. Pickens CH 12/4/61 age 21. Present through 4/62. Transferred Co. K 6/25/62. Present through 10/62. Ab. detailed Quartermaster's Dept. 11-12/62. Present 3-6/63. Ab. on horse detail 8/16-31/63. Present 9-10/63. Driving ambulance in Quartermaster's Dept. 11/10/63-2/64. Ab. on horse detail 4/1-30/64. On duty at bridge of Ashley River 7-8/64. d. at home Pickens, S. C. on rolls 9-10/64. Left widow.

BLACK, JOHN GAILLIARD. Pvt., Co. F. b. Cherokee Co. 7/42. Teacher, age 18, Harmony, York Co. 1860 census. Enl. Pickens Dist. 4/8/62. Ab. on leave New Road Station 4/30/62. NFR. However, enl. Co. D, 7th S. C. Cav. Charleston 1/17/62. Detailed in QM Dept. 1/63-4/64. Transferred Hart's S. C. Bty. 11/23/64. NFR. Gd. Charleston Medical College 1868. M. D., York Co. M. D., Blacksburg, Cherokee Co. 1880 & 1900 censuses. d. 3/14/15. Bur. Mt. View Cem., Blacksburg, S. C.

BLACK, JOHN LOGAN. Colonel. b. Cherokee, York Dist. 7/12/30. Att. USMA Class 1854. Appointed Lt. Col. 1st S. C. Bn. Cav. 10/31/61. Appointed Colonel 1st S. C. Cav. when increased to a regiment 1/25/62. Signed for 85 lbs. fresh beef, 47 lbs. salt beef, 476 lbs. bacon, 586 1/2 lbs. flour, 100 lbs. corn meal, 373 lbs. rice, 116 lbs. coffee, 511 lbs. sugar, 79 lbs. candles, 20 lbs. soap, 193 lbs. lard, 46 quarters of salt and 34 gallons of molasses 3/62. Left S. C. in 10/62 with 13 wagons & exchanged mules at Charlotte, N. C. enroute to Staunton, Va. Present 9-12/62 and on Dumfries Raid. However, ab. at CM Fredericksburg 12/31/62. Present Brandy Station 6/9/63 & WIA (head) Upperville 6/21/63. Present until WIA (right hand) Brandy Station 8/6/63 while commanding Brigade. Ab. wounded through 8/31/63. Present 9-10/63 and Mine Run 11/63. Ab. on leave for 30 days 1/31/64. Presence or absence not stated 3-4/64, however regiment ordered to S. C. 3/24/64. Dismounted men & baggage sent by railroad. Commanding East Lines, James Island 5/21-6/30/64 & 7-12/64. However, ab. on leave for 15 days 8/15/64. Present 1-4/65. Commanding at Charlotte, N. C. 4/1/65 and commanding regiment 4/10/65. NFR. Mining magnate and planter, Blacksburg, S. C. postwar. d. Camp Cherokee near Blacksburg, S. C. 3/25/02. Bur. Aimwell Cem., Ridgeway, S. C.

BLACKMAN, JERRY. Pvt., Co. G. Enl. Abbeville 3/15/62. Present as camp guard through 4/30/62. Ab. on picket 5-6/62. Present 7-8/62. Present sick in camp 9-10/62. Ab. sick and left in private house near Staunton, Va. 11/19/62. NFR. d. of consumption Staunton hospital 3/63.

BLACKWELL, L. L. Pvt., Co. A. b. N. C. circa 1832. Enl. Abbeville 6/25/62. Ab. detached duty 7-8/62. Present 9-12/62. Ab. detached duty 3-4/63. Present through 8/31/63. Ab. on horse detail to S. C. 9-10/63. Present dismounted 10/20-11/20/63. Present 1-2/64. On duty Green Pond 3-4/64. Dismounted since 1/25/64. Present 5-12/64. NFR. Miner, age 38, Eureka, Nevada 1870 census.

BLAKE, THOMAS. Pvt., Co. I. WIA Battery Wagner, S. C. 10/21/64. NFR

BLAKE, WALTER. Pvt. Co. A. b. Charleston, S. C. 8/41. On postwar roster. Reenlisted Co. K, 4th S. C. Cav. Grahamsville 10/8/62 age 23. Present until ab. detailed with Gen. Walker 12/1762 for 15 days. Present until detailed with Ge. Walker 4/28/63. Ab. on leave for 30 days 10/25/63. Present 12/63 until ab. on horse detail 8/11/64 for 40 days. Present 9-10/64. Paroled 4/65. d. Charleston 12/5/79.

BLASSINGAME, BENJAMIN FRANKLIN. 1stLt., Co. F. b. Anderson Co. 12/13/34. Farmer. Enl. Butler Guards of Greenville and present Bull Run on pension application. Enl. Co. F, 1st S. C. Cav. Pickens CH 12/4/61 age 37 as Pvt. Present until on picket Jacksonborough 3-4/62. Present 5-6/62. Elected 2ndLt. 7/12/62. On detached duty 9-12/62. Present 3-8/31/63 & commanding Co. Present 9-10/63. Elected 1stLt. 11/1/63. Commanding Co. 11/16/63-8/64. Signed for 4 axes, 4 tents, 6 curry combs & 2 buckets 1/31/64. Present 9-12/64. Present Wadesboro, S. C. 1865. NFR. Farmer, Brushy Creek, Anderson Co. 1866-1900. d. Greenville Co. 8/30/08. Bur. Springwood Cem.

BLASSINGAME, WYNN GOWEN. Pvt. Co. F. b. S. C. 1/13/36 or 5/6/38. Enl. 4th S. C. Inf. Pickens CH 1/61. NFR. Served in Co. H, 3rd S. C. Reserves. Enl. Co. F, 1st S. C. Cav. Columbia 2/15/64. Ab. on duty Green Pond 4/20-30/64. Present 5-6/64. Ab. on duty West Lines, James Island 7-8/64. Ab. detailed to guard POW's Florence 9/1912/31/64. Escorted Gen. Walker, supply train with Major Black of Reserve Artillery train, and went with report to Hillsboro, N. C. 4/65. NFR. Farmer, Grayson, Texas 1910 census. Retired to Oklahoma City, Okla. d. 1/24/29. Bur. Lanes Chapel Memorial Cem., Oklahoma City, Okla.

BLOCH, ALFRED. Pvt., Co. H. b. S. C. 7/15/24. Blacksmith, age 36, Chester Co. 1860 census. Enl. Rock Hill 1/9/62. Present 3-4/62. Armorer for Battery at Church Flats 5-6/62. Ab. on sick leave 7-8/62, dismounted. Present dismounted 9-10/62. Ab. on duty as Blacksmith 11-12/62, dismounted. Ab. on horse detail to S. C. 3/26-4/30/63. Ab. detailed to horse recruiting station 6-8/63. Present 9-12/63. Ab. on leave 2/21/28/64 & 3/31-4/19/64. Present 5-6/64. Ab. sick in Charleston hospital 8/22-31/64. Present 9-10/64. Detached Pocotalio 10-12/64. NFR. Farmer, Catawba, York Co. 1880 census. d. 1/13/92. Bur. Catawba Bapt. Ch. Cem., York Co.

BLOCKER, JAMES ALFRED RALPH. Pvt., Co. I. b. Colleton Dist. circa 1840. Enl. Co. I, 11th S. C. South Edisto Island 6/17/61 age 23. Discharged 11/25/61. Enl. Co. I, 1st S. C. Cavalry Parker's Ferry 4/3/62. Present through 6/62. Ab. on guard duty 7-8/62. Present 9-10/62. Present sick in quarters 11-12/62. Present 3-4/63. Present through 10/63. Ab. sick with chronic bronchitis in Richmond hospital 11/30-12/15/63. Ab. at horse recruiting station 2/4/64. Ab. on leave 4/26-30/64. Present 5-6/64. Discharged Charleston for phthistis pulmonalis 7/20/64. Age 24, 5'9", dark complexion, dark hair, dark eyes, Farmer. Applied for pension Shelby Co., Tenn. circa 1900. d. 5/31/11. Bur. Elmwood Cem., Shelby Co., Tenn.

BLOOMINGBURG, RICHARD THOMAS. Pvt., Co. I. b. Colleton Dist. circa 1846. Res. Waltersboro, Colleton Dist. 1860 census. Enl. Parker's Ferry 4/3/62. Present until detailed as Courier 5-6/62. Present sick in quarters 7-8/62. Present 9-12/62. Present as Teamster 3-4/63. Ab. on extra duty at recruiting station for horses 8/21/63-2/64, dismounted since 1/1/64. Ab. on leave 4/7-30/64. Present 5-10/64. AWOL 11-12/64. NFR. Farmer, Collins, Colleton Co. 1880 census.

BOGGS, JOHN CALVIN. 2ndSgt., Co. F. b. Salubity Springs, Pickens Dist. 5/6/31. Farmer, Anderson Dist. 1860 census. Enl. Pickens CH 12/4/61. Present until sick in quarters 3-4/62. Present 5-7/62. Ab. on detached service with horses from Summerville, S. C. to Staunton, Va. 9-10/62. Present 11/62-6/63. Ab. detached with horses through 8/31/63. Present 9/63-12/64. NFR. Farmer, Bushy Creek, Anderson Co. 1870 & 1880 censuses. d. Slabtown, Anderson Co. 5/22/93 Bur. Slabtown Cem.

BOLICK, WASHINGTON A. Pvt., Co.'s D & K. b. Calaw, N. C. 1835. Blacksmith. Enl. Chester CH 5/16/62. Present through 9/1/62. Detailed as Nurse in Amherst Co. 9-10/62. Present 11-12/62. Ab. sick with chronic rheumatism in Charlottesville hospital 2/19-5/18/63. Ab. sick and Nurse in Columbia, S. C. hospital 9/20/63-4/30/64. Detailed to work in S. C. Railroad shops 5-8/64. Age 29, 5' 11 1/2", light complexion, brown hair, blue eyes, Mechanic. Transferred Co. I, 6th S. C. Inf. 9/1/64. KIA in Va.

BOLICK, WILLIAM A. Pvt., Co. K. b. S. C. circa 1835. Farmer, age 25, Hadsellville, Chester Dist. 1860 census. Enl. John's Island 7/12/61. Present through 9/1/62. Detailed as Blacksmith 9-10/62. Present 11-12/62. Ab. as nurse 3-4/63. Present 4/12-8/31/63. Ab. on scout 9-10/63. Ab. on scout over the Rappahannock 12/23-31/63. KIA near Brentsville 2/14/64. Bur. Arrington's Cross Roads near Brentsville, Va. Left widow & child.

BOLICK, WILLIAM DOUGLAS. Pvt., Co. K. b. S. C. 10/16/45. Age 14, Chester Dist. 1860 census. Enl. Columbia 4/19/64. Present through 6/64. Ab. detailed driving cattle for Quartermaster, Florence, S. C. 7-12/64. Paroled Greensboro, N. C. 4/26/65. d. Fairfield, S. C. 12/6/90. Wife receiving pension, Blair, Fairfield Co. 9/27/19.

BOONE, JOSEPH HALL WARING. Pvt., Co. A. b. S. C. 1840. On postwar roster. Reenlisted Co. K, 4th S. C. Cav. Pocotaligo 6/16/63 age 23. Present until ab. on leave for 30 days 12/7/63. Promoted 3rdSgt.

Present until KIA Trevillian 6/11/64. Bur. St. Paul's Ch., Charleston, S. C.

BOSTICK, BENJAMIN R. Pvt., Co. A. b. S. C. 12/6/46. On post war roster. Reenlisted Co. K, 4th S. C. Cav. Mc Phersonville 7/10/62. Present until detailed by Gen. Walker 12/23/63 for 10 days. Ab. sick 1/7/64. Present until ab. on horse detail 8/11/64 for 40 days. DOW's received 10/1/64 at Wilson, N. C. hospital 10/5/64.

BOSTICK, EDWARD. Pvt., Co. K. b. S. C. circa 1837. Enl. 3/25/62 age 35. Promoted Captain, 21st S. C. Inf. 12/17/62 but not on muster rolls of that unit.

BOSTICK, L. A. Pvt., Co. A. DOW's 9/29/63 on post war roster.

BOSTICK, LUTHER ROBERT. Pvt., Co. A. b. S. C. 1844. On postwar roster. Reenlisted Co. K, 4th S. C. Cav. Grahamville 3/25/62 as 4thCpl. Present until promoted 5thSgt. 1/1/63. Present until ab. on leave for 30 days 12/17/63. Present until captured Cold Harbor 5/30/64. Sent to Pt. Lookout. Exchanged 3/15/65. NFR. d. 1910. Bur. Black Swamp Meth. Ch., Hampton Co., S. C.

BOUCHILLION, BENJAMIN FRANKLIN. No. Co. b. Abbeville Co. 2/14/44. Not on muster rolls. d. 12/14/10. Bur. Old Jacksonville Cem., Jacksonville, Fla. Served on wife's pension application from Marion Co., Fla.

BOURNE, E. H. Pvt., Co. B. b. S. C. 1/1/28. Carpenter, Clinton 1860 census. Enl. Clinton 8/22/61 age 33. Present through 4/63. WIA 7/8/63. Ab. wounded in Staunton hospital 8/31/63. Furloughed from Richmond hospital for 30 days 9/26/63. Ab. on wound furlough through 10/63. Present 11/63-12/64. NFR. d. 12/13/89. Bur. Clinton City Cem.

BOWEN, E. H. Pvt., Co. B. Enl. Laurens Dist. age 30 on postwar roster. Alive 1898.

BOWEN, WILLIAM REESE. Pvt. & Bugler, Co.'s F & K. b. Pickens Dist. 10/25/35. Farmer, Pickens Dist. 1860 census. Enl. Co. F, Pickens CH 4/27/62. Transferred Co. K as Bugler 6/25/62. Present through 4/63. Transferred to Band 6/1/63. Ab. sick with acute diarrohea in Charlottesville hospital 9-10/63. Present 11/63-12/64. NFR. Farmer, Post Oak, Johnson Co., Mo. 1910 census. d. Chilhowee, Johnson Co., Mo. 5/28/14. Bur. George's Creek Bapt. Ch., Pickens Co., S. C.

BOWERS, HENRY U. Pvt., Co. I. b. S. C. circa 1838. Farmer, age 22, Colleton Dist. 1860 census. Enl. Parker's Ferry 4/3/62. AWOL through 4/3/63. However, had enlisted Co. B, 3rd S.C. Cav., Jacksborough 3/20/62. Present until ab. sick at home 7-8/62. Present 9/628/63. Ab. sick in hospital 9-10/63. Present 11/636/64. Present sick in quarters 7-8/64. Ab. on sick leave 9-10/64. AWOL 11-12/64. NFR.

BOWERS, JOHN W. Pvt., Co. I. b. S. C. 1838. Enl. Parker's Ferry 4/3/62. Present through 6/62. AWOL 8/25-31/62. Ab. sick in hospital 9-10/62. Present sick in camp 11-12/62. Ab. sick in hospital 8/31-10/31/63. In hospital under charges 12/27-31/63. Sick in Richmond hospital with scurvy & pneumonia 1-2/64. Furloughed for 30 days 2/28/64. Transferred 1st S. C. Inf. 5/18/64. NFR. d. Newberry, S. C. 1/14/22.

BOWERS, ROBERT M. Pvt., Co. I. b. Fairfax, Barnwell Dist. 3/12/31. Enl. Parker's Ferry 4/3/62. AWOL through 4/30/62. Enl. Co. D, 17th Bn. S. C. Cav. Cheraw 4/2/62. Present through 10/63. Deserted to the enemy Charleston 3/22/65. Age 33, 5'10", dark complexion, dark hair, grey eyes. Took oath and released. Farmer, Allendale, Barnwell Co. 1870 census. d. Barnwell, S. C. 12/13/07.

BOYD, ARCHIBALD B. Pvt., Co. G. b. S. C. circa 1839. Enl. Abbeville CH 3/15/62, age 23. Sick in camp through 4/30/62. Present 5-8/62. Ab. sick Summerville hospital 10/1-31/62. Present 11/62-10/63. Ab. sick with "Intermittent Fever" in Richmond hospital 12/20/63-1/20/64. Furloughed for 30 days. Present 3-10/64. Detached at Pocotaligo 11-12/64. NFR. Living Abbeville Co. 1902.

BOYD, JOHN THOMAS. 3rdCpl., Co. H. b. S. C. 2/17/28. Enl. Rock Hill 1/9/61 age 34 as Pvt. NFR until on picket Jacksonborough 3-4/62. Present 5-6/62. Ab. on sick leave 7-8/62. Present 9-10/62. On picket 11-12/62. Present 3-8/63. Promoted 4thCpl. Present 9-10/63. Ab. on horse detail 12/22/63-2/64. Present 3-6/64. Promoted 3rdCpl. Present 7-12/64. Paroled Greensboro, N. C. 4/26/65. Farmer, Catawba, York Co. 1880 census. d. Lesslie, York Co. 12/8/99. Bur. Neelys Creek Presb. Ch. Cem., York Co.

BOYD, WILLIAM FRANKLIN. Pvt., Co. I. b. Fort Mill, York Dist. 9/8/46. Student, age 14, York Co. 1860 census. Enl. James Island 9/29/64. Present through 12/64. Paroled Charlotte, N. C. 6/65. Farmer, Fort Mill, York Co. d. York Co. 1/14/24. Bur. Flint Hill Bapt. Ch., York Co.

BOYKINS, J. Pvt., Co. F. Captured and d. of disease Ft. Delaware, no dates.

BOYLE, WILLIAM AUGUSTUS. Pvt., Co. A. b. S. C. 8/10/39. On postwar roster. Enl. Co. C, 11th S. C. Inf. 7/6/61 as 1stSgt. Promoted 2ndLt. 9/61. Not reelected 5/3/63. Reenlisted Co. K, 4th S. C. Cav. Elected Lt. of Arty. NFR. d. 12/13/23. Bur. St. Paul's Episcopal Ch., Dorchester Co., S. C.

BRABHAM, HENRY JASPER, SR., Capt., of Brabham's Co. b. S. C. 8/6/17. Not on muster rolls. Pension application only record. d. Allendale Co. 8/21/19. Bur. South End Cem., Bamberg Co. S. C.

BRADDY, RUSSELL F. Pvt., Co. E. b. S. C. circa 1832. Res. Orangeburg Dist. 1850 census. Enl. Ft. Motte 10/26/61 age 26. Present through 6/6/62, however, ab. on leave for 5 day to visit his wife in St. Mathews Parish 1/12/62. d. of disease Green hospital, Charleston 8/13/62, en route to Summerville hospital.

BRADFORD, WILLIAM. Pvt., Co. G. b. S. C. 1840. Served in Washington Infantry 1860-61. Enl. Abbeville CH 1/8/62. d. of disease in Adams Run hospital 3/24/62.

BRADY, WILLIAM R. Pvt., Co. B. b. S. C. 8/45. Enl. Abbeville CH 5/26/63. Present through 8/63. Ab. detached 9-10/63. AWOL 12/31/63. Present 1-2/64. AWOL 4/26-30/64. Present 5-8/64. Ab. on detail 9-10/64. Present 11-12/64. Paroled Greensboro, N. C. 4/26/65. Farmer, Raymond, Hinds Co., Miss. 1870-1920 censuses.

BRAILSFORD, ROBERT MC LEOD. Pvt., Co. A. b. S. C. 11/7/26. M. D., age 32, Charleston 1860 census. Enl. Charleston 1/3/62. Returned to Charleston 1/27/62. Served as Surgeon. d. Charleston 11/6/81. Bur. St. Marks Episc. Ch., Sumter Co.

BRANDENBURG, JOHN. Pvt., Co. E. Res. Orangeburg Dist., age 20. DOW's received near Boonsboro, Md. 7/6/63.

BRANDENBURG(H), JOHN B. Pvt., Co. E. b. S. C. 7/41. Enl. Ft. Motte 10/26/61. Present through 4/63. WIA (shoulder blade) Boonsboro Gap, Md. 7/6/63 and captured. Ab. wounded at home 9-10/63. Paroled Hillsboro, N. C. 4/65. Farmer, Pine Grove, Orangeburg Co. 1880 & 1900 censuses. d. Converse, S. C. 11/11/09. Bur. St. Matthews Luth. Ch. Cem.

BRANDENBURG(H), JOHN M. 1stCpl., Co. B. b. S. C. 3/40. Enl. Ft. Motte 9/26/61 age 19. Promoted 3rd Cpl. Present through 6/62. Promoted 2ndCpl. Present 7-8/62. On picket 9-10/62. Present 11-12/62. Promoted 1stCpl. Present 3-12/63, but WIA Boonsborough Gap, Md. 7/6/63. Present 1-2/64, dismounted 2/1/64. Ab. on horse detail 3/31-4/30/64. Present through 12/64. NFR. Farmer, Pine Grove, Orangeburg Co. 1900 census.

BRANDENBURG(H), LOUIS ASBURY. Pvt., Co. E. b. Lonestar, Orangeburg, Co. S. C. circa 1840. Student, Orangeburg Co. 1860 census. Enl. Ft. Motte 10/26/61. Present through 12/31/61. Ab. sick at home 3/30-4/30/62. Present 5-6/62. On picket Lawach's 7-8/62. Present 9-12/62. Ab. sick in Lynchburg hospital 1/21-4/30/63. WIA Boonsborough Gap, Md. 7/6/63. Ab. wounded & sick with abcess in Richmond hospital 9/23-10/63. Present 11-12/63, dismounted since 8/12/63. Present 1-2/64. Detailed Green Pond 4/26-30/64. Present 5-8/64. Ab. with wagons 9-10/64. Present 11-12/64. Paroled Hillsboro, N. C. 4/65. d.

1908. Bur. Santee 1st Baptist Church, Orangeburg Co.

BRANDENBURG(H), THOMAS PERRY. 5thSgt., Co. E. b. S. C. 1/5/41. Res. Orangeburg Dist. 1850 census. Enl. Ft. Motte 10/26/61 age 20 as Pvt. Present through 12/31/61. Promoted 2ndCpl. Sick in camp 3-4/62. On picket on Charleston & Savannah RR at Adams Run 5-6/62. Present 7-8/62. Promoted 1stCpl. Sick in camp 9-10/62. Present 11-12/62. Promoted 5thSgt. Present 3-6/63 & Gettysburg. Ab. on detail to S. C. for a horse 8/18-31/63. Present 9/63-2/64. Ab. on horse detail 3/31-4/30/64. Present 5-6/64. Present as 1stCpl. 7-8/64. On picket Pocotaligo Bridge 9-10/64. Present 11-12/64. Paroled Hillsboro, N. C. 4/65. d. St. Matthews, Calhoun Co. 5/20/91. Bur. St. Matthews Luth. Ch. Cem.

BRANTON, HENRY H. Pvt., Co. I. b. Colleton Dist. 12/40. Shingle getter, age 21, Colleton Dist. 1860 census. Enl. Parker's Ferry 4/3/62. On picket Jacksonborough through 4/30/62. Present 5-6/62. Ab. on guard duty 7-8/62. Present 9-10/62. Ab. on scout 11-12/62. Ab. sick with diarrohea in Farmville hospital 3/18-4/20/63. Furloughed for 40 days. AWOL 7/16-8/31/63. Present 9/63-12/64. NFR. Farmer, age 59, Montgomery Co., Texas 1900 census.

BRANTON, JOHN ISAAC. Pvt., Co. I. b. Colleton Dist. circa 1833. Res. Colleton Dist. 1860 census. Enl. Sheridans's Co., 11th S. C. 10/15/61. NFR. Reenlisted Co. I, 1st. S. C. Cav. Parker's Ferry 4/3/62. Present through 6/62. Ab. on guard duty 7-8/62. Present 9/62-12/63. Ab. on leave 2/12/-28/64 & 3/26-30/64. Present 5-12/64. Captured Adams Run 4/19/65. Age 35, 5'6", dark complexion, hazel eyes, black hair. NFR.

BREAKER, C. H. 4thSgt., Co. I. Enl. Parker's Ferry 4/3/62 as Pvt. Ab. 4/29-30/62. Present 5-6/62. Ab. on sick leave 8/27-30/62. Ab. sick with "Haemopfysis" in Richmond hospital 10/22-11/6/62. Present through 12/62. Promoted 4thSgt. Present 3/63-8/64. Ab. detached at Headquarters, East Line, James Island 9/19-10/31/64. Present 11-12/64. NFR.

BREAKER, H. M. Pvt., Co. I. b. circa 1845. Enl. Parker's Ferry 4/3/62. Present through 10/62. On picket 11-12/62. Present sick in camp 3-4/63. WIA Brandy Station 6/9/63. DOW'S, age 18.

BREWTON, ISSAC H. Pvt., Co. B. b. S. C. Enl. Spartanburg Co. age 25 on postwar roster. Alive 1898.

BREWTON, JAMES JONAS H. M. Pvt., Co. B. b. Spartanburg Dist. 3/36. Farmer, age 24, Ceishville, Spartanburg Dist. 1860 census. Enl. Clinton 8/22/61 age 25. Present through 4/62. Ab. on sick leave 5-6/62. Present sick in camp 7-8/62. Present 9-10/62. On picket Ellis Ford 11-12/62. Ab. detailed as Druggist, Richmond 3/63-3/16/64. Ab. on horse detail 3/31-4/30/64. Present 5-6/64. Ab. on leave 8/21-31/64. Present 9-10/64. Ab. sick with "Febris Remittens" in Raleigh hospital 2/18-25/65. NFR. M. D., Floresville, Tex. 1900 census. Retired, age 74, Bexar, Texas 1910 census.

BREWTON, THOMAS P. Cpl., Co. B. b. Spartanburg Dist. 1839. Farmer, age 24, Spartanburg Dist. 1860 census. Not on muster rolls. WIA Boonsboro, Md. 7/63 on postwar roster. d. by 1898.

BRICE, CHARLES STRONG. Pvt., Co. C. b. S. C. 10/19/31. Lawyer, age 28, Charleston 1860 census. Enl. Co. F, 6th S. C. Inf. Chester CH 4/11/61 age 29. Detailed Brigade Quartermaster Dept. 8/25-10/31/61. Present 11-12/61. Discharged 3/31/62. Enl. Co. C, 1st S. C. Cav. Chester 4/21/64. Ab. detailed Brigade Headquarters 7/28-12/31/64. Paroled Hillsboro, N. C. 4/65. Lawyer, age 38, Chester CH 1870 census. d. 10/28/78 age 46. Bur. Evergreen Cem., Chester, S. C.

BRICE, JAMES ALEXANDER. Pvt., Co. D. b. S. C. 11/22/39. Enl. Co. H, 6th S. C. Inf. 3/24/62. WIA (leg) & captured 5/31/62. Exchanged 8/5/62. Present 9/62-2/63. Hired a substitute and discharged 4/12/63. Enl. Co. D, 1st S. C. Cav. Fairfield, S. C. 1/29/64. Detailed Green Pon 4/26-30/64. Present 5-8/64. Ab. detailed to guard POW's Florence 9/18-12/31/64. NFR. Merchant, Winnsboro 1880 census. Cotton buyer, Winnsboro 1900 census. d. Winnsboro 1909. Bur. Bethel A. R. P. Ch. Cem., Winnsboro, S. C.

BRICE, WILLIAM WATT. Pvt., Co. D. b. Fairfield Dist. 3/4/46. Enl. Fairfield 3/12/64. Present through 4/30/64. Ab. sick in Fairfield Dist. 5/17-6/30/64. Present 7-8/64. Ab. detailed to guard POW's Florence

9/18-12/31/64. Paroled Hillsboro, N. C. 4/65. Farmer, age 24, Winnsboro, 1870 census. Confederate Veteran living Chester Co. 1904.

BRIDGE, ALFRED LOWERY ANDERSON. Pvt., Co. I. b. S. C. circa 1844. Enl. Parker's Ferry 4/3/62. AWOL 4/27-30/62. Present 5-6/62. Ab. on sick leave 8/27-31/62. Present 9-12/62. Ab. sick in hospital 3-4/63. Present 5-6/63. Ab. on horse detail 8/18-31/63. Present 9-10/63. Ab. guarding wagon yard 11-12/63. Present 1-2/64. Ab. detailed Green Pond 4/26-30/64. Dismounted 2/15-5/1/64. Present 5-10/64. Ab. detailed at Pocotaligo 10-12/64. NFR. Farmer, Sheridan, Colleton Co. 1880-1920 censuses. Bur. Ackerman Cem., Colleton Co., no dates.

BROCK, CHESLEY H. Pvt., Co. F. b. S. C. 1820. Enl. Pickens CH 12/4/61 age 40. Present through 12/31/61. Ab. with Quartermaster wagons 3-4/62. On picket Camp Means 5-6/62. Present 7-10/62. Ab. detailed as nurse in hospital 11-12/62. Present 3-10/63. Ab. on horse detail 12/22/63-2/64. Ab. on leave 4/27-30/64. Present 5-6/64. On duty Dells Bluff 8/29-12/31/64. Discharged Charleston 1864 on postwar roster. NFR. Farmer, Grayson Co., Texas 1880 census.

BROGAN, MARTIN. Pvt., Co.'s C & K. b. S. C. circa 1835. Enl. Co. C Hamburg 8/27/61. Present through 12/31/61. On picket Bear Bluff, Wadmalaw Island 1/62. Transferred Co. K 6/27/62. Present 9-10/62. Captured Fredericksburg 12/13/62. Age 29, 5'7", light complexion, blue eyes, dark hair. Paroled for exchange 12/17/62. On parole Richmond 12/31/62. Ab. sick in hospital 1/15-8/31/63. Ab. sick at home 9-10/63. Transferred Co. B, Lucas's 1st S. C. Bn. Arty. 5/16/64 but not on muster rolls of that unit. Actually transferred to 20th S. C. Inf. Joined 6/3/64. Present through 10/64. Paroled Augusta, Ga. 6/6/65. Alive 1898.

BROOKER, EDWARD P. Lt., Co. I. b. Barnwell Dist. 9/18/25. Not on muster rolls. Enl. Co. E, 11th S. C. Inf. Store Clerk, Saluda, Edgefield Co. 1870 census. Agent, Gins & Engines, Williston, Barnwell Co. 1880 census. d. 12/14/94 (10/30/93 on military tombstone). Bur. Tabernacle Bapt. Ch. Cem., Aiken Co., S. C.

BROOKS, DAVID D. Pvt., Co. A. b. S. C. 8/35. Farmer, age 28, Abbeville Dist. 1860 census. Enl. Abbeville CH 8/20/61 age 29. On picket Bears Island 11-12/61. Present 3-6/62. Present sick in camp 7-8/62. Present 9-12/62 & 3-4/63. Ab. sick with fever in Staunton hospital through 8/31/63. Ab. detached in charge of recruiting horses 9/63-2/64. Ab. on leave 4/29-30/64. Present 5-12/64. Farmer, Washington Co., Ga. 1900 & 1910 censuses. d. 3/26/16. Wife receiving pension Plum Branch, McCormick Co.

BROWN, ALFRED. Teamster, Co. A. (Colored). Enl. 11/7/61. NFR.

BROWN, ANGUS P. Captain, Co.'s C & K. b. S. C. 1826. Enl. Co. C Hamburg 8/27/61 as 2ndLt., age 34. Ab. detailed on recruiting duty 12/17-31/61. Present Wadmalaw Island 1/62. Promoted Captain of Co. K 6/26/62. Present through 9/62. Present sick in camp 11-12/62. Ab. on sick leave 1-4/63. WIA & captured Upperville 6/21/63. Sent to Old Capitol. Transferred Johnson's Island. Transferred to Baltimore 2/9/64. Exchanged 3/10/64. Ab. on paroled through 4/64. Ab. on leave 5-6/64. Issued forage for 56 horses 7/2/64. Signed for 18 pair of shoes 7/8/64. Issued forage for 48 horses 7/21/64. Present 7/31/64. Present Charleston 8/64 with 3 officers & 48 men present & 9/64 with 3 officers and 41 men present. Present with 1 officer and 40 men 10/64, however, ab. on leave for 7 days 10/3/64. Present 11/64 with 1 officer & 64 men present. Signed for 12 jackets, 16 pr. pants, 23 shirts, 7 pr. drawers, 1 pr. socks, 28 pr. shoes, 25 blankets & 1 cap 11/23/64. Signed for 5 blankets, 4 pr. shoes, 4 jackets & 2 shirts 11/25/64. Present 12/64 with 1 officer & 34 men present. Present 2/65 & paroled Greensboro, N. C. 4/26/65. Farmer, Edgefield Co. 1870 census. Nightwatchman, age 84, Columbia 1910 census. d. 7/1/19. Wife receiving pension Columbia 10/22/19.

BROWN, ELIJAH W. Pvt., Co. F. b. S. C. 2/30/39. Merchant, age 21, Anderson Dist. 1860 census. Enl. Co. A, 1st S. C. Inf. (Hagood's) Coles Island 3/26/62. Present until hired J. D. Patrick as substitute and discharged 4/26/62. Enl. Co., F, 1st S. C. Cav. Columbia 3/1/64. On duty Green Pond 4/20-30/64. Present 5-6/64. Ab. on horse detail 8/10-31/64. Present 9-10/64. Ab. on leave 12/23-31/64. NFR. Clerk in store, age 31, Varens, S. C. 1870 census. Merchant, age 41, Anderson, S. C. 1880 census. d. 3/7/91. Bur. Silver Brook Cem., Anderson Co.

BROWN, J. NESBIT. Pvt., Co. B. b. Enoree, Spartanburg Dist. 1/17/39. Enl. Clinton 8/22/61 age 21. Ab. sick 10/31/61. Present 11/61-6/62. Ab. on leave 7-8/62. Present 9-10/62. Ab. with wagons hauling forage 11-12/62. Ab. sick in hospital 3-4/63. Returned to duty from Farmville hospital, 5/9/63. Ab. on horse detail to S. C. 8/18-31/63. Present 9-10/63. Present 11-12/63, horse died. Ab. on leave 1/18-2/28/64 & on horse detail 3/31-4/30/64. Present 5-10/64. On duty Pocotaligo 11-12/64. Paroled Greensboro, N. C. 4/26/65. Receiving pension Enoree, Spartanburg Co. 3/25/19.

BROWN, JOHN. Pvt., Co. G. b. S. C. circa 1842. Enl. Abbeville CH 3/15/62. Present through 10/62. Ab. on scout 11-12/62. Present through 8/31/63. Ab. on detached service 9-10/63. AWOL 12/30-31/63. Present 1-12/64. NFR.

BROWN, SAMUEL EDWARD. Pvt., Co. G. b. S. C. 1/13/29. Enl. Abbeville Ch 3/15/62. Present sick in quarters through 4/30/62. Ab. on picket 5-6/62. Present 7-10/62. Ab. on scout 11-12/62. Present 3-12/63. Detailed with Provost Guard Hanover Junction 1/2-2/28/64. Present 3-10/64. Ab. sick in hospital 11-12/64. NFR. d. 11/29/80. Bur. Long Cane Presb. Ch. Cem., McCormick Co., S. C.

BROWN, WILLIAM. Teamster, Co. A. (Colored). Enl. 11/7/61. NFR.

BROWNLEE, JAMES HARVEY. Pvt., Co. A. b. Abbeville Co. 1843. Enl. age 19 on postwar roster. Living Dallas, Tex. 1880. Living Abbeville Co. 1902.

BULLOCK, JOHN RICHARD. Pvt., Co. A. b. S. C. 3/10/44. Enl. Abbeville CH 4/24/61. Present 11/61-12/64. NFR. Farmer, Greenwood, Abbeville Co. 1870-1910 censuses. d. 2/15/18. Bur. Coronaca Cem., Greenwood.

BULLOCK, L. O. Pvt., Co. I. Enl. Adams Run 4/28/62. Ab. on sick leave 8/31-10/31/63. Transferred to Ward's S. C. Bty. 5/13/64. NFR.

BULLOCK, ZADOCK Q. Pvt., Co. I. b. Robeson Co., N. C. 7/4/37. Farmer, age 22, Robeson Co., N. C. 1860 census. Enl. Co. G, 9th S. C. Bn. for 12 months Blanton's Cross Roads 1/1/62, age 22. NFR. Enl. Co. I, 1st S. C. Cav. Culpepper CH 5/27/63. WIA and in Richmond hospital 7/29-8/8/63, furloughed for 30 days. Present 1-4/64. Transferred to Ward's S. C. Bty. 5/10/64. Present through 12/64. Paroled Greensboro, N. C. 5/1/65. Farmer, Green Sea, Horry Co. 1880 census. d. Horry Co. 8/24/18. Bur. Williiams Cem., Horry Co. S. C.

BURBRIDGE, JOHN WILLIAM. Pvt., Co. I. b. S. C. 4/10/15. Enl. Parker's Ferry 4/3/62. Ab. on leave 4/24-6/30/62. Present 5-10/62. Ab. on scout 11-12/62. Ab. sick with chronic dysentery in Richmond hospital 3/14-4/30/63. Present until ab. sick with acute diarrohea 7/18-8/18/63. Present through 12/63. Ab. on horse detail to S. C. 1/26-2/28/64. On duty Green Pond 4/16-30/64. Present 5-8/64. AWOL 10/27-12/31/64. NFR. d. 4/12/91. Bur. Live Oak Cem., Colleton Co.

BURDEN, J. G. Pvt., Co. G. On postwar roster.

BURDETT, GEORGE F. Pvt., Co. G. b. Abbeville Co. circa 1842. Enl. Columbia 3/25/64. Present through 8/64. Ab. detached Anderson, S. C. 10/27-31/64. Ab. sick in hospital 12/30-31/64. Paroled Greensboro, N. C. 4/26/65. Living Abbeville Co. 1902.

BURDETT, HENRY KELSEY. Pvt., Co. A. b. S. C. 5/15/27. Enl. Abbeville CH 8/13/61. Present through 10/31/61. Ab. on duty Bear Island 11-12/61. Present 5/62-6/6/63. Ab. on horse detail 8/18-31/63, dismounted since 8/1/63. Present 9/63-2/64. Ab. on horse detail 4/15-30/64. Present 5-6/64. Present in arrest 10/15-31/64. Detached at Pocotaligo 12/10-31/64. NFR. Laborer, Calhoun, Abbeville Co. 1900 census. Applied to enter Old Soldiers' Home, Columbia from Abbeville Co. 4/30/09. Age 83, 5'8", Farmer. Entered 4/3/09 age 86. d. McCormick Co. 6/4/18. Bur. Willington Cem., McCormick Co.

BURKE, R. Pvt., Co. F. Enl. Pickens CH 12/4/64. Detailed in Commissary Dept. through 12/31/64. NFR. May have served in Co. H, 25th S. C. Inf.

BURKETT, ----------. Pvt., Co. I. Not on muster rolls. KIA Brandy Station 6/9/63 on casualty list.

BURKETT, WILLIAM. Pvt., Co. E. b. S. C. circa 1843. Enl. Ft. Motte 10/26/61 age 17. Present through 6/62. Ab. as Courier, Parker's Battery 7-8/62. Present 9/62-4/63. Ab. detailed with Signal Corps through 10/63. Ab. detached horse recruiting station 10/26/63-2/64. Ab. on leave 4/2-5/5/64. Present 5-10/64. Ab. detached Green Pond 12/10-31/64. Paroled Hillsboro, N. C. 4/65. School Teacher, age 27, Walker Co., Ala. 1870 census. Wife receiving pension Richland Co. 1919.

BURNETT, BARNWELL RHETT. Pvt., Co. A. b. S. C. 1844. On post war roster. Reenlisted Co. K, 4th S. C. Cav. Grahamville 3/25/62 age 17. Present until ab. on horse detail 8/11/64 for 40 days. AWOL 9/30-10/31.64. NFR. d. 1919. Bur. St. Peters Epis. Ch., Charleston.

BURNETT, HENRY DE SAUSSURE 'HARRY.' Pvt. Co. A. b. S. C. 1842. On post war roster. Reenlisted Co. K, 4th S. C. Cav. Grahamville 3/25/62. Present until detached by Gen. Walker 12/17/62 until ab. on leave 1/30/63 for 30 days. Transferred to Co. C 3/7/63. Present until detailed as Forage Master for Regt. 12/1/63. Transferred 3rd S. C. Cav. 1/15/64. Detached to Lt. R. Jones's Section of Horse Arty. & promoted Sgt. Present through 10/31/64. NFR. d. 1916. Bur. Grahanville Cem., Jasper Co., S. C.

BURNETT, WILEY. Pvt., Co. G. b. N. C. 1840. Enl. Abbeville CH 4/12/62. age 21. Present 5-10/62. Ab. on picket 11-12/62. Present 3-8/63. Ab. on detached service 9-10/63. Ab. on leave 21 days 12/19/63. Present 1-2/64. Ab. on horse detail 4/15-30/64. Present 5-8/64. Ab. detailed guarding POW's Florence 9/10-12/31/64. Paroled Augusta, Ga. 5/18/65. Farmer, Sandy Creek, Franklin Co., N. C. 1880 census. Living Abbeville Co. 1902. d. Sandy Creek, N. C. 12/5/26. Bur. Burnett family Cem.

BURNS, WILLIAM. Pvt., Co. D. d. 5/4/62. Bur. Richmond.

BURNS, WILLIAM H. (1). Pvt., Co. E. b. S. C. 1847. Student, age 13, Ft. Motte, Greenville Dist. 1860 census. Enl. date unknown. Captured Bentonsville, N. C. 3/6/65. Sent to New Bern & Pt. Lookout. Released 6/4/65.

BURNS, WILLIAM H. (2). Pvt., Co. C. b. S. C. 1843. On postwar roster. d. Greenville, S. C. 1885.

BURRELL, BRIGHT RUSSELL, JR. Pvt., Co. F. b. Macon Co., N. C. 1/4/30. Farmer, Mocassi, Raburn Co., Ga. 1860 census. Enl. Pickens CH 6/2/62. Present through 10/62. On picket 11-12/62. Present 3-4/63. Present Gettysburg. Ab. on detached service 8/31/63. Present 9-12/63. Ab. guarding private property 1 mile form camp 1/15-2/28/64. Ab. on leave 4/27-30/64. Present 5-12/64. NFR. Farmer, Clayton, Raburn Co, Ga. 1870 census and Mocassi. Raburn Co., Ga. 1880 census. d. Pine Mountain, Raburn Co. 8/05. Bur. family cem., Raburn Co. Brother of Butler Burrell.

BURRELL, BUTLER. Pvt., Co. F. b. Macon Co., N. C. 11/30/26. Enl. Pickens CH 12/4/61 age 35. Present through 4/62. On picket Camp Means 5-6/62. Present 7-10/62. On picket 11-12/62. Present 3-4/63. KIA Gettysburg near Huntersville, Pa., between Round Tops and the wheat field 7/3/63. His brother Bright witnessed his death. Left widow. Reburied. Burrell Cem., Pine Mountain, Ga. after the war.

BURRELL, SAMUEL PERCY C. Pvt., Co. F. b. Macon Co., N. C. 12/2/32. Enl. Pickens CH 6/1/62. Present through 4/63. Ab. sick in hospital 8/5-31/63. On detached service 9-10/63. AWOL 10/22/63-2/64. Transferred Capt. Hammond's Co., 25th S. C. Inf. 4/27/64 because he was dismounted. Not on rolls of that unit. Listed as on picket 11-12/64 on rolls Co. F, 1st S. C. Cav. NFR. Farmer, Raburn Co., Ga. 1880 census. d. Raburn Co., Ga. 4/18/08. Bur. Macedonia Missionary Bapt. Ch., Raburn Co., Ga.

BURRIS, LEVI J. Pvt., Co. F. b. S. C. circa 1844. Age 6, Anderson Dist. 1850 census. Enl. Columbia 5/1/63. Ab. on detached service 8/18-31/63. Present 9-12/63, however, driving Ordnance wagon 12/27-31/63. Present 3-12/64. NFR. Farmer, Pendelton, Anderson Co. 1870 census. Wife receiving pension Anderson Co. 1919.

BURTON, JOHN N. Pvt., Co. G. b. S. C. 1840. Enl. Abbeville CH 9/1/63. Present through 10/63. Ab. on horse detail 12/22/63-2/64. Ab. detached service at Green Pond 4/16-30/64. Present 5-6/64. Present sick in camp 7-8/64. Ab. sick James Island hospital 10/6-31/64. Ab. on sick leave 12/13-31/64. Paroled Greenville, S. C. 5/23/65. Living Abbeville Co. 1902.

BURTON, WILLIAM L. Pvt., Co. G. b. S. C. 1836. Enl. Abbeville CH 2/8/62 age 20. On picket Jacksonborough 3-4/62. Ab. sick in hospital 6/9-30/62. Present 7/62-8/31/63. Ab. on detached service 9-10/63. Present 11/63-12/64. NFR. d. 2/22/67. Bur. Little River Bapt. Ch. Cem., Abbeville Co.

BUSH, JEFF G. Pvt., Co. K. b. Barnwell Dist. circa 1840. Res. Barnwell Dist. 1850 census. Enl. Coles Island 4/9/62, joined from Co. G, 2nd S. C. Arty. Present 7/62-5/63. Ab. sick 7-8/63. Ab. on sick leave 9-10/63. Present 11/63-8/64. Detailed to Military Telegraph Office, Charleston. d. Charleston hospital 10/5/64 age 18. Effects: 75 cents & sundries. Bur. Magnolia Cem., Charleston.

BUSH, JOHN. Pvt., Co. H. Res. Edgefield Dist., age 18. d. Charleston, S. C. 1864-65.

BUTLER, HEZEKIAH H. Pvt., Co. F. b. S. C. circa 1843. Res. Edgefield Dist. 1860 census. Enl. Pickens CH 12/4/61 age 18. Present through 4/62. On picket Camp Means 5-6/62. Present 7/62-4/63. KIA Kelley's Mills or Brandy Station 6/9/63.

BUTLER, JOHN WHITMILL. 2ndSgt., Co. C. b. Beach Island, S. C. 12/17/39. Student, age 21, Calhoun, Ga. 1860 census. Enl. Hamburg 8/27/61 age 21 as 1stCpl. Ab. detached Wadmalaw Island 11-12/61. Promoted Sgt. Ab. on sick leave from a fall from a horse 5-8/62. AWOL 9-10/62. Promoted 2ndSgt. Ab. sick 11-12/62. Discharged 4/23/63. Enl. Co. H, 2nd S. C. Arty. 6/16/64. Present through 6/30/64. NFR. Farmer, Gaines Co., Ga. 1870 & 1880 censuses. d. Shannon, Ga. 11/1/96. Bur. Pleasant Valley North Bapt. Ch. Cem., Rome, Ga.

BYERS, DAVID T. Pvt., Co. H. b. S. C. circa 1847. Res. York Dist. Enl. Columbia 11/6/64. Exchanged Company by order Gen. Taliaferro 12/15/64. Present through 12/31/64. NFR.

BYNUM, BENJAMIN B. Pvt., Co. C. b. Tennessee circa 1823. Enl. Co. A, 15th S. C. Inf. 1861. Discharged 1862. Enl. Co. C, 1st S. C. Cav. Richmond 10/9/62 as substitute for John N. Mims, age 40. Present through 10/31/62. AWOL 11/62-3/63. Dropped as a deserter. NFR. Farmer, Coryell, Texas 1880 census.

CADDEN, LEWIS. Pvt., Co. I. b. S. C. circa 1838. Age 12, Lancaster Dist. 1850 census. Enl. Parker's Ferry 4/3/62. AWOL through 4/30/62. NFR. Enl. Co. H, 2nd S. C. Troops (6 months unit) Colleton Dist. 8/1/63. Ab. 11/63. NFR. Conscripted 11/27/63. Assigned to Parker's S. C. Bty. 12/17/63. Present through 4/64. Ab. on leave 6/24-30/64. Present 7-12/64. NFR. Res. Abbeville Co. 1870 census.

CALDWELL, DAVID. Pvt., Co. A. Enl. Abbeville CH 8/20/61. Discharged 9/5/61. He served again in some capacity with the South Carolina Cavalry as he wrote about their activities in 1864 in the postwar.

CALDWELL, ROBERT CALVIN. Pvt., Co. D. b. S. C. 1844. Enl. Co. A, 10th Bn. S. C. Cav. as 1stSgt. 1/20/62. Ab. on special duty 2/62. NFR. Enl. Co. D, 1st S. C. Cav. Chester CH 1/29/64. Present 3-6/64. Ab. on horse detail 8/8-31/64. Present 9-12/64. Paroled Hillsboro, N. C. 4/65. Farmer, Caldwell, Newberry Co. 1880 census. d. Newberry, S. C. 1/17/20. Wife receiving pension 2/28/38.

CALDWELL, WILLIAM J. Pvt., Co. A. b. S. C. circa 1839. Enl. Abbeville CH 8/13/61. NFR. Enl. Co. A, 10th Bn. S. C. Cav. 1/20/62. Discharged 6/3/62. Reenl. Co. D, 1st S. C. Cav. Chester CH 1/29/64. Present through 4/64. Ab. on horse detail 6/26-30/64. Present 7-8/64. Ab. sick Chester Dist. 10/21-31/64. Ab. guarding pontoon bridge 12/29-31/64. Paroled Hillsboro, N. C. 4/65.

CALDWELL, WILLIAM K. P. Pvt., Co. B. b. S. C. 5/26/26. Enl. Columbia 2/1/64. Present 3-4/64. Ab. sick Charleston hospital 6/1-7/31/64. Ab. sick leave at home 8/29-31/64. Present 9-10/64. Detailed Pocotaligo 12/2-31/64. Paroled Hillsboro, N. C. 4/65. d. 1/1/79. Bur. Nazareth Presb. Ch. Cem., Spartanburg Co.

CALLEAN, JOHN. Pvt., Co. I. Pvt., Co. I. On post war roster.

CALLEAN, RICHARD. Pvt., Co. I. On post war roster.

CALLER (COLLER), H. S. Pvt., Co. F. On post war roster.

CAMPBELL, JOHN M. Pvt., Co. A. b. Ga. Enl. Abbeville CH 12/23/61. On duty Wilson's Point 3-4/62. Present 5-6/62. Ab. sick Summerville hospital 7/-8/62. Present 9-10/62. On scout 11-12/62. Ab. sick with chronic diarrohea in Staunton hospital 3/14/63. Transferred to Lynchburg hospital. d. there 4/10/63. Effects: sundries. Left widow. Bur. Old City Cem., Lynchburg, Va.

CAMPBELL, JOHN W. Pvt., Co. I. b. S. C. 1839. Res. Abbeville Dist. 1850 census. Enl. Parker's Ferry 4/3/62. On picket Jacksonborough 4/30/62. Present 5-6/62. Ab. on sick leave 8/25-31/62, and in Summerville hospital. Present 9-10/62. On picket 11-12/62. Present 3-8/63. Ab. sick with "Int. Fever" in Richmond hospital 10/3-11/25/63. Present 11/63-2/64. Ab. on detail Green Pond 4/16-30/64. Present 5-6/64. Ab. on Courier duty 7-8/64. Present 9-12/64. NFR.

CANADAY, JACOB L. C. Pvt., Co. I. Enl. Capt. Sheridan's Co., 11th S. C. Inf. Fenwick Island 10/15/61. Ab. sick 12/27-31/61. NFR. Enl. Co. I, 1st S. C. Cav. Parker's Ferry 4/3/62. Present through 6/63, however, ab. sick with chronic bronchitis in Charlottesville hospital 5/23-7/29/63. Ab. on horse detail 8/18-31/63. AWOL 9-10/63. Present 11/63-2/64. Ab. detached Green Pond 4/16-30/64. Remounted 5/1/64. Present through 11/64. AWOL 12/15-31/64. NFR. Living Adams Run, Colleton Co. 1890.

CANNON, WILLIAM CLARKE. Pvt., Co. D. b. New Prospect, Spartanburg Co. 8/24/41. Att. Wofford College 1860-61. Enl. Co. I, 13th S. C. Inf. Spartanburg CH 8/27/61. Present through 10/61. Ab. sick at home 2/28/62. Discharged for ill health 5/2/62. Enl. Co. D, 1st S. C. Cav. by 5/29/63 in postwar account. "Returned to Virginia and served a few months with the 1st South Carolina Cavalry. Health failed again and sent home. Served in the "Home Reserves." Merchant, Spartanburg 1895. d. Spartanburg 6/5/21. Bur. Oakwood Cem.

CANTRELL, DUKE WILLIAM. Pvt., Co.'s D & K. b. Pickens Dist. 3/18/41. Enl. Co. D Pickens CH 12/4/61. Present through 4/62. Transferred Co. K John's Island 6/26/62. Present through 12/62. Ab. on detached service 4/10-30/63. Ab. on detached service 7/11-8/31/63. Ab. on horse detail 9-10/63. Present 11/63-2/64. On duty at bridge over Ashley River 7-8/64. On duty Bee's Ferry 9-10/64. On picket Chambers Ford 12/29-31/64. Paroled Raleigh, N. C. 4/65. Farmer, Eastatoe, Pickens Co. 1910 census. Receiving pension Pickens Co. 4/15/19. d. 12/8/19. Bur. Mountain Grove Bapt. Ch. Cem., Pickens Co.

CAPEHART, HAMILTON BERNARD. Pvt., Co. F. b. S. C. circa 1835. Enl. Pickens CH 12/4/61. Present through 12/31/61. Discharged for disability Adams Run 2/6/62. NFR. Painter, age 45, DeKalb Co., Tenn. 1880 census.

CAPMAN, J. A. Pvt., Co. F. On post war roster.

CARLILSE, JAMES COZBY. Pvt., Co. G. b. S. C. 12/1/38. Farmhand, age 22, Lowndesville, Abbeville Dist. 1860 census. Enl. Co. D, 7th S. C. Inf. Abbeville CH 4/15/61 age 23. Present through 8/61. Promoted 4thSgt. 8/2/61. Present 9-10/61, promoted 3rdSgt. Present 11/61-2/62. Promoted 2ndSgt. 2/2/62. Elected 1stLt. 5/13/62. Ab. sick 7/23/62. Furloughed 30 days. Ab. sick 8/21-31/62. Returned to duty 9/24/62. Furloughed 40 days. Ab. sick with scurvy & fever 12/1-31/62. Present 1/63-6/64. Resigned 8/11/64. Reenl. Co. G, 1st S. C. Cav. Richmond 8/25/64. Present through 12/64. NFR. Policeman, Atlanta, Ga. 1900 census. Retired, Atlanta 1920 census. d. Atlanta, Ga. 11/5/27.

CARLILSE, WILLIAM H. 1stLt., Co. E. b. S. C. 9/7/39. Enl. Co. H, 21st S. C. Inf. Georgetown 1/1/62.

Transferred Co. E, 1st S. C. Cav. 5/1/62. Present 5-12/62. Elected 2ndLt. 2/25/63. Present through 6/63. Present in arrest 8/6-10/31/63. Elected 1stLt. 11/11/63. Present through 8/64. Commanding Company 9-10/64. Paroled Greensboro, N. C. 5/1/65. M. D., Lamar Co., Texas 1880 census. M. D., Antlers Co., Okla. 1910 census. d. Antlers, Okla. 10/22/22. Bur. Restland Cem., Lamar Co., Tex.

CARNES, WILLIAM B. Pvt., Co. I. b. S. C. 8/45. Student, age 15, Bishopville, Sumter Dist. 1860 census. Enl. James Island 6/16/64. Present through 8/64. On picket 9-10/64. Ab. on sick leave for 30 days 12/11/64. NFR. Store Clerk, Bishopville, S. C. 1870 census. Merchant & Farmer, Bishopville, S. C. 1900 census.

CARR, A. JOSEPH. 1stSgt., Co. E. b. Poplar, Orangeburg Dist. 1828. Farmer, age 32, Orangeburg Dist. 1860 census. Enl. Ft. Motte 10/26/61 age 35 as 4thSgt. On picket White Point through 12/31/61. Promoted 1stSgt. Ab. on sick leave 3/31-4/30/62. Present 5-10/62. On scout 11-12/62. Ab. sick in Lynchburg hospital 3-4/63. Present 5-6/63. Ab. on horse detail 8/18-31/63. Present 9/63-10/64. Ab. detached at Pocotaligo 12/10-31/64. Paroled Hillsboro, N. C. 4/65. Farmer, age 52, Poplar, Orangeburg Co. 1880 census. d. Charleston 6/27/88.

CARR, JEFFERSON RANDOLPH. Pvt., Co. I. b. Baltimore, Md. 9/11/47. Served in Quartermaster's Dept., Richmond. Enl. Co. A, 25th Bn. Va. Local Defense Troops 1/20/63. Transferred Co. I, 1st S. C. Cav. 4/20/63. NFR. Farmer, Henrico Co., Va. 1870 census. Clerk to Sheriff, Baltimore 1880 census. d. 3/9/98 age 51. Bur. Loudon Park Cem., Baltimore.

CARR, WILLIAM HENRY. 3rdCpl., Co. E. b. S. C. 10/2/30. Enl. Ft. Motte 10/26/61 as Pvt. On picket White Point through 12/31/61. Present 4-6/62. Ab. on sick leave 8/16-31/62. Present 9-12/62. Promoted 4thCpl. Present 3-8/63. Promoted 2ndCpl. Present 9-12/63, dismounted since 12/9/63. Present dismounted 1-2/64. Ab. detached Green Pond 2/26-4/30/64. Present 5-6/64. Present as 3rdCpl. 7-10/64. Detailed as guard Wappo Bridge 12/29/64-2/65. Paroled Hillsboro, N. C. 4/65. d. 3/3/15. Bur. St. Stephens Episcopal Chl. Cem., Berkeley Co. Wife receiving pension St. Stephens, Berkeley Co. 10/27/19.

CARROLL, JAMES L. Pvt., Co. G. b. S. C. 1812. Laborer, Dorns Gold Mine, Abbeville Dist. 1860 census. Enl. Legares Point 9/10/62 as substitute for W. G. Neil. Present through 10/62. Ab. sick in Richmond hospital 12/23-31/62. KIA or died of disease 1863 on postwar roster.

CARSON, JAMES. Pvt., Co. D. b. S. C. 1815. Enl. Chester CH 9/10/61 age 44. Ab. on sick leave 10/31/61. Present 11/61-4/62. Discharged 5/18/62. Reenl. Co. A, 2nd S. C. Arty. 3/1/63. Present through 12/63. Ab. sick in hospital 3-4/64. Ab. on sick leave 5-5/64. Present 7-10/64. Paroled Greensboro, N. C. 5/1/65. Farmer, age 55, West of the Brazos, Texas 1870 census.

CARSWELL, E. W. Pvt., Co. E. UDC records are only information that this man served.

CARTER, A. G. Served on Chester Co. Confederate Veterans list, 1904.

CARTER, B. Musician., Co. C. b. St. Clair Co., Ala. 1820. Farmer, Graniteville, Edgefield Dist. 1860 census. Enl. Hamburg 8/27/61 as 40 as a Musician. Present through 8/61. Present as Saddler 11-12/61. Ab. on extra duty for 60 days repairing saddles & bridles for the company 3-6/62. Discharged for disability 7/21/62. Age 43, 5'10", ruddy complexion, blue eyes, red hair, Farmer. Enl. Co. D, 5th S. C. Reserves 11/10/62 for 90 days. Present sick in camp 12/31/62. Served through 2/15/63. NFR. Carpenter, Shaws, Edgefield Co. 1880 census.

CARTER, FREEMAN W. Pvt., Co. C. b. S. C. circa 1840. Farmer, age 20, Edgefield Dist. 1860 census. Enl. Hamburg 8/27/61 age 22. Present through 12/31/61 and 5-6/62. Ab. on sick leave 7/25-31/62. Ab. sick in Manchester hospital 10/17-31/62. Present 11-12/62. Ab. detailed as Nurse, Orange CH 3/63-1/64. Ab. sick in hospital 2/28-29/64. AWOL 4/1-30/64. Transferred Co. A, Lucas's 15th Bn. S. C. Arty. 5/16/64. Present through 12/64. Captured Calhoun, S. C. 3/16/65. Sent to New Bern & Pt. Lookout. Released 6/26/65. 5'8 3/4", light complexion, brown hair, hazel eyes. Farmer, Gregg, Aikens Co. 1880 census.

CARTER, JAMES M. Pvt., Co. D. b. S. C. circa 1843. Res. Darlington Dist. 1860 census. Enl. Chester CH 1/30/64. Present until ab. sick in Columbia hospital 4/28-30/64. Present 5-10/64. Ab. detached Coosawahatchie 12/10-31/64. Paroled Hillsboro, N. C. 4/65. Farmer, age 37, Shiloh, Sumter Co. 1880 census. Confederate Veteran living Chester Co. 1904.

CARTER, J. Z. Pvt., Co. I. Enl. Parker's Ferry 4/3/62. On picket Jacksonborough through 4/30/62. Present 5-8/62. Ab. sick with chronic rheumatism in Richmond hospital 9-11/1/62. d. Stevensburg, Va. 1/20/63.

CARY, JOHN L. Pvt., Co. C. b. Edgefield Dist. 12/29/27. Enl. Wadmalaw Island 3/17/62. Present through 6/62. Ab. on sick leave 8/18-31/62. Present 9/62-4/63. WIA (slightly in the arm) Brandy Station 6/9/63. Ab. on detached service 8/8-31/63. Present 9-12/63. Ab. on leave 1/29-2/18/64. AWOL 2/19-29/64. Present 3-6/64. Ab. on detail at District Headquarters 7/28-8/31/64. Orderly, Colonel Black 9-12/64. Paroled Greensboro, N. C. 5/1/65. Received pension Putnam Co., Fla. d. 7/14/06. Bur. Old City Cem., Jacksonville, Fla.

CARY, WILLIAM H. Pvt., Co. C. b. S. C. 6/20/36. Holsterer, age 24, Barnwell Dist. 1860 census. Enl. Wadmalaw Island 3/18/62. Ab. on sick leave 5-6/62. Present 7/62-10/63. Ab. on leave 12/21-31/63. Ab. sick in hospital with bronchitis 2/8-5/1/64. Present6-8/64. Secretary for Colonel Black 9-12/64. Present Fayetteville, N. C. 4/65. NFR. Farmer, Windsor, Ark. 1900 census. d. 1/21/05. Bur. Old White Pond Cem., Aiiken Co. Wife receiving pension White Pond, Aiken Co. 9/8/19.

CASTON, ALBERT D. Pvt., Co. I. b. S. C. circa 1840. Res. Colleton Dist. 1850 census. Enl. James Island 6/30/62. Present 7-8/62. Ab. sick Summerville hospital 9-10/62. Present sick in camp 11-12/62. d. of pneumonia in Lynchburg hospital 2/18/63. Bur. Old City Cem., Lynchburg, Va.

CATHCART, ----------. Pvt., Co. F. d. Charleston, S. C. 2/14/65.

CATHCART, ELIAS J. Pvt., Co. B. b. Spartanburg Dist. 9/25/35. Enl. Clinton 8/22/61 age 25. Ab. sick 10/31/61. On picket White Point 11-12/61. Present as Teamster 3-6/62. On picket Little Brittan 7-8/62. Ab. on march with horses from Summerville, S. C. to Staunton, Va. 9-10/62. Present as Teamster for Company 11-12/62. Present 3-6/63. Ab. on horse detail 8/18-31/63. Present 9-12/63. Ab. detailed horse recruiting camp 1/9-2/29/64. Present 3-10/64. Ab. detached Pocotaligo 12/2-31/64. Paroled Greensboro, N. C. 5/1/65. Farmer, Woodruff, Spartanburg Co. 1870-1900 censuses. d. Roebuck, Spartanburg Co. 11/8/09. Bur. Unity Bapt. Ch. Cem.

CATHCART, JAMES H. Pvt. Co. b. b. S. C. circa 1840. Age 20, Richland Dist. 1860 census. Enl. Spartanburg CH 1/26/64. Ab. sick 3/15-31/64. Present 5-8/64. On picket 9-10/64. Present 11-12/64. d. of disease at home in Spartanburg Dist. 6/15/65.

CATHCART, JOHN C. Pvt., Co. B. b. Spartanburg Dist. 11/29/39. Age 20, Reidsville, Spartanburg Dist. 1860 census. Enl. Clinton 8/22/61 age 22. Ab. sick 10/31/61. Present 11/61-5/62. Ab. on leave 7-8/62. Present 9-10/62. Ab. with wagon train hauling forage 11-12/62. Present 3-4/63. Ab. detailed recruiting horses in Nelson Co., Va. 7/4-12/31/63. Present 1-10/64. On detached duty Pocotaligo 11-12/64. NFR. Farmhand, Youngs, Laurens Co. 1870 census. Farmer, Reidsville, Spartanburg Co. 1880 census. d. 10/25/87. Bur. Mountain View Bapt. Ch. Cem., Spartanburg Co., S. C.

CATHCART, ROBERT A. Pvt., Co. B. b. S. C. circa 1836. Res. Spartanburg Dist. Enl. Camp Tripp 4/29/62. Present until KIA in skirmish at Walpole Gate, John's Island 6/7/62, age28.

CATTLES, JACOB J. Pvt., Co. I. b. S. C. 1838. Enl. Parker's Ferry 4/3/62. On picket Jacksonborough 4/30/62. On picket John's Island Ferry 5-6/62. AWOL 8/28-31/62. Ab. sick in Richmond hospital 10/13-11/5/62. Furloughed for 40 days. Present 3-4/63. Ab. sick in hospital through 8/31/63. AWOL 9-12/63. Present 1-2/64. Ab. sick in Lynchburg hospital 3/64. Ab. on horse detail 4/16-31/64. Present 5-6/64. Present sick in camp 7-8/64. Present 9-10/64. Ab. on horse detail for 30 days 12/28/64. NFR.

Teamster, Sheridan, Colleton Co. 1900 census. Bur. Ackerman Cem., Red Oak, Colleton Co., no dates.

CHALMERS, ROBERT L. 2ndSgt., Co. A. b. Newberry Dist. circa 1837. Res. Newberry Dist. 1850 census. Att. Erskine College 1859-60. Enl. Abbeville CH 8/13/61 age 23, as 3rdSgt. Present through 12/62. Promoted 2ndSgt. Ab. on detail 3-4/63. d. of disease Culppeper CH 6/16/63. Effects: Sundries. Left widow.

CHAMBERS, JOHN L. Pvt., Co.'s K & D. b. S. C. circa 1836. Enl. Co. A, 6th S. C. Inf. Chester CH 4/11/61 age 25. Present through 12/61. Ab. sick with "Febris Intermittens & debility" 3/24-4/12/62. Discharged 4/30/62. Enl. Co. K, 1st S. C. Cav. Chester CH 3/2/63. Transferred Co. D 6/12/63. Present through 12/63. Ab. recruiting horses 1/9-2/29/64. Present 3-12/64. Paroled Hillsboro, N. C. 4/65. Merchant, age 44, Chester CH 1880 census.

CHANDLER, E. R. Pvt., Co. A. Enl. Columbia 3/31/64. On duty Green Pond through 4/30/64. Ab. on leave 6/27-30/64. Ab. on sick leave 10/21-31/64. AWOL 12/21-31/64. Ab. sick in Charlotte, N. C. hospital with debilitas 2/15-3/23/65. Sent to Chester hospital with "Heamsplysis" 3/24/65. NFR.

CHAPMAN, GEORGE I. Pvt., Co. F. b. S. C. 6/10/37. Res. Pickens Dist. Enl. Pickens CH 11/7/63. Present through 11/30/63, no horse. Present 1-2/64. On duty Green Pond 4/29-30/64. Present 5-6/64. Ab. sick in hospital 6/29-8/31/64. Present 9-12/64. Paroled Charlotte, N. C. 5/3/65. Age 25, 5'8", light complexion, grey eyes. d. Pickens Co. 4/7/70. Bur. Boggs-Garner Cem.

CHAPMAN, ISAAC A. Pvt., Co. F. b. S. C. 1822. Farmer, age 28, Eastern Dist., Pickens Dist. 1850 census. Enl. Pickens CH 12/4/61 age 40. Present through 12/31/61. d. of disease Pickens Dist. 4/16/62. Left widow.

CHAPMAN, JACOB MATHEW. Pvt., Co. B. b. N. C. 1840. Enl. Co. G, 1st North Carolina Cav. Asheville, N. C. 5/20/61 age 21, res. Henderson Co., N. C. Present through 10/62. Ab. with fracture of left humerus 11/23/62-2/5/63. Present until transferred Co. B, 1st S. C. Cav. 12/5/63. Present through 8/64. Confined in arrest 10/15-31/64. Present 11-12/64. NFR. Farmer, age 44, Virginia, Iowa 1880 census. Alive 1898.

CHEATHAM, WILLIAM JASPER. Pvt., Co. G. b. Edgefield Dist. 6/22/29. Farmer, age 30, Warrenton, Abbeville Dist. 1860 census. Enl. Co. B. 19th S. C. Inf. Abbeville CH 12/19/61. Ab. sick Abbeville CH 1/26-2/28/62. Ab. sick Atlanta, Ga. 4/14-30/62. Discharged Tupelo, Miss. 7/30/62. Age 32, 5'4 1/2", sallow complexion, light hair, light eyes, Farmer. Enl. Co. G, 1st S. C. Cav. Abbeville CH 7/1/63. WIA Brandy Station 8/1/63. DOW's Richmond hospital 8/9/63. Bur. Hollywood Cem. Effects: $53.00 & sundries. Left widow & 2 children. Horse killed Brandy Station 8/1/63. Wife paid $350.00.

CHEEK, JAMES WILLIAM. Pvt., Co. K. b. Laurens Dist. 12/17/33. Farmer, Monroe, Laurens Co. 1860 census. Enl. Columbia 12/23/63. Ab. sick in Laurens Dist. through 12/31/63. AWOL 1-2/64. Ab. detached Green Pond 4/16-30/64. d. Laurens Dist. 9/9/64 age 32. Bur. Raburn Creek Bapt. Ch. Cem., Laurens Co. Left widow. Twin brother of William A. Cheek.

CHEEK, WILLIAM ABRAM. 1stCpl., Co. B. b. Laurens Dist. 12/17/33. Enl. Clinton 8/22/61 age 28 as 2ndCpl. Ab. sick 10/31/61. On picket Laroches 12/31/61. Present 3-4/62. Promoted 1stCpl. Ab. on sick leave 5-8/62. Ab. on horse detail 10/1/62 and ab. sick 10/31/62. AWOL 11-12/62. Present 3-4/63, reduced to Pvt. Ab. sick in Columbia, S. C. hospital 5-8/63. Ab. sick 9-10/63. AWOL 11-12/63. Present 1-2/64. Ab. sick with partial paralysis in Richmond hospital 3/7-4/1/64. Ab. on horse detail 4/2-30/64. On duty Dill's Bluff 6/5-30/64. Ab. as nurse waiting on brother, McLeod's hospital 7-8/64. Ab. on sick leave 9/9-12/31/64. NFR. Farmer, age 47, Dial, Laurens Co. 1880 census. d. Princeton, S. C. 7/14/04. Bur. Princeton Bapt. Ch. Cem.

CHESNUT, ALEXANDER MELTON. Pvt., Co. A. b. Horry Dist. circa 1837. Laborer, Kingston, Horry Dist. 1860 census. Enl. Co. G, 10th S. C. Inf. 9/4/61 age 26. Ab. sick 6/62. Discharged for chronic diarrohea 7/27/62. Age 27, 5'8", light complexion, dark blue eyes, Farmer. Enl. Co. A, 1st S. C. Cav. Columbia 3/31/64. On duty Green Pond 4/1-30/64. Present 5-6/64. Ab. guarding POW's Florence 9/17-10/31/64.

Present 11-12/64. NFR. Farmer, Conway, Horry Co. 1880 census. d. 1937. Bur. Poplar Meth. Ch., Conway, S. C.

CHISHOLM, JOHN MAXWELL. Pvt., Co. A. b. S. C. 9/18/39. On postwar roster. Reenlisted Co. K, 4th S. C. Cav. Grahamville 3/25/62. Present until ab. on sick leave 10/1-31/62. Present 11/62 until ab. sick leave 12/7-31/63. Present until ab. oh horse detail 8/31/64 for 40 days. WIA date and place unknown. Ab. wounded 10/31/64. Paroled 4/65. d. 1/22/98. Bur. Magnolia Cem., Charleston, S. C.

CHISHOLM, WILLIAM D. Pvt., Co. D. b. Barnwell Dist. circa 1843. Enl. Chester CH 9/10/61. Present through 12/31/61. Ab. detailed Charleston 4/29-30/62. Ab. detailed Brigade Commissary Dept. 5-6/62. Present 7-10/62. Ab. detailed Commissary Dept. 11/62-8/31/63. Ab. sick with diarrohea in Richmond hospital 10/9/63. Furloughed for 30 days 10/11/63. Ab. sick Chester CH 10/31-12/31/63. Present, detailed in Regimental Commissary Dept. 1-4/64 & in Brigade Commissary Dept. 6-12/64. Paroled Hillsboro, N. C. 4/65.. d. Barnwell Co. circa 1880.

CHISHOLM, WILLIAM S. Pvt., Co. D. b. Chester Dist. circa 1825. Enl. Camp Ripley 1/10/62. Ab. on sick leave 3/30-4/30/62. Ab. sick Chester Dist. 6/10-30/62. Discharged 8/19/62. Age 35, 6', fair complexion, blue eyes, dark hair, Farmer. Reenlisted by 1865. Paroled Hillsboro, N. C. 4/65. Miller, age 45, Lewisville, Chester Co. 1870 & 1880 censuses. d. there 7/80.

CLARK, ALEXANDER. Pvt., Co. D. Res. Chester Dist. KIA Bentonsville, N. C. 3/19/65 on casualty list.

CLARK, J. W. Pvt., Co. A. On post war roster.

CLARK, JAMES. Pvt., Co. I. b. S. C. 11/6/26. Res. Charleston. Enl. Parker's Ferry 4/3/62. Ab. on sick leave 4/19-30/62. Present 5-10/62. Ab. on scout 11-12/62. Present 3-6/63. Ab. on horse detail 8/18-31/63. AWOL 9-10/63. Present in arrest Castle Thunder, Richmond undergoing sentence of CM 3/20-11/30/64, dismounted since 7/19/63. Released 12/1/64. Age 40, 5'5", dark complexion, light hair, blue eyes. Deserted to the enemy Charleston 3/6/65. Took oath and released 3/15/65. Farmer, age 46, York Co. 1870 census. d. 1906.

CLARK, JOHN M. Pvt., Co. A. b. S. C. 7/16/30. On postwar roster. Reenlisted Co. G, 5th S. C. Inf. 5/13/62 as 1stCpl. Present until reduced to Pvt. by 7/64. Present through 2/65. Paroled Burkeville Junction 4/11-17/65. d. 3/12/05. Bur. Philadelphia Meth. Ch., York Co., S. C.

CLARK, L. E. Pvt., Co. A. On postwar roster.

CLARKE, THOMAS ELI. Pvt., Co. A. b. Anderson Co.,S. C. 6/26/35. Overseer, Abbeville Dist. 1860 census. Enl. Abbeville CH 8/13/61 age 26. On duty Bear Island 11-12/61. Present 3/62-4/63. Ab. sick in Lynchburg hospital 5-6/63. Present 9-12/63. Detailed Provost Guard, Fredericksburg 1-2/64. Ab. sick Charlotte, N. C. hospital 3/30-6/30/64. Present 7-12/64. NFR. Farmer, Center, Oconee Co. 1870 & 1880 censuses. Farmer, Fork, Anderson Co. 1900 census. d. 1908. Bur. Fair Play Cem., Oconee Co.

CLARKSON, ROBERT. Pvt., Co. A. b. S. C. 1844. Enl. 11/7/61. Left 2/7/62 (underage). May have served later in Holcombe Legion Cav., but not on muster rolls. d. LaGrange, Fayette Co., Texas 1940.

CLARKSON, THOMAS BOSTON, JR. Pvt., Co. A. b. Richland Co. circa 1834. Enl. 11/22/61. Discharged, end of enlistment 2/7/62. Reenlisted Co. H, 2nd S. C. Cav. Columbia 4/1/64. Present through 12/64. Ab. sick in Charlotte, N. C. hospital with "Febris. Int." 4-14/65. Clergyman, Richland Co. 1870 & 1880 censuses. d. Gaffney, S. C. 1/23/82.

CLARY, THOMAS J. Cpl., Co. A. b. S. C. Laurens Dist. 4/2/38. Farmer, Fraziersville, Abbeville Dist. 1860 census. Enl. Abbeville CH 8/13/61 age 26 as Cpl. Ab. sick 8/23-10/31/61. Ab. sick in hospital 11-12/61. Enl. Co. F, 1st S. C. Inf. (Hagood's) Greenville 8/20/61. No record until Ab. sick in Richmond hospital 10/27/62. Present 11-12/62. Promoted 4thSgt. Present 3-6/63. Promoted 3rdSgt. Ab. sick with typhoid fever in hospital 7/19-9/11/63. Ab. on leave 11-12/63. Promoted 2ndSgt. WIA Wilderness 5/6/64. Ab.

wounded through 2/65. NFR. d. 7/19/10. Bur. Orange Hill Cem., Williston, Fla.

CLAWSON, THOMAS WILLIAMS. Pvt., Co. I. b. S. C. 6/8/45. Enl. Co. A, State Cadets Bn. Charleston 7/10/63. Present through 9/63. NFR. Enl. Co. I, 1st S. C. Cav. Columbia 5/19/64. Detailed, Headquarters, East Lines, James Island 5/20-8/31/64. Ab. on sick leave 10/24-31/64. Detailed, Headquarters, East Lines, James Island 12/28-31/64. Paroled Charlotte, N. C. 5/23/65. Lawyer, York Co. 1870-1910 census. d. 11/20/13. Bur. Rose Hill Cem., York Co.

CLAY, HENRY. Pvt., Co. G. Paroled Augusta, Ga. 5/29/65.

CLAY, WILLIAM ALEXANDER. Pvt., Co. G. b. Abbeville Dist. 8/15/33. Farmer, age 23, Abbeville Dist. 1860 census. Enl. Camp Jackson 7/1/62. Ab. sick in Charleston hospital 7/30-8/31/62 & 9/13-10/31/62. Present 11-1/62 & 3-6/63. Ab. on horse detail 8/31/63-10/31/63. AWOL 12/30-31/63. Ab. detailed as Teamster for Captain Williams 1/2-12/29/64. Ab. on horse detail 4/1-30/64. Present 5-12/64. NFR. d. 8/22/13. Bur. Love Cem., St. Lucia Co., Fla. Marker in Lebanon Presb. Ch. Cem., Abbeville.

CLAYTON, ALFRED T. Captain, Co. F. b. S. C. circa 1829. Farmer, Pickens Dist. 1860 census. Enl. Pickens CH 12/4/61 age 32, as 1stLt. Present through 6/62. Ab. on sick leave 7-8/62. Present 9-10/62. On picket duty Richards Ford 11-12/62. Commanding Co. 3-4/63. Signed for 13,392 lbs. corn & 15,624 lbs. hay 3/31/63. Signed for 28,016 lbs. corn & 33,152 lbs. hay 4/63. Present through 8/31/63. Signed for 3 tent flys, 20 pr. horse shoes, 1 horse shoe anvil & 1 book 5/25/63. Signed for 6,487 lb. corn & 33,152 lb. hay 6/63. Signed for 5 jackets & 5 caps 8/26/63. Signed for 15 caps, 27 jackets, 16 pr. pants and 52 shirts 8/24/63. Signed for 14,180 lbs. corn, 14,148 lbs. hay, 13 pr. boots, 15 pr. socks, 3 blankets, 59 pr. drawers, 15 pr. pants & 2 overcoats 9/8/63. Reported only 21 horses present 9/30/63. Ab. on leave 9-10/63 on rolls. Signed for 8,832 lbs. corn & 10, 304 lbs. hay 10/63. Elected Captain 11/1/63. Signed for 14,800 lbs. corn & 16,562 lbs. hay 2/64. Ab. on leave 2/26-29/64. Signed for 18 pr shoes, 5 blankets, 2 fly tents, 1 camp kettle, 14 pr. pants, 1 overcoat, 1 axe, 4 shirts, 4 jackets, 2 pr. drawer, 1 Army tent, 14,086 lbs. corn & 16,562 lbs. hay 3/31/64. Captain, 1 Lieutenant & 19 NCO's present 1-3/64. On detached duty Pickens CH 4/30/64. Present sick in camp 6/30-7/1/64. Signed for 14 caps, 20 jackets, 20 pr. pants, 10 cotton shirts, 10 pr. drawers, 4 pr. shoes & 15 pr. socks at Green Pond 5/20/64. Signed for 55 pr. pants, 32 jackets, 2 pr. socks, 28 shirts, 21 pr. drawers, 17 caps, I shovel, 6 Army tents, 2 camp kettles & 7 mess pans 6/30/64. Signed for 21,208 lbs. corn & 24, 238 lbs. hay 7/31/64. Ab. on leave 8/31/64 & 9/16-25/64. Signed for 1 jacket, 5 pr. pants, 5 shirts, 8 pr. drawers, 6 pr. shoes, 8 pr. socks & 1 wall tent 9/30/64. Present 10/64-1/31/65. NFR. d. 1884. Some of his papers misfiled with J. O. Clements.

CLEMENTS, ISAAC D. Pvt., Co. D. b .S. C. circa 1841. Enl. Camp Johnson, Chester 10/8/61. age 19. Present through 6/62. Ab. on sick leave Rutherford Co., N. C. 8/9-31/62. Present 9/62-4/63. Ab. sick in hospital 8/5-31/63. Ab. on sick leave through 12/31/63. Present 1-2/64. Ab. on horse detail Rutherford Co., N. C. 4/25-30/64. Present 5-10/64. Present, driving ambulance for Regt.'l. QM Dept. 11/17-12/31/64. Paroled Greensboro, N. C. 5/1/65.

CLEVELAND, ROBERT LEWIS. Pvt., Co. B. b. Laurens Dist. 9/13/32. Farmer, Laurens Dist. 1860 census. Enl. Clinton 8/22/61 age 30. Ab. sick 10/31/61. Present 11/6-4/62. Ab. detailed as Courier for Colonel Dunovant at Church Flats 5-6/62. Present 7-8/62. Hired S. T. Prior as substitute and discharged 9-10/62. Reenl. Columbia 2/1/64. Ab. on leave 4/26-30/64. Present 5-6/64. Ab. on horse detail 8/10-31/64. Present 9-12/64. NFR. Farmer, Scuffeltown, Laurens Co. 1880 census. d. 3/23/95. Bur. Bethany Presb. Ch. Cem., Clinton, S. C.

CLIFTON, JESSE ALEXANDER. Pvt., Co. D. b. Chester Dist. 9/26/45. Age 14, Clinton 1860 census. Enl. Camp Johnson, Chester 10/4/61 age 16. Present through 6/62. Present sick in camp 7-8/62. Present 9/62-6/63. Ab. on horse detail 8/18-31/63. Present 9-12/63, dismounted 90 days. "Entered Yankee lines & captured Gen. Meade's horse hitched outside his tent. Received a deep saber cut on side of neck and another on the corner of an eye and a bullet just over his heart which was never removed." Present 1-10/64. Ab. detached Coosawhatchie 12/10-31/64. Paroled Hillsboro, N. C. 4/65. Att. UVa. Methodist Minister in postwar. Clergyman, Sumter, Sumter Co. 1900 census. d. Marion, S. C. 6/14/06.

CLINE, CHARLES P. Pvt., Co. H. b. S. C. 1841. Enl. Co. E, 17th S. C. Inf. Columbia 3/26/61. WIA 2nd Manassas 8/30/62. Present through 2/63. AWOL 5/16-8/1/63. Present sick in camp 11-12/63. Ab. sick in hospital 1-2/64. Transferred Co. H, 1st S. C. Cav. 3/12/64. Ab. detailed at Green Pond 4/16-30/64. NFR. d. York Co. 10/11/15.

COBB, AUGUSTUS BURT. 4thSgt., Co. A. b. S. C. 9/2/45. Merchant's Clerk, age 15, Abbeville 1860 census. Enl. Abbeville CH 8/13/61 age 19. Ab. on sick leave 8/23-10/31/61. Ab. on duty Bear Island 11-12/61. Present 3-10/62. Ab. in arrest 11-12/62. CM'd 3/17/63. Present 5-6/63. Ab. on horse detail 8/18-31/63. Present 9-12/63, promoted 2ndCpl. Present 1-2/64, however, ab. sick with debility in Richmond hospital 2/20-3/14/64. Ab. on horse detail 4/24-30/64. Present 5-10/64, promoted 4thSgt. Detached at Pocotaligo 12/10-31/64. NFR. House Painter, Mason Co., Texas 1910 census.

COBB, JAMES EDMUND. Sgt., Co. A. b. Abbeville, S. C. 11/8/42. Silversmith. Enl. Abbeville CH 3/19/62 as Pvt. Present through 7/8/62. Promoted 4thCpl. Present 9/62-6/63, however, ab. sick with acute diarrohea in Farmville hospital 6/10-7/1/63. Ab. on horse detail 8/18-31/63. Promoted 1stCpl. & Sgt. d. Abbeville of dysentery 9/23-24/63. Bur. Sharon Meth. Ch. Cem., Abbeville, S. C.

COBB, JOHN G. Pvt., Co. H. b. S. C. 1815. Enl. Rock Hill 1/9/61 age 46. On picket Simons Bluff 3-4/62. Present 5-6/62. Present sick in camp 7-8/62. AWOL 9/62-4/63. Present 1-2/64. Detailed 3/29/;64. Present 4/64. Transferred Co. B, Lucas' 15th Bn. S. C. Arty. 5/13/64. Joined 5/20/64. Present through 8/64. AWOL 9/11-12//64. CM'D 12/24/64. NFR. d. King's Mountain, York Co. 3/26/81.

COCHRAN, THOMAS W. Pvt., Co. A. Enl. age 30 on postwar roster. On Abbeville Co. roster 1902.

COLCOCK, RICHARD HUTSON. Captain, Co. A. b. S. C. 8/18/23. On postwar roster. Reenlisted Co. K, 4th S. C. Cav. Grahamville 3/25/62 as 1stLt. Elected Captain 12/12/62. Present until ab. on leave 1/16/63 for 18 days. Present until ab. on leave 4/28/63 for 10 days. Present until ab. on leave 12/22/63 for 10 days. Present until ab. sick in Limestone Springs, S. C. hospital 8/9/64. Ab. on sick leave 10/21/64 for 30 days. NFR. d. 9/15/01. Bur. Stoney Creek Cem., Beaufort Co. S. C.

COLCOCK, THOMAS HUTSON. Pvt., Co. A. b. S. C. circa 1837 Gd. South Carolina College 1859. On post war roster. Reenlisted Co. K, 4th S. C. Cav. Grahamville 3/25/62. On duty as Secretary 9/62-1/63. Detailed by Gen. Walker 2/16/63. Appointed Adjutant, 3rd S. C. Cav. 5/2/63. Present until ab. on leave 8/20/64 for 20 days. Paroled Appomattox CH 4/9/65. Cotton Broker, Charleston 1870-1894. Bur. Evergreen Cem., Ocala, Fla., no dates.

COLE, IRA F. Pvt., 1st S. C. Cav. b. Ga. circa 1840. Age 26, Ashland, Ala. 1870 census. Confederate Veteran living Chester Co. 1904.

COLE, MORGAN S. Pvt., Co. F. b. Pickens Dist. 1827. Farmer, Tunnel Hill, Pickens Dist. 1860 census. Enl. Pickens CH 12/4/61 age 35. Present through 8/62. Ab. sick with diarrohea in Richmond hospital 10/24-26/62. Transferred Farmville hospital with lumbago. Furloughed for 60 days 12/20/62. In Farmville hospital with chronic Nephritis & general emaciation 1/21/63. In Richmond hospital with Nephritis 10/4-31/63. Discharged 11/63. Age 37, 6', dark complexion, dark hair, dark eyes, Farmer. Returned to duty from Richmond hospital 1/25/64. NFR. Farmer, Cherokee Co., Texas 1870 & 1880 censuses. d. Alto, Texas 12/31/06.

COLEMAN, WALTER FRANCK MARION. Pvt., Co.'s K & D. b. S. C. 9/21/46. Student, age 13, Chester 1860 census. Enl. Co. C, Manigault's Bn. Arty. 3/11/62. Present through 5/31/62. NFR. May have been discharged for underage. Enl. Co. K, 1st S. C. Cav. Columbia 4/18/64. Ab. on horse detail through 4/30/64. Ab. as Courier for Signal Corps, Charleston 6/25-30/64. Ab. on leave from Charleston hospital 7-8/64. Transferred Co. D 10/1/64. Detached as Coosawhatchie 12/10/-31/64. Paroled Smithfield, N. C. 4/26/65. d. Chester, S. C. 12/25.95. Bur. Pleasant Grove Presb. Ch. Cem., Chester. Wife receiving pension Ridgway, Fairfield Co. 9/20/19.

COLLIER, WILLIAM OSCAR. Pvt., Co. E. b. Orangeburg Dist. 11/27/27. Enl. Branchville 10/26/61. On picket White Point through 12/31/61. Present 3-4/62. Present sick in camp 5-6/62. On scout Edisto Island 7-8/62. Present 9-10/62. On scout 11-12/62. Present 3-8/63. Detailed at recruiting station for horses 9-10/63. Present 11-12/63. Detailed at regimental recruiting station for horses 1-2/64. Detached at Green Pond 4/15-30/64. Present 5-10/64. Detached at Pocotaligo 12/10-31/64. WIA in N. C. 1865. Paroled Greensboro, N. C. 4/29/65. Farmer, Providence, Orangeburg Co. 1870 census. d. Orangeburg Co. 8/30/76.

COMBS, JOHN T., JR. Pvt., Co. G. b. S. C. circa 1847. Age 3, Edgefield Dist. 1850 census. Enl. age 18 on postwar roster.

COLN, ALEXANDER GASTON. Pvt., Co. D. b. N. C. 1833. Enl. Chester 9/1/62. Ab. detailed as Teamster for QM Dept. of regiment 9/62-2/64. Ab. on leave Chester Dist. 4/22-30/64. Present 5-12/64. Paroled Hillsboro, N. C. 4/65. d. 1904. Bur. Evergreen Cem., Chester Co., S. C.

COLN, IRA S. Pvt., Co. D. b. S. C. 3/25. Enl. Chester CH 9/10/61 age 36. Present through 12/61. AWOL 1/62. Present 3-8/62. AWOL Chester Dist. 9/27-30/62. Present detailed as Blacksmith in QM Dept. 11/1/62-2/64. Ab. on leave Chester Dist. 4/23-30/64. Present 5-6/64. AWOL & ab. sick in Chester Dist. 7/3-8/31/64. Ab. sick Chester Dist. 10/21-31/64. AWOL 11/12-12/31/64. Paroled Greensboro, N. C. 5/1/65. Farmer, Baton Rouge, Chester Co. 1880-1910 censuses.

COLN, W. ADDISON. Pvt., Co. D. b. S. C. circa 1837. Enl. Camp Means, Chester CH 4/17/62, age 25. Present through 8/62. Ab. sick Chester Dist. 10/2-12/31/62. Present 3-10/63. Detailed as Teamster in Brigade QM Dept. 12/14/63-2/64. Ab. on leave in Chester Dist. 4/28-30/64. Ab. in Brigade/McLoud hospital, James Island 5/27-12/31/64. Paroled Hillsboro, N. C. 4/65.

COMBS, JOHN T., JR. Pvt., Co. G. b. S. C. 1847. Age 3, Edgefield Dist. 1850 census. Enl. age 18 on postwar roster.

CONE, JOSEPH HAMILTON. Pvt., Co. I. b. Colleton Dist. 3/3/43. Res. Ridgeville, Colleton Dist. Enl. Parker's Ferry 4/3/62. Present through 4/63. WIA (right leg) & captured Brandy Station 6/9/63. In Alexandria hospital 6/10/63. Transferred Baltimore hospital. Exchange 8/24/63. In Williamsburg hospital same day. Transferred Farmville hospital 8/28/63. Furloughed for 30 days 9/11/63. Ab. on wound furlough through 4/64. Ab. sick in hospital 7/21-12/31/64. Deserted to the enemy Charleston 3/22/65. Age 21, 5'6", dark complexion, dark hair, dark eyes. Took oath and released. d. 11/17/90.

CONNARD, M. Pvt., Co. C. Not on muster rolls. Captured near Columbia 2/19/65. POW through 2/28/65. NFR.

CONNER, JOHN. Pvt., Co. G. b. S. C. circa 1846. Enl. Abbeville CH 2/10/62. On picket Jacksonborough 3-4/62. Discharged 6/16/62, probably for underage. Served later in 1st S. C. Arty. Farmhand, Fulton, Miss. 1870 census.

CONNOR, WILLIAM D. 3rdLt., Co. E. b. S. C. circa 1838. Enl. Branchville 10/26/61 age 20 as Pvt. Promoted 2ndSgt. Present through 12/31/61 & 3-6/62. Reduced to Pvt. 8/13/62. On detached duty Gordonsville through 8/31/62. Present sick in camp 9-10/62. On scout 11-12/62. Elected 3rdLt. 3-4/63. Ab. sick with "Febris Int." through 9/16//63. Furloughed for 30 days 10/1/63. Present 11-12/63. Ab. in charge of horse recruiting station for the regiment 1/27-2/29/64. Present 3-6/64. Detached at Green Pond 5/18-6/20/64. Ab. sick in Charleston hospital 8/19-31/64. Present 9-10/64. NFR. Wife applied for admission to Old Soldiers' Home, Columbia from Dorchester Co. 4/25/1951.

COOK, JAMES. Pvt., Co. A. b. Woodruff, Spartanburg Dist.10/45. Not on muster rolls. Paroled Hartsville, Ga. 5/19/65. Farmer, Reidsville, Spartanburg Co. 1870 census. Farmer, Marion, S. C. 1910 census. Living Bethea, Dillon Co. 1920.

COOK, JOHN E. Pvt., Co. D. b. S. C. 1842. Enl. Camp Ripley 1/17/62. Present 3-6/62. Ab. on leave Chester 8/9-31/62. Present 9-10/62. Present sick in camp 11-12/62. Present 3-8/63. Ab. sick in Richmond hospital with "Int. Fever" 10/19-31/63. Ab. on horse detail Chester Dist. 12/22/63-2/64, dismounted 72 days. Present 3-10/64. Ab. guarding Pontoon Bridge 12/29-31/64. Paroled Hillsboro, N. C. 4/65. Farmer, Blackstock, Chester Co. 1870 census and Rock Mills, Anderson Co. 1880 census.

COONER, E. J. Pvt., Co. E. b. S. C. circa 1830. Farmer, age 30, Colleton Dist. 1860 census. Enl. Branchville 10/26/61 age 34. Killed by Barr (a person) at Branchville 10/29/61.

COPELAND, GEORGE DUCKETT. Pvt., Co. B. b. S. C. 10/24/46. Enl. Columbia 4/11/64. Present through 12/64. Paroled Greensboro, N. C. 5/1/65 as Courier. d. 12/15/07. Bur. Clinton City Cem.

COPELAND, JOHN HENRY. 2ndLt., Co. B. b. S. C. 1839. Enl. Clinton 8/22/61 age 22 as 1stCpl. Ab. sick 10/31/61. Promoted 3rdSgt. 12/31/61. Present 3-10/62. On raid with General Hampton 11-12/62. Promoted 2ndSgt. Present 3-10/63. Ab. on horse detail 12/22-31/63. Ab. on leave 4/26-30/64. Present 5-8/64. Elected 2ndLt. 9/15/64. Ab. guarding POW's Florence 9/16-10/64. Commanding Co. 11-12/64 at Florence. NFR. d. of pneumonia Richmond, Va. 12/30/06. Body brought back to Laurens Co. by mule teams and buried Hurricane Bapt. Ch., Laurens Co. Wife receiving pension Clinton, Laurens Co. 9/10/19.

COPELAND, JOHN HOLLAND. Pvt., Co. B. b. S. C. 6/10/39. Not on muster rolls. d. 12/20/06. Bur. Duncan's Creek Presb. Ch. Cem., Laurens Co. May be confused with John Henry Copeland above.

COPELAND, LEISEL WATSON "WATTS." Pvt., Co. B. b. S. C. 3/14/25. Farmer, Laurens Co. 1860 census. Enl. Co. D, 9th S. C. Reserves (90 days) Clinton as 2ndLt. 11/17/62. Present through 2/63. NFR. Enl. Co. B, 1st S. C. Cav. Columbia 3/3/64. Ab. on leave 4/27-30/64. Present 5-12/64. NFR. Farmer, Jacks, Laurens Co. 1870 & 1880 censuses. d. 12/27/94. Bur. Duncan's Creek Presb. Ch., Laurens Co.

COPELAND, ROBERT JAMES. Pvt., Co. B. b. Clinton, Laurens Dist. 9/19/47. Enl. Co. E, 4th S. C. Bn. Res. 9/1/64. NFR. On postwar roster Co. B, 1st S. C. Cav. Paroled 4/26/65. Farmer, Laurens Co. 1880 & 1900 censuses. Receiving pension Clinton 10/17/19 and Rockledge, Brevard Co., Fla. 1920. d. Clinton 5/26/21. Bur. Clinton City Cem.

CORBETT, JOHN H. Pvt., Co. G. b. Orangburg Dist. 3/26/26. Enl. Abbeville CH 1/2/62. Present through 6/62. Ab. sick 7-8/62. AWOL 9-10/62. On picket 11-12/62. Ab. sick in hospital 3-4/63. Present through 8/31/63. Ab. on detached service 9-10/63. AWOL 12/30-31/63. Present 1-2/64. Transferred Co. B, Lucas' 15th Bn. S. C. Arty. 5/13/64. Joined 5/20/64. Present through 12/31/64. Admitted Pettigrew Hospital, Raleigh, N. C. 4/17/65. Paroled 4/20/65. d. Orangeburg, S. C. 3/9/28.

CORNWELL, JOHN BENNETT. 1stSgt., Co. D. b. S. C. 7/26/36. Farmer, age 24, Chester Dist. 1860 census. Enl. Chester CH 9/10/61 age 25. Ab. on leave 12/21-31/61. Present 3-4/62. Ab. on leave Chester Dist. 5/13-31/62. Discharge for disability 8/8/62. Age 26, 5'10", fair complexion, blue eyes, dark hair, Farmer. May have served in Co. D, 3rd S. C. Reserves. Farmer, Blackstock, Chester Co. 1870 & 1880 censuses. d. 7/25/89. Bur. Woodward Bapt. Ch. Cem., Chester Co.

CORNWELL, JOHN DAVID. b. S. C. 8/27/40. Res. Chester Dist. 1860 census. Enl. Columbia 2/5/64. Ab. detached at Green Pond 4/18-30/64. On picket Moorland's Warf, Charleston 6/25-30/64. Ab. sick in Charleston hospital 7-8/64. Present 9-10/64. Detached at Green Pond, Dist. Headquarters 12/28-31/64. NFR. Member, York Co. Survivors Association 8/26/80. d. 12/23/08. Bur. Fishing Creek Presb. Ch. Cem., Chester Co.

CORNWELL, JOHN PRATT. Pvt., Co. D. b. S. C. 6/1/40. Res. Chester Dist. 1860 census. Enl. Chester CH 7/14/62. Ab. on sick leave Chester Co. 8/19/62. Ab. sick with acute dysentery in S. C. hospital, Manchester, Va. 10/4-11/62. d. 10/12/62. Bur. Pleasant Grove Presb. Ch. Cem., Chester Co.

CORRETHERS, J. K. Pvt., Co. H. Enl. Columbia 3/19/64. Present through 6/64. Ab. sick in hospital 8/12-10/31/64. AWOL 11-12/64. NFR.

COTHRAN, E. TOBIAS. Pvt., Co. A. b. S. C. circa 1840. Enl. Abbeville CH 8/13/61 age 21. Present through 10/31/61. Ab. sick in hospital 11-12/61. Present, sick in camp 3-4/62. Present 5-6/62. Ab. on sick leave 7-10/62. Present 11/62-6/63. Ab. on horse detail 8/18-31/63. AWOL 9-10/16/63. Ab. sick with "Int. Fever" in Richmond hospital 12/17/63-2/22/64. Present through 2/29/64. Detached at Green Pond 3-4/64. Present 5-6/64. Ab. on leave 10/31/64. Detached at Pocataligo 12/10-31/64. NFR. Farmer, Talbert, Edgefield Co. 1880 census.

COVERT, H. C. Pvt., Co. A. b. Fla. circa 1831. Merchant, age 29, Charleston 1860 census. Enl. 11/7/61. Discharged by Surgeon 11/27/61. NFR.

COWAN, A. T. Pvt., Co. A. Enl. by 8/61. Not on muster rolls 10/11/61.

COWAN, FRANK E. Pvt., Co. A. b. S. C. 7/6/30. Enl. Abbeville CH 8/13/61. Present until ab. sick 11/10-12/31/61. Present 3-12/62. Ab. sick 3-4/63. Present through 12/63. Ab. detached to recruit horses 1/19-2/29/64. Ab. on horse detail 4/15-30/64. Present 5-12/64. Paroled as Teamster Greensboro, N. C. 5/1/65. Farmer, Magnolia, Abbeville Co. 1880 & 1900 censuses. d. 2/9/08. Bur. Salem Cem., Abbeville. Wife received pension Calhoun Falls, Abbeville Co.

COWAN, SAMUEL FAIR KIRKSEY. Pvt., Co. F. b. S. C. 10/3/36. Enl. Pickens CH 12/4/61 age 25. Present through 4/62. Ab. nurse in hospital 5-6/62. Present 7-10/62. Ab. nurse in hospital 11-12/62. Present 3/634/64. Present sick in camp 5-6/64. Present 7-12/64. Paroled 1865 on postwar roster. Farmer, Keowee, Oconee Co. 1870 & 1910 censuses. d. 10/7/12. Bur. Bethel Presb. Ch. Cem., Walhalla, Oconee Co.. Wife receiving pension Walhalla, Oconee Co. 10/13/19.

COWARD, JAMES W. Pvt., no company. Not on muster rolls. Receiving pension Madison Co., Tenn. 1907.

COX, ALEXANDER P. Pvt., Co. C. b. S. C. circa 1816. Enl. Hamburg 8/27/61 age 45. Ab. sick in hospital 11-12/61. Ab. detached at Church Flats 5-6/62. Present 7-10/62. On picket 11-12/62. Captured Bell's Ford 12/31/62. Sent to Old Capitol. Paroled for exchange 2/5/63. Paroled POW 3-4/63. Exchanged. Ab. sick at home 5/1-9/30/63. AWOL 12/20-31/63, lost cartridge box, fined $4.50. Captured Ellis Ford on Rapidan 3/5/64. Sent to Old Capitol & Ft. Delaware. d. of pneumonia Ft. Delaware 1/27/65. Bur. Finn's Point, Nat. Cem., N. J.

COX, GEORGE W. Pvt., Co. A. b. S. C. 1/42. Enl. Abbeville CH 6/29/62. Present through 8/62. AWOL 9-10/62. On picket 11-12/62. Present 3-5/63. WIA & captured Brandy Station 6/9/63. Sent to Old Capitol. Exchanged 7/63. Ab. on horse detail 8/18-31/63. Present 11-12/63, dismounted 7/9-10/17/63. Present 1-4/64, remounted 4/15/64. Present 5-6/64. Detached at Pocataligo 12/10-31/64. Paroled Augusta, Ga. 5/28/65. Farmer, age 28, Brushy Creek, Anderson Co. 1870 census. Farmer, Darcusville, Pickens Co. 1900 & 1920 censuses.

COX, MALCOLM LUTHER. Sgt., Co. A. b. S. C. circa 1837. Farmer, age 23, Abbeville Dist. 1860 census. Enl. Abbeville CH 8/13/61. Present through 6/62. Ab. on sick leave 7-8/62. Present 9/62 until WIA (saber cut to head) & captured Gettysburg 7/5/63. Sent to Ft. Delaware. 7/30/63. d. of imflamation of the lungs 6/30/64. Bur. Finn's Point, N. J. Nat. Cem.

COX, S. L. Pvt., Co. A. Missing in action Boonsboro, Md. 6/9/63 on record of events for Company A. Not on muster rolls.

CRADDOCK, WILLIAM ARCHIBALD. Pvt., Co. B. b. S. C. circa 1824. Enl. Clinton 8/22/61 age 37. Ab. sick 10/31/61. Present 11-12/61. Ab. on sick leave 1/62. Present sick in camp 3-4/62. d. at home in Laurens Dist. 6/25/62 age 36. Bur. Bethany Presb. Ch. Cem., Clinton, S. C.

CRAIG, GEORGE W. Pvt., Co. K. b. S. C. 1844. Farmhand, age 16, York Dist. 1860 census. Enl. Adams Run 9/12/62. Ab. on detached service through 10/31/62. Present 11-12/62. Ab. on detached service 3-8/15/63. Ab. on horse detail 8/16-31/63. Ab. on sick leave 9-10/63. Ab. sick York Dist. 11/1/63. AWOL 12/31/63. Ab. sick in South Carolina hospital 1-2/64. Ab. in prison Columbia, S. C. 4/14-5/31/64. Ordered transferred Co. A, 20th S. C. Inf. 5/16/64. Present 6/64. Ab. sick with phthisis pulmonalis in Richmond hospital 7/22/64. Furloughed for 30 days 8/6/64. NFR.

CRAWFORD, JAMES A. 2ndCpl., Co. A. b. S. C. 8/44. Enl. Abbeville CH 12/23/61, age 23 as Pvt. Present through 6/62. Ab. on sick leave 7-8/62. Present 9-10/62. On picket 11-12/62. Present 3-12/63, promoted 4thCpl. Present 1-2/64. Ab. on leave 4/24-30/64. Present 5-8/64, promoted 2ndCpl. Present 11-12/64. NFR. Farmer, Gaines, Anderson Co. 1870 census. d. 9/25/71. Bur. Lebanon Presb. Ch. Cem., Abbeville Co.

CRAWFORD, JAMES ALEXANDER, Pvt., Co. A. b. S. C. 8/6/45. On postwar roster. Farmer, Greenwood, S. C. 1900-1910 censuses. Farmer, Gastonia, N. C. 1920 census. d. Greenwood, S. C. 3/28/27. Bur. Rock Church Cem.

CRAWFORD, JAMES ANDREW. Pvt., Co. G. b. S. C. 1840. On postwar roster. d. 9/25/71. Bur. Lebanon Presb. Ch., Abbeville.

CRAWFORD, JAMES R. Pvt., Co. G. b. S. C. 12/14/30. Enl. Abbeville CH 3/15/62. Ab. sick in hospital through 4/62. Present 5/62-4/64. AWOL 4/15-6/30/64. Present 7-12/64. NFR. d. 5/24/06. Bur. Lower Cane Presb. Ch. Cem., McCormick Co., S. C.

CRAWFORD, JOHN ALEXANDER. Pvt. & Musician, Co. G. b. S. C. 6/6/40. Clerk, age 20, Columbia, Richlands Dist. 1860 census. Enl. Abbeville Ch 1/9/62 as Pvt. Present through 6/63. Ab. on horse detail 8/31/63. Present 9/63-6/64. Appointed Musician in Regimental Band 8/3/64. Present with Band though 12/64. NFR. Miller, Palestine, Ark. 1880 census. Cotton Ginner, Kaufman, Texas 1900 census. d. Kaufman, Texas 6/12/19.

CRAWFORD, JOHN I. 4thSgt., Co. A. b. Ala. circa 1842. Enl. Abbeville CH 8/13/61 age 20 as Pvt. On duty Bears Island 11/11-12/31/61. Present 3-6/63. Ab. on leave 7-8/62. Ab. detached duty 9-10/62. Ab. on scout 11-12/62. Promoted Cpl. Present 3-4/63. WIA (upper part of thigh -flesh) Brandy Station 6/9/63. Ab. wounded in Richmond hospital through 8/31/63. Promoted 4thSgt. Furloughed for 60 days 9/17/63. Ab. wounded 11/63-2/64. Present 3-8/64. Ab. on horse detail 8/3-31/64. Discharged and placed on retired list 9/2/64. NFR. d. 3/27/67. Bur. Lebanon Presb. Ch., Abbeville Co.

CRAWFORD, ROBERT W. 2ndLt., Co. D. b. S. C. 12/2/38. Enl. Chester CH 9/10/61. Commanding Co.10/31/61. Present 11/61-1/62. Ab. on sick leave 4/3-30/62. Present James Island 6/8/62. Ab. sick in Chester Dist. 6/14-7/31/62. Commanding Co. 8-10/62. Thrown from a horse and dragged to his death near Staunton, Va. 11/7/62. Bur. Cedar Shoals Presb. Ch., Chester Co.

CREIGHTON, ALBERT N. Pvt., Co. D. b. S. C. circa 1834. Age 16, Chester Dist. 1850 census. Enl. Chester CH 9/10/61 age 21. Present 11/61-10/62. Ab. on scout 12/25-31/62. Present 3-4/63. Ab. on horse detail 8/18-31/63 & 10/23-31/63. Present 11/63-12/64. Paroled Hillsboro, N. C. 4/65.

CREIGHTON, JOHN MC PHERSON. Pvt., Co. A. b. S. C. 1825. Enl. 11/7/61. Detailed by Gen. Evans 2/7/62. NFR. Enl. Co. E, 6th S. C. Reserves 1862-1863. Enl. Walpole's Co., Stono Scouts 11/30/63. Enl. Co. K, 5th S. C. Cav. 1863-64. Paroled 5/65. Lawyer, Newport, R. I. 1875. Retired, age 45, Newport, R. I. 1880 census.

CREWS, THOMAS BISSELL. Captain, Co. A. b. Rutherford Co., N. C. 6/7/32. Moved to Laurens 1849. Clerk & Printers Apprentice. Moved to Atlanta and then Abbeville CH. Asst. Editor & Publisher of "Abbeville Damer," 1850 census. Enl. Abbeville CH 8/13/61 age 24, as 1stSgt. Ab. on leave 10/31/61. Present 11/61-10/62. On scout 11-12/62. Elected 2ndLt. Present 3-8/1/63. Signed for 23,328 lbs. corn & 26,040 lbs. hay 7/31/63. Signed for 4,228 lbs. corn & 21,266 lbs. hay 8/31/63. Ab. on detached service 8/31/63, however, signed for 13 pr. shoes 8/11/63, 125 jackets, 94 shirts, 49 pr. socks & 99 pr. drawers 8/19/63, & 19,916 lbs. corn & 23,022 lbs. hay 8/31/63. Promoted 1stLt. Present 9-10/63. Signed for 12,430 lbs. corn & 12,440 lbs. hay 9/30/63. 106 horses present 10/5/63. Signed for 1,672 lb. corn & 1,489 lbs. hay, 31 horses present 10/6/63. Signed for 1,208 lbs. corn & 1,664 lbs. hay 10/7/63 & 104 horses present. Signed for 265 lbs. corn & 375 lbs. hay 10/20/63 & 116 horses present. Commanding Co. 11-12/63. Signed for 24,800 lbs. corn & 25,930 lbs. hay 12/30/63, 50 horses present. 2 Lt.s & 30 men present 1-3/64. Signed for 1 jacket, 2 pr. shoes & 1 pr. pants 3/1/64, however reported ab. on leave 2/21-4/30/64. Ab. on leave 6/27-7/31/64.Signed for 44 pr. pants, 49 jackets, 11 pr. socks, 26 shirts, 42 pr. drawers, 22 caps, I shovel, 8 Army tents, 3 camp kettles & 7 mess pans 6/30/64. Present 9-10/64, however, AWOL 9/16-30/64. Signed for 1 jacket, 5 pr. pants, 6 cotton shirts, 7 pr. drawers, 6 pr. shoes & 6 pr. socks 9/30/64. Promoted Captain & commanding Co. to end of war in postwar account. Editor & publisher, "Laurens Herald" 1865-1895. Served in State Leg. 2 terms. Living Laurens 1910 census. d. 5/28/11. Bur. Laurens Cem.

CRIMMINGER, RUFUS. Pvt., no company. d. 8/15/63. Bur. Gordonsville Exchange Hotel Hospital, Gordonsville, Va. & buried there.

CRISWELL, DAVID PRESSLEY. Pvt., Co. A. b. S. C. 4/11/28. Painter, Abbeville Dist. 1860 census. Enl. Abbeville CH 8/13/61 age 28. Ab. sick in hospital 11-12/61. Present 3-6/62. Present sick in camp 7-8/62. Present 9/62-10/64. Detached Pocotaligo 12/10-31/64. Paroled Augusta, Ga. 5/29/65. Farmer, Columbia Co., Fla. 1870 census. d. Lake City, Columbia Co., Fla. 1/23/79.

CROOK, R. L. Pvt., Co. K. b. S. C. 3/1/26. Enl. Charleston 11/3/64. Courier, Stocks Causeway 12/29-31/64. Paroled Charlotte, N. C. 5/22/65. D. 5/20/80. Bur. Harmony Bapt. Ch. Cem., Chester Co.

CROSBY, HESACHIEA H. Pvt., Co. I. b. S. C. 10/43. Enl. Johns Island 6/30/62. Present until ab. on leave 8/27-31/62. Ab. Summerville hospital 9-10/62. AWOL 11/62-2/64. NFR. Probably dropped as a deserter. Farmer, Bell, Colleton Co. 1900 census.

CROW, H. B. Pvt., Co. F. b. S. C. circa 1834. Enl. Pickens CH 12/4/61 age 27. Present through 4/62. Ab. detached Church Flats 5-6/62. Present 7-10/62. On pickets 11-12/62. Present 3-4/63. Captured Upperville 6/21/63. Sent to Old Capitol. Exchanged 7/24/63. d. of typhoid fever in Staunton hospital 8/13/63. Bur. Thornrose Cem., Staunton, Va. Effects: 1 coat, 1 jacket, 1 blanket, 1 pr. shoes & sundries. Left widow.

CROW, JAMES PINCKNEY. Pvt., Co. B. b. S. C. 7/4/38. Laborer, Spartanburg Dist. 1860 census. Enl. Clinton 8/22/61 age 22. Present through 8/31/61. Ab. detached Adams Run through 12/31/61. Present 3-4/62. Ab. sick in Church Flats hospital 5-6/62. d. in hospital Church Flats 7/4/62. Bur. Cedar Shoals Bapt. Ch. Cem., Spartanburg Co.

CROW, JOHNATHAN JANTRY. Pvt., Co. F. b. S. C. 1822. Enl. Pickens CH 12/4/61 age 40. Present through 4/62. Ab. sick in hospital 5-6/62. Present 7/62-2/64. Transferred Capt. Hands Co., 25th S. C. Inf. 4/30/64. Joined 5/13/64. NFR. Farmer, age 50, Keowee, Oconee Co. 1870 & 1880 censuses. Living Cherokee Co., S. C. 1900.

CROW, JOHN F. Pvt., Co. F. b. Spartanburg Dist. 10/19/45. Enl. circa 1/64, when present. AWOL 5/6-9/4/64. Present 11-12/64, dismounted since 4/1/64. Paroled as Teamster Greensboro, N. C. 5/1/65. Miller, Glen Springs, Spartanburg Co. 1870 & 1900 censuses. Farmer, Fair Forest, Spartanburg Co. d. Spartanburg 11/20/03.

CROW, LABARON SIMPSON. 3rdSgt., Co. B. b. S. C. 10/11/37. Enl. Clinton 8/22/61 age 22, as 4thCpl. Present through 4/62, promoted 3rd Cpl. Present 5-6/62, promoted 2ndCpl. On picket Little Brittian 7-8/62. Present 9-10/62. On picket Ellis Ford 11-12/62. Promoted 4thSgt. Present 3-8/63. Ab. on horse detail 9-10/63. Returned to duty 12/6/63. Present through 2/64. On detached duty Green Pond 4/16-30/64. Present 5-6/64. Promoted 3rdSgt. Present 7-8/64. Ab. sick McLoud's House hospital 10/24-31/64. Present 11-12/64. NFR. Farmer, Cross Anchors, Spartanburg Co. 1880 census. d. 11/28/17. Bur. Cedar Shoals Bapt. Ch. Cem., Hobbyville, Spartanburg Co., S. C.

CROW, RANDOLPH C. Pvt., Co. B. b. 1803. Not on muster rolls. d. 3/31/71. Bur. Crow family cem., Franklin Co., Ga. Tombstone only record of service.

CROW, ROBERT CUNNINGHAM. Pvt., Co. B. b. near Enoree, S. C. 2/26/40. Farmer. Enl. Camp Ripley 1/20/62. Present through 6/62. Ab. as Courier at Railroad Station 7-8/62. Ab. on march with horses from Summerville, S. C. to Staunton, Va. 9-10/62. Present 11/62-3/63. WIA (left hip) 7/863 at Boonsboro Gap, Md. Ab. wounded in Staunton hospital through 8/31/63. Ab. on detached service 9-10/63. Ab. sick in hospital 11-12/63. Returned to duty 1/16/64. On duty Hanover Junction, horse died 12/30/63. Remounted 4/15/64. Present through 8/64. Present sick in camp 9-10/64. Present 11-12/64. Paroled Greensboro, N. C. 4/26/65. Farmer, Woodruff, Spartanburg Co. 1900 & 1920 censuses, d. Spartanburg 11/3/22. Bur. Bethel Ch. Cem.

CULLER, HENRY L. 2ndCpl., Co. E. b. Lexington, S. C. 1840. Farmhand, age 20, Lexington Dist. 1860 census. Enl. Ft. Motte 10/26/61 as Pvt. On picket White Point 12/31/61. Present 3-6/62. On scout Edisto Island 7-8/62 & appointed 3rdCpl. 8/13/62. Present sick in camp 9-10/62. Present 11-12/62, promoted 2ndCpl. Present 3-4/63. WIA (left hip & abdomen) Gettysburg 7/3/63. DOW's in Cavalry Camp hospital 7/9/63, age 23.

CUMMINGS, THADEUS J. Pvt., Co. I. b. Sumter, S. C. 5/5/46. Served in Co. K, 3rdBn. Palmetto Arty. Enl. James Island 6/16/64. Present sick in camp through 6/30/64. Present 9-12/64. Served to 4/14/65. d. Oswego, Sumter Co. 11/16/16. Wife receiving pension Oswego, Sumter Co. 4/26/19.

CUNNINGHAM, JOHN. Pvt., Co. E. b. Hancock Co., Ga. 11/23/18. Enl. Charleston 2/4/62. Ab. detached at Brigade Headquarters 3-4/62. Discharged by order Gen. Evans being a member of the State Legislature 6/25/62. Age 43, 5'5", fair complexion, blue eyes, light hair, Lawyer. d. 3/10/93. Bur. Cunningham Memorial Cem., Laurens Co., S. C.

CUSHMAN, ELBERT B. Pvt., Co. C. b. Vermont circa 1841. Enl. Co. H, 7th S. C. Inf. Aiken 6/4/61. Present until ab. sick in hospital 10/26-31/61. Discharged Richmond 12/23/61. Enl. Co. C, 1st S. C. Cav. Wadmalaw Island 4/14/62. Present sick in camp 5-6/62. Ab. on sick leave 7/5-12/31/62. Ab. on detached service 3/63-4/64. Transferred Co. A, Lucas's 15th Bn. S. C. Arty 5/16/64. Present through 6/64. KIA near Battery Pringle 7/1/64 in engagement with gunboats.

CUSHMAN, J. C. Pvt., Co. C. b. S. C. circa 1840. Not on muster rolls. Transferred Captain Ayers Co., Lucas' 15th Bn. S. C. Arty. 5/10/64. Not on muster rolls of that unit. d. 3/8/15 age 75. Bur. Darien Bapt. Ch. Cem., Aiken Co.

CUSHMAN, JABEZ B. Pvt., Co. C. b. S. C. 5/10/39. Enl. Wadmalaw Island 3/19/62. Present through 12/62. Ab. sick at home 3-4/63. Ab. sick in Richmond hospital 7/20-10/6/63. Ab. recruiting horses 11/20/63-2/64. Present 3-10/64. Present as Wagonmaster 11-12/64. Paroled Augusta, Ga. 5/20/65. Planter, Aiken, S. C. 1900 census. d. 3/7/05. Bur. Levels Bapt. Ch. Cem., Aiken Co.

DAILEY, CHARLES S. 2ndLt., Co. E. b. S. C. circa 1840. Res. Orangeburg Co. Enl. age 21. Appointed Asst. Surgeon 1/63. NFR.

DAILEY, FRANCIS B. Sgt., Co. E. Res. Orangeburg Co. Enl. by 1862. Discharged for disability 1862.

DANDRIDGE, WILLIAM NATHANIEL. Pvt., Co. I. b. S. C. 5/7/48. Res. Colleton Dist., 1860 census. Enl. Columbia 4/12/64. Detached Green Pond through 4/30/64. Present 5-8/64. Ab. sick in hospital 10/20-31/64. Present 11-12/64. NFR. Student, Grover, Colleton Co. 1870 census. Farmer, Sheridan, Colleton Co. 1900 census. Farmer, Grover, Colleton Co. 1910 census. Living Cottageville, Colleton Co. 1930 census. d. Grover, Colleton Co. 3/16/32. Bur. Cottageville Cem.

DANIELS, COMMODORE. Pvt., Co.'s C & K. Res. Barnwell Dist. Enl. Hamburg 8/28/61 age 40. Present until ab. on extra duty as Nurse in hospital 11-12/62. Transferred Co. K, 6/27/62. Present through 8/62. Ab. detached service as Farrier 9-10/62. Present sick in camp 11-12/62. Present 3-4/63. Ab. sick with debility & neuralgia in Staunton hospital 7/1/63 until d. near Staunton 6/64. However, listed as a Farmer, Williamsburg, Sumter Co. 1870 census. May have been discharged because of age and illnesses.

DARBY, CHARLES SINKLER. 2ndLt., Co. E. b. S. C. 7/14/37. Gd. Medical College of Charleston 1860. M.D., age 22, Orangeburg Dist. 1860 census. Enl. Ft. Motte 10/26/61 as 1stSgt., age 27. Promoted Brevet 2ndLt. 11/1/61. Commanding picket White Point 10/26-12/31/61. Ab. on leave 1-2/62. Present 3-6/62. Commanding Company 7-8/62. Acting Adjutant of Regiment 11-12/62, however, resigned 11/13/62. Appointed Asst. Surgeon from South Carolina 4/21/63. Assigned to Hood's Escort 5/30/64, S. D. Lee's Escort 8/31/64, & Hood's Escort 11/64. Assigned to Tupelo, Miss. hospital 11/31/64-1/18/65. Assigned to Gen. Wood's Escort. NFR. M. D., Pine Grove, Orangeburg Co. 1870 census and Fairfield Co. 1880 census. d. 5/11/94. Brother of Francis R. Darby.

DARBY, FRANCIS BRAMAR. 1stSgt., Co. E. b. S. C. 1812. Farmer, Orangeburg Dist. 1860 census. Enl. Ft. Motte 10/26/61, age 49, as 2ndSgt. Promoted 1stSgt. Ab. on sick leave 12/8-31/61. NFR. Probably discharged.

DARBY, FRANCIS RANDOLPH. Pvt., Co. E. b. S. C. circa 1845. Student, age 15, Orangeburg Dist. 1860 census. Enl. Ft. Motte 10/26/61 age 18. Present through 4/62. Discharged for underage, age 15, 6/26/62. 5'11", dark complexion, dark hair, dark eyes, Student. Enl. Marion, S. C. Arty. 3/19/63. Present through 10/63. Admitted to Arsenal Academy (Citadel) Columbia, S. C. 1/25/64. Present until ab. sick 3-6/4/64. Present until ab. on leave 10/22-30/64. Present 11-12/64. NFR. Brother of Charles S. Darby.

DARBY, JIM. (Colored). Servant, Captain Trezevant on pension application. Res. Ft. Motte, S. C. 5/1/23.

DARBY, JOHN B. Pvt., Co. E. b. S. C. circa 1822. M. D. Enl. Ft. Motte 10/26/61 age 37. Present through 12/31/61. Ab. on leave 1/623. Present 3-4/62. Ab. sick at home 5/4-6/30/62. Present sick in camp 7-8/62. Ab. on orders of Gen. Evans 9-10/62. AWOL 11-12/62. Transferred Charleston Light Dragons (Co. K, 4th S. C. Cav.) 3/26/63. Joined 6/1/63. Present through 8/63, however, ab. sick 8/13-11/63. Present 12/63. Present sick in camp 1-2/64. Detailed 7/14/64. Ab. sick 8/31/64. Ab. sick at home 9-10/64. NFR. Farmer, age 48, Lyons, Orangeburg Co. 1870 census, and Halleysville, Chester Co. 1880 census.

DAVENPORT, LUDIE P. Pvt., Co.'s B & K. b. Waterloo, Laurens Co. 3/25/25. Res. Laurens Dist. Enl. Co. A, 3rd S. C. Bn. Inf. Columbia 12/5/61. Discharged 6/3/62. Enl. Co. B, 9th S. C. Reserves Clinton as 1stLt. 2/14/63. Hired as substitute and discharged. Enl. Co. K, 1st S. C. Cav. Columbia 12/23/63. Ab. on leave through 1/64.Transferred Co. B 2/18/64. Present until ab. on leave 4/26-30/64. Present 5-8/64. Ab. sick at home 9/26-10/31/64. AWOL 11/28/-12/31/64. Paroled Greensboro, N. C. 4/26/65. 6', dark complexion, dark eyes, dark hair. d. 10/13/02. Bur. Waterloo Cem., Laurens Co.

DAVIDSON, JOHN D. Pvt., Co. D. Enl. Chester CH 1864. Present 11-12/64. Paroled Hillsboro, N. C. 4/65.

DAVIS, ALFRED. 4thCpl., Co. I. b. S. C. circa 1832. Enl. Parker's Ferry 4/3/62. On picket Jacksonboro through 4/30/62, & John's Island 5-6/62. Ab. on sick leave 8/27-31/62. Ab. sick and died at a private house near Staunton, Va. 11/29/62 age 30.

DAVIS, EDWARD A. 1stCpl., Co. I. b. Abbeville Dist. 1816. Enl. Parker's Ferry 4/3/62 as Pvt. On picket Jacksonborough through 4/30/62. & John's Island Ferry 5-6/62. Ab. as Courier for Captain Ryan 7-8/62. Present 9-10/62, promoted 2ndCpl. On scout 11-12/62, promoted 1stCpl. Present 3-8/63. Ab. sick with chronic rhumatism in Charlottesville hospital 9/2-11/15/63. Ab. detailed horse recruiting station 12/22-31/63. Ab. on leave 3/14-31/64. Present 5-8/64. Deserted from Camp, Mt. Pleasant 8/31/64. Reduced to Pvt. Deserted from guard tent 12/29/64. NFR. Farmer, age 59, Centerville, Anderson Co. 1870 census and age 69, 1880 census.

DAVIS, HENRY. Cook. Colored. Served as Cook and Servant for John W. Lyles. Receiving pension Fairfield Co. 1919.

DAVIS, W. J. Pvt., Co. G. b. S. C. circa 1838. Clerk, age 22, Chester, 1860 census. Enl. Abbeville CH 1/8/62, age 24.. Ab. on sick leave 4/26-30/62. On picket 5-6/62. Present 9-12/62. Ab. on detached service

3-4/63. Present through 10/31/63. Ab. sick with acute diarrohea in hospital 12/28-31/63 & 1-2/64. Furloughed for 30 days 1/20/64. AWOL 4/27-30/64. Present 5-8/64. Present on detail 9-10/64. Present 11-12/64. NFR. On Abbeville veterans roster 1902.

DAVIS, W. R. Pvt., Co. A. b. S. C. circa 1826. Enl. 1861 age 35. Reenlisted Co. K, 4th S. C. Cav. KIA Armstrong's Mill, Va. 10/1/64.

DAVIS, WILLIAM. Pvt., Co. B. b. S. C. circa 1831. Enl. Clinton 8/22/61. Ab. sick 10/31/61. Present 3-6/63. Present sick in camp 7-8/62. Present 9/62-8/31/63. Horse killed 9/13/63. Ab. on horse detail 10/31-12/31/63. Present 1/1-6/64. AWOL 8/28-31/64. Present 9-12/64. NFR. Farmer, age 39, Fair Forest, Spartanburg Co. 1870 census. Entered Old Soldier's Home, Columbia from Anderson Co. 916/09. NFR.

DEALE, JAMES ALEXANDER. Pvt., Co. G. b. S. C. 11/26/44. Farmhand, age 16, Abbeville Dist. 1860 census. Enl. Abbeville CH 1/8/62. On picket Simmons House 3-4/62. Present 5-10/62. On scout 11-12/62. Ab. on guard duty 3-4/63. Present through 8/31/63. Ab. sick in hospital 9-10/63. Furloughed from hospital for 40 days 12/5/63. Present 3/64. Ab. on leave 4/25-30/64. Present 5-10/64. Detached Pocotaligo 12/10-31/64. Paroled Augusta, Ga. 5/18/65. Farmer, Hibler, Edgefield Co. 1880 & 1900 censuses. d. Greenwood Co. 6/28/21. Bur. Bethany Ch. Cem., McCormick Co., S. C.

DEAN, EDWARD F. Pvt., Co. G. b. S. C. circa 1846. Age 14, Edgefield Dist. 1860 census. Enl. 9th S. C. Inf. 7/4/61 age 16. Present through 12/61. Transferred Co. D, 6th S. C. Inf. Discharged for under age 5/15/62. Enl. Co. G, 1st S. C. Cav. Camp Jackson, Legres Point 7/1/62. Present through 8/31/63, however, ab. sick in hospital 6/17/63. Horse killed 6/30/63. Ab. sick with hernia in hospital 9-10/27/63 & 1-2/64. Furloughed to Abbeville for 30 days 1/21/64. Present 3-12/64. NFR. Farmer, age 24, Hinds Co., Miss. 1870 census.

DE CARADENC, FRANCOIS "FRANK." Pvt., Co. C. b. S. C. 1841. Gd. Citadel 1860. Recommended for appointment as Lt. 3/11/61. Enl. 7th S. C. Inf. WIA & captured Gainesville, Va. but not on muster rolls. Enl. Co. C, 1st S. C. Cav. Wadmalaw Island 3/18/62. Present through 6/62. Ab. on sick leave 8/17-31/62. Applied for appointment as 2ndLt. in Regular Artillery 8/62. Ab. sick in Richmond hospital with typhoid fever 10/18/62. Deserted from hospital 11/15/62. Served as a Scout. d. of Catarrah in Staunton hospital 12/4/62 age 21. Bur. St. Lawrence Cem., Charleston, S. C. Brother of John Antonio De Caradenc.

DE CARADENC, JOHN ANTONIO. Pvt., Co. C. b. Barrwell Dist. 1845. Res. Johnson's, S. C. Enl. Charleston 11/14/61 age 16, as substitute for William Bennett. Age 16, 5'6", dark complexion, black hair, black eyes, Student. Detached on Wadmalaw Island through 12/61. Ab. on leave 5-6/62. Present 7-8/62. Ab. sick in Richmond hospital with typhoid fever 10/19/62. d. Winder hospital, Richmond 11/3/62 age 16. Bur. St. Lawrence Cem., Charleston, S. C. Brother of Francois De Caradenc.

DENDY, CHARLES NEWTON. Pvt., Co. G. b. S. C. 1846. Student, age 14, Abbeville Dist. 1860 census. Enl. Columbia 4/14/64. Present through 6/64. Present sick in camp 7-8/64. Ab. sick in Charleston hospital. Furloughed for 60 days 9/19-10/31/64. Detached Pocotaligo 12/10-31/64. Paroled Greensboro, N. C. 4/26/65. In Old Soldier's Home, New Orleans, La. 1920 age 75.

DESEL, CHARLES M. Pvt., Co. A. b. S. C. 7/10/29. On post war roster. Reenlisted Co. K, 4th S. C. Cav. Grahamville 3/25/62. Present until detached with Gen. Beauregard 1/20/63 as Clerk in Charleston Arsenal until dropped from rolls 4/9/64. Sent to Trans- Mississippi Dept. Paroled Cheraw, S. C. 7/15/65. Moved to Galveston, Texas. d. Dickinson, Galveston Co., Texas 9/15/10. Bur. Dickinson Cem.

DESEL, JOHN B. Pvt. Co. A. b. S. C. circa 1840. On post war roster. Reenlisted Co. K, 4th S. C. Cav. Grahamville 3/25/62 age 22. Present until ab. on sick leave 6/18/62 for 15 days. Present until ab. on leave 8/15/62 for 20 days. Present 9/62 until WIA & captured Cold Harbor 5/30/64. Sent to Lincoln Hospital, Washington, D. C. Sent to Elmira. Exchanged 11/15/64. Paroled 4/65. Farmer, Brazoria, Texas 1870 census.

DE SHIELDS, AUSBURN YOUNG. Pvt., Co. B. b. Greenville Dist. 12/12/40. Enl. Clinton 8/22/61 age 20. On picket Welborn Bluff 11-12/61, however, ab. sick 12/31/61. Present 3-6/62. Present as Courier for Col. Twiggs 6-7/62. Horse died 7/25/62. Ab. on march with horses from Summerville, S. C. to Staunton, Va. 9-10/62. Present 11/62-8/31/63. Ab. on horse detail 9-10/63, sold horse 9/3/63. Ab. 11-12/63, remounted 12/3/63. Ab. detailed recruiting camp for horses 1-2/64. Present 3-12/64. Paroled Greensboro, N. C. 4/26/65. Farmer, Reidsville, Spartanburg Co. 1900 census & Chick Springs, Greenville Co. 1920 census. d. Greenville, S. C. 1/21/22. Bur. Wood Chapel Meth. Ch., Spartanburg Co.

DICKEY, WILLIAM O. Pvt., Co. B. b. S. C. circa 1846. Age 14, Greenville Dist. 1860 census. Enl. Co. F, 3rd S. C. Reserves for 90 days Greenville 11/13/62. Present through 12/31/62. NFR. Enl. Co. B, 1st S. C. Cav. Columbia 2/5/64. On duty Green Pond 4/26-30/64. Present 5-10/64. Ab. on sick leave 11/5-12/31/64. NFR. Alive 1898.

DICKS, ENOCH E. 1stSgt., Co. C. b. Beech Island, S. C. 11/42. Student, age 17, Edgefield Dist. 1860 census. Enl. Hamburg 8/27/61 as 3rdSgt. Ab. on leave 12/26-31/61. Present 5-6/62, promoted 2ndSgt. Present 7-8/62, promoted 1stSgt. Present 9/62-4/63. Ab. sick with debility in Charlottesville hospital 5/31-7/29/63. Ab. on horse detail 8/15-31/63. Ab. sick at home 9-12/63. Present 1-12/64. NFR. Farmer, Augusta, Ga. 1870 census. d. 1875. Bur. family cem., Beech, Island, S. C.

DILL, BART. Pvt., Co. A. Enl. Abbeville CH 3/14/62. Present through 6/62. Present sick in camp 7-8/62. Present 9/62-8/31/63, however, WIA & ab. on leave 9-12/63. Present 1-12/64. NFR.

DINKINS, WINFIELD SCOTT. Pvt., Co. I. b. S. C. circa 1848. Enl. Co. B, 5th Bn. S. C. Reserves 10/6/64. Enl. Co. I, 1st S. C. Cav. by 1865. Paroled Greensboro, N. C. 4/26/65. Farmhand, age 24, Sumter, Sumter Co. 1870 census. Farmer, age 32, Swimming Pens, Sumter Co. 1880 census. Stenographer for Ice Factory, Sumter Co. 1910 census. d. 7/29/13. Wife receiving pension Sumter, S. C. 5/15/19.

DIXON, JAMES HOLMES. Pvt., Co. I. b. S. C. 1/7/49. Student, age 11, Sumter, Sumter Dist. 1860 census. Enl. 2/65 on wife's pension application. NFR. Farmer, age 21, Bishopville, Sumter Co. 1870 census. d. Columbia, S. C. 9/20/07. Wife receiving pension Columbia 3/4/20.

DIXON, WILLIAM K. Pvt., Co. I. b. S. C. circa 1846. Res. Sumter Dist. 1860 census. Enl. Co. K, 3rd S. C. Palmetto Arty. 11/1/63. Present through 4/64. Transferred Co. K, 1st S. C. Cav. 6/16/64 James Island. On Courier duty through 6/30/64. Present sick in camp 7-8/64. Ab. on sick leave for 60 days 12/26/64. NFR. Merchant & Farmer, age 24, Bishopville, Sumter Co. S. C. 1870 census.

DOBEY, WILLIAM E. Pvt., Co. C. b. S. C. 1/6/20. Enl. Columbia 3/30/64. Detached at Green Pond 4/16-30/64. Ab. sick in hospital 6/4-7/31/64. Detailed Enrolling Dept., Edgefield Dist. 8/25-11/13/64. Present through 12/31/64. NFR. Farmer, Butlers, Edgefield Co. 1870 & 1880 censuses. d. 5/5/04. Bur. Sweetwater Cem., Edgefield Co.

DOBY, JOHN. Pvt., Co. C. Enl. Aiken Dist. age 35 on post war roster. Alive 1898.

DOD, J. Pvt., Co. E. On postwar roster. Res. Orangeburg Co.

DODD, W. FRANK. Pvt., Co. E. Conscript assigned 7/12/64. Present sick in camp & dismounted 7/31/64. Present dismounted 9-10/64. Ab. detailed to work at Stone Bridge 12/3-31/64, dismounted. Paroled 4/16/65. d. 2/23/94. Wife receiving pension West Union, Oconee Co. 9/23/19.

DODDS, H. M. On Confederate Veterans list Chester Co. 1904.

DODDS, REUBEN MOSE. Pvt., Co. D. b. Chester Dist. 6/4/42. Enl. Chester CH 9/10/61 age 17. Present through 10/62. Ab. on scout 12/25-31/62. Present through 2/64. Ab. on leave 4/28-30/64. Present 5-10/64. Present in arrest by order Col. Black 11-12/64. Paroled Hillsboro, N. C. 4/65. Farmer, Baton Rouge, Chester Co. 1910 census. d. Chester 6/12/18. Bur. Chester Co.

DOHERTY, HENRY. Pvt., Co.'s C & K. b. Boston, Mass. 3/42. Enl. Co. C Kaolin 8/27/61. Present through 12/31/61. On picket Bear Island 1-2/62. Transferred Co. K 6/27/62. Present through 4/63. Ab. detailed to work in crockery ware factory 5/24/63-4/64. AWOL 5-6/64. Ab. detailed permanently at Kaolin factory 7-8/64. NFR. Moved to Richmond Co., Ga. 5/2/65. Potter, age 36, Philadelphia, Pa. 1870 census & 1880 census. Living Atlantic City, N. J. 1890 census. d. there 1904.

DOLIN, MICHAEL. Pvt., Co. E. b. circa 1829. Enl. Adams Run 3/20/62, age 31, res. Orangeburg Dist. Present through 6/62. Ab. on sick leave 8/14-31/62. Ab. sick Manchester hospital 9-10/28/7/62. Furloughed for 30 days 10/28/62. Ab. on scout 12/62. Present 3-8/63. Present as Teamster 9/63-2/64, dismounted. Ab. on horse detail 3/31-4/30/64. Present 5-10/64, dismounted 10/20-31/64. On detail Pocotoligo 12/10-31/64, dismounted. Paroled Hillsboro, N. C. 4/65. Alive 1898.

DORSEY, CHARLES B. Pvt., Co. D. Enl. Chester CH 7/18/62. Present through 10/63, however, ab. sick with "Fever Tertiana" in Richmond hospital 7/2-30/63. Ab. on horse detail 12/22/63-2/64, dismounted 1 month, however, ab. sick with chronic diarrohea in Richmond hospital 11/30-12/22/63. Present 3-4/64. Present sick in camp 5-6/64. Present 7-12/64. Paroled Hillsboro, N. C. 4/65.

DOUGLASS, GEORGE A. 3rdSgt., Co. G. b. S. C. circa 1840. Res. Abbeville Dist. 1860 census. Enl. Abbeville CH 1/8/62 as 1stCpl. Ab. sick in hospital 3-4/62. On picket 5-6/62. Present 7-8/62. Present 9-10/62, promoted 4thSgt. 9/15/62. Present 11-12/62, promoted 3rdSgt. 11/27/62. Present 3-6/63. Ab. on horse detail 8/31/63. Present 9/63-2/64. Ab. on leave 4/27-30/64. Present 5-6/64. Ab. on detached service 8/25-31/64. Present 9-10/64. Detached at Pocotaligo 12/10-31/64. NFR.

DOUGLASS, JOHN C. Pvt., Co. A. b. S. C. circa 1842. Enl. Abbeville CH 3/19/62. Present through 6/62. Present sick in camp 7-8/62. Present 9/62-6/63, dismounted 5 days. Ab. on horse detail 8/18-31/63. Ab. on sick leave 9-10/63. Ab. sick in hospital 1/17-2/29/64, dismounted since 9/25/63. Present 1-5/64. Transferred Co. H, 2nd S. C. Arty. 6/23/64. Joined James Island 7/1/64. Present through 12/64. NFR. Laborer, Atlanta, Ga. 1880 census.

DOUGLASS, JOSEPH. Pvt., Co. H. b. S. C. circa 1845. Age 14, Chesterfield Dist. 1860 census. Enl. Columbia 3/30/64. Present through 4/64. Present sick in camp 5-6/64. Ab. sick in hospital 8/11-31/64. Ab. sick 9-10/64. AWOL 11-12/64. NFR. Mason, age 65, New Orleans, La. 1910 census.

DOUGLASS, THOMAS D. Pvt., Co. G. b. S. C. circa 1839. Enl. Abbeville CH 3/15/62. Ab. sick in hospital, Adams Run until d. of disease there 4/1/62 age 25.

DOY (DAY), W. C. Pvt., Co.'s D & K, 1st S. C. Cav. On post war roster.

DRAYTON, CHARLES ELLIOTT ROWLAND. Pvt., Co. A. b. Dorchester Co. 5/26/36. Gd. U. Va. 1855. Enl. 11/2/61. Left end of enlistment 2/7/62. Enl. Co, A, Rutledge Mounted Rifles, 7th S. C. Cav. at Pocotaligo 3/26/62. Captured near Old Church, Va. 5/30/64. Exchanged 9/30/64. NFR. Farmer, Swimming Pens, Sumter Co. 1870 census & Aikens, S. C. 1880 census. d. Aiken Co. 2/7/88. Bur. St. Thadeus Cem.

DRENNAN, HORACE. Asst. Surgeon. b. S. C. 1/3/36. Gd. Jefferson Med. College, Philadelphia 1859. Enl. Co. C, 7th S. C. Inf. Abbeville CH 4/15/61 age 24 as Pvt. Present through 9/61. Ab. sick with "Ictues" in Danville hospital 1/24/62. Sent to Warrenton, Va. hospital. NFR. Enl. Co. G, 1st S. C. Cav. Abbeville CH 3/7/63 as Pvt. Ab. on detached service 3-4/63. Acting Asst. Surgeon 6/9-8/31/63 & 9/63-2/64. Appointed Asst. Surgeon PACS from S. C. 7/26/64. Ab. on leave 8/27-31/64. Present 9-10/64. Surgeon in charge of hospital Walhalla, S. C. 1/10-3/3/65. NFR. M.D., age 44, Cedar Springs, Abbeville Co. 1870 census. and Verdery, Greenwood Co. 1900 & 1910 censuses. d. Verdery, S. C. 9/8/11. Bur. Cedar Spring Cem., Greenwood Co., S. C.

DRIGGERS, DANIEL F. Pvt., Co. I. b. S. C. 1825. Res. Colleton Dist. 1850 census. Enl. Parker's Ferry 4/17/62. Ab. on leave 4/27-31/62. Detailed as Nurse 5-6/62. Courier for Capt. Ryan 7-8/62. Ab. sick with rheumatism in Richmond hospital 10/18-11/11/62. Present until ab. sick with pneumonia in Richmond hospital 1/6/63. Ab. sick in hospital 3-4/63. Ab. on horse detail 4/6-6/6/63, dismounted since 4/20/63. Ab. sick in Staunton hospital 5-6/63-2/64. Present 5-10/64. AWOL 12/11-31/64. Deserted to the enemy Charleston 2/18/65. Age 47, 5'8", dark complexion, dark eyes, black hair, took oath and released 3/15/65. Farmer, St. Paul's Parish, Colleton Co. 1870 census. Farmer, St. James, Goose Creek, Charleston 1880 census. Res. Charleston 1889.

DROZE, THOMAS N. 3rdCpl., Co. I. b. S. C. 2/10/34. Enl. Parker's Ferry 4/3/62 as Pvt. Present through 4/30/62. Ab. on sick leave 8/2-31/62. Ab. sick Chester hospital 9/62-4/63. Present through 8/31/63. Ab. on detail recruiting horses 9/63-2/64. Detached Green Pond 4/16-30/64. Present 5-8/64, promoted 4thCpl. Present 9-10/64, promoted 3rdCpl. Present 11-12/64. NFR. d. 4/25/18. Bur. Old Midway Ch. Cem., Bamberg Co., S. C.

DRUMMOND, IRA L. Pvt., Co. B. b. Spartanburg Dist., S. C. 1847. Enl. Columbia 4/14/64. Detached Green Pond 4/26-30/64. Present 5-12/64. Discharged, Yorkville 4/28/65 on pension application. Farmer, Woodruff, Spartanburg Co. 1870 & 1880 census. Grocer, Woodruff, Spartanburg Co. 1900 census. d. 1903.

DRUMMOND, SEABORN SIMPSON. 4thCpl., Co. B. b. near Woodruff, Spartanburg Dist. 9/20/44. Res. Spartanburg Dist. 1860 census. Enl. Adams Run 9/28/62 as Pvt. Present through 8/31/63. Ab. on horse detail 9-10/63. Present 12/6-31/63. Detailed horse recruiting camp 2/4-29/64. Detached Green Pond 4/16-30/64. Present 5-8/64, promoted 4thCpl. Present 9-12/64. Paroled Yorkville 4/28/65 on pension application. Farmer, Woodruff, Spartanburg Co. Applied for pension 4/9/19. d. Spartanburg 7/30/22. Bur. Bethel Cem., Spartanburg Co.

DU BOSE, SAMUEL F. Pvt., Co. K. b. S. C. circa 1845. Paroled Raleigh, N. C. 4/65. Farmer, postwar. d. Oktibbeha, Miss. 1/80.

DUFFIE, J. E. Pvt., Co. D. Captured and died of disease David's Island, N. Y. 8/30/63.

DUFFORD, EPHRAIM. Pvt., Co. E. Res. Orangeburg Dist. Paroled Hillsboro, N. C. 4/65.

DUFFORD, EUGENE. Pvt., Co. E. b. France 1821. Minister, Pine Grove, Orangeburg Dist. 1860 census. Enl. Ft. Motte 10/26/61 age 40. Present through 12/61. Ab. on leave 8 days 1/62. AWOL 3-4/62. Present 5-6/62. On duty "Pine Berry" 7-8/62. Present 9-10/62, however, AWOL 2 weeks in Sept. Ab. sick in hospital 11-12/62. Furloughed from Farmville hospital 12/21/62 for 30 days. Ab. sick with chronic rheumatism 4/21-5/21/63. Present through 12/63, dismounted 8/15-12/4/63. Ab. on leave Milford Station 2/27-29/64. Present 3-12/64. NFR. Minister, Pine Grove, Orangeburg Co. 1870 census. Widow receiving pension Union, Tenn. 1914.

DUKE, RANSOM. 4thSgt., Co. F. b. S. C. 9/11/29. Mechanic, age 32, Pickens CH 1860 census. Enl. Pickens CH 12/4/61 age 25, as Pvt. Present through 4/62, promoted 4thCpl. Present 5-6/62. Presence or absence not stated 7-8/62. Ab. on detached service 9-10/62. Promoted 3rdCpl. Present 11-12/62, promoted 2ndCpl. However, ab. sick with primary Syphilis in Richmond hospital 12/1762-3/1/63. Present 3-4/63, promoted 1stCpl. Present 5-8/63. Ab. on detached service 9-10/63. Present 11/63-2/64. On duty Green Pond 4/16-30/64. Present 5-8/64, promoted 4thSgt. Present 9-12/64. NFR. Blacksmith, Pickens CH 1880 census. d. 3/7/01. Bur. Tabor Meth. Ch. Cem., Pickens Co. Brother of Russell Duke.

DUKE, RUSSELL. Pvt., Co. F. b. S. C. 1840. Mechanic, age 23, Pickens CH 1860 census. Enl. Pickens CH 12/4/61 age 22. Present through 8/62. Ab. on detached service 9-10/62. Ab. sick in Richmond hospital 11-12/62. Present 3-4/63. Captured Upperville 6/21/63. Sent to Old Capitol. Exchanged. Ab. on detached service 8/13-31/63. Present 9-10/63. Ab. on horse detail 12/22/63-2/64. AWOL 4/27-30/64. Present 5-8/64. Ab. detailed guarding POW's Florence 9/17-12/31/64. Paroled as Courier, Greensboro, N. C. 5/1/65. Blacksmith, Pickens CH 1870 & 1880 censuses. d. 7/30/94. Bur. Tabor Meth. Ch. Cem., Pickens Co. Brother of Ransom Duke.

DUKES, JAMES J. 3rdCpl., Co. E. b. S. C. circa 1840. Enl. Adams Run 1/8/62. Present 3-4/62. Present sick in camp 7-8/62. On duty Pine Berry 7-8/62. Present sick in camp 9-10/62. Present 11-12/62. Captured on scout Catlett's Station 3/30/63. Paroled 4/2/63. Returned to duty 4/23/63. Present through 8/31/63. Promoted 3rdCpl. Present 9-10/63. Ab. detached horse recruiting station 12/23/63-2/64. Ab. on leave 4/15-30/64. Present 5-6/64, reduced to 4thCpl. Present 7-10/64. Detailed as guard Wappoo Bridge 12/29-31/64. Paroled Hillsboro, N. C. 4/65. Farmhand, age 30, Calvary, Colleton Co. 1870 census.

DUKES, THADDEUS C. S. Pvt., Co. E. b. Orangeburg, S. C. 1845. Enl. Co. I, 2nd S. C. Inf. (2nd Palmetto) 12/61 as Cpl. NFR. Enl. Co. E, 1st S. C. Cav. Adams Run 4/28/62. Present through 6/62. Ab. sick at home 8/12-31/62. Present 9-10/62. Discharged Staunton, Va. 11/2/62 for disabilty. Age 18, 5', sallow complexion, grey eyes, dark hair, Farmer. Enl. Co. B, 20th S. C. Inf. 3/1/63. Present through 4/63. Ab. on leave 20 day 6/14/63. AWOL 7/14-8/31/63. Ab. sick 9-10/63. Present 11-12/63. AWOL 1/28-2/25/654. Present 3-6/64. Paroled Greensboro, N. C. 5/2/65 as 1stCpl. Farmer, age 24, Branchville, Orangeburg Co. 1870 census.

DUNCAN, DARLING P. ALLEN. 3rdLt., Co. 's C & K. b. S. C. circa 1836. Res. Barnwell Dist. Enl. Co. C Camp Johnson 11/1/61 as Pvt. Detailed recruiting service 12/17-31/61. Detailed with Col. Black at Camp Ripley 1/62. Present as Bugler 5-6/62. Transferred Co. K 6/27/62. Promoted 1stSgt. Present 6-10/62. On picket Richards Ford 11-12/62. Commanding Co. 3-4/63. Ab. sick with hemorrhoids in Williamburg hospital 5/6-7/9/63. Resigned 7/14/63 from chronic cystitis disease of the bladder. NFR. M.D., Burke Co., Ga. 1880 census.

DUNCAN, H. AUGUSTUS. Pvt., Co. K. b. S. C. circa 1843. Student, age 15, Barnwell Dist. 1860 census. Enl. Co. A, 2nd S. C. Arty Ft. Johnson 6/16/64. Present 7-8/64. Transferred Co. K, 1st S. C. Cav. James Island 8/15/64. Present until ab. on sick leave 10/15-31/64. Present 11-12/64. Paroled Raleigh, N. C. 4/65.

DUNLAP, LAWSON H. Pvt., Co. D. b. York Dist. 12/35. Farmhand, age 25, York Dist. 1860 census. Enl. Chester CH 9/10/61 age 25. Present through 8/62. Present sick in camp 9-10/62. Ab. sick in Staunton & Manchester hospitals 11/19-12/5/62. Furloughed for 35 days. Present through 10/63. Ab. on horse detail 12/22/63-2/64. Present 3-8/64. Detailed as Overseer of slaves near Stoney, S. C. 9/7-10/31/64. Present 11-12/64. Paroled Hillsboro, N.C. 4/65. Farmer, Chester Co. 1870 census, and Bethesda, York Co. 1880 & 1900 censuses. Farmer, Union, Union Co., 1910 census.

DUNLAP, NICHOLAS W. Pvt., Co. D. b. Fairfield Dist. 1843. Enl. Chester CH 9/10/61 age 21. Present through 6/62. Ab. on sick leave Chester CH 7/30/62. Present 9-10/62. On scout 12/25-31/62. Detailed as Teamster 2-3/63. Present through 8/31/63. Ab. sick with "Int. Febris" in Richmond hospital 12/10/63-1/4/64. Present through 2/64. Detached Green Pond 4/16-30/64. Present 5-6/64. On picket 7-8/64. Present 9-10/64. Detached Coosawhatchie 12/10-31/64. Paroled Hillsboro, N.C. 4/65. Engineer, Charleston, S. C. 1890.

DUNLAP, WILLIAM THOMAS. Pvt., Co. D. b. S. C. 1843. Age 16, Saluda, Abbeville Dist. 1860 census. Enl. Chester CH 9/10/61 age 19. Pressent through 8/62. Present sick in camp 9-10/62. Present 11-12/62. d. of disease Chester, S. C. 3/18/63.

DUNLOP, F. N. Pvt., Co. H. On post war roster.

DUNLOP, THOMAS N. 4thCpl., Co. H. b. S. C. circa 1837. Res. York Dist. 1860 census. Enl. Rock Hill 1/9/61 age 24, as Pvt. Present through 6/62. Ab. on sick leave 7-8/62, horse died 8/9/62. Present as Teamster for 2nd N. C. 9-10/62. Driving wagon for QM 11-12/62. Ab. on horse detail 2/9-28/63. Present 4/30-8/31/63. Ab. on horse detail 9-10/63. AWOL 1/1-2/29/64, however, returned to duty from Richmond hospital 1/28/64. Present 3-4/64. Ab. on leave 4/25-5/4/64. Present through 6/64, promoted 4thCpl. Present 7-10/64. Detached at Pocotaligo 12/10-31/64. NFR. Barkeeper, York Co. 1870 census.

DUPONT, BOHUN CARTER. Pvt., Co. A. b. Beaufort Co. 3/26/44. Age 17, St. Luke's, Beaufort Co. 1860 census. Att. S. C. College 1860-61. On post war roster. Reenlisted Co. K, 4th S. C. Cav. Grahamville 3/24/62 age 18. Present until ab. on leave 2/28/64 for 10 days. Present until detailed as Brigade Mail

Courier 9/20-10/31/64. NFR. Trucker, Lake City, Fla. 1900. Farmer, Lemon City, Fla. 1920 census. d. Dade Co., Fla. 1929. Brother of Theodore B. Dupont.

DUPONT, THEODORE DEHON. Pvt., Co. A., 1st S. C. Cav. b. Ga. 8/2/36. On post war roster. Reenlisted Co. K, 4th S. C. Cav. Grahamville 5/12/62. Transferred Co. B, 2nd S. C. Cav. Present until detached with Gen. Walker 10/20/62. Present until ab. on sick leave 8/25/63. Present 9-12/63. Dropped from rolls 4/9/64, reduction of company. NFR. d. 4/15/94. Bur. Oakview Cem., Albany, Ga.

DURANT, THOMAS MC CREA. Pvt., Co. A. b. Sumter Dist.1839. Clerk, Sumter Dist. 1860 census. Enl. Co. D, 2nd S. C. Inf. 4/8/61. Resigned 5/13/62. On post war roster Co. A, 1st S. C. Cav. Reenlisted Co. K, 4th S. C. Cav. Pocotaligo 2/13/63. Present until ab. on sick leave 10/10/63-2/64. Present 3-10/64. NFR. d. Sumter Co. 1867.

DYE, L. T. Pvt., Co. K. Enl. Columbia 4/11/64. Present through 4/30/64. Detailed as Courier for Signal Corps, Charleston 6/25-30/64. Detailed driving cattle for QM Dept. 7-12/64. Paroled Greensboro, N. C. 4/26/65.

DYE, WILLIAM C. 1stCpl., Co.'s D & K. b. S. C. 1837. Enl. Chester CH 10/10/61 age 23, as Pvt. Present through 6/62. Transferred Co. K 6/25/62. Present through 12/62, promoted 3rdCpl. Ab. detached Gordonsville 4/10-5//63, promoted 2ndCpl. Present through 8/31/63 & 9/63-2/64. Ab. on leave 4/28-30/64. Promoted 1stCpl. Present 5-8/64. Detailed as Courier, Charleston 9-12/64. Paroled Greensboro, N. C. 5/2/65. Farmer, Halsellville, Chester Co. 1870 & 1880 censuses. d. 4/1/90. Wife receiving pension Shelton, Fairfield Co. 10/7/19.

EADES, MILES A. Pvt., Co. F. b. Pickens Dist. 1843. Farmer, age 18, Pickens Dist. 1860 census. Enl. Colleton Dist. 4/8/62. On picket Jacksonborough through 4/30/62. Present 5-8/62. Ab. on detached service 9-10/62. Present 11-12/62. Ab. detached service 3-4/63. Present through 8/31/63. Captured on picket Robinson's River 10/18/63. Sent to Old Capitol. Transferred Pt. Lookout. In Pt. Lookout hospital with "variolodi" 1/13/65. Escaped 3/24/65. Surrendered Washington, D. C. 4/6/65. Farmhand, age 37, Wallingford, Hardin Co., Ky. 1880 census. d. Rule, Haskell Co., Texas 9/7/07.

EARLE, JAMES WASHINGTON. 2ndLt., Co. F b. S. C. 7/10/27. Farmer, Pickens Dist. 1860 census. Enl. Co. F Pickens CH 12/4/61 age 36, as 2ndLt. Present through 4/62. Ab. on sick leave 6/23-7/13/62. Ab. on sick leave 7-8/62. Resigned for ill health 9/20/62. Reenlisted Co.K, 2nd S. C. Cav. Culpeper CH 8/6/63 as Pvt. Present through 6/64. Detailed to arrest deserters in Anderson Dist. 8/9/64. Detail extended 9/15/64. NFR. Living Savannah, Anderson Co. 1880 census. d. 2/1/01.

EATON, EPHRAM B. Pvt., Co. K. b. S. C. circa 1843. Enl. Co. B, 7th S. C. Cav. Pocotaligo 3/31/63. Transferred Capt. Kirk's Squadron S. C. Cav. 2/25/64. NFR. Enl. Co. K, 1st S. C. Cav. Columbia 5/18/64. Courier, Telegraph Office, Charleston 5/25-30/64. Present 7-8/64. Ab. on leave for 15 days 10/28/64. On picket Chatman's Ford 12/21/31/64. NFR. Farmer, Garvin, Anderson Co. 1870 & 1880 censuses.

EDWARDS, EPAMINODAS E. Pvt. Co. G, 1st S. C. Cav. Enl. Co. D, 7th S. C. Inf. 4/15/61 age 26, Teacher. Discharged for disability 4/20/62. Enl. Co. G, 1st S. C. Cav. Abbeville CH 9/1/62. Present through 11/62. Ab. sick with scabies in Staunton hospital 12/31/62. Present 3-4/63. Ab. sick in Richmond hospital 6/30-7/18/63. Furloughed for 30 days. Ab. on horse detail 8/31/63. Present 9-12/63. Ab. on leave 2/13-29/64. Present 3-6/64. Present sick in camp 7-8/64. Present 9-10/64. Detached Pocotaligo 12/10-31/64. Paroled Greensboro, N. C. 4/16/65. Farmer, Abbeville Co. postwar. d. 8/23/05. Bur. Sharon Meth. Ch. Cem.

EDWARDS, JOHN J. Pvt., Co. H. b. S. C. 10/28/43. Served briefly in Co. H, 1st S. C. Cav. 1861. Enl. Co. A, 17th S. C. Inf. 12/27/61. Surrendered Appomattox CH 4/9/65/. d. 7/31/17. Bur. Neely's Reformed Ch. Cem., York Co.

EDWARDS, JOHN L. 1stCpl. & Musician, Co. H. b. S. C. 1836. Enl. Rock Hill 1/9/61 age 26, as Pvt. Promoted 1stCpl. On picket Jacksonborough 3-4/62. Present 5-6/62. Ab. on sick leave 7-8/62. Present sick in camp 9-10/62. AWOL 11-12/62. Present 3-4/63. Transferred to Adjutant as Bugler 6/1/63. Present through 10/63. Ab. on horse detail 12/22-31/63 & 1-2/64. Present 3-12/64. NFR. Farmer, age 34, Lansford, Chester Co. 1870 census.

EDWARDS, M. LEWIS. Pvt., Co. G. b. S. C. circa 1841. Enl. Abbeville CH 1/8/62. Ab. sick in hospital 3-4/62. Ab. sick 5/11-6/11/62. AWOL 6/12-30/62. Present 7-10/62. On picket 11-12/62. Present 3-8/63, however WIA & in Richmond hospital 7/20-8/863 & on horse detail during this period. Present 9-12/63. Detailed horse recruiting station 1/20-2/29/64. AWOL 4/27-30/64. Present 5-12/64. Ab. sick with scabies in Charlotte hospital 2/9/65. NFR.

ELLIOTT, THOMAS O. Pvt., Co. A. b. S. C. circa 1841. On post war roster. Reenlisted Co. K, 4th S. C. Cav. Grahamville 3/25/62 age 26. Present until ab. on leave 10/28/62 for 15 days. Present until ab. sick 6/22/63-10/31/64. Paroled 4/65. M. D., age 41, Charleston 1880 census. d. Charleston 11/3/91 age 50.

ELLIS, JESSE C. 4thCpl., Co. A. b. S. C. circa 1837. Farmer, ag 23, Abbeville Dist. 1860 census. Enl. Abbeville 8/13/61 age 23. Present through 12/61. d. of disease Wadmalaw Island 6/15/62.

ELMORE, ALBERT RHETT. Pvt., Co. A. b. S. C. 10/23/43. Res. Richland Dist. Att. S. C. College 1860-61. On post war roster. Enl. Co. C, 7th S. C. Bn. Inf. 3/20/62. Appointed Sgt. Major 5/27/62. Transferred Co. K, 4th S. C. Cav. 6/26/63. Joined Green Pond 8/1/63. Present until ab. on leave 2/24/64 for 20 days. Present until ab. on horse detail 8/11/64 for 40 days. Present 10/31/64. Paroled 4/65. Railroad Agent in post war. d. Gainesville, Fla. 3/11/15. Bur. Evergreen Cem., Gainesville, Fla.

ENLOE, WILLIAM B. Pvt., Co. H. b. S. C. 1844. Student, age 15, Yorkville, 1860 census. Enl. Rock Hill 1/2961 age 18. Present through 4/62. Ab. as baggage guard Church Flats 5-6/62. Present as Wagoner for QM Dept. 7-8/62. Present 9-10/62. On raid 11-12/62. Present 3-4/63. Ab. detached with wagon train 4/30-8/31/63. Ab. detailed to recruiting station for horses 8/31-12/31/63. Present 1-2/64. Ab. on leave 3-4/64. Transferred C. S. Navy 5/13/64. NFR. Stationary Engineer, age 52, Hartshorne, Cherokee Ind. Terr. 1900 census.

ERVIN, JAMES. Pvt., Co. H. b. S. C. Res. York Dist. d. age 27. Not on muster rolls.

ERWIN, D. H. or W. Pvt., Co. G. On postwar roster.

ERWIN, JAMES A. Pvt., Co. H. b. York Dist. circa 1826. Farmer, age 34, York Dist. 1860 census. Enl. York Dist. 2/10/64. Present through 8/25/64. Sick in camp 8/26-31/64. Present 9-10/64. Detached Pocotaligo 12/10-31/64. NFR. Farmer, Bethesda, York Co. 1870 & 1880 censuses. d. 7/9/01. Wife receiving pension Rock Hill, York Co. 10/24/19.

ERWIN, JOHN W. Pvt., Co. G. b. S. C. circa 1845. Enl. Camp Jackson 7/1/62. Present through 8/62. Present, dismounted 9-10/62. Present 11/62-12/63. Detailed Commissary Dept. 2/29/64. Present 3-12/64. NFR. Laborer, age 35, York Co. 1880 census. On Abbeville veterans roster 1902.

ERWIN, THOMAS. Pvt., No Co. B. S. C. 1823. Bur. Lebanon Presb. Ch. Cem., Abbeville. U. D. C. records.

ERWIN, WILLIAM JOHN. Pvt., Co. G. b. Ireland 8/21/34. Enl. Abbeville CH 1/8/62. On picket Jacksonborough 3-4/62. Present 5/52-10/63. On horse detail 12/22/63-2/64. Detached Green Pond 4/27-30/64. Ab. on leave 6/22-30/64. Present 7-12/64. NFR. d. 6/12/14. Bur. Mt. Zion Presb. Ch., Anderson Co., S. C.

EUBANK, JOHN S. Pvt., Co. C. b. S. C. 1847. Age 13, Barnwell Dist. 1860 census. Not on muster rolls. d. Cheraw, S. C. 2/15/65 age 22.

EUBANKS, WILLIAM JETER. Pvt., Co. C. b. Union, S. C. 5/3/45. Enl. Columbia 3/30/64. Ab. on horse detail 4/16-30/64. Present 5-8/64. Ab. sick in hospital 9/29-10/31/64. Present 11-12/64. NFR. d.

Campobello, S. C. 7/30/13. Bur. Motlowe Creek Bapt. Ch. Cem., Spartanburg Co.

EVANS, CHARLES D. Pvt., no Co. b. 10/20. Not on muster rolls. Enl. Co. F, 8th S. C. Inf. 3/14/63. Present until ab. sick in Charlottesville hospital 6/9-8/18/63. Present until WIA Chickamaunga 9/20/63. Present 11-12/63. Surrendered Appomattox CH 4/9/65. d. 12/16/23. Bur. Lebanon Presb. Ch. Cem, Abbeville Co. UDC records only citation for 1st S. C. Cav.

EVANS, JAMES. Pvt., Co. A. b. Belfast, Ireland 9/24/42. Enl. Abbeville CH 8/13/61 age 18. Present through 12/61. Nurse in hospital, Adams Run 1/62. Present 3-6/62. Ab. on leave 7-8/62. Present 9/62-4/63. Ab. detached recruiting horses 5-8/63. Present 11-12/63, dismounted since 10/30/63. Present 1-12/64. NFR. Farmer, age 67, Abbeville Co. 1910 census. d. 9/21/10. Bur. Lebanon Presb. Ch., Abbeville Co.

EVANS, JAMES. W. Pvt., Co. A. On postwar roster. Reenlisted Co. K, 4th S. C. Cav. by 1864. Captured Hawes' Shop 5/28/64. Sent to Pt. Lookout. Exchanged 3/15/65. Paroled 4/65.

EVANS, JOHN. Pvt., Co. A. b. Abbeville Dist. 4/15/40. Enl. Abbeville CH 12/23/61. Present until discharged for "General debility, Access of the lungs" 4/28/62. Age 21, 5'8", dark complexion, black eyes, brown hair, Student. Reenlisted Columbia 1/7/64. Present through 12/64. Paroled Charlotte, N. C. 4/65 on pension application. Res. Bordeaux, Abbeville Co. 1880 census. Receiving pension Abbeville Co. 9/8/19.

EVANS, WILLIAM K. Pvt., Co. B. b. S. C. 1837. Enl. Clinton 8/22/61 age 24. On detached service Camp Bee through 10/31/61. Present 3-4/62. AWOL 5-8/62. Present 11-12/62, however, transferred Co. D, 3rd S. C. Bn. Inf. 6/1/62. Age 27, 5' 9", light complexion, brown hair, hazel eyes. Present through 12/62, promoted Cpl. 9/15/62. Present until detailed on South Carolina Rail Road as Mechanic by War Dept. 11/5/63 to end of war.

EVERETT, JOHN D. 1stSgt., Co. C. b. S. C. 1830. Farmer, age 29, Edgefield Dist. 1860 census. Enl. Hamburg 8/27/61 age 31. Ab. on sick leave 12/3-31/61. On sick leave 1/62. Discharged for disability 8/28/62. NFR. Applied to enter Old Soldier's Home, Columbia from Aiken Co. 11/23/10. d. 1911. Bur. Hammond Cem., Beech Island, S. C.

FAIRBAIRN, GEORGE W. Pvt., Co. B. b. S. C. circa 1836. Res. Laurens Dist. Enl. Clinton 8/22/61 age 22. Ab. sick 10/31/61. Present in arrest 3-4/62. Ab. in confinement 7-8/62. Present 9/62-4/63. On detached service with horses in Nelson Co., Va. 3-10/63. Present 11-12/63. Ab. on leave 2/11-29/64 & 4/26/-30/64. Present 5-6/64, detailed in Commissary Dept. 6/15-30/64. &9-12/64. NFR. Merchant, age 34, Hunter, Laurens Co. 1870 census. Married Phillips, Ark. 1878. Alive 1888.

FAIRLY, WILLIAM H. Pvt. Co. A. b. S. C. circa 1837. Law Student, age 23, Laurens Dist. 1860 census. On post war roster. Reenlisted Co. K, 4th S. C. Cav. Grahamville 3/25/62 age 25. Present until ab. on leave 6/22/63 for 15 days. Present until reached with Gen. Beauregard as Clerk 2/19-4/8/64. Present until KIA Trevillian 6/11/64. Bur. Magnolia Cem, Charleston.

FARRINGTON, ORSON J. Quartermaster Sgt., Co. G. b. N. Y. 1833. Tailor, Abbeville 1860 census. Res. Columbia. Enl. Abbeville CH 1/8/62. Present 3-4/62. Ab. in Charleston Acting Commissary for Company 5-6/62. Present 7-8/62. Present as Company Commissary Sgt. 9-10/62. Ab. on detached service as Tailor for Company & then AWOL 11-12/62. Ab. sick with chronic rheumatism in Richmond hospital 2/16-5/26/63, no horse. Ab. sick in Richmond hospital 8/31/63-4/64. Transferred Co. B, Lucas's 15th Bn. S. C. Arty. 5/13/64. Joined 5/20/64. Present through 12/64, promoted QM Sgt. 11/27/64. Paroled Raleigh, N. C. 4/15/65. Sent to Ft. Monroe. Took oath and transportation furnished to Albany, N. Y. Tailor, Albany, Ga. 1870 & 1880 censuses. Paid taxes there 1882.

FARRIS, ARTHUR B. Pvt., Co. H. b. King's Mountain, York Dist. 1/10/35. Enl. Rock Hill 1/9/61 age 27. Present through 12/62. Transferred Co. C, 5th S. C. Cav. in exchange for A. A. Rose 1/29/63. Present through 8/63. Ab. sick with "Febris Remittens" in Episcopal Church Hospital, Williamsburg 9/28/63. Transferred to Richmond hospital. WIA (slight) 5/6/64. Ab. sick with abcess on head in Richmond hospital

6/20/64. Transferred Huguenot Springs Hospital 7/8/64. Present 9/64-2/65. Paroled Appomattox CH 4/9/65. Farmer, King's Mountain, York Co. 1870 & 1880 censuses. d. 12/10/80. Bur. Enon Bapt. Ch. Cem., York Co.

FAUST, HENRY MEDICUS. Asst. Surgeon. b. Barnwell Dist. 2/8/35. Appointed Asst. Surgeon PACS from South Carolina 3/5/62. Appointed Asst. Surgeon 1st S. C. Bn. Cav. 4/18/62 to rank from 2/24/62. Present through 5/31/63. Ab. sick at home 8/31/63. Assigned to 5th S. C. Cav. 11/28/63. Detached at Adams Run by order Gen. Butler 4/7/64. Ab. sick leave for 30 days 8/8/64. Ab. sick 9-10/64. Present Stony Creek 11-29-30/64. Ab. sick 12/31/64. NFR. M. D., Barnesville, S. C. 1870 census. d. 11/9/97. Bur. Springtown Bapt. Ch., Bamberg Co.

FAYSSOUX, TEMPLAR SHUBLICK. Captain & Quartermaster. b. S. C. circa 1825. Gd. West Point 1839. Served Ft. Moultrie, S. C. 1856-57. Appointed Lt. of Cavalry 4/11-17/61. Served in DeSaussure's Sqd., S. C. Cav. Appointed Jr. 1stLt., Co. A, 15th S. C. Arty. 6/5/61. Present until promoted Captain 10/10/61. Present until appointed Captain & Quartermaster, 1st S. C. Cav. 12/15/61. Failed to give bond and dropped 2/24/62. Appointed 1stLt. of Artillery 11/6/62. Appoint Captain & Ordnance Officer, Evan's S. C. Brigade 12/28/62. Transferred to Elliott's S. C. Brigade 7/5/64. Surrendered Appomattox CH 4/9/65 as J. S. Fayaoux, Captain & Ordnance Officer, Wallace's S. C. Brigade. d. near Philadelphia, Pa 5/8/29

FELDER, SEBASTIAN FLETCHER. 3rdSgt., Co. E. b. Lone Star, Calhoun Dist. 3/20/34. Farmer, age 30, Orangeburg Dist. 1860 census. Enl. Ft. Motte 10/26/61. Present through 6/62. Present sick in camp 7-8/62. Present 9/62-8/63. Hired his stepson, L. W. Weeks as substitute, not yet 18 years old 9-10/63 and discharged. Enl. Hart's S. C. Arty. 2/15/64. Present until ab. on horse detail 4/1-30/64. Paroled Greensboro, N. C. 5/1/65. Farmer, Pine Grove, Orangeburg Co. 1880 census. d. Lone Star, Calhoun Co. 10/16/91.

FELTS, JOHN S. Pvt., Co. H. b. S. C. circa 1837. Farmhand, age 23, York Dist. 1860 census. Enl. Rock Hill 1/18/61 age 22. On picket Jacksonborough 3-4/62. Present 5-6/62. Present sick in camp 7-8/62. Ab. detailed to drive horses from Summerville, S. C. to Staunton, Va. 9-10/62. On raid 11-12/62. Present 3-6/63. Ab. on horse detail 8/18-31/63. Present 9/63-2/64. Ab. on leave 3-4/64. Present 5-10/64. Detached Pocotaligo 12/10-31/64. Paroled Greensboro, N. C. 4/26/65. Retired, age 72, Charlotte, N. C. 1910 census. Mill Worker, Charlotte, N. C. 1915.

FENLEY, ELISHA or ELIHU. Pvt., Co. G. b. S. C. circa 1815. Enl. Rock Hill 1/8/62. Present through 6/62. Ab. on leave 7 days 8/27-31/62. Present 9-12/62. Present sick in camp 3-4/63. Ab. sick with variola in Farmville hospital 5/2/63. d. 5/5/63 age 47. Probably buried in unmarked grave in Farmville Confederate Cem.

FERGUSON, ABRAM T. Pvt., Co. H. b. Chester, S. C. 1/2/47. Age 11, Chester Dist. 1860 census. Enl. Columbia 4/15/64. Detached Green Pond 4/16-30/64. Present 5-8/64. On picket 9-10/64. Detached Pocotaligo 12/10-31/64. NFR. Farmer, age 32, Lansford, Chester Co. 1880 census. d. Lancaster, S. C. 6/26/12.

FERGUSON, ALFRED. Pvt., Co. K. Enl. Pickens CH 12/4/61. Not on muster rolls On pay rolls 11/1-12/31/63. Paid 11/6/63. Paroled Raleigh, N. C. 4/65.

FERGUSON, BENJAMIN. Pvt., Co. I. b. S. C. circa 1839. Enl. Parker's Ferry 4/3/62. On picket Jacksonborough through 4/30/62. Present 5-6/62. Present sick in camp 7-8/62. Present 9-10/62. On scout 11-12/62. Present 3/63-2/64. Ab. on leave 4/22-5/3/64. Present sick in camp 5-6/64. On courier duty 7-8/64. On picket 9-10/64. Present 11-12/64. NFR. Farmer, age 41, Marshall, Ala. 1880 census. d. in Ala. between 1880 & 1900.

FERGUSON, GEORGE P. 1stSgt., Co. B. b. S. C. circa 1834. Enl. Clinton 8/22/61 age 27 as 2ndSgt. On detached service Camp Bee 10/31/61. Present 11-12/61, promoted 1stSgt. Present 3-6/62. Ab. on sick leave 7-8/62. Detailed to march horses from Summerville, S. C. to Staunton, Va. 9-10/62. Present 11/62-6/63. Ab. on detached service in S. C. 8/18-31/63. Present 11-12/63, however, ab. on horse detail

12/29/63-2/64. Present 3-8/64. Ab. sick McLeod hospital, James Island & furloughed 10/14/64. Present 11-12/64. NFR. Paid taxes in S. C. 1865-66. Entered Old Soldier's Home, Raleigh, N. C. from Gaston Co., N. C. d. there 11/2/11 age 86.

FERGUSON, JAMES TURNER. Pvt., Co. K. b. Chester Dist. 9/43. Enl. Pickens CH 6/25/62. Ab. on sick leave 7-8/62. Present 7-12/62. Ab. detached 3-4/63. Present 5-6/63. Ab. on horse detail 8/16-31/63. AWOL 9-10/63. WIA Stevensburg on post war roster. Present 11/63-4/64. On picket Bee's Ferry 6/25-30/64. Present 7-8/64. On picket Bee's Ferry 9-10/64. Present 11-12/64. Paroled Raleigh, N. C. 4/65. Farmer, age 56, Catawba, York Co. 1900 census. d. York Co. 8/3/07.

FERGUSON, JOHN A. Pvt., Co.'s F & K. b. Pickens Dist. 7/41. Enl. Pickens CH 12/4/61 age 20. Present through 4/62. Transferred Co. K 6/25/62. Present through 8/62. Ab. on detached service 9-10/62. Present 11/62-6/63. Ab. on horse detail 8/16-31/63. Present 9/63-2/64. Ab. on leave 4/27-31/64. Present 5-6/64. On picket Bee's Ferry 7-8/64. Ab. sick Charleston hospital and furloughed for 60 days 10/24/64. AWOL 12/25-31/64. NFR. Farmer, Pickens Co. 1870-1900 censuses. d. 5/31/06. Bur. Clinton City Cem.

FERGUSON, JOSEPH JAMES (or James Joseph). Pvt., Co. I. Enl. James Island 6/30/62. Present 7-10/62. Ab. sick in Richmond hospital 11/21-12/7/62 & 3-4/63. Present through 2/64. Ab. on horse detail 4/16-30/64. Present sick in camp 5-6/64. Present 7-10/64. AWOL 12/11-31/64. Transferred 2nd S. C. Arty. 2/4/65 NFR.

FERGUSON, RICHARD C. Pvt., Co. B. b. S. C. circa 1832. Res. Laurens Dist. Enl. Clinton 8/22/61. On picket White Point 11-12/61. Present 3-6/62. On picket Little Brittan 7-8/62. Present sick in camp 9-10/62. Ab. sick with pleurisy in Richmond hospital 11/12-12/7/62. Ab. sick with Variola in Richmond hospital 12/14/62. d. of chronic diarrohea 12/19/62 age 30. Effects: $26.75.

FERGUSON, STEPHEN R. Pvt., Co. D. b. S. C. 11/26/16. Enl. 9/10/61. Detailed as Farrier. Ab. sick at home 12/23/61. Accidentialy wounded on furlough. Discharged for disability 3/27/62. NFR. Farmer, age 66, Lewisville, Chester Co. 1880 census. d. 11/29/07. Bur. Old Hopewell Bapt. Ch., Cem., Chester Co.

FERGUSON, WILLIAM. Pvt., Co. I. b. S. C. 1843. Enl. John's Island 6/30/62. Ab. on leave 8/21-31/62. Present 9/62-8/31/63. Ab. on horse detail 9-10/63 & 12/24-31/63, however, ab. sick in Richmond hospital 12/31/63-2/12/64. Present 3-6/64. On courier duty 7-8/64. Present 9-12/64. Paroled April, 1865 on wife's pension application. Farmer, age 26, Pickens Co. 1870 census. Farmhand, age 56, Rafen Creek, Sumter Co. 1900 census. d. 12/24/00. Wife receiving pension Cottageville, Colleton Co. 10/20/19.

FERGUSON, WILLIAM M. Pvt., Co. K. b. S. C. 6/30/16. Enl. Pickens CH 6/25/62. Ab. on sick leave 9-10/62. Present 11-12/62. Ab. on detached service 3-4/63. Present through 8/31/63. Ab. on horse detail 9-10/63. AWOL 12/23-31/63. Detailed horse recruiting camp 1/10-2/29/64. Present 3-6/64. On picket Bee's Ferry 7-8/64. Ab. sick Charleston hospital and furloughed for 20 days 10/24/64. AWOL 12/25-31/64. Paroled Raleigh, N. C. 4/65. d. 6/11/87. Bur. Old Hopewell Bapt. Ch. Cem., Chester Co.

FERGUSON, WILLIAM T. Pvt., Co. H. b. S. C. 1/5/46. Age 13, Chester Dist. 1860 census. Enl. Columbia 4/15/64. Detached Green Pond 4/16-30/64. Present through 12/64. NFR. Farmer, age 23, Lansford, Chester Co. 1870 census. Farmer, age 52, Navarro Co., Texas 1900 census. d. there 11/3/19. Bur. Oakwood Cem., Navarro Co., Texas.

FIELD, OLIVER PERRY. 4thSgt., Co.'s F & K. b. Pickens, S. C. 8/28/41. Enl. Pickens CH 12/4/61 age 17 as Pvt. Present until transferred Co. K 6/25/62. Promoted 1stCpl. Ab. on sick leave 9/1-12/31/62. Ab. on detached service 4/8-30/63. WIA Upperville 6/21/63. WIA Gettysburg 7/3/63. Ab. wounded in hospital through 8/31/63. Ab. on leave for 30 days from hospital 9/19/63. Ab. wounded through 12/63. In S. C. hospital 1-2/64. Present 3-4/64. Promoted 4thSgt. Present 5-8/64. Commanding Company & on duty at Department Headquarters, Charleston 9-10/64. Present 11-12/64. 27 men present in Company 11/64. Served in Co. K to 4/65 on pension application. Farmer, Pickens Co. 1880 & 1900 censuses. Receiving pension Pickens, S. C. 4/17/19. d. Greenville, S. C. 3/2/00. Bur. Mountain Grove Church, Pickens Co.

FINLEY, ELIHU. Pvt., Co. G. Enl. age 24 on postwar roster. Reportedly d. of small pox Farmville hospital circa 4/20/63. Bur. Confederate Cem.

FINLEY, WILLIAM GREEN. Pvt., Co. H. b. N. C. 6/12/31. Enl. Rock Hill 1/9/61 age 30. Present 3-4/62. Ab. on sick leave 6/23-30/62. Present 7-8/62. Detailed to drive wagon from Summerville, S. C. to Staunton, Va. 9-10/62. Ab. sick 11-12/62. Ab. sick with debility in Charlottesville hospital 1/2-14/63. Ab. on horse detail 2/216/30/63. Ab. detached with wagon train through 8/31/63. Detailed horse recruiting camp 9/63-2/64. Present 3-12/64. NFR. Farmer, age 37, York Co. 1880 census. d. 6/17/83. Bur. Bethel ARP Ch. Cem., York Co.

FISHBURN, EDWARD B. Pvt., Co. A. b. S. C. 11/40. Enl. 11/21/61. Detailed to Gen. Evans as Guide 2/2/62. Reenlisted Co. K, 4th Va. Cav. 1/9/64. Transferred 4/9/64. NFR. Married 1873. Contractor, Charleston 1880. Contractor, Summerville Dist., Dorchester Co. 1900. Living Charleston 1910.

FITZSIMMONS, PAUL G. Pvt., Co. A. b. Charleston 10/5/34. Att. S. C. College 1859-60. On post war roster. Reenlisted Co. K, 4th S. C. Cav. Grahamville 3/25/62 age 34. Ab. detached with Gen. Pemberton until discharged for disability 12/8/62. Appointed Lt., 21st S. C. Inf. on postwar roster. d. 4/12/71. Bur. Winyah Cem., Georgetown, S. C.

FLENNIKEN, DAVID REID. Pvt., Co. D. b. S. C. 3/10/45. Student, age 15, Chester Dist. 1860 census. Enl. Chester 6/18/62. Present until ab. sick at home 8/14-31/62. Present 9-10/62. Ab. as Regimental Mail Carrier to Culpeper CH 11-12/62. Present 3-4/63. WIA (arm - flesh wound) Gettysburg, Pa. 7/3/63. Ab. on horse detail 8/3-31/63. Present 9/63-6/64. Courier for Major Walker 7-12/64. Paroled Greensboro, N. C. 4/15/65 on pension application. Grocer, Winnsboro, S C. 1870 census. Grocery Broker, Columbia, S. C. 1900 census. Receiving pension Columbia 3/21/19. d. 9/18/23. Bur. 1st Presb. Ch. Cem., Columbia.

FLUD, DANIEL. Pvt., Co. A. b. 3/16/18. M. D. On post war roster. Reenlisted Co. K, 4th S. C. Cav. Grahamville 3/25/62. Present until detached with Surgeon 11-12/62. Present until discharged for overage 7/1/63. d. Summerville, S. C. 3/21/96. Bur. St. Paul's Epis. Ch., Charleston.

FOGLE, HENRY W. Pvt., Co. E. b. S. C. 1838. Farmer, age 22, Orangeburg Dist. 1860 census. Enl. Ft. Motte 10/26/61. On picket White Point 12/31/61. Present 3-4/63. Present sick in camp 5-6/62. On scout to Edisto Island 7-8/62. Present 9/62-6/63. Ab. on horse detail 8/18-31/63. AWOL 10/25-11/22/63, dismounted since 9/1/63. Present 1-2/64. Ab. on horse detail 4/13-5/13/64. Present through 10/64. Detailed as guard Wappo Bridge 12/29-31/64. Paroled Hillsboro, N. C. 4/65. Farmer, age 34, Pine Grove, Orangeburg Co. 1870 census. d. 11/73 age 35.

FORD, GEORGE. Pvt., Co. K. b. Chester Dist. circa 1840. WIA Upperville 6/26/63 and discharged on post war roster. Res. Chester Co.

FORD, JOHN WESLEY. Pvt., Co. A. b. S. C. 2/7/40. Enl. Abbeville CH 8/13/61 age 29. On duty Bear Island through 12/31/61. Present3-6/62. Ab. on sick leave 7-8/62. Present 9/62-8/31/63, however, ab. sick with "hypertrophy of the heart" in Farmville hospital 6/10-8/17/63. Present 9-12/63. Ab. on horse detail 2/29-4/29/64. Present 5-12/64. NFR. Farmer, King's Mountain, N. C. 1870 & 1880 censuses. Receiving pension Townsville, Anderson Co. 1901. Retired, King's Mountain, N. C. 1910 & 1920 census. d. Cleveland, N. C. 11/11/22.

FORD, W. B. Pvt., Co. K. b. S. C. 8/14/41. Enl. 4/27/63 on Mississippi pension application. Served to 5/4/65. d. Water Valley, Miss. 1/22/25. Bur. Goshen Cem., near Water Valley, Miss.

FORD, WILLIAM B. Pvt., Co. K. b. S. C. 8/8/47. Enl. Ridgeway 2/22/64. Ab. detached Green Pond 4/18-30/64. Present 5-6/64. Ab. sick in Charleston hospital 8/27-31/64. On duty Bee's Ferry 9-10/64. Detailed driving wagon Green Pond 12/24-31/64. Paroled Raleigh, N. C. 4/65. Farmer, Yabobusha Co., Miss. 1910 & 1920 censuses. May be same as W. B. Ford, above.

FOREMAN, A. Pvt. Co. K. On post war roster.

FOREMAN, BARNEY. 4thSgt., Co.'s C & K. b. Barnwell Dist. 11/22/28. Planter, Barnwell Dist. 1860 census. Enl. Co. C Hamburg 8/27/61 age 32 as 4thSgt. On picket Bear Bluff, Wadmalaw Island 1/62. Transferred Co. K 6/27/62 and reduced to Pvt. Present through 10/62. Present sick in camp 11-12/62. Present 3-4/63. Ab. sick with pleurisy in Farmville hospital and returned to duty 5/10/63. Present through 8/31/63. Ab. on horse detail 9-10/63. AWOL, Barnwell Dist. 12/23-31/63. Present in arrest 1-2/64. Ab. detached Green Pond 4/18-30/64. Present 5-6/64. Ab. on leave 9-10/64, however, transferred Co. A, 2nd S. C. Arty. 9/1/64. Transferred back to Co. K, 1st S. C. Cav. 2/11/65. NFR. Farmer, Sleepy Hollow, Barnwell Co. 1870 & 1880 censuses. d. 7/18/92. Bur. Jackson Memorial Cem., Aiken Co., S. C.

FOREMAN, ISAAC WILLIAM. Pvt., Co. K. b. S. C. circa 1844. Age 16, Barnwell Dist. 1860 census. Enl. date unknown. Paroled Raleigh, N. C. on post war roster. Res. Barnwell Co. Farmer, Sleepy Hollow, Aiken Co. 1880 census. d. Augusta, Ga. 2/20/87.

FOREMAN, JACOB J. Pvt., Co. K. b. S. C. circa 1844. Student, age 14, Barnwell Dist. 1860 census. Enl. John's Island 6/26/62. Present 7-10/62. On picket Richards Ford 11-12/62. Present 3-4/63. WIA (gunshot wound to brain) Funkstown, Md. 7/8/63. Left wounded Williamsport, Md. Since died in Hagerstown, Md. hospital 7/11/63 age 18. Marker in Jackson Memorial Cem., Jackson, S. C.

FOREMAN, JESSE JOSEPH, JR. Pvt., Co.'s C & K. b. S. C. 10/28/39. Res. Barnwell Dist. Enl. Co. C Hamburg 8/27/61, age 21. Detached Wadmalaw Island 11-12/61 & Bear Bluff 1/62. Transferred Co. K 6/27/62. Present 7-8/62. Ab. on detached service 9-10/62. Present 11/62-8/31/63. Ab. on horse detail 9-10/63. AWOL 12/23-31/63. Ab. on horse detail 2/64. Ab. sick with debility in Richmond hospital 3/4-15/64. Furloughed for 30 days. On detached duty Green Pond 8/18-31/64. Present 5-10/64. Courier, Tar Bluff 12/21-31/64. Captured Mineral Springs, S. C. 3/5/65. Sent to New Bern & Point Lookout. Released 6/27/65. 5'9", fair complexion, brown hair, hazel eyes. Farmer, Sleepy Hollow, Barnwell Co. 1870 census. d. 3/6/01. Bur. Wesley Meth. Ch. Cem., Aiken Co., S. C. Wife receiving pension Kathwood, Aiken Co. 11/25/19. Brother of Joseph W. Foreman.

FOREMAN, JOSEPH W. Pvt., Co. K. b. S. C. circa 1847. Student, age 13, Barnwell Dist. 1860 census. Enl. Columbia 4/23/64. Detached Green Pond through 4/30/64. Present 5-6/64. On duty bridge on Ashley River 7-8/64. Detailed in Commissary Dept. Hqtrs., Charleston 10/7-31/64. On picket Chatman's Ford 12/21-31/64. NFR. Farmer, age 33, Sleepy Hollow, Barnwell Co. 1880 census. Brother of Jesse J. Foreman.

FOREMAN, L., Pvt., Co. K. On postwar roster.

FORGEY, MANOAH DAVIS. Pvt., Co. A. b. S. C. 1841. Medical Student, Laurens Dist. 1860 census. Enl. 2nd Co. G, Col. Gault's 12th Ark. Inf. Adelphia, Ark. 7/27/61. Present until captured Port Hudson, La. 7/9/63. Paroled 7/12-13/63. Transferred Co. A, 1st S. C. Cav. 7/1/64. Present until detailed guard POW's Florence 9/29-12/31/64. Appointed Surgeon, C. S. A. NFR. Farmer, age 43, Grayson Co., Texas 1880 census. d. Bells, Texas 1905.

FORRESTER, THOMAS PRESSLEY. Pvt. Co. A. b. S. C. 1838. Age 22, Spartanburg Dist. 1860 census. Enl. 11/7/61. Discharged for disability 11/21/61. Enl. Co. E, 23rd Ga. Inf. 12/6/61, as Sgt. Elected 2ndLt. 12/16/61. Promoted 1stLt. Captured Chancellorsville 5/3/63. Sent to Old Capitol. Exchanged 1863. Present 1-2/64. Ab. sick with chronic diarrhea in Richmond hospital 6/30/64. Furloughed to Atlanta, Ga. for 30 days. Paroled in Texas 1865. Receiving pension in Ga. 1917.

FORTESQUE, THOMAS J. Pvt., Co. G. b. S. C. circa 1843. Res. Abbeville Dist. 1850 census. Enl. Abbeville Dist. 5/12/62. On picket duty through 6/62. Present 7-10/62. On scout 11-12/62. Present 3-12/63. Ab. on leave 2/13-29/64. Present 3-6/64. Present sick in camp 7-8/64. Detailed as Supt. of Negroes working on James Island 9/8-10/31/64. Present 11-12/64. Present through 4/65 on wife's pension application. Married Abbeville 1866. d. 11/11. Wife receiving pension Abbeville 9/8/19.

FOWLER, THOMAS W. 4thSgt. Co. B. b. Cross Anchor, S. C. 1/7/33. Res. Spartanburg Dist. Enl. Clinton 8/22/61 age 27. d. in Columbia, S. C. hospital 10/15/61 age 30.

FOX, ISAIAH IRVIN. 3rdLt., Co. I, b. Walterboro, Colleton Dist. 4/5/41. Student, age 19, Walterboro, Colleton Dist. 1860 census. Enl. Parker's Ferry 4/3/62 as 2ndSgt. Ab. sick at home through 4/30/62. Present 5-10/62. On scout 11-12/62. Promoted 3rdLt. Present 3-8/63. Ab. sick Gordonsville hospital 9-10/63. On picket 11-12/63. Present 1-8/64, however, detached at Green Pond 4/16-30/64. Signed for 13 pr. shoes, 3 blankets, 2 fly tents, 10 pr. pants, 1 axe, 21 shirts, 1 jacket & 1 pr. drawers 3/31/64. Ab. sick 10/15-31/64. Present in arrest 11-12/64. NFR. Att. U. of S. C. 1868-70. Lawyer, Walterboro, 1870 census. d. there 4/18/78. Bur. Live Oak Cem., Waterboro, S. C. Brother of John R. P. Fox.

FOX, JOHN RICHARD PERRY. Captain, Co. I. b. Cottageville, Colleton Co. 11/16/26. Enl. Captain Sheridan's Co., 11th S. C. Inf. as 2nd Lt. Fenwick Island 10/15/61. Ab. sick at home 12/16-31/61. Company disbanded. Enl. Co. I, 1st S. C. Cav. Parker's Ferry 4/3/62 as Captain. Signed for 2 wall tents, 20 Army tents, 12 camp kettles, 5 hatchets, 30 mess pans & 5 axes 4/14/62. 42 horses present Adams Run 4/62. Present through 6/62. 65 horses present 5/62. Ab. on sick leave 8/22-31/62. Present 9-10/62. Ab. sick in Richmond hospital 11/62-4/63. WIA (right arm) & captured Brandy Station 6/9/63. In Alexandria hospital & sent to Old Capitol. Transferred Ft. McHenry, Johnson's Island & Pt. Lookout. Exchanged 5/8/64. In Richmond hospital with old wound 5/9-11/64. Signed for 3 tent flies 5/25/64. Ab. sick through 8/64. 61 horses present 8-9/64. Present 9-10/64. Signed for 8 pr. shoes 9/30/64. 40 horses present 9-10/64. Ab. before Medical Board 11-12/64. Transferred to Invalid Corps 1/16/65 & assigned to S. C. Reserves. NFR. Farmer, Collins, Colleton Co. 1880 census. d. 6/1/96. Bur. Fox Cem., Colleton Co.

FRANKLIN, CHARLES E. Pvt. & Bugler, Co. B. b. S. C. 12/4/34. Enl. Clinton 8/22/61. Present until ab. on leave 12/27-31/61. Present 3-6/62. Ab. on sick leave 7-8/62. Present 9-10/62. Appointed Bugler. Present 3-4/63. Transferred to Band 6/1/63. WIA (left shoulder) & captured Gettysburg 7/3/63. Exchanged and in Williamsburg hospital 8/29/63. Transferred to Petersburg hospital 9/4/63. Ab. wounded through 10/63. Present 11-12/63, however, ab. wounded and at home 12/7-31/63. Present 1-2/64, returned to Co. B from Band 2/1/64. Present 5-8/64, however, in Summerville hospital 6/30/64. Ab. on leave 10/21-31/64. Present as Bugler for Company 11-12/64. NFR. Merchant, Hunter, Laurens Co. 1870 census and Clinton, Laurens Co. 1880 census. Fruit Grower, Kissimmee, Fla. 1900 census. Farmer, Osceola, Kissimmee, Fla. 1910 census. d. Kissimmee, Fla. 10/27/12. Bur. Rose Hill Cem.

FRANKLIN, CHARLES R. 2ndLt., Co. B. b S. C. circa 1831. Enl. Clinton 8/22/61. Ab. sick 10/31/61. Present 11-12/61. Present in arrest 3-4/62, since 12/18/61. CM'd. 4/20/62. Cashiered 5/20/62. NFR. Paid Taxes in S. C. 1866, as retailer Cigar Maker. d. 10/22/12. Bur. Rose Hill Cem., Osceola Co., Fla.

FRANKLIN, GEORGE W. Pvt., Co. B. b. S. C. 4/40. Res. Laurens Dist. Enl. Clinton 8/22/61. Ab. sick 10/31/61. Present 11/61-8/62. Ab. on march with horses from Summerville, S. C. to Staunton, Va. 9-10/62. Present 11/62-8/31/63. Ab. on horse detail 10/23-12/31/63. Present 1-2/64. AWOL Columbia, S. C. 3-4/64. Present 5-8/64. Deserted 9/12/64. Deserted to the enemy New Bern, N. C. 3/65. Sent to Annapolis, Md. 3/9/65. Took oath and sent to Baltimore 3/11/65. Stonecutter, Newberry Co. 1870 census. Wagon Maker, Hazelwood, Chesterfield Co. 1880 census. Wheelwright, Columbia, S. C. 1900 census. Applied for pension 1916.

FREDERICK, ALFRED J. 1stLt., Co. E. Enl. Ft. Motte 10/26/61. Ab. recruiting 12/26-31/61. Ab. sick 2/62. Present 3-4/62. Ab. sick leave 5/24-8/30/62. Ab. on detached service 9-10/62. Resigned for phthisis pulmonalis 12/9/62. NFR. Paid taxes in S. C. 1866.

FREEMAN, BARNEY LYNN. Pvt., Co. F. b. S. C. 11/6/32. Enl. Pickens CH 12/4/61 age 30. Present through 6/62. Ab. sick Summerville hospital 7-10/62. Ab. sick at home 11-12/62. Present 3-4/63. Ab. recruiting horses through 8/31/63. Present 9-10/63. Ab. recruiting station for horses 12/22/63-2/64. Present 3-4/64. On duty West Lines, James Island 8/30/64. Present 9-12/64. NFR. d. 2/18/76. Bur. Glassy Mountain Meth. Ch. Cem., Pickens Co.

FREEMAN, BENTON STRAIN. Pvt., Co. F. b. S. C. 5/26/37. Enl. Co. H, 4th S. C. Inf. as 1stSgt. Pickens Dist. 4/14/61. Present until ab. sick in hospital 11-12/61. Present through 12/11/62. Regiment disbanded. Enl. Co. F, 1st S. C. Cav. Pickens CH 2/20/63. Present through 4/63. Present until WIA (date & place unknown) & admitted Jackson hospital, Richmond 8/2/63. Furloughed for 30 days 8/15/63. Ab. on detached service 9-10/63. Present 11/63-6/64. On duty Paden's Hqtrs. 8/26-31/64. Present 9-12/64. NFR. Farmer, Darcusville, Pickens Co. 1900 & 1910 censuses. d. Greenville, S. C. 4/17/21. Bur. Graceland Cem.

FREEMAN, DAVID COLUMBUS. Pvt., Co. F. b. S. C. 12/22/36. Farmer, Pickens Dist. 1860 census. Enl. Pickens CH 12/4/61. Present through 12/31/61. Discharged for disability by Surgeon 4/20/62. NFR. Farmer, Dacusville, Pickens Co. 1880 census. d. Easley, S. C. 4/12/97. Bur. Cross Roads Bapt. Ch., Pickens Co.

FREEMAN, DAVID CROCKETT., Pvt., Co. F. b. S. C. 1839. Farmer, age 23, Pickens Dist. 1860 census. Enl. Pickens CH 12/4/61. Present through 12/31/61. Discharged for disability 4/20/62. d. 1863. Bur. Freeman Cem., Collinsville, Ala.

FREEMAN, J. G. Pvt., Co. F. b. Rutherford Co., 12/17/36. Enl. Pickens CH 4/21/62. Present through 8/62. Ab. on detached service 9-10/62. Present 11-12/62. Ab. on detached service 3-7/63. WIA Brandy Station 8/1/63. Ab. on detached service 8/31/63. Present 11-12/63. Ab. driving cattle for Major Beggs, Brigade Commissary Officer since 7/1/63-2/64. On duty Green Pond 4/25-30/64. Present 5-12/64, horse killed 10/5/64. NFR. Farmer, age 43, Windsor, Aiken Co. 1880 census. d. Logan's Store, Rutherford Co., N. C. 1/11/15.

FREEMAN, JOHN. Pvt., Co. C. b. S. C. circa 1825. Enl. age 35 from Aiken Dist. on postwar roster. Farmer, Darlington Co. 1880 census. Alive 1898.

FREER, JAMES HAMILTON. Pvt., Co. A. b. S. C. circa 1832. On post war roster. Reenlisted Co. K, 4th S. C. Cav. Grahamville 3/25/62 age 30. Present until ab. on sick leave 8/28/62 for 10 days. Present until detached with Gen. Walker as Sutler 6/11/63-10/31/64. Paroled 4/65. Merchant, Summerville 1880 census. d. John's Island 2/22/00.

FRIERSON, AUGUSTUS CONVERSE. Pvt., Co. A. On postwar roster. Reenlisted Co. K, 4th S. C. Cav. 4/1/63. Captured Cold Harbor 5/30/64. Sent to Pt. Lookout. d. of disease 7/29/64. Bur. Confederate Cem., Point Lookout, Md.

FRIERSON, JAMES JULIAN. Pvt., Co. A. b. Sumter Co. 1/19/32. Planter & Lawyer, St. Helena Island, Beaufort Co. 1860 census. On post war roster. Reenlisted Co. K, 4th S. C. Cav. Grahamville 3/25/62. Present until detached in A. A. Gen.'s Office, Gen. Beauregard 9/25/62-8/63. Present until dropped 4/9/64, reduction of company. May have served in 23rd or 25th S. C. Inf.'s. d. Clarendon Co. S. C. 8/1/99.

FRITH, WILLIAM HENRY. Pvt., Co. G. b. S. C. 2/22/38. Enl. Abbeville CH 1/8/62. Present through 4/63. Present dismounted 5-6/63. Ab. on horse detail 8/31/63, horse killed 8/1/63. Paid $250.00. Present 9-10/63. On picket Chancellorsville 12/31/63. Ab. recruiting station for horses 1/13-4/30/64. Present 5-10/64. Detached Pocotaligo 12/10-31/64. NFR. Farmer, Cedar Springs, Abbeville Co. 1880 & 1900 censuses. Receiving pension 3/31/19. d. Abbeville 3/7/22. Bur. Old Bethia Cem., Abbeville Co.

FULLER, HENRY M., JR. Pvt., Co. A. b. S. C. 6/19/35. Planter & Lawyer, St. Helena Island, Beaufort Co., 1860 census. On post war roster. Reenlisted Co. K, 4th S. C. Cav. Grahamville 3/25/62. Present until transferred Co. F, 11th S. C. Inf. 1/1/63, but not on rolls of that unit. Detailed to make salt. Transferred from Co. A, 3rd S. C. Cav. to Co. K, 4th S. C. Cav. 5/1/63. Present until ab. detached as Clerk for Captain Samuel Lowndes, AAG 7/1-12/31/63. Present 1/64-3/24/64 & 8/15/64. NFR. d. 9/23/90. Bur. Burnt Churchyard Cem., Beaufort Co.

FURGUSON, DAVID G. Pvt., Co. A. Enl. 11/12/61. Detailed Special Duty 2/2/62. NFR.

GAILLARD, EDWARD T. Pvt., Co. A. b. S. C. 4/6/37. On post war roster. Reenlisted Co. K, 4th S. C. Cav. Grahamville 3/25/62. Present until ab. sick 5/15-6/30/62. Discharged for disability 7/24/62. 5'6 1/2", pale complexion, black hair, blue eyes, Clerk. d. 12/11/05. Bur. Magnolia Cem., Charleston.

GALLIARD, W. F. Pvt., Co. F. b. S. C. circa 1839. Enl. Pickens CH 12/4/61 age 22. Ab. on leave through 12/31/61. Present 3-4/62. Discharged for disability 5/31/62.

GAINES, ANDY WALKER. Pvt., Co. G. b. Anderson Dist., S. C. 12/16/36. Res. Abbeville Dist. 1860 census. Enl. Abbeville CH 3/15/62. On service with the wagons 3-4/62. Ab. sick in hospital 6/9-8/31/62. Present 9-10/62. Ab. sick in Richmond hospital 12/23-31/62. d. Strasburg, Va. 2/14/63 age 28. Left widow.

GAINES, L. PRESSLEY. Pvt., Co. G. b. S. C. circa 1839. Student, age 21, Anderson Dist., 1860 census. Enl. Abbeville Ch 3/15/62. Present through 3/31/62. Ab. sick in hospital 6/9-30/62. Present 7-12/62. d. of disease Strasburg, Va. 1/27/63 age 24, res. Abbeville Dist.

GAINES, TANDY. Pvt., Co. G. Res. Abbeville Dist., age 28. d. of pneumonia, Strasburg, Va. 1/7/63.

GALL, CHARLES. Pvt., Co. I, WIA (slightly) near Culpepper CH 9/13-14/63 on Record of Events for Co. I. Not on muster rolls.

GALLAHER, W. J."KOON", Pvt., Co. C. b. S. C. circa 1840. Enl. Kaolin 8/27/61 age 21. Ab. sick in hospital 12/31/61. Discharged by Surgeon 5/27/62. May have reenlisted as paroled Augusta, Ga. 5/19/65. Alive 1898.

GALLOWAY, JOHN W. Pvt., Co. I. b. S. C. circa 1842. Farmer, age 19, St. Paul's Parish, Colleton Dist. 1860 census. Enl. Parker's Ferry 4/3/62. On fatigue duty with wagons 4/30/62. On picket John's Island Ferry 5-6/62. Present 7-10/62. On duty with wagons 11-12/62. Present 3-4/63. Ab. sick in hospital 8/31/63. Ab. on horse detail 9-10/63. AWOL 12/30-31/63. Present 1-2/64. WIA (fractured right arm) 3/6/64. Ab. wounded in Richmond hospital 3/20-27/64. Ab. sick leave 4/26-30/64. Ab. sick in hospital 7/7-12/31/64. Deserted to the enemy Charleston, S. C. 3/6/65. Age 24, 5'3", light complexion, light hair, blue eyes, res. Charleston. In Hilton Head hospital with scurvy 5/7-11/65. NFR. Farmer, age 38, Pine Grove, Edgefield Co. 1880 census.

GALPHIN, STEPHEN S. Pvt., Co. C. b. S. C. 12/44. Res. Edgefield Dist. Enl. Hamburg 8/27/61 age 18. Present through 12/31/61. On picket Bear Bluff, Wadmalaw Island 1/62. Discharged by Surgeon 3/22/62. Reenlisted in Va. 3/1/63. Ab. on horse detail 8/15-31/63. Present 9-10/63. Ab. on horse detail 12/22/63-2/64. Present 3-4/64. Ab. on detached service 5/29-6/30/64. Present 7-8/64. Ab. on leave 60 days 9/1-10/31/64. Detached Pocotaligo 12/10-31/64. NFR. Farmer, Hammond, Edgefield Co. 1870 census. Farmer, Midville Village, Burke Co., Ga. 1900 census.

GAMBLE, JULIAN. Pvt., Co. K. Enl. by 11/64. On picket duty Chatman's Ford 12/31/64. NFR.

GAMMON, MICHAEL JOHN. Pvt., Co. A. b. N. Y. 1845. Res. Charleston. Served in Arsenal Bn., State Cadets as 4th Class Pvt. in Co. B. Enl. by 1865. Deserted to the enemy Charleston 2/18/65. Age 20, 5'6", dark complexion, black hair, dark eyes. Took oath and discharged 3/13/65. Marble & Stone business, Charleston postwar. d. 1883.

GANT, RUTHERFORD. Pvt., Co. F. Res. Pickens Co. 1850 census. Enl. Co. H, 3rd S. C. Reserves and served 11/24-12/31/62. Enl. Co. F, 1st S. C. Cav. Columbia 2/9/64. On duty Green Pond 4/29-30/64. Present 5-8/64. AWOL 9/10/64 - now driving wagon for AQM of Regt. 10/31/64. Present 11-12/64, dismounted since 7/16/64. Admitted Charlotte hospital with debilities 4/26/65. Patient paroled from Charlotte hospital 5/12/65. NFR.

GARRICK, DAVID B. Pvt., Co. E. b. S. C. 1831. Enl. 10/26/61 age 30. Present through 12/31/61. Ab. sick in Adams Run hospital 3-4/62. Ab. on sick leave 4/30-6/30/62. Present 7/62-10/63. Detailed horse recruiting station 12/23/63-2/64. Detached at Green Pond 4/16-30/64. Present 5-12/64. Paroled Hillsboro, N. C. 4/65.

GARLAND, EDWARD. Pvt., Co. F. Enl. by 1865. Captured and died Point Lookout, Md. 5/31/65.

GARRETT, G. Pvt., Co. C. b. S. C. circa 1830. Farmer, age 30, Pond Springs, Ga. 1860 census. Enl. Aiken Dist. by 1862 age 30. WIA in Va. Discharged by Surgeon 7/28/62, all on postwar roster. Farmer, age 40, Ft. Worth, Texas 1870 census.

GARRETT, HENRY C. Pvt., Co. C. b. Edgefield Dist. circa 1837. M.D, age 23, Collins, Edgefield Dist. 1860 census. Enl. Co. I, 7th S. C. Inf. Edgefield Dist. 4/15/61. Present through 7/1/61. Discharged for "inability" 7/11/61. Age 24, 5'11", light complexion, blue eyes. Enl. Co. C, 1st S. C. Cav. Wadmalaw Island 5/17/62. Ab. attending the sick at Church Flats through 6/62. Discharged by Surgeon 7/29/62. NFR. Farmer, age 35, Edgefield Co. 1870 census.

GARRISON, WILLIAM F. Pvt., Co. H. b. S. C. 10/14/41. Enl. Rock Hill 3/17/62. On picket Rantole Bridge 3-4/62. Present 5-6/62. Ab. on sick leave 7-8/62. Ab. sick with "Intermitten Fever' in Richmond hospital 10/22-11/6/62. Ab. sick with acute bronchitis in Richmond hospital 11/21/62-3/1/63. Ab. on horse detail 3-4/63 & 8/31/63. Ab. recruiting horses 9-10/63. Transferred Co. E, 17th S. C. Inf. 12/17/63. Ab. sick in Richmond hospital. 12/63-1/64. Joined 2/27/64. Present 3-4/64. Ab. sick 5-6/64. Paroled Hillsboro, N. C. 4/65. Farmer, Pennington, Bradley Co., Ark. 1870 census. Farmer, Rockwall, Texas 1880 census. d. Garland, Dallas Co., Texas. 10/6/82. Bur. Pleasant Valley Cem., Garland, Tex.

GASTON, E. LEANDER. Pvt., Co. D. Enl. Camp Means 4/17/62. On picket Bear Bluff through 4/30/62. On picket Townsend's House 5-6/62. Horse killed John's Island 6/7/62. Paid $150.00. Present 7-10/62. Ab. driving QM wagon 11/62-4/63. Present 5-6/63. Ab. on horse detail to Chester, S. C. 8/18-31/63. Detailed Regimental QM 10/28-12/10/63. Ab. on leave 12/11-22/63, however, listed as ab. sick in Richmond hospital 11-12/63. (Probably furloughed from there). Present 1-2/64. Ab. on leave 4/23-30/64. Present 5-12/64. Paroled Hillsboro, N. C. 4/65. Bur. Bullock's Creek Ch. Cem., York Co., no dates.

GATCH, JOHN DAVID. Pvt., Co. I. b. Charleston Dist. 1845. Enl. 4/3/62, but not accepted by mustering officer. (under age). Enl. Co. B, 3rd S. C. Cav. Pocotaligo 3/20/63. Present through 4/63. Paid as Company Farrier 5/63-10/64. Present sick in camp 11-12/64. NFR. Farrner, Pocotaligo, Hampton Co. 1880 census. d. Columbia, S. C. 1/2/06. Bur. Zahler Cem., Yemassee, Hampton Co., S. C.

GEORGE, DEDRICK LEWIS. Pvt., Co. I. b. S. C. 6/13/45. Enl. Columbia 3/9/64. Detailed Green Pond 4/16-30/64. Present 5-12/64. NFR. Farmer, Hollow Creek, Lexington Dist. 1870 & 1900 censuses. d. 11/9/01. Bur. George Cem., Chapin, Lexington Co.

GEORGE, JOSEPH B. 2ndLt., Co. I. Res. Stewart Co., Tenn. Enl. 3/28/61. NFR. Enl. Co. D, 3rd Tenn. Cav. (Forrest's) Memphis 2/27/62. Present 3-4/62. Promoted 2ndLt. 7/6/62. Present through 10/62. NFR until deserted Nashville, Tenn. 2/1/65. 5'9", fair complexion, brown hair, grey eyes. Took oath 2/15/65. NFR.

GERALD, WILLIAM E. Pvt., Co. D. b. Horry Dist. 1841. Age 19, Kinston, Horry Dist. 1860 census. Enl. by 1863. KIA Gettysburg 7/3/63.

GIBSON, JOSHUA. Pvt., Co. G. b. Chesterfield Dist. circa 1841. Age 19, Chesterfield Dist. 1860 census. Enl. Edgefield 5/30/64. Present through 8/64. Detailed 9-10/64. Present 11-12/64. NFR. Living Abbeville Co. 1902.

GILL, CHARLES. (1). Pvt., Co. I. Enl. Parker's Ferry 4/3/62. On picket Jacksonborough through 4/30/62. On picket John's Island 5-6/62. Present 7-8/62. Ab. sick in Richmond hospital 9-10/62 & Lynchburg hospital 11-12/62. Present 3-8/63. WIA (slightly) Culpepper CH 9/13-14/63. Present sick in camp through 10/63. On picket 11/63-2/64. Ab. on leave 4/22-5/2/64. Present through 10/64. Detached Pocotaligo

12/10-31/64. NFR. Farmer, Concord, Sumter Co. 1880 census. d. Charleston 1/9/05. Bur. Magnolia Cem.

GILL, CHARLES. (2). Pvt., Co. I. Res. Colleton Dist. Enl. by 1864. d. Columbia, S. C. 2/15/64 age 30. Bur. Magnolia Cem., Charleston 1/9/65.

GILL, JAMES ARCHIBALD. Pvt., Co. H. b. S. C. 12/18/44. Enl. Rock Hill 1/9/61. On picket Jacksonborough 3-4/62. Present 5-6/62. Ab. sick in hospital 7-8/62. Detailed to drive horses from Summerville, S. C. to Staunton, Va. 9-10/62. Present 11-12/62. Present sick in camp 3-4/63. Present 6/23-8/31/63 & 9-12/63. Ab. recruiting station for horses 1-2/64, however, ab. sick with bronchitis in Charlottesville hospital 2/19-3/9/64. Detached Green Pond 4/16-30/64. Present 5-6/64. Present sick in camp 7-10/64. Detached as Courier 11-12/64. NFR. Farmer, York Co. 1870 census. Farmer, Alvord, Wise Co., Texas 1900 census. Farmer, Clay, Texas 1910 census. Farmer, Ft. Stephens, Okla. 1920 census. d. York Co., S. C. 11/27/33. Bur. Ebenezer Presb. Ch. Cem., York Co.

GILLIAM, JAMES MASON. Pvt., Co. A. b. Saxaphaw, Alamance Co., N. C. 8/20/20. Enl. Columbia 2/1/64. Present through 6/64. Captured John's Island 7/2/64. Exchanged Port Royal Ferry 8/16/64. AWOL 10/24-31/64. Present 11-12/64. NFR. Farmer, Abbeville Co. postwar. d. 10/26/91. Bur. Sharon Meth. Ch., Abbeville Co.

GILLISON, JOHN M. Pvt., Co. K. b. S. C. 10/29/23. Enl. Columbia 3/1/64. Ab. sick 4/18-30/64. Present 7-8/64. Ab. sick leave 10/10-12/31/64. NFR. Res. Saluda, Edgefield Co. 1870 census and Senaca, Oconee Co. 1880 census. d. Oconee Co. 2/12/09. Bur. Richland Presb. Ch. Cem.

GILMER, ROBERT ALEXANDER. 1stSgt., Co. F. b. S. C. 8/23/37. Enl. Pickens CH 12/4/61. Present through 10/62. Ab. sick in Lynchburg hospital 11-12/62. Present 3/63-12/64. NFR. Miller, Tugaloo, Oconee Co. 1880-1900 on cenuses. d. Tuccoa, Ga. 7/31/05. Bur. Hopewell Meth-Epis. Ch., Oconee Co., S. C.

GILMER, WILLIAM J. G. Pvt., Co. A. b. Abbeville Dist. circa 1837. Enl. Abbeville CH 3/19/62. Present through 4/62. Ab. sick Summerville hospital 5-6/62. Present sick in camp 7-8/62. Present 9-12/62. Discharged from Staunton hospital 4/4/63 for "Syphilis Consectiva." Age 26, 5'8", fair complexion, blue eyes, dark hair, Clerk. Enl. Co. F, 2nd S. C. Inf. (Palmetto Regiment) Greenwood 3/1/64. Present through 6/64. Captured Halltown, Va. 8/26/64. Sent to Old Capitol and Camp Chase. d. of pneumonia 3/3/65. Bur. Camp Chase, Ind., Nat. Cem.

GLADDEN, ELIHU. Pvt. & Musician, Co. A. b. S. C. circa 1831. Res. Chester Dist. Enl. Co. C, 6th S. C. Res. Chester 11/18/62 as 2ndLt. Disbanded 1/1/63. Enl. Co. L, 5th S. C. State Troops 8/15/63. Disbanded 11/63. Enl. Co. A, 1st S. C. Cav. Columbia 2/15/64 as Pvt. Transferred to Regimental Band 4/30/64. Present through 6/64. Ab. on leave by Medical Board 9/1064 for 60 days. Paroled Hillsboro, N. C 4/65. Farmer, Rossville, Chester Co. 1880 census.

GLADDEN, JESSE ALEXANDER. Chief Musician. b. Fairfield Dist. 3/33. Music Teacher, age 27, Fairfield Dist. 1860 census. Enl. Co. G, 6th S. C. Summerville 6/15/61. Transferred to Regiment Band 6/30/61. Present through 8/61. Ab. sick with chronic bronchitis in Charlottesville hospital 10/8-31/61. Furloughed for 30 days 11/5/61. In same hospital with lung disease 12/20-29/61. Discharged for disability 12/30/61. Enl. Co. D, 1st S. C. Cav. at Camp Means 4/20/62 as Pvt. Present through 4/30/62. On picket Townsend's House 5-6/62. Present 7-10/62. Present as Bugler 11/62-4/63. Transferred Regimental Band 6/1/63. Present 9/63-8/64. Ab. on leave 10/31/64 for 10 days. Present 11-12/64. Paroled Hillsboro, N. C. 4/65. Farmer, age 37, Winnsboro, Fairfield Co. 1870 census. d. Fairfield Co. 2/24/97 age 63. Bur. Bethesda Meth. Ch., Chester Co.

GLENN, FRANK D. Pvt., Co. G. Enl. age 18 on Abbeville Co. roster 1902.

GLENN, H. J. Pvt., Co. B. b. circa 1816. Methodist Episcopal Minister, age 44, Laurens Dist., 1860 census. Enl. 8/22/61 age 45. d. Laurens Dist. 10/27/61.

GLENN, JAMES A. 2ndCpl., Co. H. b. S. C. 1838. Student, age 23, Cherokee, Ala. 1860 census. Enl. Rock Hill 3/20/62 as 4thCpl. On picket Jacksonborough through 4/62. Present 5-6/62. Ab. sick 7-8/62. Detailed to drive horses from Summerville, S. C. to Staunton, Va. 9-10/62. Ab. on raid 11-12/62. Present through 8/31/63. Promoted 3rdCpl. Present 9-10/63. Ab. on horse detail 12/22/63-2/64. Present 3-6/64. Promoted 2ndCpl. Present 7-12/64.

GLENN, JUDGE. Pvt., Co. B. b. circa 1821. Res. Laurens Dist. Not on muster rolls. d. Columbia 6/15/61 age 40. Bur. Bramlett Meth. Ch. Cem., Laurens. Co.

GLENN, S. D. Pvt., Co. G. b. S. C. 6/15/21. Res. Laurens Dist. 1860 census. Enl. Co. A, 9th S. C. Reserves (90 days) as 1stLt. 11/17/62. Disbanded 2/14/63. Enl. Co. A, 6th S. C. Cav. Laurens Dist. 8/27/63. Ab. on sick leave 8/30/63. Never mustered in. Claimed exemption as Blacksmith. Enl. Co. G, 1st S. C. Cav. Columbia 8/8/64. Present through 12/64. NFR. d. 4/13/96. bur. Friendship Presb. Ch. Cem., Laurens Co.

GLOVER, DURANT. Pvt., Co. C. b. S. C. circa 1842. Farmer, age 18, Edgefield Dist. 1860 census. Enl. Hamburg 8/27/61. Detached Wadmalaw Island through 12/31/61. Present 5-6/62. Ab. sick 7/25-8/31/62. Present 9-12/62. On detached service 3-4/63. Present 5-6/63. Ab. on horse detail 8/15-31/63. Present 9-10/63. Ab. on horse detail 12/22/63-2/64. Ab. on leave 4/16-5/3/64. Present through 10/64. Detached Pocotaligo 12/10-31/64. Farmhand, age 28, Hamburg, Edgefield Co. 1870 census. Farmer, Hammond, Aiken Co. 1880 census. Alive 1898.

GLOVER, FLORENCE. 2ndLt., Co. E. b. S. C. 1843. Student, age 17, Orangeburg Dist. 1860 census. Enl. Coosawhatchie 11/14/61 as Pvt. Present through 12/31/61. Promoted 4thSgt. Present sick in camp 3-4/62. Present 5-6/62. Ab. on sick leave 8/27-31/62. Promoted 3rdLt. Present through 12/31/62, promoted 2ndLt. 11/28/62. Present 3-10/63. Ab. on leave 12/27-1/20/64. Present through 2/64. A detached Green Pond 4/16-30/64. Present 5-8/64. Signed for 36 caps, 39 pr. pants, 31 shirts, 36 pr. drawers, 3 pr. sock and 12 pair horse shoes 5/20/64. On picket Pontoon Bridge 9-10/64. In charge of guard, Wappo Bridge, James Island 12/20/64-1/20/65. NFR. Age 27, no occupation, Amelia, Orangeburg Co. 1870 census.

GLOVER, WILEY C. Pvt., Co. C. b. S. C. 11/25/24. Enl. Wadmalaw Island 5/20/62. Present sick in camp 6/30/62. Ab. on sick leave 7/12-8/31/62. Ab. sick at home 9-12/62. Present 3-8/63. WIA (minnie ball right arm) near Pony Mountain 9/13/63. Ab. wounded in Richmond hospital until furloughed for 30 days 9/23/63. Ab. wounded through 2/64. Present 3-6/64. Placed on retired list for total disability 7/5/64 and transferred to Invalid Corps. NFR. d. 8/5/96. Bur. Glover Cem., Aiken Co., S. C.

GOLDEN, O. P. Pvt., Co. G. Enl. by 1863. WIA (right leg) 9/20-10/19/63. DOW's in post war account.

GOLDING, REUBEN L. Saddler, Co. A. b. Laurens Dist. circa 1832. Harness Maker, age 28, Edgefield Dist. 1860 census. Enl. Abbeville CH 8/13/61 age 30 as Pvt. Ab. sick in hospital 12/31/61. Discharged for hand and wrist injuries 4/24/62. Age 35, 5'8", fair complexion, hazel eyes, black hair, Saddler. Paid 9/1/62. Assigned to Guard, Augusta, Ga. 5/4/63. NFR.

GOOD, J. G. Pvt., Co. F. Res. Pickens Co. Alive circa 1900 on postwar roster.

GOOLEY, JOHN G. Blacksmith, Co. F. b. Pickens Dist. circa 1799. Enl. Pickens CH 12/4/61 age 31?. Present through 12/31/61. Ab. with QM wagon 3-4/62. Present 5-6/62. Discharged for disability 7/21/62. Age 63, 6', fair complexion, gray eyes, dark hair, Farmer. NFR.

GORDON, ALEXANDER BURGESS. Pvt., Co. A. b. S. C. 10/6/31. On post war roster. Reenlisted Co. K, 4th S. C. Cav. Grahamville 3/25/62. Present until ab. on leave 4/27/63 for 15 days. Present until detached with Col. Rhett 10/27/63 for 15 days. Present until ab. on leave 2/25/64 for 40 days. Captured Trevillian 6/11/64. Sent to Pt. Lookout. Exchanged 3/11/65. Paroled 4/65. Carpenter, Charleston 1870 census. d. 2/9/83. Bur. Gordon Cem., Cross Anchor, Spartanburg Co.

GORDON, JAMES P. Pvt., Co. A. b. S. C. 6/38. Farmhand, age 24, Abbeville Dist. 1860 census. Enl. Abbeville CH 8/13/61 age 22. Ab. on sick leave 10/31/61. Present 11/61-4/62. NFR. Reported serving in McGowan's Brigade 1864. NFR. Retired, age 75, Lone Cane, Abbeville Co. 1910 census.

GOW, E. J. Pvt., Co. F. Enl. by 1862. WIA (leg amputated) 9/20/62. NFR.

GRAHAM, J. R. Sgt., Co. H. Enl. Rock Hill 2/10/62. On picket Rantole Bridge 3-4/62. Present 5-6/62. On picket 7-8/62. Detailed to drive horses from Summerville, S. C. to Staunton, Va. 9-10/62. Present 11-12/62, reduced to Pvt. Present 3-6/63. Ab. on horse detail 8/18-31/63. Present 9-10/63. Transferred Co. C, 1st N. C. Cav. 11/2/63. Present through 8/64. Ab. on detached service 10/14-31/64. Present 11-12/64. Paroled Burksville, Va. 4/22/65.

GRAHAM, JOHN W. Pvt., Co. E. b. Cherokee Co., Ala. 12/42. Enl. by 1864. Captured Cherokee Co., S. C. 7/12/64. Sent to Nashville & Louisville, Ky. Sent to Camp Douglas 7/28/64. Took oath and enlisted Co. K, 6th U. S. Vols. 3/4/65. Age 22, 5'10", dark complexion, hazel eyes, auburn hair, Farmer. Present through 7/65. Escort for Paymaster to Ft. Laramie 9-12/65. On detached duty with artillery at Bishop Ranch, Nevada Territory 1-4/66. Transferred QM Dept. & promoted Cpl. 10/6/66. Assigned to Ft. Reamy, Nevada Territory. NFR. Farmer, Hickory Ridge, Monroe Co., Ark. 1870 census. Farmer, Rose Creek, Perry Co., Ark. 1900 census. Res. Canon, Fremont Co., Nev. 1910 census.

GRANT, JAMES BENSON. Pvt., Co. F. b. S. C. 7/25/46. Enl. Pickens Dist., Militia 1858. 5' 11", dark complexion, hazel eyes, dark hair, Farmer. Enl. Columbia 7/16/64. Present through 8/64. Ab. sick 10/16-31/64. Present 11-12/64. Paroled Greensboro, N. C. 4/26/65. Farmer, Keowee, Oconee Co. 1880 census. Miller, Keowee, Oconee Co. 1900 census. d. 4/1/08. Bur. Whitmire Meth. Ch. Cem., Oconee Co., S. C. Wife receiving pension Salem, Oconee Co. 10/27/19.

GRAVES. GEORGE CRAWFORD. Pvt., Co. G. b. S. C. 4/10/45. Student, age 15, Abbeville Dist. 1860 census. Enl. Co. B, Battalion of State Cadets, Charleston 7/10/63. Transferred Arsenal Academy for duty 8/17/63. NFR. Enl. Co. G, 1st S. C. Cav. Columbia 9/16/64. Present through 12/64. Captured in hospital, Raleigh, N. C. 4/13/65. Paroled 5/5/65. Farmer in postwar. Receiving pension Calhoun Falls, Abbeville Co. 4/7/19. Retired, age 75, Magnolia, Abbeville Co. 1920 census. d. 6/2/23. Bur. Latimer, S. C.

GRAY, H. JUDGE. Pvt., Co. B. Res. Laurens Co. Enl. 1861 age 35. d. at home 1861 on post war roster.

GRAY, J. BERRY. Pvt., Co. B. b. S. C. circa 1833. Farmer, age 27, Laurens Dist. 1860 census. Enl. Clinton 8/22/61 age 28. Present through 10/61. On picket Edisto River 11-12/61. Present 3-7/62. Ab. on guard Staunton, Va. 9-10/62. Teamster hauling forage 11/62-10/63. Detached with Infirmary Corps 11/63-2/64. Detached Green Pond 4/26-30/64. Present 5-10/64. Detached Pocotaligo 12/2-31/64. NFR. Farmer, age 37, Youngs, Laurens Co. 1870 census. Farmer, age 47, Washburn, Logan Co., Ark. 1880 census. d. by 1898.

GRAY, JOEL HARVEY. Pvt., Co. B. b. S. C. 1823. Enl. Camp Trapp 4/1/62. On picket Hallover Bridge through 4/30/62. Ab. sick Church Flats hospital 5-6/62. Ab. on sick leave 7-8/62. Ab. on detail driving horses from Summerville, S. C. to Staunton, Va. 9-10/62. Present 11/62-4/63. WIA & horse killed Brandy Station 6/9/63. Ab. wounded in Richmond hospital until furloughed for 30 days 8/19/63. Ab. wounded through 10/63. AWOL 12/1/63-2/64. Detached Green Pond 4/16-30/64. Present 5-8/64. Ab. detached guarding POW's Florence 9/16-10/31/64. Detached Pocotaligo 12/2-31/64. NFR. Farmer, Laurens Co. 1870 census. d. 5/17/07 age 84. Bur. Dial Meth. Ch. Cem., Laurens Co. Wife receiving pension Gray Court, Laurens Co. 1919.

GRAY, P. LEANDER. Pvt., Co B. b. S. C. circa 1847. Age 3, Laurens Co. 1850 census. Married Laurens Co. 1860. Enl. age 17 on post war roster. Alive 1898.

GREEN, IRVING. 1st. S. C. Cav. Not on muster rolls. d. date unknown. Bur. Oakwood Cem., Richmond.

GREEN, JOHN S. Pvt., Co. K. b. Barnwell Dist. circa 1841. Enl. John's Island 6/25/62. Present through 12/62 & 7-8/63. WIA (left side) Upperville 6/21/63. Ab. wounded in Richmond hospital until discharged for disability 8/6/63. Age 22, 5'10", light complexion, brown eyes, dark hair, Student. NFR. Farmer, age 39, Sleepy Hollow, Aikens Co. 1880 census.

GREEN, NATHANIEL A. 3rdLt., Co. B. b. Laurens Dist. 12/31/19. Enl. Clinton 8/22/61 age 41, as 3rdSgt. Ab. sick 10/31/61. Present 11/61-4/62. Promoted 2ndSgt. Present 5-12/62. Promoted 3rdLt. Present 3-12/63. On Provost Guard, Hanover Junction 2/1-29/64. Present 3-4/64. 2 officers & 29 m3n present 3/64. On detached service Charleston 6/18-7/31/64. Resigned for chronic rheumatism 8/13/64 age 46. NFR. Farmer, age 60, Clinton, Laurens Co. 1880 census. d. 11/28/93. Bur. Clinton City Cem.

GREEN, THOMAS W. Pvt., Co. C. b. S. C. circa 1828. Enl. Camp Butler 9/13/61 age 33. Present through 12/31/61. Present 5-6/62. Ab. sick 7/27-31/62. AWOL 9-10/62. Ab. on sick leave 10/1/62-8/31/63. NFR. Alive 1898.

GREGORIE, ALEXANDER FRAISER. Pvt., Co. A. b. S. C. 7/1/24. On post war roster. Reenlisted Co. K, 4th S. C. Cav. Grahamville 3/25/62. Present until AWOL 11-12/63. Present until ab. on horse detail 8/11/64 for 40 days. Ab. on sick leave 10/4/31/64. NFR. d. 1/14/04. Bur. Stoney Creek Cem., Beaufort Co.

GREGORIE, ISAAC MC PHERSON. Pvt., Co. A. b. S. C. 1/15/25. M. D. On post war roster. Reenlisted Co. K, 4th S. C. Cav. Grahamville 3/25/62. Present until appointed Surgeon of regiment 2/13/63. Present through 10/31/64. Paroled 4/65. d. 9/27/91. Bur. Stoney Creek Cem., Beaufort Co.

GREGORIE, WILLIAM DOUGLAS, JR. Pvt., Co. A. b. S. C. 10/10/25. On post war roster. Reenlisted Co. K, 4th S. C. Cav. Grahamville 3/25/62. Present until transferred Col. Colcock's 8th S. C. Bn. Cav. 8/5/62. Also listed as serving in Co. H, 3rd S. C. Cav. but not on rolls. d. 12/4/00. Bur. Stoney Creek Cem., Beaufort Co.

GRESSETT, MARION S. 3rdSgt., Co. E. b. Orangeburg Dist. 1837. Att. College, Ebenezer, York Dist. Teacher. Enl. Branchville 10/26/61 age 20 as Pvt. Promoted 1stCpl. & present through 12/31/61. On picket at Railroad 3-4/62. Present 5-8/62, promoted 4thSgt. 8/13/62. Present sick in camp 9-10/62. Present 11-12/62. Promoted 3rdSgt. & ab. guarding baggage Gordonsville 3-4/63. Present Brandy Station, Upperville, Louisa CH and Gettysburg in postwar account. Ab. sick in hospital 8/31/63. Present 9-12/63. Detached with disabled horses at recruiting station 2/1-29/64. Detached Green Pond 4/16-30/64. Present 5-6/64. Present sick in camp 7-10/64. Present 11-12/64. Paroled Hillsboro, N. C. 4/65. Gd. Medical College of S. C. in Charleston 1869. M. D., Branchville through 1900. d. there 5/19/08.

GRIFFIN, ANDERSON. Pvt., Co. F. b. Pendleton Dist. 1821. Enl. Colleton Dist. 6/16/62 as substitute for W. G. Blassingame 6/16/62 age 42. Present through 10/62. On scout 11-12/62. Ab. sick 3-4/63. WIA (right arm amputated) Upperville 6/21/63. Ab. wounded in Richmond hospital until retired 9/15/64. Assigned to Invalid Corps 10/26/64. NFR. No occupation, age 58, Pickens CH 1880 census. d. Pickens Co. 1887.

GRIFFIN, JAMES B. Pvt., Co. D. b. S. C. circa 1836. Student, age 24, Lowriesville, Chester Dist. 1860 census. Enl. Co. E, 6th S. C. Inf. Chester Dist. 4/11/61 age 25. Ab. sick most of time until discharged 4/25/62. Enl. Co. D, 1st S. C. Cav. Chester CH 7/28/62. Present through 10/62. On scout 12/25-31/62. Present through 8/31/63. Ab. on horse detail 10/23-12/31/53. Present 1-6/64. Present sick in camp 7-8/64. Ab. sick in Summerville hospital 9-10/64. Present 11-12/64. Paroled Hillsboro, N. C. 4/65. Farmer, age 43, Chester Co. 1880 census.

GRIFFIN, THOMAS. Pvt., Co. F. Enl. Columbia 2/15/64. On detached duty Green Pond 4/25-30/64. Present 5-12/64. NFR.

GRIFFIN, W. M. Pvt., Co. G. Enl. 1864. Present 11-12/64. NFR.

GRIMES, ANDREW JACKSON. Pvt., Co. E. b. Orangeburg Dist. circa 1843. Enl. Adams Run 1/8/62. Present sick in camp 3-4/62. Ab. on sick leave 5/10-6/30/62. AWOL 7-8/62. Ab. sick in Summerville hospital 9/30/62-6/63. Ab. sick in Columbia hospital 10/26-31/63. Ab. sick with gonorrhea in Richmond hospital 12/20/63-2/64. Age 30, 5'10", dark complexion, grey eyes, dark hair. Returned to duty 3/28/64. Ab. sick in Columbia hospital 4/64. Present 5-6/64. Detailed to guard POW's Florence 9/17-12/31/64. Paroled Hillsboro, N. C. 4/65. d. Hillsborough Co., Fla. 1/22/79. Wife receiving pension Polk Co., Fla. circa 1900.

GRIMES, LEWIS. Pvt., Co. E. b. S. C. circa 1837. Enl. Adams Run 1/8/62. Present through 6/62. Ab. on sick leave 7-8/62. Ab. sick in Summerville hospital 9-10/62. On picket 11-12/62. Present until captured Brandy Station 8/1/63. Sent to Old Capitol. Transferred Pt. Lookout. Exchanged 4/27/64. Ab. sick with debility in Richmond hospital 5/1/64. Furloughed for 30 days 5/6/64. Present 7-8/64. Detailed guarding POW's Florence 9/17-12/31/64, dismounted since 10/9/64. Paroled Hillsboro, N. C. 4/65. Farmer, age 42, Polk Co., Fla. 1885. d. Hillsborough Co., Fla. 10/22/98 on wife's pension application.

GROVES, GEORGE G. Pvt., Co. I. Enl. 4/3/63. AWOL through 4/30/62. NFR. Enl. Co. B, 3rd S. C. Cav. on roster but not on muster rolls. NFR.

GRUBER, JOHN WILLIAM. Pvt., Co. I. b. S. C. 6/24/46. Enl. Columbia 4/16/64. On detail Green Pond 4/26-30/64. Present 5-8/64. Ab. on sick leave 9/14-10/31/64. Detached at Pontoon Bridge 12/29-31/64. NFR. Farmer, Sheridan, Colleton Co. 1870-1900 censuses. d. 3/28/18.

GUYTON, JAMES C. Pvt., Co. C. b. Pickens Dist. 1845. Enl. Aiken 8/27/61 age 18. Present through 12/61/61. On picket Bear Bluff 1/62. On detached service Church Flats 5-6/62. Ab. on sick leave 7/12-8/31/62. Present 9/62-4/63. Ab. sick in Petersburg hospital 5/5-11/63. Present through 8/31/63. Ab. sick in Charlottesville hospital with typhoid fever 10/17-12/15/63. Ab. on horse detail 12/22/63-2/64. Ab. on horse detail 4/1-5/1/64. Present 5-8/64. Courier for Gen. Taliaferro 7/2-10/31/64. Detached Telegraph Office 12/28-31/64. NFR. Bookkeeper, age 25, Edgefield Co. 1870 census. d. Pickens, S. C. 3/6/30.

HADDON, THOMAS LUTHER. Pvt., Co. G. b. S. C. 4/10/47. Enl. age 18 on postwar roster. Farmer, Long Cane, Abbeville Co. 1900 census.   d. 11/3/05. Bur. Lindsay Cem., Abbeville Co.

HAFHER, JOHN ANDREW. Pvt., Co. K. b. Chester Dist. 7/20/46. Enl. Columbia 4/18/64. Present through 4/30/64. On picket fortifications of Charleston 6/29-30/64. Present 7-8/64. d. of yellow fever in Charleston hospital 10/28/64. Effects: $11.60 and sundries worth $42.10 given to Capt. Brown. Bur. Pleasant Grove Presb. Ch. Cem., Chester Co.

HAIR, HENRY M. Pvt., Co.'s C & K. b.Blacksville, Barnwell Dist. 4/26/41. Farmer, age 20, Edgefield Dist. 1860 census. Enl. Hamburg 8/27/61 age 21. Detached Wadmalaw Island 11-12/61. Transferred Co. K 6/27/62. Present through 10/62. On scout 11-12/62. Present 3-6/63. Captured Upperville 6/21/63. Sent to Old Capitol. Exchanged by 8/63. Ab. on horse detail 8/16-31/63. Present 9-10/63. Detailed with Provost Guard, Fredericksburg 11/9-12/31/63. Detailed horse recruiting camp 1/25-2/29/64. Present 2-6/64. On duty Telegraph Office 7/8/64. Present 9-11/64. Courier, Dist. Headquarters, Charleston 12/15-31/64. Paroled Raleigh, N. C. 4/65. Farmer, Butler, Edgefield Co. 1870 census. Cotton Mill Worker, Charleston 1900 census. d. 6/6/04. Bur. Magnolia Cem., Augusta, Ga.

HALE, MOSES THOMAS. Pvt., Co. A. Enl. by 10/11/61 on muster rolls. NFR.

HALL, P. Pvt., Co. A. Enl. Co. A, 19th Bn. S. C. Cav. 7/11/62. Present through 12/64. Transferred Co. A, 1st S. C. Cav. 1865. Paroled Augusta, Ga. 5/19/65.

HALL, SAMUEL B. Pvt., Co. H. Enl. Rock Hill 3/19/62. On picket Jacksonborough through 4/62. Present sick in camp 7-8/62. Present 9/62-4/63. WIA & captured Gettysburg 7/6/63. Sent to David's Island, N. Y. Exchanged by 8/63. In Richmond hospital 8/28/63 until furloughed for 30 days 9/7/63. On leave from Parole Camp -10/63. Present 11/63-2/64. Ab. on leave 4/27-6/64. Ab. sick in Charleston hospital

8/1810/31/64. AWOL 11-12/64. NFR.

HALL, WILLIAM S. 2ndCpl., Co. I. b. S. C. 1820. Enl. Parker's Ferry 4/3/62 as Pvt. On picket Jacksonborough through 4/62. Ab. on leave 5-6/62. Present sick in camp 7-8/62. Present 9/62-4/63. Promoted 4thCpl. Present through 8/31/63. Promoted 3rdCpl. Present 9-12/63. Ab. on leave 2/12-29/64. Detached Green Pond 4/16-30/64. Present 5-6/64. Promoted 2ndCpl. Present 7-12/64. Paroled Augusta, Ga. 5/18/65. Farmer, age 41, Lamar, Randolph Co., Ala. 1870 census. Confederate Veteran living Chester Co. 1904. d. 8/7/12. Bur. Evergreen Cem., Chester.

HALLEY, J. WILLINGHAM. 1stLt., Co. K. Res. Chester Dist. d. date and place unknown.

HAMILTON, A. Pvt., Co. C. Enl. Aiken Dist. age 18 on post war roster. Alive 1898.

HAMILTON, JOHN D. Pvt., Co. C. b. S. C. 8/45. Enl. Columbia 6/29/64. Present through 12/64. NFR. Farmer, age 37, Bullock's Creek, York Co. 1880 census. Grocer, age 54, Sharon Town, York Co. 1900 census.

HAMILTON, MC DUFFIE. Pvt., Co.'s F & K. b. S. C. circa 1840. Enl. Pickens CH 12/4/61 age 21. Present through 12/31/61. On picket Simmons House 3-4/62. Transferred Co. K 6/27/62. Present through 6/63. Ab. on horse detail 8/16-31/63. Present 9/63-2/64. Ab. on horse detail 8/29-31/64. Ab. sick in Charleston hospital 10/15-31/64. Ab. on sick leave for 60 days 11/7-12/31/64. Paroled Raleigh, N. C. 4/65. Res. Pickens Co. 1870. d. between 1870-80/ Bur. West View Cem., Easley, S. C. on wife's pension application.

HAMLIN, ALBERT BARTON. Pvt., Co. A. b. S. C. 11/14/40. Res. Abbeville Dist. 1860 census. Enl. Abbeville CH 3/19/62. Ab. on sick leave 4/15-30/62. Discharged 5/30/62. NFR. Farmer, Abbeville Co. 1900 census. Stock Dealer. d. Abbeville Co. 11/29/18. Bur. Melrose Cem.

HAMLIN, JAMES R. 3rdSgt., Co. F. b. S. C. circa 1837. Enl. Pickens CH 12/4/61 age 34 as Cpl. Promoted 4thSgt. Present through 12/31/61. Ab. on leave 3-4/62. On duty Church Flat 5-6/62. Ab. detached 7-10/62. Present 11/62-4/63. Promoted 3rdSgt. Captured Upperville 6/21/63. Sent to Old Capitol. Exchanged 8/63. Ab. detached 8/18-31/63. AWOL 9-10/63. Present 11-12/63. Ab. sick with Bronchitis in Richmond hospital 1/8-3/28/64. Present 4/64, dismounted since 1/1/64. Present 5-6/64. Present sick in camp 8/27-31/64. Present 9-12/64. NFR. Married Greenville Co. 1875. Living in Missouri circa 1900 on postwar roster.

HAMLIN, JOHN MARCUS. 3rdCpl., Co. F. b. Anderson Dist. 5/28/38. Enl. Pickens CH 5/15/62 as Pvt .On picket Camp Means through 6/62. Present 7-8/62. On detached service 9-10/62. On scout 11-12/62. Present 3-8/31/63. On detached service 9-10/63. Present 11-12/63. Detailed horse recruiting station 1/20-2/29/64. Detached Green Pond 4/16-30/64. Present 5-6/64. Detached horse recruiting camp 8/23-31/64. Promoted 3rdCpl. Present 9-12/64. NFR. Farmer, Williamston, Anderson Co. 1870 census. Farmer, Bervard, Translyvania Co., N. C. 1900-1920 censuses. d. there 5/25/25.

HAMMOND, WILLIAM P. Pvt., Co. H. b. S. C. 12/25/39. Enl. Rock Hill 3/19/62. Ab. with wagons through 4/62. Present 5-10/62. On raid 11-12/62. Ab. on horse detail 2/21-3/30/63. AWOL 7/16-10/31/63. Present 11/63-2/64. Detached Green Pond 4/16-30/64. Present 5-12/64. NFR. d. York Co. 11/17/71.

HANCOCK, D. O. 2ndLt., Co. C. b. Edgefield Co. 12/14/33. Enl. Hamburg 8/27/61 age 24, as Pvt. Present through 10/31/61. Ad. detached Wadmalaw Island 11-12/61.Present 5-6/62. Elected 2ndLt. and ab. on sick leave 7-8/31/62. Present 9-10/62. On picket 11-12/62. Ab. sick with typhoid fever in Charlottesville hospital 8/3/-9/21/63. Present 10/63. Commanding Co. 11-12/63. 32 horses present 12/31/63. Present sick in camp 1-2/63. Present 3-8/64. 1 officer & 30 men present 1-3/64. 1 officer & 23 men present 3/31/64. Signed for 1 jacket & 1 pr. shoes Columbia 4/10/64. Signed for 7 caps, 18 jackets, 32 pr. pants, 15 shirts, 16 pr. drawers, 4 pr. shoes & 10 pr. socks 5/20/64, Commanding Co. Signed for 15 pr. pants, 29 jackets, 3 pr. socks, 24 cotton shirts, 15 pr. drawers & 8 caps James Island 6/30/64. Commanding Co. 9-10/64. Detached Pocotaligo 12/10-31/64. NFR. Paroled Greensboro, N. C. 4/26/65 on pension application, farmer. Columbus Co. Ga. 1880 census. Receiving pension Augusta, Ga. 1907.

HANCOCK, J. Pvt., Co. G. On post war roster. Reenlisted Co. H, 7th S. C. Cav. Richland Co. 7/14/63. Present until captured New Kent CH 8/29/63. Exchanged. Issued clothing 11/15/64. NFR.

HANKINSON, ALFRED G. Pvt., Co. C. b. Millbrook, Aiken Dist. 1846. Enl. Camp Johnson 10/8/61 age 17. Present through 12/31/61. On picket Bear Bluff, Wadmalaw Island 1/62. Present 5-6/62. Ab. on sick leave 7/27-30/62. Present 9/62-8/31/63. Ab. on horse detail 9-12/63. On detached service 2/1-29/64. Ab. on horse detail 4/12-5/1/64. Present through 8/64. Detailed to drive cattle 9/22-12/31/64. NFR. d. Aiken, S. C. 6/12/13.

HANKINSON, JOHN NEAL. 3rdLt., Co. C. b. Aiken, S. C. 2/22/43. Gd. Yorkville Academy. Farmer. Served briefly in Co. E, 1st S. C. Inf. (Hagood's). Enl. Camp Butler 10/8/61, age 21, as Pvt. Detached Wadmalaw Island 11-12/61. Present 5-6/62. Appointed 3rdLt. 7/7/62. Ab. on sick leave 8/13-31/62. Present 9-12/62. Signed for feed for 136 horses Staunton 10/21/62. Present Brandy Station (both fights) & Gettysburg. Signed for feed for 25 horses in the field 7/3/63. Signed for 4 pr. shoes 8/4/63. Paid $1.00 each for shoeing horses 8/11/63. Ab. sick with "Febris Conmumo" in Richmond hospital until furloughed for 30 days 9/21/63. Ab. sick leave 9-10/63. Present 11/63-8/64, however, detached Green Pond 4/29-30/64. Signed for 17 pr. shoes, 7 blankets, 13 pr. pants, 4 shirts, 1 pr. socks, 5 jackets, 3 pr. drawers 4 tent flys, 1 spider, 3 camp kettles, 1 mess pan, & 1 axe 3/31/64. Signed for 1 spade, 2 axes, 5 camp kettles & 13 mess pans 6/30/64. On detached service guarding POW's Florence 9/17-12/31/64. Rejoined at Fayetteville, N. C. 4/65. Paroled Aikenville, S. C. Surveyor, Windsor, Aiken Co. 1880 census. Civil Engineer. Clerk of Court Aiken Co. 1888-1896. Farmer, Windsor, Aiken Co. 1920 census. d. Aiken, S. C. no date. Bur. family cem., Aiken Co.

HANKINSON, RICHARD J. Pvt., Co. C. b. S. C. 1845. Res. Barnwell Dist. 1860 census. Enl. Hamburg 8/17/61 age 18. Ab. on sick leave 12/21-31/62. On picket Bear Bluff, Wadmalaw Island 1/62. Present 3-6/62. Discharged by Surgeon 7/1/62 (Probably for underage). Reenlisted Columbia 3/5/63. Present 5-6/63. Ab. on horse detail 8/15-31/63. Present 9-10/63. Ab. on horse detail 1/22/63-2/64. Ab. on horse detail 4/16-5/3/64. Present through 8/64. On guard Dill's Bluff for Capt. Marshall 8/22-12/31/64. Paroled Augusta, Ga. 5/19/65. d. Barnwell, S. C. 3/12/15.

HANKINSON, RICHARD R. Pvt., Co. C. b. Edgefield Dist. circa 1839. Enl. Wadmalaw Island 4/3/62. Ab. on sick leave 5-8/31/62. Present 9-12/62. Ab. sick with chronic rheumatism White Sulphur Springs hospital 1/15-9/11/63. Present through 10/63. Ab. sick with "cele of the nose" in Richmond hospital 11/20/63-3/28/64. Age 35, 5'6", light complexion, blue eyes, light hair, Farmer 2/3/64. Detached Green Pond 4/16-30/64. Present 5-12/64. NFR. Farmer, Aiken Co. 1880 census. Alive 1898.

HAPPOLDT, JOHN PHILLIP. Pvt., Co. A. b. S. C. 9/25/30. Farmer, Charleston Co., 1860 census. On post war roster. Reenlisted Co. K, 4th S. C. Cav. Grahamville 3/25/62. Present until detached with Gen. Walker 12/6-31/62. Present until ab. on sick leave 2/22/63. Present until detailed in Railroad Workshops 6/1/63-4/64. Transferred to Kanapaux's Bty. 4/15/64, but not on rolls of that unit. d. Charleston 8/26/05. Bur. Magnolia Cem

HARDEN, ADAM WALKER. Pvt., Co.'s K & D. b. Chester Dist. 3/3/47. Student, age 13, Chester 1860 census. Enl. Columbia 4/18/64. Present through 6/64. Ab. sick in Charleston hospital & furloughed 7-8/64. Transferred Co. D 10/1/64. Present 11-12/64. Paroled Hillsboro, N. C. 4/65. Farmer, Chester, Chester Co. 1880-1920 censuses. d. 9/1/20. Bur. Capers Chapel Bapt. Ch., Chester Co. Brother of Benton F, Ebenezer L. & William A. Harden.

HARDEN, ANDREW JACKSON. Pvt., Co. D. b. S. C. 5/1/31. Miller, age 29, Chester Co. 1860 census. Enl. Chester CH 9/10/61. Ab. sent for baggage 12/31/61. On picket Bear Bluff 3-4/62. Present 5-6/62. Ab. sick 7-8/62. Ab. sick in Summerville hospital 9/21-12/31/62. Present 3-4/63. Ab. sick with rheumatism in Richmond hospital 8/26/63-4/11/64. Furloughed for 60 days. Present 6/64. On retired list 8/23/64. Surrendered Greensboro, N. C. 4/26/65. Farmer, York Co. 1870 census. d. Chester Co. 1/1/97. Bur. Unity Bapt. Ch. Cem.

HARDEN, BENTON FLETCHER. Pvt., Co. D. b. Chester Dist. 10/44. Enl. Camp Means 4/4/62. On picket Bear Bluff through 4/20/62. Present 5-6/62. Courier for Col. Dunovant 7-8/62. Present 9-10/62. Ab. sick in hospital 11/26-12/31/62. Present 3-10/63. Ab. on horse detail 12/22/3-2/64. Present 3-6/64. Present sick in camp 7-8/64. Present 9-10/64. Detached Coosawhatchie 12/10-31/64. Ab. sick with debilitis in Charlotte hospital 2/2-8/65. Paroled Greensboro, N. C. 4/26/65. Farmer in postwar. d. Weston, Texas 4/23/10. Bur. Cottage Hill Cem., Chester Co., S. C. Brother of Adam W., Ebenezer L., & William A. Harden.

HARDEN, EBENEZER "EBB." L. Pvt., Co. D. b. Chester Dist. circa 1838. Enl. Co. A, 6th S. C. Inf. Chester CH 4/11/61 age 24. Discharged for phthisis pulmonalis or "Haemopthsis" Centerville, Va. 2/8/62, age 23. 6', sallow complexion, hazel eyes, dark hair, Farmer. Reenlisted Co. D, 1st S. C. Cav. Chester CH 5/15/62. On picket Townsend's House through 6/30/62. Ab. on sick leave 8/1-3/62. Present 9-10/62. Present, nursing sick in camp 11-12/62. Present 3-4/63. Ab. sick in hospital 6/1-30/63. Horse killed 8/1/63. Paid $450.00. Ab. sick with "hemorages" in Richmond hospital 8/31-9/30/63, furloughed for 30 days. Ab. on sick leave 12/1-31/63. Ab. sick in Columbia, S. C. hospital l2/2-15/64. Discharged for disability 2/16/64. Age 36, 5'11", dark complexion, hazel eyes, dark hair, Farmer. NFR. Brother of Adam W., Benton F. & William A. Harden.

HARDEN, JOHN H. Pvt., Co.'s D & K. b. Chester Dist. 2/7/40. Enl. Co. D Chester CH 9/10/61, age 25. Ab. guarding baggage at Depot 11-12/61. Present 3-4/62. Ab. on sick leave 5/13-6/30/62. Discharged 8/8/62. Age 26, 5'10", fair complexion, light hair, blue eyes, Farmer. Reenlisted Co. K 32/63. Transferred Co. D 6/1/63. Present 5-6/63. Transferred to Regimental Band 7/1/63. Present 9-10/63. Ab. sick with "Henriat Febris Remittens" in Richmond hospital 11/30-12/22/63. Ab. on horse detail 12/23/63-2/64. Present in Co. D 3-7/64. Ab. on horse detail 8/9-31/64. Present 9-12/64. Paroled Greensboro, N. C. 4/26/65. Farmer, age 60, Drew, Bartholmew Co., Ark. 1900 census. d. Dermott, Chicot Co., Ark. 9/3/07.

HARDEN, JOSEPH NEELY. Pvt., Co.'s K & D. b. Chester Co. 8/11/46. Age 13, Chester Co. 1860 census. Enl. Co. K, Columbia 4/18/64 age 17. Present until transferred Co. D 5/10/64. Present through 12/64. Paroled Hillsboro, N. C. 4/65. Farmer, age 32, Chester Co. 1880 census. d. Chester Co. 2/18/03. Bur. Capers Chapel Cem.

HARDEN, MILES. Pvt., Co.'s C & K. b. Edgefield Dist. circa 1828. Farmer, age 32, Beech Island, Edgefield Dist., 1860 census. Enl. Co. C Wadmalaw Island 4/3/62 age 33. Transferred Co. K 6/27/62. Discharged for physical disability 8/21/62. Age 33, 5'10", dark complexion, grey eyes, dark hair, Farmer. NFR. Farmer, Hammond, Edgefield Co. 1880 census. d. by 1898. Bur. Springwood Cem., Greenville, S.C.

HARDEN, WILLIAM A. Pvt., Co. D. b. Chester Dist. 1842. Age 18, Chester, 1860 census. Enl. Camp Ripley 2/16/62. Present through 4/62. On picket Townsend's House 5-6/62. Courier, Col. Dunovant 7-8/62. Present 9-10/62. Ab. on scout 12/25-31/62. Present 3-6/63. Ab. on horse detail 8/18/31/63. Present 9/63-12/64. Ab. sick with "Febris Int." in Charlotte hospital 4/20/27/65. Paroled Greensboro, N. C. 5/1/65. Farmer, age 28, Baton Rouge, Chester Co. 1870 census. Alive 1909.

HARLEY, ALONZO MONCRIEF. Pvt., Co.'s C & K. b. S. C. 9/27/43. Enl. Co. C Hamburg 8/27/62. Detached Wadmalaw Island through 12/61. On picket Bear Bluff, Wadmalaw Island 1/62. Present 5-6/62. Transferred Co. K 7/1/62. Present through 8/31/63, however, ab. sick with "Affiction of Kidney" in Farmville hospital 6/10-720/63. Ab. on horse detail 9-10/63. AWOL Barnesville, S. C. 12/23-31/63. Present in arrest 1/6-2/64. Ab. on horse detail 3-4/64. Ab. on leave 6/20/7/31/64. AWOL 8/29-31/64. Present 9-12/64. Paroled Raleigh, N. C. 4/65. d. Jackson, Ala. 6/17/11. Bur. Jackson Memorial Park Cem., Jackson, Ala.

HARLEY, CALLAWAY A. Pvt., Co. E. b. S. C. circa 1834. Res. Orangeburg Dist. Enl. Co. C, 2nd S. C. Arty. Orangeburg 3/19/62 as 3rdLt. Resigned for "incepient TB" 6/26/63. Enl. Co. E, 1st S. C. Cav. 3/29/64. Ab. detached Green Pond 4/16-30/64. Ab. on sick leave 6/15-8/31/64, furloughed from Charleston hospital 6/25-9/24/64 AWOL 10/24-31/64. Detached Pocotaligo 12/10-31/64. Captured Salisbury, N. C. 4/12/65. Sent to Nashvlle and Camp Chase. Released 6/13/65. Age 31, 5'7", florid complexion, dark hair, grey eyes. Paid taxes in S. C. 1866. Bur. Sunnyside Cem., Orangeburg, S. C., no marker.

HARLESTON, EDWARD, JR. Pvt., Co. A. b. S. C. 20/20/35. On post war roster. Reenlisted Co. K, 4th S. C. Cav. Grahamville 3/25/62 as 1stSgt. Present until elected 3rdLt. 1/63. Present until detailed as Acting Adjutant at Green Pond 5-6/63. Present until ab. on leave 2/23/64 for 20 days. Present until commanding company 8/31/64. Commanding Squadron 9-10/64. Ab. sick with "splenitus" in Danville hospital 11/29-12/30/64. Sent from Lynchburg to Lancaster Co., S. C. 12/31/64. Paroled 4/65. d. 10/30/91. Bur. Strawberry Chapel Cem., Berkeley Co.

HARLESTON, JOHN E. Pvt., Co. A. On post war roster. Reenlisted Co. K, 4th S. C. Cav. Pocotaligo 12/1/62. Present until appointed 2ndLt., 1st S. C. Arty. DOW's 11/24/63.

HARMON, FREDERICK. Pvt., Co. I. b. S. C. 10/45. Enl. Columbia 4/11/64. Present through 7/64. Detailed to Commissary Dept. 8/8/64-12/31/64. NFR. Farmer, age 54, Richland Co. 1900 census. d. Columbia, S. C. 8/15/15. Bur. St. John's Ch., Lexington Co, S, C.

HARPER, JAMES B. R. Pvt., Co. H. b. S. C. 3/30/35. Enl. Rock Hill 10/31/63. Ab. on horse detail 12/22/63-2/64. Present 3-8/64. Detailed guarding POW's Florence 9-12/64. Paroled as Courier, Greensboro, N. C. 5/1/65. Farmer, Bethel, York Co. 1900 census. d. 10/8/04. Bur. Bethel Presb. Ch., York Co.

HARRIS, DANIEL G. Pvt., Co. C. b. Ridgeville, S. C. circa 1846. Enl. Columbia 5/5/64. Present through 8/64. Ab. on horse detail for 15 days 10/27/64. Detached Pocotaligo 12/1--31/64. NFR. d. Charleston, S. C. 11/13/81. Bur. Ridgeville, S. C.

HARRIS, N. SPENCER. Pvt., Co. B. b. S. C. circa 1829. Enl. Co. I, 3rd S. C. Inf. Clinton 4/14/61 as 2ndLt. age 32. Present through 12/61. Not reelected 65/13/62. Enl. Co. B, 1st S. C. Cav. Columbia 4/18/63. Present through 10/63. Ab. sick with bladder irritation in Richmond hospital 11/30-12/23/63. Furloughed for 30 days. Present 2/64. Detached Green Pond 4/26-30/64. Present 5-6/64. Ab. on horse detail 8/10-31/64. Present 9-10/64. Detached Pocotaligo 12/2-31/64. NFR. d. by 1898. Bur. Clinton City Cem., no dates.

HARRIS, THOMAS D. Pvt., Co. I. b. S. C. circa 1840. Enl. Co. I, 5th S. C. Inf. York 4/13/61 age 21. Present through 12/61. NFR. Enl. Co. G, Palmetto Sharpshooters York 4/13/62. NFR. Enl. Co. G, 25th Bn. Va. Local Defense Troops 5/5/62. Transferred Co. I, 1st S. C. Cav. 5/17/63. Joined Culpeper CH 7/10/63. Ab. on horse detail 8/18-31/63. Present 9-10/63. Ab. sick with chronic diarrohea & lumbago in Richmond hospital 11/28-2/64. Ab. sick 3/30-4/30/64. Present 5-6/64. Ab. on horse detail 8/10-31/64. Ab. on leave 10/31/64. AWOL 11/21-12/31/64, said to be in Lunatic Asylum. NFR.

HASTE, G. D. Pvt., Co. K. Paroled Greensboro, N. C. 5/1/65 as Courier.

HAVERLY, B. B. Pvt., Co. A. Enl. Abbeville CH 5/13/63. Present through 6/63. Ab. on horse detail 8/18-31/63. Present 9-10/63. Ab. on horse detail 12/23/63-2/64. Ab. on leave 4/29-30/64. Present 5-12/64. Paroled Greensboro, N. C. 5/1/65.

HAWKINSON, JOHN N. SR. Pvt., Co. C. b. White Pond, Aiken Dist. 2/22/40. Not on muster rolls. Served 6/61-4/65 on pension application. Paid taxes in S. C. 1866. Farmer, Aiken Co. 1900 census. Applied for pension from White Pond, Aiken Co. 4/29/19.

HAYNES, HARRISON. Blacksmith, Co. F. b. S. C. 1822. Enl. Colleton Dist. 4/8/62 as Blacksmith/Artificer for Company. Present through 4/30/62. Ab. sick 6/23-7/8/62. Ab. on leave 9-10/62. AWOL 10/17-12/31/62. Ab. sick in with "Enysipelus" in Farmville hospital 4/25-5/29/63. Present through 4/64. Present sick in camp 6/25-30/64. Ab. sick in hospital 7/10-12/31/64. NFR. Farmer, Keowee, Oconee Co. 1880 census. d. 1895. Bur. New Hope Bapt. Ch., Senaca, S. C.

HAYS, WILLIAM H. Pvt., Co. H. b. S. C. 1844. Enl. Rock Hill 6/25/64. Present through 12/64. Paroled Charlotte, N. C. 5/22/65. Farmhand, age 26, Hillsboro, Marion Co. 1870 census. Farmer, age 57, Manning, S. C. 1900 census. Farmer, age 63, Buford, Lancaster Co. 1910 census.

HAYWARD. J. K. Pvt., Co. A. On post war roster. Also listed in Co. K, 4th S. C. Cav. & Co. H, 3rd S. C. but not on rolls of either unit.

HAZARD, PATRICK. Pvt., Co. A. b. Ireland circa 1819. Enl. Abbeville CH 8/13/61 age 42. On duty Bear Island 11-12/61. Present 3-4/62, horse died 4/17/62. Ab. on horse detail 5-6/62. Present 7-10/62, dismounted. Present 11/62-10/63. On picket 12/30/63-1/2/64. Detached, recruiting horses 1/19-2/29/64. Present 3-10/64. Detached Pocotaligo 12/10-31/64. Discharged 1/11/65. Enl. Co. F, 1st Confederate Bn. Deserted 3/24/65. Sent to Washington, D. C. Took oath and sent to Pittsburg, Pa. 5/18/65.

HAZARD, THOMAS. Pvt., Co. A. b. Ireland circa 1831. Res. Abbeville Dist. Enl. Abbeville CH 8/13/61 age 30. On duty Bear Island 11-12/61. Present 3/62-8/31/63. Ab. on detail Brigade Headquarters 9-10/63. Ab. detailed horse recruiting station 11-12/63. On picket 2/27-29/64. Present 3-6/64. On picket 10/30-31/64. Present 11-12/64. Deserted to the enemy James Island 1/6/65. Age 40, 5'10", light complexion, blue eyes, light brown hair. Farmer, Monroe Co., N. Y. 1870 census.

HEATH, GEORGE DAWSON. b. 5/19/46. On Chester Co. Confederate Veterans list 1904. Res. Chester 1910 census. d. Chester 5/6/14. Bur. Evergreen Cem.

HEATH, WILLIAM MOULTON. Pvt., Co. A. b. Canada East 1824. Enl. Abbeville CH 8/13/61 age 35. On duty Bear Island 11-12/61 and Bear Bluff 3-4/62. Present 5-6/62. Present sick in camp 7-8/62. Ab. sick 9-10/62. On picket 11-12/62. Present 3-10/63. Ab. sick with acute diarrohea in Richmond hospital11/7/12/1/63. Furloughed for 30 days 12/2/63. Ab. sick 1-2/64. Present 3-4/64. Appointed Bonded Agent in QM Dept. & Purchasing Agent for Greenville Dist. at Greenville CH 5/24/64. NFR. Farmer, age 44, Grant, Edgefield Dist. 1870 census. Farmer, Collier, Edgefield Co. 1880 & 1890 censuses. d. 3/18/92. Bur. Antioch Bapt. Ch., Edgefield Co.

HEISE, JOHN EDWARD. Pvt., Co. G. b. Columbia, S. C. 11/46. Enl. Columbia 11/4/64. Transferred to Regimental Band 12/1/64. Present through 12/31/64. Paroled as Courier Greensboro, N. C. 5/1/65. d. Columbia, S. C. 8/3/06. Bur. Elmwood Memorial Gardens, Columbia.

HEMMINGER, JOHN OLIVER. Pvt., Co. G. b. S. C. 3/19/46. Enl. Abbeville CH 2/1/64. Detached Green Pond 4/27-30/64. Present 5-10/64. Detached Pocotaligo 12/10-31/64. Paroled Washington, Ga. 5/7/65. Farmer, Calhoun Mills, Abbeville Co. 1870-1900 censuses. d. 1/4/02. Bur. Guthrie Cem., McCormick Co. Wife receiving pension Campobello, Spartanburg Co. 11/1/19.

HENDERSON, JAMES. Captain & Quartermaster. b. S. C. circa 1831. Farmer, age 29, Gowenville, Greenville Co. 1860 census. Enl. Co, I, Cobb Ga. Legion Atlanta, Ga. 1/16/63. Present through 4/63. Transferred 1st S. C. Cav. as Acting QM 10/23/63. Signed for 1 wool shirt, 6 pr. socks, 7 jackets, 9 pr. pants, 1 overcoat, 7 shirts, 7 pr. drawers, 2 blankets, 2 pr. shoes 1/29/64. Recommended by Col. Black. Appointed Columbia 4/1/64. Ordered to Richmond 4/27/64. Present 5-6/64. Signed for 7 mules, 2 horses, 2 wagons, 4 sets wagon harness, 2 m. of buttons, 4 sets lead harness, 7 halters & 2 wagons saddles. Ab. detailed at Summerville, S. C. by order Gen. Hardee 9-10/64. Present 11-12/64. Paroled Augusta, Ga. 5/24/65.   d. S. C. 1895, res. Ga.

HENDERSON, WILLIAM PINCKNEY. Pvt., Co. B. b. S. C. 5/42. Enl. Clinton 8/22/61 age 20. Ab. sick 10/31/61. On picket Larochers 12/31/61. On picket Hallover Bridge 3-4/62. Present 5-6/62, horse killed in skirmish 6/7/62. On duty Edisto Island 7-8/62. Present 9/62-4/63. MIA through 10/31/63. Apparently captured & exchanged. Present 11-12/63, lost horse 7/7/63, remounted 12/23/63. Present 1-8/64. Deserted 9/12/64. NFR. Laborer, age 59, Mt. Vernon, Fairfax Co., Va. 1900 census. d. District of Columbia 11/19/06.

HENDERSON, WILLIAM R. Blacksmith, Co. B. b. S. C. 1826. Negro. Res. Spartanburg Co. Enl. Camp Tripp 5/5/62. Present through 8/62. Ab. on march with horses from Summerville, S. C. to Staunton, Va. 9-10/62. Ab. on detached service as Blacksmith 11-12/62. Present 3-4/63. Ab. on horse detail 8/18-31/63. Present 9-12/63. Present as Blacksmith for Company 1-2/64. Ab. on horse detail 4/24-30/64. Present 5-6/64. Ab. sick in Charleston hospital 8/25-10/31/64. d. of disease in Summerville hospital 11/20/64 age 38.

HENDRIX, FIELD E. Pvt., Co. F. b. S. C. circa 1834. Enl. Co. I, 4th S. C. Inf. Pickens Dist. 4/14/61. Present through 6/31/61. Discharged for "Effects of Rubeola" 9/16/61. Age 27, 5'5", fair complexion, blue eyes, yellow hair, Farmer. Enl. Co. F, 1st S. C. Cav. Pickens Dist. 4/8/62. Present through 4/30/62. On picket Camp Means 5-6/62. Ab. sick with chronic bronchitis in Summerville hospital 7-8/62. Discharged for disability 10/18/62. NFR.

HENNEMAN, JOHN ADAM. Pvt., Co. B. b. Bavaria, Germany 1836. Res. Charleston. Enl. Co. A, Holcombe Legion, Adams Run 12/28/61 as 2ndSgt. Ab. sick leave 2/19-3/5/62. Present through 4/62. Elected 3rdLt. 5/1/62. Under arrest Adams Run 6/62. WIA & in Charlottesville hospital 8/27/62. Transferred to Lynchburg hospital with rheumatism 9/18/62. Resigned 2/16/63. Enl. Co. B, 1st S. C. Cav. Columbia 3/3/64. Present through 8/64. AWOL 10/25-31/64. Detailed Courier, McLeod Hospital 11/26-12/31/64. Deserted to the enemy Charleston 2/18/65. Took oath 4/1/65. Age 30, 5'6", light complexion, dark eyes, dark hair. Sent to Hilton Head & New York City 4/7/65. Jeweler, age 34, Spartanburg 1870 census and age 44, 1880 census. d. 1891. Bur. Oakwood Cem., Spartanburg.

HENNIES (HEINS on muster rolls), HENRY J. Pvt., Co. H. b. S. C. 8/18/46. Enl. Columbia 7/14/64. On picket 7-8/64. Detailed as Mechanic, Engineering Dept., James Island 8/31/64. Present sick 9-10/64. Detached as Carpenter 12/7-31/64. NFR. Merchant, Columbia 1900 census. d. 11/6/20. Bur. Ebenezer Luth. Ch., Columbia.

HERBERT, J. Y. Pvt., Co.'s F & K. b. Pickens Dist. Enl.Co. F Pickens Dist. 4/22/62. Present through 4/30/62. Transferred Co. K 6/25/62. Present through 10/62. Ab. on scout 11-12/62. Present 3-4/63. Captured Upperville 6/21/63. Sent to Old Capitol. Exchanged by 8/63. Detached in QM Dept., Cavalry Corps at Orange CH 8-12/63. Present 1-8/64. Died of disease in Charleston hospital 9/26/64. Effects: $56.00 & sundries.

HEWITT, EDWARD L. Pvt., Co.'s D & K. b. Catawba Co., N. C. circa 1827. Enl. Co. D Chester CH 9/10/61 age 33. Present through 4/62. Transferred Co. K 6/25/62. Present through 8/62. Ab. on detached service 9-10/62. Ab. sick in Staunton hospital 12/1/62-4/63. Discharged for chronic rheumatism 8/11/63. Age 35, 5'7", dark complexion, dark eyes, dark hair, Farmer. NFR.

HEYWARD, THOMAS SAVAGE. Pvt., Co. A. b. S. C. 10/39. Age 10 Beaufort Co. 1850 census. Enl. 11/7/61. Discharged 2/7/62. Reenlisted Co. H, 3rd S. C. Cav. Charleston 3/12/62 as 3rdSgt. Promoted 3rdLt. Present sick 8/31/62. Present 9/62-3/63. Ab. sick leave 3/8-4/30/63. Ab. detailed McPhersonville 8/63. Commanding Company 9-10/63. Commanding 29th Bn. Ga. Cav. 8/31/64. Paroled Bennett House, N. C. 4/26/65. Bank Officer, Charleston 1880 census. Bookkeeper, Charleston 1900 census. d. 1901. Bur. St. Michaels Ch. Cem., Charleston.

HIGGINS, JOHN ADDISON. Pvt., Co. F. b. S. C. 6/30/41. Farmer, age 19, Pickens Dist. 1860 census. Enl. Pickens CH 12/4/61 age 22. Present through 4/62. Discharged for disability 5/10/62. NFR. Farmer, Easley, Pickens Co. 1880-1900 censuses. d, Easley, S. C. 3/13/18. Bur. West View Cem., Easley, S. C.

HIGHTOWER, BENJAMIN B. Pvt., Co. C. b. Edgefield Dist. 2/43. Student. Enl. Wadmalaw Island 2/1/62. Present through 6/62. Ab. on sick leave 8/14-31/62. Present 9-12/62. Discharged 2/20/63. Age 19, 5'8", dark complexion, dark hair, dark eyes. Reenlisted Columbia 4/13/64. Ab. detached Green Pond 4/21-30/64. Ab. detailed in QM Dept. 5-6/64. On duty bridge over Ashley River 7-8/64. On duty at Signal Corps Station 9-10/64. Present 11-12/64. Present Fayetteville, N. C. 4/16/65. NFR. Farmer, Windsor, Aiken Co. 1880-1900 censuses. d. 5/10/03. Bur. Old White Pond Cem., Aiken Co. Wife receiving pension Aiken Co. 9/25/19.

HILL, ANDREW MALONE. Pvt., Co. A. b. S. C. 3/42. Farmhand, age 18, Walker Co., Ala. 1860 census. Enl. Abbeville CH 3/19/62. Present through 4/62. Ab. detached 5-6/62. Ab. on sick leave 7-8/62. Present 9/62-4/63. Ab. sick with "scroful diatlesis, cervical abscess, debility & chronic diarrohea" in Farmville hospital 6/11-8/31/63. Furloughed for 50 days. Ab. sick with "nephritis" in Columbia, S. C. hospital through 5/64. Ab. detached James Island 8/3-12/31/64. NFR. Farmer, Madison Co., Tex. 1900 & 1910 censuses. d. 1916. Bur. Sharon Meth. Ch. Cem., Abbeville Co., S. C.

HILL, ASAPH. Pvt., Co. F. b. S. C. 1840. Res. Pickens Dist. Enl. Pickens CH 12/4/61. Ab. sick in Adams Run hospital until died of disease 3/21/62.

HILL, CHARLES ALLEN. Pvt., Co. E. b. Cottageville, Colleton Co. 2/23/46. Student, age 12, Colleton Dist. 1860 census. Enl. 1864 on roster but not on muster rolls. Paroled Greensboro, N. C. 5/1/65. Farmer, Sheridan, Colleton Co. 1900 census. Receiving pension Cottageville, Colleton Co. 9/24/19. d. there 6/28/31. Bur. Fox Cem. Son of John R. Hill.

HILL, JAMES W. Pvt., Co. I. b. S. C. 1/25/47. May have served in 5th S. C. Reserves. Enl. Columbia 4/12/64. Detached Green Pond 4/20-30/64. Present 5-8/64. Ab. on sick leave 10/21-31/64. Detached Pocotaligo 12/29-31/64. NFR. Farmer, Gray, Edgefield Co. 1870 & 1880 censuses. d. 12/19/18. Bur. Cottageville Cem., Colleton Co.

HILL, JOHN RISHER. 1stLt., Co. E. b. Colleton Dist. 6/15/15. Farmer, Colleton Dist. 1860 census. Enl. Adams Run 3/19/62 as Pvt. Present through 12/62, promoted 1stLt. 11/28/62. Present through 6/63, signed for forage for 43 horses 3/31/63. Present through 8/31/63, however, WIA (head) & horse killed Gettysburg 7/3/63. Paid $150.00 for horse. Ab. on leave 9-10/63. Commanding Co. 11/63-2/64. 32 horses present 12/31/63. Present 3-4/64. Signed for for forage for 14 horses 3/14/64. Signed for 36 pr. horse shoes 4/27/64. Signed for 21 pr. shoes, 3 blankets, 14 pr. pants, 1 overcoat, 3 shirts, 4 jackets, 3 pr. drawers, 1 tent fly & 3 Army tents 3/31/64. Present sick in camp 5-8/64. Ab. sick with "Febris chron. termitt" in Charleston hospital 9/19/64. Present through 12/64. Paroled Hillsboro, N. C. 4/65. Farmer, Sheridan, Colleton Co. 1880 census. d. Round O, Colleton Co. 11/22/91. Bur. Fox Cem., Colleton Co. Father of Charles A. Hill.

HILL, L. W. Pvt., Co. E. Enl. Captain Sheridan's Co., 11th S. C. Inf. Fenwick Island 10/15/61. Present through 12/61. Company disbanded. Enl. Co. E, 1st S. C. Cav. Adams Run 3/19/62. Present through 6/62. Courier, General's Headquarters 7-8/62. Present sick in camp 9-12/62. Present 3/63-2/64. Ab. on leave 4/26-5/6/64. Present sick in camp through 10/64. Present 11-12/64. Paroled Hillsboro, N. C. 4/65.

HILL, LAWRENCE HENRY. Pvt., Co. I. b. S. C. circa 1842. Student, age 18, Colleton Dist. 1860 census. Enl. Co. I, 6th S. C. Inf. Summerville 7/10/61 age 19. Ab. sick sick in hospital 8/31/61. Ab. sick with measles, mumps, & typhoid fever in Charlottesville hospital 10/3-11/18/61. Present through12/31/61. NFR. Probably discharged. Enl. Co. I, 1st S. C. Cav. by 4/65. Paroled Hillsboro, N.C. 4/65 on wife's pension application. Farmer, age 28, Sheridan, Colleton Co. 1870 census. d. 5/28/79. Wife receiving pension Dorchester Co. 10/10/19.

HILL, LEONARD. Pvt., Co. B. b. S. C. 1833. Res. Laurens Dist. 1860 census. Enl. Clinton 8/22/61 age 28. Present through 12/61. On picket Hallover Bridge 3-4/62. Present 5-6/62. Ab. on sick leave 7-8/62. Ab. driving horses from Summerville, S. C. to Staunton, Va. 9-10/62. Present 11/62-8/63. Ab. on horse detail 8/18-31/63. Present 9-10/63. Ab. recruiting station for horses 11/8/63-2/64. Ab. on horse detail

4/18-30/64. Present as Farrier 5-12/64. Paroled Hillsboro, N. C. 4/65. d. by 1898.

HILL, RICHARD K. Pvt., Co. F. b. S. C. 1/24/38. Enl. Pickens CH 12/4/61. Present through 8/62. Ab. sick 9-10/62. Present 11/62-8/31/63. On detached service 9-10/63. Present 11-12/63. Ab. horse recruiting station 2/64. Detached Green Pond 4/16-30/64. Present 5-12/64. Paroled Greensboro, N. C. 4/26/65. Farmer, Easley, Pickens Co. 1880 & 1900 censuses. d. 7/28/00. Bur. West End Cem., Easley, S. C.

HILL, RICHARD S. Pvt., Co. E. Enl. Capt. Sheridan's Co., 11th S. C. Inf. 10/15/61 as Sgt. Presence or absence not stated through 12/31/61. Company disbanded. Reenl. Co. E, 1st S. C. Cav. Adam's Run 2/25/62. Present sick 3-4/62. Courier, General's Headquarters 7-8/62. Present 9-10/62. On picket 11-12/62. Ab. sick with debilities in Charlottesville hospital 2/20-3/14/63. Present through 8/63. Ab. on detail with disabled horses 8/31/63. Detailed to Regimental Band 9-10/63. Present 11/63-12/64. Paroled Hillsboro, N. C. 4/65.

HILL, ROSWELL A. Pvt., Co. E. b. Pickens Dist. 12/6/41. Student, age 18, Pickens Dist. 1860 census. Enl. date unknown. d. Adams Run 12/19/62.

HILL, THOMAS H. Pvt., Co. E. b. S. C. circa 1841. Enl. Adams Run 2/28/62. Present 3-4/62. Present sick in camp 5-6/62. Ab. on sick leave 7-8/62. Present 9-10/62. On scout 11-12/62. AWOL 3-4/63. Present 5/63-2/64. Ab. on leave 4/29-5/12/64. Present through 8/64. Detailed guarding POW's Florence 9/17-12/31/64. Paroled Hillsboro, N. C. 4/65. Farmer, age 29, Darlington Co. 1870 census.

HILL, WILLIAM B. Pvt., Co. I. b. S. C. circa 1841. Enl. Sheridan's Co., 11th S. C. Inf. 10/15/61. Present through 12/31/61. Company disbanded. Enl. Co. I, 1st S. C. Cav. Parker's Ferry 4/3/62. AWOL through 2/64, however, enl. Co. C, 5th S. C. Cav. 4/14/63. Present through 10/63. Issued clothing 12/25/64. NFR.

HILTON, JOHN H. Pvt., Co. D. b. S. C. circa 1843. Res. Lancaster Co. On postwar roster. Farmer, age 37, Stilesboro, Ga. 1880 census. Res. Attalla Town, Ala. 1900 census. Res. Gaden, Ala. 1910 census. d. Elowah, Co. Ala. 3/26/11. Bur. Forest Cem., Gadsden, Ala.

HILTON, MILES. Pvt., Co. D. b. S. C. 1823. Enl. Lancaster, S. C. 6/26/64. Present sick in camp through 7/31/64. Present 9-10/64. Detailed guarding stores on Savannah Railroad 12/28-31/64. NFR. d. 2/22/06. Bur. Tennessee Colony Cem., Anderson Co., Texas.

HINE, MICHAEL. Pvt., Co. A. b. Germany 8/43. Enl. by 1864. Captured John's Island 7/2/64. Sent to Ft. Pulaski 10/25/64. "Desires to take oath." NFR. Farmer, Sherman, Michigan 1900 & 1910 censuses.

HIOTT, JAMES EZEKIEL. Pvt., Co. E. b. Colleton Dist. 3/3/47. Enl. 4/12/64. Detached Green Pond 4/16-30/64. Present through 12/64. Paroled Greensboro, N. C. 5/1/65. Receiving pension Round O, Colleton Co. 9/8/19. d. Charleston 3/3/29. Bur. Maple Cane Bapt. Ch. Cem., Cottageville, Colleton Co.

HIPP, ROBERT H. Pvt., Co. I.b. S. C. 4/10/33. Enl. Co. I, 3rd S. C. Inf. Clinton 7/1/62. Present through 12/63. NFR. Enl. Enl. Co.I, 1st S. C. Cav. Columbia 2/1/64. On detail Green Pond 4/16-30/64. On Courier duty 7-8/64. Present sick in camp 9-10/64. Detailed Pocotaligo 12/10-31/64. Ab. sick with debility from pneumonia in Raleigh, N. C. hospital 2/18-21/65. Paroled Greensboro, N. C. 4/28/65. d. 5/13/65. Bur. Hipp Cem., Lafayette Co., Miss.

HITT, MARTIN G. Pvt., Co. C. b. S. C. 1842. Farmer, age 18, Edgefield Dist. 1860 census. Enl. Hamburg 8/27/61 age 18. Detached Wadmalaw Island through 12/31/61. Present 5-6/62. Ab. on sick leave 8/8-31/62. Present 9/62-8/63. Ab. on horse detail 10/23-12/31/63. Ab. sick in hospital 1/64. Present 2/29/64. Ab. on horse detail 4/19-5/4/64. Present through 8/64. Detailed driving cattle 9/19-12/31/64. Paroled Augusta, Ga. 5/23/65. Farmer, age 67, Wise, Edgefield Co. 1910 census.

HOFF, ROBERT WILLIAM, JR. 3rdSgt., Co. I. b. Tarlton, Colleton Co. 4/18/35. Carpenter, age 23, Colleton Co. 1860 census. Enl. Parker's Ferry 4/3/62 as 3rdSgt. On picket Jacksonborough through 4/30/62. Present 5-6/62. Present sick in camp 7-8/62. Ab. sick with Lumbago in Richmond hospital 10/22-11/6/62. Ab. on sick leave through 12/62. Present 3-6/63. Ab. on horse detail 8/18-31/63. AWO 9/28-12/31/63. Reduced to Pvt. Present 1-2/64. Listed as age 29, 5'10", light complexion, blue eyes, dark hair, Farmer on list 3/31/64. Detached Green Pond 4/16-30/64. Present 5-8/64, promoted 3rdCpl. Detached Eastern Lines, James Island as Carpenter 10/6-12/31/64, reduced to 4thCpl. NFR. Farmer, age 45, Sheridan, Colleton Co. 1880 census. d. Colleton Co. 7/2/04.

HOGAN, JAMES F. Pvt., Co. G. b. S. C. circa 1840. Enl. Abbeville CH 1/8/62 age 22. Present sick in camp 3-4/62. Present 5-10/62. On scout 11-12/62. AWOL 3-4/63. Present 5-6/63. Ab. on horse detail 8/18-31/63. Present 9-10/63. Detailed as Teamster, Brigade Commissary Officer 11/15/63-2/64. Ab. on horse detail 4/4-30/64. Transferred Co. B, Lucas's Bn. S. C. Arty. 5/13/64. Joined 5/20/64. Present through 12/64. Paroled 4/65.

HOGAN, JOHN H. Pvt., Co. G. b. Abbeville Dist. circa 1828. Enl. Abbeville CH 1/8/62. Present through 4/62. Ab. on sick leave 6/2-30/62. Present 7-10/62. On scout 11-12/62. Ab. sick in hospital 3-4/63. Present through 8/31/63. On detached service 9-10/63. Ab. sick in hospital 12/28-2/64. Ab. sick in Columbia, S. C. hospital 4/16-30/64. d. in Ladies hospital, Columbia, S. C. 5/12/64 age 36. Left widow.

HOLDEN, JACKSON B. Pvt., Co. F. b. S. C. circa 1831. Enl. Pickens CH 12/4/61 age 30. On picket Jacksonborough through 12/31/61. NFR. Enl. Co. C, Moore's Bn. S. C. Volunteers. Transferred Co. E, 1st S. C. Cav. 4/17/62. On picket Camp Means 4/30/62, dismounted since 4/5/62. Present 7-8/62. Ab. on detached service 9-10/62. On scout 11-12/62. Present 3/63-2/64. Pay stopped for loss of 1 English repeating pistol value $70.00. Present 3-8/64. d. Garodines Station on Northeast Railroad of fever 10/15/64.

HOLDEN, RICHARD J. Pvt., Co. F. b. S. C. 1823. Enl. Pickens CH 12/4/61. Present through 12/31/61. On picket Simmons House 3-4/62. Present 5-6/62. Ab. on leave 7-8/62. Present 9/62-4/63. Ab. wounded in Richmond hospital 8/8/63. Furloughed for 30 days 8/24/63. Present 10/63. Ab. on horse detail 12/22/63-2/64. Ab. on horse detail 4/20-30/64. Present 5-6/64. Present sick 8/25-31/64. Present 9-12/64. Paroled as Teamster Greensboro, N. C. 5/1/65. d. 1873. Bur. Holden Cem., Rayburn Co., Ga.

HOLDER, BENJAMIN LUTHER. 1stLt., Co.'s F & K. b. S. C. 5/23/35. Res. Pickens Dist. 1860 census. Enl. Pickens CH 12/4/61 as 2ndLt. Present through 4/62. Transferred Co. K 6/25/62. Promoted 1stLt. Resigned 10/18/62. NFR. Minister, Pumpkintown, Pickens Co. 1880 census. d. Pickens Co. 1/29/18. Bur. Secona Bapt. Ch. Cem., Pickens Co.

HOLDER, TYRA JAMES H. 1stLt. Co. K. b. S. C. 1831. Age 28, Pickens Dist. 1860 census. Resigned. Married Pickens Co. 1864 age 33. Married Pickens Co. 1869 age 38. Farmer, Pumpkintown, Pickens Co. 1870 census. d. Pickens Co. 5/14/24.

HOLBROOK, WILLIAM YANCEY. Pvt., Co. G. b. Ga. 3/25/38. Farmer, Carnesville, Ga. 1860 census. Enl. Co. B, 15th Ga. Inf. 7/14/61. Discharged 3/22/62. Enl. Co. G, 1st S. C. Cav. Carnesville, Ga. 3/20/63. Ab. sick with dispepsia in Richmond hospital 7-9/63. Furloughed for 30 days 9/23/63. Returned to duty 11/10/63. Ab. sick in hospital 11/20/63-2/64. Detached Green Pond 4/17-30/64. Present 5-8/64. AWOL 10/12-31/64. Ab. sick in hospital 12/14-31/64. Captured & paroled Hartwell, Ga. 5/20/65. Farmer, Gunnell, Franklin Co., Ga. 1900 census. d. Carnesville, Ga. 12/26/07. Bur. Bold Springs Cem.

HOLLADAY, ALPHEUS J. Pvt., Co. I, b. Clarendon Dist. 10/4/46. Enl. John's Island 9/29/64. Present through 12/64. Present in N. C. 4/15/65. NFR. Farmer, Friendship, Clarendon Co. 1880-1900 censuses. Res. Limestone, Orangeburg Co. 1910 census. d. 9/14/19. Wife receiving pension Orangeburg Co. 12/16/19.

HOLLAND, BLUFORD RUSSELL. 1stCpl., Co. B. b. Laurens Dist. 8/14/39. Age 21, no occupation, Scuffletown, Laurens Co. 1860 census. Enl. Camp Johnson 8/25/61 as Pvt. Ab. sick 10/31/61. Ab. sick in hospital 12/31/61. On picket Hallover Bridge 3-4/62. Present 5-6/62, promoted 4thCpl. Present 7-8/62. On picket 9-10/62. On picket Ellis's Ford on Rappahannock River 11-12/62, horse killed12/30/62. Paid $200.00. Promoted 2ndCpl. Present 3-12/63. Ab. recruiting camp for horses 1/9-2/29/64. Present 3-8/64. Promoted 1stCpl.   Present sick in camp 9-10/64. Present 11-12/64. NFR. Farmer, age 31, Scuffletown, Laurens Co. 1870 census. d. Laurens Co. 3/21/75. Bur. Leesville Cem., Clinton, S. C.

HOLLAND, EDWIN CLIFTON. Pvt., Co. A. b. S. C. 8/18/28. On post war roster. Reenlisted Co. K, 4th S. C. Cav. Grahamville 3/25/62. Present until WIA Pocotaligo 10/22/62. Present until promoted 1stCpl. 1/1/63. Present until promoted 5thSgt. 6/63. Present until detailed with Gen. Butler 7/18-8/31/64. Promoted 4thSgt. Detailed with Brigade QM 10/15-31/64. NFR. Chief Clerk, Planter's & Merchants Rice Mill, Charleston 1899. d. 7/4/02.

HOLLAND, GREENBURY R. Pvt., Co. B. b. S. C. 2/7/35. Res. Laurens Dist. 1860 census. Enl. Clinton 8/22/61 age 26. Ab. sick 10/31/61. Ab. nurse in hospital 12/31/61. On picket Hallover Bridge 3-4/62. Present 5/62-8/31/63. On detached duty 9-10/63. Ab. horse recruiting station 10/27-12/31/63. Present 1-2/64. Detached Green Pond 4/16-30/64, remounted 4/12/64. Present as Bugler 5-10/64. Detached Pocotaligo 12/2-31/64. NFR. d. Laurens Co. 6/15/65 age 28. Bur. Holland Cem. No. 2, Laurens Co.

HOLLAND, JOHN A. Pvt., Co.'s C & K. b. S. C. circa 1838. Farmer, age 22, Laurens Dist. 1860 census. Enl. Wadmalaw Island 3/18/62. Transferred Co. K 6/27/62. Transferred back to Co. C 7/8/62. On duty with QM Dept. 7-8/62. Present 9/62-4/63. Horse killed Funktown, Md. 7/9/63. Ab. detached 8/8-31/63, however, ab. sick with "Int. fever" in Richmond hospital 8/30-11/22/63. Furloughed for 30 days. Ab. on horse detail 2/22-29/64. Detached Green Pond 4/16-30/64. Present 5-8/64. Detailed as Overseer for Lt. Hall on James Island 10/6-31/64. Present 11-12/64. NFR.

HOLLAND, P. G. Pvt., Co. A. Enl. 11/7/61. Returned to Carolina Light Inf., Co. K, 1st S. C. Inf. (McCray's) 11/30/61, but not on muster rolls of that unit.

HOLLAND, W. H. Pvt., Co. I. b. S. C. 1/5/45. Enl. Culpeper CH 5/27/63. Present through 10/63. On picket 11-12/63. Detailed horse recruiting station 1/20-2/29/64. Detached Green Pond 4/16-30/64. Present sick in camp 5-8/64. Present 9-12/64. Paroled Raleigh, N. C. 4/65. d. Spartanburg 2/2/20. Bur. Arkwright Cem.

HOLLEY, JOHN T. Pvt., Co. K. b. S. C. 12/44. Enl. Culpeper CH 5/26/63. Present 5-6/63, dismounted. Ab. on horse detail 8/16-31/63. Present 9-10/63. Ab. on horse detail 12/23/63-2/64. Ab. on sick leave 4/7-30/64. Ab. sick in Columbia, S. C. hospital 5-6/64. Present 7-8/64. Detailed to go home for provisions 10/19-31/64. Present 11-12/64. Paroled Raleigh, N. C. 4/65. Farmer, age 25, Cross Hill, Laurens Co. 1870 & 1880-1900 censuses. Retired, Pleasant Hill, Oglethorpe Co., Ga. 1920 census.

HOLLEY, TURNER WILLINGHAM. 1stLt., Co.'s D & K. b. S. C. 1820. Farmer, age 40, Chester Dist. 1860 census. Enl. Chester CH 9/10/61 as 3rdSgt., age 41. Ab. on sick leave 10/31/61. Present 11/61-4/62. Transferred Co. K 6/25/62 and elected 2ndLt. Present 7-8/62. Promoted 1stLt. Commanding Company 9-10/62. On scout 11-12/62. Commanding company 2/26/63. Ab. sick with "Icterus" in Richmond hospital 4/63. Transferred Columbia, S. C. hospital 4/18/63. Commanding Company 5-6/63. Signed for 40 pr. pants, 41 jackets, 1 pr. shoes, & 4 pr. socks 8/23/63, commanding company. Present 9-10/63. On picket Goldvein Fork on Rapidan River 11-12/63. On picket Chancellorsville 1-2/64. Commanding Company 3-6/64. Signed for forage for 25 horses 5/28/64. Signed for forage for 15 horses & 40 cotton shirts, 42 pr. drawers, 31 caps, 6 pr. boots, 10 pr. socks, 13 jackets, 9 pr. pants 6/30/64. On duty bridge at Ashley River 7-8/64. Signed for forage for 55 horses 7/10/64. Signed for 14 pr. drawers, 5 shirts, I overcoat, 2 pr. pants & 4 caps 9/8/64, commanding company.   d. of yellow fever in Charleston hospital 10/4/64. Effects: $396.00 & sundries.

HOLMES, THOMAS G. Pvt., Co. A. b. Charleston 1836. On post war roster. Reenlisted Co. K, 4th S. C. Cav. Grahamville 3/35/62 age 25. Present until detailed as Asst. Clerk, C. S. Stables, Charleston.

5/15/62-2/26/64. KIA Hawes Shop 5/28/64.

HOLMES, WILLIS. Pvt., Co. C. b. S. C. 3/21/27. Res. Edgefield Dist. Served in Co. K, 5th S. C. Reserves (90day unit) 11/2062-2/10/63. Enl. Co. C, 1st S. C. Cav. Columbia 4/5/64. Detached Green Pond 4/16-30/64. Present 5-12/64. Present Raleigh, N. C. 4/65. d. Edgefield Co. 4/3/05. Bur. family cem., Edgefield Co.

HOLT, A. B. 1stLt., Co. K. b. 6/21/33. On roster but not on muster rolls. d. 4/19/04. Bur. Westview Cem., Fulton Co., Ga.

HOOD, HENRY T. Pvt., Co. F. b. N. C. circa 1847. Enl. by 1865. Captured Greenville, S. C. 5/24/65 and paroled. Farmer, age 33, Ebenezer, York Co. 1880 census.

HOOD, ISREAL MC DANIEL, SR. Pvt., Co. D. b. S. C.5/7/24. Enl. Chester CH 9/2/64. On picket through 10/31/64. Present 11-12/64. NFR. d. 2/9/97. Bur. Pleasant Grove Presb. Ch. Cem., Chester Co.

HOOD, ISREAL MC DANIEL, JR. Pvt., Co. D. b. Chester Dist. 8/22/46. Enl. by 1865. Paroled Greensboro, N. C. 4/26/65. Store Keeper, Chester 1880 census. County Auditor, Chester Co. 1910 census. d. 6/25/15. Bur. Evergreen Cem., Chester. Wife receiving pension Chester 1919.

HOOD, W. HARRIS "HARVEY." Pvt., Co. D. b. S. C. 1838. Enl. Camp Ripley 1/17/62. Present through 4/62. On picket Townsend's House 5-6/62. Ab. on sick leave 8/9-31/62. Present 9-10/62. On picket 12/29-31/62. Ab. on horse detail 3/4-4/30/63. Present through 8/31/63. Ab. detailed Regimental Commissary 10/23/63-4/64. Ab. detailed Brigade Commissary Dept. 6/10-12/31/64. Paroled Greensboro, N. C. 4/26/65. d. Chester Co. 5/2/00. Wife receiving pension Chester 10/8/19.

HOPE, ROBERT H. Pvt., Co. H. b. Mecklenburg Co., N. C. 1820. M. D., age 41,York Dist. 1860 census. Enl. Rock Hill 1/9/61 age 43. NFR. M.D., age 50, York Co. 1870 census.

HOPKINS, JAMES. b. S. C. 5/30. On postwar roster. Reenlisted Co. K, 4th S. C. Cav. Mc Phersonville 7/6/62. Present until WIA (left shoulder) Pocotaligo 10/22/62. Ab. wounded through 4/63. Present 5/63 until ab. on leave 2/21/64 for 20 days. Present until captured Cold Harbor 5/30/64. Sent to Pt. Lookout. Exchanged 3/11/65. NFR. Farmhand,Swing Pens, Sumter Co. 1870 census.

HORNE, ROBERT LEANDER. Pvt., Co. A. b. S. C. 10/4/42. Enl. Columbia 3/64. AWOL since assigned to Regiment on rolls through 12/31/64. Ab. on sick leave from Charlotte, N. C. hospital 2/14-3/28/65. NFR. Farmer, Langsford, Chester Co. 1870 census and Lewisville, Chester Co. 1880 census. Cattle Dealer, Chester Co. 1900 census. Retired, Chester Co. 1910 census. d. 9/14/28. Bur. Evergreen Cem.

HORN, SAMUEL A. 4thCpl., Co.'s C & K. b. Maine 5/42. Kiln Worker, age 20, Edgefield Dist. 1860 census. Enl. Co. C Camp Johnson 10/17/61 as Pvt., age 21. Present through 12/61.Transferred Co. K 6/27/62. Present 7-8/62. Ab. detached service 9-10/62. Present 11-12/62. Promoted 4thCpl. Present 3-4/63, detailed to Telegraph Office, Richmond. Ab. detailed to work in crockery factory and reduced to Pvt. 5/63-4/64. AWOL 5-6/64. Permanently detailed to Kaolin Works 7-8/64. NFR. Farmer, age 38, Hammond, Aiken Co. 1880 census and 1900 census. Retired, age 73, Columbia 1910 census.

HORSEY, FREDERICK K. 1stLt., Co. I. b. S. C. circa 1835. Age 22, Charleston, 1860 census. Enl. Parker's Ferry 4/3/62 as 2ndLt. Present through 4/30/62. Ab. on sick leave 5-6/62 & 8/25-31/62. Ab. on leave 9-10/62. Promoted 1stLt. Present 11-12/62. Commanding Company 3/63-12/63. Signed for forage for 46 horses 6/30/63. Signed for 13 caps, 25 jackets, 12 pr. pants & 42 pr. shoes 8/24/63. Signed for 2 jackets, 8 caps & 28 shirts 9/6/63. Signed for 13 pr. boots, 13 pr. shoes, 6 pr. socks, 5 blankets, 37 pr. drawers, 16 pr. pants, 1 overcoat & 10 shirts 9/8/63. Signed for forage for 92 horses 9/26/63. Signed for forage for 31 horses 10/23/63. Ab. on detached service 1/13/64. Present 3-4/64. Commanding Company 5-8/64. Signed for 39 pr. drawers, 13 pr. shoes, 35 pr. socks, 26 caps, 41 jackets, 39 pr. pants & 32 cotton shirts Green Pond 5/20/64. Signed for 10 caps, 19 jackets, 14 pr. pants, 25 cotton shirts, 14 pr. drawers, 3 pr. sock, 3 Army tents, 5 camp kettles, 13 mess pans, 2 axes & 1 spade 6/30/64. Signed for forage for 59 horses 8/1/64. Ab. on sick leave 10/25-31/64. Commanding Company 11-12/64. NFR. Merchant, age 35, Charleston 1870 census, age 45, 1880 census.

HOUCK, CALVIN W. Pvt., Co. E. b. S. C. circa 1826. Enl. Ft. Motte 10/26/61 age 35. Present through 12/31/61. Present sick in camp 3-4/62. Ab. sick at home 5/15-6/30/62. Courier at General's Headquarters 7-8/62. Present 9/62-4/63. Ab. sick in hospital 8/10-31/63. Present 9/63-2/64. Detached Green Pond 4/26-30/64. Present 5-12/64. Paroled Hillsboro, N. C. 4/65. Farmer, age 57, Orange, Orangeburg Co. 1880 census.

HOUSER, JAMES W. Pvt., Co. H. b. N. C. 1843. Enl. Rock Hill 1/30/62 age 19. On picket Jacksonborough 3-4/62. Present 5-10/62. On picket 11-12/62. Ab. driving ambulance 3-4/63. Present through 8/31/63. Present 9/63-2/64. Ab. on leave 3-4/64. Transferred Co. C, 1st S. C. Inf (Butler's) 5/3/64. NFR. Laborer, Golconda, Pope Co., Ill. 1880 census.

HOUSER, L. C. Pvt., Co. H. b. N. C. 12/36. Enl. Rock Hill 1/9/62. Detailed as Wagoner in QM Dept. 3-8/62. Present 9-10/62, dismounted. Present 11/62-4/63. Ab. detailed to hospital to wait on the sick 8/10/63-2/64. Returned to duty 3/23/64. Ab. on leave 4/24-5/24/64. Present 5-10/64. Detached Pocotaligo through 12/31/64. NFR. Farmer, age 63, Prentriss, Miss. 1900 census.

HOUSTON, ALEXANDER RICHARD. Pvt. Co. G. b. S. C. 1846. Res. Abbeville Dist. 1860 census. Enl. Abbeville CH 3/15/62. Ab. on detached service through 4/62. Hired Robert Martin as substitute and discharged 5/15/62. Appointed Regimental Commissary Officer. NFR. Farmer, age 23, Richmond Co., Ga. 1870 census. Warf Keeper & Keeper, Toll Bridge, Augusta, Ga. 1877-90. d. by 1902.

HOUSTON, ROBERT A. Pvt., Co. C. b. S. C. circa 1837. Clerk, age 23, Hamburg, Edgefield Dist. 1860 census. Enl. Wadmalaw Island 5/17/62. Present through 8/31/63. Ab. sick with "Secondary Syphilis" in Richmond hospital 10/17-11/20/63. Present 11/63-10/64. Detached Pocotaligo 12/10-31/64. NFR. Merchant, age 34, Augusta, Richmond Co., Ga. 1870 census. Alive 1898.

HOWARD, JAMES BUSH. 4thCpl., Co.'s C & K. b. Edgefield Dist. 1/40. Enl. Hamburg 8/27/61 as Pvt., age 20. Transferred Co. K 6/27/62. Promoted 4thCpl. Ab. sick at home through 9/1/62. AWOL 9-10/62. Present 11-12/62. Discharged for "Phthis Pulmonalis" 1/1/63. Age 22, 6', dark complexion, hazel eyes, dark hair, Farmer. Paid 1/6/63. NFR. Farmer, age 60, Johnson, Smith Co., Ga. 1900 census. Age 80, Blue Springs, Pulaski Co., Ga. 1920 census. d. by 1927 on wife's pension application.

HOWARD, JESSE. Pvt., Co. C. b. S. C. 7/35. Res. Edgefield Dist. 1850 census. Enl. Camp Butler 9/17/61. Ab. sick in hospital 11-12/61. Accidentialy wounded by pistol and ab. on leave 1/62. Discharged 4/9/62. NFR. Laborer, age 64, Wahee, Marion Co. 1900 census. d. by 1910.

HOWARD, JOHN A. Pvt., Co. C. b. S. C. circa 1832. Enl. Hamburg 8/27/61 age 27. Present until ab. sick in hospital 11-12/61. Present 5-12/62. Ab. sick with "secondary syphilis" in Richmond hospital 3-4/63. Present through 8/31/63. Captured near Madison CH 9/23/63. Sent to Old Capitol & Pt. Lookout. Exchanged 5/64. Ab. sick with debilities in Richmond hospital 5/8-7/7/26/64. Present 8/64. Ab. detailed to guard POW's Florence, S. C. 9/17-12/31/64. NFR. Alive 1898.

HOWARD, WILLIAM S. Pvt., Co. C. b. S. C. circa 1845. Enl. Columbia 1/31/64. Ab. on horse detail 4/4-5/5/64. Present through 12/64. NFR. Bookkeeper, age 65, Philadelphia, Pa. 1910 census.

HOWELL, JESSE M. Pvt., Co. A. b. S. C. 1836. On post war roster. Reenlisted Co. C, 7th S. C. Bn. Inf. and elected 1stLt. 12/31/61. Present until ab. on leave 9-10/62. Present until resigned 11/11/63. Reenlisted Co. K, 4th S. C. Cav. Charleston 1/24/64. Ab. on leave 2/25/64 for 40 days. Captured Cold Harbor 5/30/64. Sent to Pt. Lookout. Exchanged 11/1/64. Paroled Greensboro, N. C. 5/5/65. d. 1882. Bur. Elmwood Mem. Gardens, Columbia, S. C.

HOWELL, LEWIS ROBERT CARN. Pvt., Co. E. b. St. George, S. C. 12/12/22. Farmer, age 38, Branchville, Orangeburg Dist. 1860 census. Enl. Branchville 10/26/61. Ab. recruiting duty 12/28-31/61. Present sick in camp 3-4/62. Ab. on leave 6/6-30/62. On scout Edisto Island 7-8/62. Present 9/62-8/31/63. Ab. on horse detail 10/24-12/31/63. Present 1-10/64. Detached Pocotaligo 12/10-31/64. NFR. Farmer, Cow Castle, Orangeburg Co. 1880 census. d. 10/22/98. Bur. Indian Fields Meth. Ch. Cem., Rosinville, S. C.

HOWLAND, HENRY A. Pvt., Co. G. b. Ga. 1847. Student, age 13, Elberton, Ga. 1860 census. Enl. Legare's's Point 8/5/62 as substitute for W. D. Mars. Present through 10/62. On scout 11-12/62. AWOL 2/21/63-2/64. Ab. in arrest Columbia 4/1030/64. Transferred Co. A, Lucas' Bn. S. C. Arty. 5/13/64. Joined 5/16/64. Present through 8/64. AWOL 10/25-12/31/64. NFR. Brother of Lewis Howland.

HOWLAND, LEWIS. Pvt., Co. G. b. Ga. 1845. Student, age 15, Elberton, Ga. 1860 census. Enl. Legare's Point 8/5/62 as substitute for William Tennent. Present through 10/62. Ab. on scout 11-12/62. Present 3-10/63. Detailed horse recruiting station 12/22/63-2/64. Present 3-12/64. NFR.

HOUSER, A. MONROE. Pvt., Co. H. b. Lincoln Co., N. C. circa 1838. Enl. Co. K, 1st N. C. Inf. 4/25/61. Served through 11/12/61. Regiment disbanded. Enl. Co. H, 1st S. C. Cav. Rock Hill 1/9/62 age 24. Ab. driving wagon 3-4/62. Present 5/62-8/63. Ab. on horse detail 9-10/63. AWOL 12/31/63-1/16/64. Present through 2/64. Ab. on leave 3-4/64. Transferred C. S. Navy or Co. G, 18th Ga. Inf. 5/13/64, dismounted. Age 25, 6', dark complexion, grey eyes, dark hair, Mechanic. Married in N. C. 11/15/66.

HOUSER, JAMES W. Pvt., Co. H. b. Lincoln Co., N. C. 1843. Enl. Rock Hill 1/30/62. On picket Jacksonborough through 4/62. Present 5-10/62. On picket 11-12/62. Ab. driving ambulance 3-4/63. Present through 8/31/63 & 9/63-2/64. Ab. on leave 3-4/64. Transferred Co. C,1st S. C. Inf. (Butlers) 5/3/64, dismounted. NFR. Laborer, Golconda, Pope Co., Ill. 1880 census.

HOUSER, L. C. Pvt., Co. H. b. Lincoln Co., N. C. 12/36. Enl. Rock Hill 1/9/62 age 25. Wagoner, 3-6/62. Present as Wagoner in QM Dept. 7-8/62. Present 9-10/62, dismounted. Present 11/62-4/63. Ab. detailed a nurse in Orange CH hospital 8/10/63-3/23/64. Ab. on leave 4/24-5/24/64. Present 5-10/64. Detached Pocotaligo 12/10/31/64. NFR. Farmer, age 63, Prentiss, Miss. 1900 census.

HUBBARD, HENRY H. Pvt., CO. I. b. S. C. circa 1845. Age 15, Augusta, Ga. 1860 census. Enl. Parker's Ferry 4/3/62. AWOL since 4/30/62. NFR. [May have been discharged for underage]. Reenlisted Columbia 3/1/64. Detached Green Pond 4/16-30/64. Present 5-10/64. AWOL 11-12/64. NFR. Laborer, age 24, Gragg, Edgefield Co. 1870 census.

HUDSON, A. Pvt., Co. E. Enl. 5/10/64. Present through 6/64. Ab. sick leave 8/23-31/64. Detailed as Courier 9-10/64. Present 11-12/64. NFR.

HUDSON, DANIEL. 1stLt., Co. I. b. S. C. circa 1832. Farmer, age 30, Colleton Dist. 1860 census. Enl. Parker's Ferry 4/3/62. Present through 6/62. Ab. on leave 8/23-30/62. Resigned for ill health 10/22/62. NFR. Farmer, age 36, Bowen, Colleton Co. 1870 census. Farmer, age 48, Heyward, Colleton Co. 1880 census.

HUDSON, JOHN D. Pvt., Co. D. b. S. C. circa 1837. Enl. Chester CH 9/10/61 age 24. Present through 12/61. Discharged for disability 4/1/62. Enl. Co. E, 3rd S. C. Reserves (90 days) 11/13/62. Ab. did not report on rolls through 12/31/62. NFR.

HUDSON, MAYWOOD D. Pvt., Co. E. b. S. C. circa 1843. Enl. Adams Run 3/12/62. Ab. sick at home through 4/30/62. On picket Adams Run Station 5-6/62. On scout Edisto Island 7-8/62. Present 9-10/62. Ab. sick in hospital 11-12/62. Present 3-4/63. Ab. on horse detail 8/18-31/63. Present 9/63-2/64. Detached Green Pond 4/16-30/64. Present 5-12/64. Paroled Hillsboro, N. C. 4/65. Brick Maker, age 67, Gainesville, Ga. 1910 on pension application.

HUDSON, RICHARD H. 3rdCpl., Co. I. b. Colleton Dist. circa 1844. Enl. Co. I Parker's Ferry 4/3/62 age 18. Present through 6/62. Discharged for "chronic Hepatitis" Summerville hospital 9/22/62. Age 18, 5'5", fair complexion, grey eyes, dark hair. NFR.

HUDSON, RUSSELL STOBO. Pvt., Co.'s I & E. b. S. C. 5/43. Res. Colleton Dist. Enl. Co. I Parker's Ferry 4/3/62. On picket Jacksonborough through 6/62. Transferred Co. E 6/17/62. On picket Welltown 7-8/62. Present 9-10/62. On picket 11-12/62. Present 3-6/63. Ab. on horse detail 8/18-31/63. Present 9-12/63. Ab. detailed horse recruiting station 1-2/64. Detached Green Pond 4/16-30/64. Present 5-10/64. Detached Pocotaligo 12/10-31/64. Paroled Hillsboro, N. C. 4/65. Collector, Charleston 1886. Farmer, age 57, Fulton Co., Ga.1900 census. Merchant, age 72, Cross Keys, DeKalb Co., Ga. 1910 census.

HUDSON, P. STOBO. Pvt., Co. E. Res. Colleton Dist. Enl. by 1865. Paroled Hillsboro, N. C. 4/65. Collector, Charleston, 1886.

HUGER, BENJAMIN FROST. Pvt., Co. A. b. S. C. 7/21/36. On post war roster. Reenlisted Co. K, 4th S. C. Cav. Grahamville 3/25/62 & promoted 3rdSgt. Ab. on sick leave through 6/62. Present until promoted 2ndSgt. 5-6/63. Present through 2/64. WIA (right thigh) Hawes' Shop 5/28/64. Ab. wounded in Richmond hospital until transferred Columbia, S. C. hospital 6/3/64. Ab. wounded through 10/31/64. Issued clothing 11/25/64. Paroled 4/65. Post Master, Charleston. d. 3/25/87. Bur. Magnolia Cem.

HUGHES, CICERO. Pvt., Co. A. b. S. C. 9/10/40. Enl. Abbeville CH 9/17/61. Present through 4/62. Ab. detached as Regimental Forage Master 5/62-6/63. Ab. on horse detail 8/18-31/63. Ab. detailed in QM Dept. as Forage Master 9/63-2/64. Detached Green Pond 4/64. Ab. sick 6/25-12/31/64. NFR. Constable, Abbeville Co. 1880 census. Entered Old Soldier's Home, Columbia from Abbeville Co. 7/31/09. d. 1/26/14. Bur. Long Cane Cem., Abbeville Co.

HUGHEY, WILLIAM M. Pvt., Co. A. b. S. C. circa 1839. Res. Abbevlle Dist. 1860 census. Enl. Abbeville CH 3/19/62 as Pvt. Present as Bugler through 6/62. Ab. sick in Summerville hospital 7-8/62. Ab. sick 9-10/62. On picket 11-12/62. Ab. sick in Richmond hospital 2/19/63. Transferred to Danville hospital 4/3/63. Present 5-6/63. Ab. on horse detail 8/18-31/63. Present 9-12/63. Detail with Provost Guard, Hanover Junction 2/16-29/64. Present 3-6/64. Present sick in camp 7-8/64. d. of disease in James Island hospital 11/6/64.

HUNTER, SAMUEL R. Pvt., Co. D. b. S. C. 6/16. Enl. Chester CH 9/10/61 age 43. Ab. on sick leave 10/31/61-1/62. Discharged 3/27/62. NFR. Lecturer, age 53, Baltimore 1870 census. Store Clerk, age 62, Baltimore 1880 census. Farmer, age 84, Pickens Co. 1900 census.

HURBERT, D. T. Pvt., Co. K. b. S. C. circa 1834. Res. Pickens Dist. Not on muster rolls. d. of disease in Charleston hospital 9/24/64 age 30. Bur. Magnolia Cem., Charleston.

HUSSEY, JOHN M. Pvt., Co. E. b. St. George, Colleton Dist. 1835. Farmer, Colleton Dist. 1860 census. Enl. Adams Run 3/19/.62. On picket Adams Run Station on Railroad through 4/62. Present sick in camp 5-6/62. Present 7-10/62. On scout 11-12/62. Present 3-6/63. Shot through the body Gettysburg 7/3/63. DOW's 7/13/63. Left widow with 2 children.

HUTCHINSON, JOSEPH HALL WARING. Pvt., Co. A. b. S. C. 10/17/33. On post war roster. Reenlisted Co. K, 4th S. C. Cav. Grahamville 3/25/62, as 1stCpl. Promoted 4thSgt. 4/25/62. Ab. on sick leave 5/14/62-1/63. Reduced to Pvt. Present through 10/31/64. Paroled 4/65. d. Drayton Hall 6/22/97. Bur. St. George Parish Ch., Dorchester Co.

HUTCHINSON, PHILIP HENRY. Pvt., Co. A. b. 11/7/39. Served as 1stLt., Washington Arty. 1861-62. On postwar roster Co. A, 1st S. C. Cav. Reenlisted Co. K, 4th S. C. Cav. Mc Phersonville 7/10/62. Present until ab. on sick leave 6/15/63 for 15 days. Present 7-8/63. Ab. on sick leave 9/30-10/31/63. Present through 2/64. WIA (right thigh) Hawes' Shop 5/28/64. Ab. wounded in Richmond hospital until transferred Summerville, S. C. hospital 6/3/64. Ab. wounded through 8/31/64. Present until detailed as Scout

9/20-10/31/64. Paroled 4/65. d. Summerville, S. C. 10/10/10. Bur. St. Paul's Epis. Ch., Summerville.

HUTCHINSON, THOMAS. Pvt., Co. A. b. Ireland circa 1840. Enl. Abbeville CH 8/13/61 age 21. On duty Bear Island 11-12/61. Present 3-8/62. Ab. sick 9-12/62. Present 3-12/63. Ab. on horse detail 1/4-2/29/64. Present 3-12/64. Deserted to the enemy Charleston, S. C. 2/20/65. Age 36, 5'10", light complexion, light hair, grey eyes. Took oath 3/2/65. Seems to be two different men because of ages.

HUTTO, DAVID ANDREW. Pvt., Co. K. b. S. C. 2/2/50. Enl. Columbia 8/15/64. Ab. sick in Summerville hospital 10/4-31/64. Courier, Falls Point 12/31/64. Paroled Raleigh, N. C. 4/65. Convict Guard, Elbert, Ga. 1900 census. d. 10/30/28. Bur. Blacksville, S. C. Cem.

HYDE, JAMES NEWTON. Pvt., Co. F. b. S. C. 8/24. Farmer, age 36, Pickens Dist. 1860 census. Enl. Pickens CH 12/4/61 age 40. Present through 12/31/61. Ab. in confinement Abbeville CH 3-4/62. Present 5-6/62. Ab. sick in Summerville hospital 7-10/62. Present 11/62-6/63. Ab. sick with debilitas in Richmond hospital 7/4-18/63. Ab. on horse detail 8/18-31/63. AWOL 9-12/63. Present 1-2/64, dismounted since 7/14/63. AWOL 3/29-4/30/64. Present 5-12/64. NFR. Farmer, age 56, Franklin Co., Ga. 1880 census. Retired, age 75, Gunnell, Franklin Co., Ga. 1900 census.

HYDE, JOHN STEPHENS. Pvt., Co. F. b. Equinox Mill, Anderson Dist. 10/4/45. Student, age 15, Horse Shoe, Pickens Dist. 1860 census. Enl. Pickens CH 12/4/61 age 17. Present through 4/62. On picket Camp Means 5-6/62. Ab. sick Summerville hospital 7-10/62. Present 11/62-12/63. Ab. on horse detail 1/7-2/29/64. On duty Green Pond 4/16-30/64. Present 5-12/64. NFR. Receiving pension Equinox Mill, Anderson Co. 4/28/19. Attended reunion Walhalla, S. C. 1922.

HYDE, JOSEPH A. Pvt., Co. F. b. S. C. circa 1847. Enl. Columbia 8/13/64. Present through 12/64. Admitted Raleigh, N. C. hospital 4/12/65. Captured there 4/13/65. Issued clothing there 4/18/65. NFR. Bookkeeper, age 53, Charleston 1880 census.

INNABNETT, WILLIAM MOSCOW. Asst. Surgeon. b. S. C. 1834. Appointed Asst. Surgeon PACS from S. C.12/4/62 to rank form 9/2/62. Assigned to 1st S. C. Cav. 9/19/62. Present through 10/62, however, ordered transferred to 24th S. C. Inf. 9/25/62 & Asst. Surgeon Yates to be relieved from duty with 24th S. C. Inf. and assigned to 1st S. C. Cav., but not on rolls 24th S. C. Inf. Assigned for treatment 12/62. Ordered to report to Surgeon Brodie, Dept. of S. C., Ga. & Fla. 8/3/63. Resigned at Sullivan's Island the same day. NFR. d. Denmark, S. C. 12/30/92.

IENE, MICHAEL. Pvt., Co. A. On roster but not on muster rolls. NFR.

IRWIN, JAMES WILLIAM. Pvt., Co. G. b. S. C. 11/6/30. Enl. Abbeville CH 3/15/62. Present through 10/62. Ab. sick in Richmond hospital 11/26-12/31/62. Present 3/63-8/64. On duty with Regimental QM Dept. 9-10/64. Present 11-12/64. NFR. Farmer, Bradley, Greenwood Co. 1900 census. d. 9/22/09. Bur. Horeb Bapt. Ch., Greenwood Co.

IRWIN, THOMAS A. Pvt., Co. unknown. b. Greenville Dist. 9/18/46. Student, age 13, Spartanburg Dist. 1860 census. Served in Holcombe S. C. Legion with father 1862, but not on muster rolls. Enl. 1864 in postwar account. Railroad Agent, Southern RR, Spartanburg 1880-1910 censuses. Retired, Spartanburg 1920 census. d. Spartanburg 6/30/23.

ISENHOWER, JOHN. Pvt., Co. B. b. Lincoln Co., N. C. circa 1825. Enl. Co. B, 4th S. C. Bn. Cav. Columbia 2/1/62 age 37. Present until discharged 6/24/62. Age 37, 6', light complexion, blue eyes, dark hair, Millwright. Enl. Co. B, 1st S. C. Cav. 2/65 on wife's pension application. d. 6/27/07. Bur. Mt. Olivet Ch., Cem., Fairfield Co. Wife receiving pension Winnsboro, Fairfield Co. 10/8/19.

JACKSON, --------------. Pvt., Co. C. Enl. Aiken Dist. age 30 on postwar roster. Alive 1898.

JACKSON, A. Pvt., Co. B. Res. Spartanburg Dist., age 32. d. in Pa. 1863.

JACKSON, F. B. Pvt., Co. B. Res. Spartanburg Dist. Captured and died of typhoid fever Gettysburg 8/2/63.

JACKSON, GRANVILLE T. Y. 3rdLt., Co. G. b. Tenn. circa 1838. Married Abbeville Dist. 1859. Enl. Abbeville CH 1/8/62. Signed for 20 pr. horse shoes 2/16/62. On picket Jacksonborough through 4/62, however, ab. on leave for 20 days 4/7/62. 76 horses present. Ab. on sick leave 6/16-30/62. Commanding picket on Stone River 9-10/62. Present 9-10/62. On picket 11-12/62. Ab. on sick leave 3-7-29/63. 36 horses present 3/31/63. Present 1-2/64. Ab. on leave 3-4/64, however, signed for 24 pr. shoes, 7 blankets, 4 tent flys, 1 camp kettle, 17 pr. pants, 5 shirts, 1 pr. socks 13 jackets and 3 pr. drawers 3/31/64. Present 5-10/64, however, ab. sick with "Irritable stucture of urethra" in hospital 9/19/64, furloughed for 20 days. Detached Pocotaligo 12/10-31/64. NFR. Farmer, age 35, Long Cane, Abbeville Co. 1870 census and age 45, Ninety Six, Abbeville Co. 1880 census. d. Ninety-Six, S. C. 10/22/87.

JACKSON, H. M. Pvt., Co. H. b. York Dist 1829. Farmer, age 31, York Dist. 1860 census. Enl. Rock Hill 1/9/62 age 32. Present through 6/62. Ab. on leave 7-8/62. Present 9-12/62. Ab. on leave 3-4/63. WIA (shoulder) Brandy Station 6/9/63. Ab. wounded & in hospital 6/10-7/11/63. Furloughed for 20 days. Ab. wounded & on leave through 812/31/63. Present 1-2/64. Ab. on leave 3/31-4/28/64. Present 5-6/64. On picket 7-8/64. Present 9-12/64. NFR.

JACKSON, J. ALEXANDER. Pvt., Co. B. b. S. C. circa 1833. Res. Spartanburg Dist. Enl. Clinton 9/10/61 age 28. Present through 10/31/61. Ab. on leave 12/16-31/61. Present 1-8/62. Ab. on leave 9-10/62. Ab. on raid with General Hampton & horse lost 11-12/62. Ab. on horse detail 3-4/63. Left as nurse with wounded Gettysburg 7/3/63. Captured 7/5/63. Sent to Ft. McHenry & Pt. Lookout. d. there 4/24/64 age 32. Bur. Pt. Lookout, Md. Confederate Cem.

JACKSON, R. Pvt., Co. B. Res. Spartanburg Dist., age 30. d. in Pa. 1863.

JACKSON, THOMAS ROBERT. Pvt., Co. B. b. S. C. 2/25/34. Farmer, age 24, Spartanburg Dist. 1860 census. Enl. Camp Tripp 5/19/62. Ab. on sick leave 7-8/62. Present sick in camp 9-10/62. Present 11-12/62, dismounted since 10/10/62. Ab. sick with Bronchitis in Richmond hospital until transferred Lynchburg hospital 4/21/63. WIA & captured sick with typhoid fever Gettysburg 7/3/63. DOW's 8/2/63 age 30. Bur. Gettysburg National Cem.

JACKSON, Q. A. Pvt., Co. B. On post war roster.

JAMES, JOHN F. 2ndCpl., Co. F. b. S. C. 9/39. Enl. Pickens CH 12/4/61 as Pvt., age 22. Present through 10/62. On picket 11-12/62 & 3-4/63, promoted 4thCpl. Present through 8/31/63, promoted 2ndCpl. Captured on picket Robinson River 9/23/63. Sent to Old Capitol, Pt. Lookout & Elmira. Exchanged 10/64. Furloughed from Charleston hospital for 30 days. NFR. Farmer, Hill Co., Texas 1870-1900 censuses. d. 6/22/04. Wife receiving pension Greer, Greenville Co., S. C. 1/25/27.

JAMES, SAMUEL A. Pvt., Co. F. b. S. C. 7/38. Enl. Pickens CH 12/4/61. Present through 12/31/61. On picket Simmons House 3-4/62. Present 5-6/62. Ab. on leave 7-8/62. Present 9-10/62. Detailed as wagon guard 11/62-4/63. Present 3-8/63. Ab. detached 9-10/63. Present 11/63-2/64. Detached Green Pond 4/16-30/64. Present 5-12/64. NFR. Farmer, age 43, St. Joe, Montague Co., Texas 1880 census. Carpenter, age 61, St. Joe, Montague Co., Texas 1900 census. Mason. In house for aged, age 82, Tarrant Co., Texas 1920 census.

JARRELL, WILLIAM G. 1stSgt., Co. D. b. S. C. circa 1835. Res. Newberry Dist. Enl. Chester CH 9/10/61 as 2ndSgt., age 26. Present through 6/62. Ab. on leave to Rockingham Co., N. C. 8/16-31/62. Promoted 1stSgt. Ab. sick in N. C. 9-10/62. Present 11-12/62. Ab. on horse detail 2/22-29/63. KIA near Gettysburg 7/3/63 age 28. Left widow.

JEANS, E. W. B. Pvt., Co. E. Enl. by 1864. Transferred Co. B, Lucas's Bn. S. C. Arty. 4/30/64. Not on muster rolls of that unit.

JENKINS, JOSEPH G. Pvt., Co. K. b. Greenwood, Greenwood Dist. 5/13/47. Enl. Columbia 4/146/64. Present until detail as Courier for Col. Black on James Island 5-12/64. Paroled Greensboro, N. C. 4/216/65. Bookkeeper, Camden, N. J. 1870-1900 censuses. Farmer, Greenwood, Greenwood Co. 1920 census. Receiving pension there 4/4/19. d. 1/6/24. Bur. Methodist Cem., Greenwood Co.

JOHNSEY, THOMAS JACKSON. Pvt., Co. D. b. S. C. circa 1824. Enl. Chester CH 9/10/61. Present through 6/62. Ab. on sick leave 8/9-31/62. Present 9-12/62. Present sick in camp 3-4/63. Detailed in Regimental QM Dept. 5-12/63. Ab. recruiting station for horses 1-2/64. Present 3-4/64. On retired list 6/30/64. NFR. Age 55, Chester, 1880 census.

JOHNSON, FRANCIS MARION. Pvt., Co. C. b. S. C. circa 1844. Enl. by 1865. Captured Bennettsville, S. C. 3/6/65. NFR. Farmer, age 34, Flat Creek, Lancaster Co. 1880 census.

JOHNSON, J. H. Pvt., Co. D. On postwar roster.

JOHNSON, L. T. Pvt., Co. B. b. S. C. circa 1834. Trader, age 26, Clinton, Laurens Dist. 1860 census. Enl. Co. G, 3rd S. C. Inf. 4/14/61. Ab. sick with gastritis & rheumatism in Culpepper CH & Charlottesville hospitals 7/15-7/19/61. Discharged 8/25/61. Enl. Co. B, 1st S. C. Cav. Adams Run 7/25/62. Present sick in camp through 8/62. Ab. sick in Summerville hospital 9-10/62. Ab. sick with chronic rheumatism in Richmond hospital 11/22/62-1/3/63, furloughed for 30 days. Present 3-4/63. Discharged 6/12/63. Served in Co. H, 1st S. C. State Troops 9/12-12/31/63. NFR.

JOHNSON, LEROY J. "OLD HICKORY." Captain, Co. G. b. S. C. 1836. Enl. Abbeville CH 1/8/62. 41 horses present 2/28/62. 59 horses present 3/14/62. 57 horses present 3/29/62. Ab. on sick leave 4/26-30/62. 57 horses present 4/2/62. Signed for 7 Army tents, 4 axes, & helves, 4 spades, 4 hatchets, 4 camp kettles & 2 mess pans. 4/16/62. Signed for 7 jackets, 15 caps, & 1 shirt 4/30/62. Present 5-6/62. Signed for 370 lbs. beef 5/28/62. 82 horses present 6/8/62. 76 horses present 7/3/62. Ab. on leave 8/20-31/62. Present 9-10/62. Signed for box 1,000 buck & ball cartridges cal. 68 with caps, box 1,000 Enfield cartridges with caps, 1 shovel, 7 Army tents, 3 camp kettles & 8 mess pans 9/3/62. On scout 11-12/62. Present through 7/63. 51 horses present 7/1/63. WIA Brandy Station 8/2/63, however, present through 12/64, however, ab. on leave for 15 days 9/5/64. Signed for 43 pr. pants, 41 jackets, 5 pr. shoes, 3 pr. socks, 51 cotton shirts, 46 pr. drawer & 32 caps, 6/30/64. Signed for 16 pr. pants, 21 jackets, 23 shirts, 5 caps & 1 pr. shoes 8/24/63. 34 horses present 8/31/63. 39 horses present 10/23/63. 41 horses present 8/1/64, 48 horses present 9/1/64, signed for 1 jacket, 4 pr. pants, 8 shirts, 5 pr. drawers, 6 pr. shoes, 3 pr. socks, & 1 wall tent 9/30/64 & 48 horses present 10/1/64. Signed for 12 pr. boots, 16 pr. socks, 3 blankets, 35 pr. drawers, 14 pr. pants, & 2 shirts on undated invoice. Present 4/65. NFR. d. by 1902. Bur. Sharon Meth. Ch. Cem., Abbeville Co., no dates.

JOHNSON, RICHARD. Chaplain. b. Beaufort, S. C. 11/13/09. Gd. Norwich U., Vermont, 1829, AB. Ordained Epis. Priest in S. C. 1834 & Ga. 1836. Appointed Chaplain, Hampton Legion 7/12/61. Relieved 2/5/63. Appointed Chaplain, 1st S. C. Cav. 1/8/63. Present 6/9/63. Ab. on leave 8/13/63. AWOL 10/25-12/31/63. Present 1-2/64. Ab. detached Green Pond 4/18-30/64. Present James Island 6/20/64. Ab. on leave 6/23/64 for 7 days. Present 8/25/64. Present 9-12/64, however, ab. on leave 9/12/64 for 15 days. Ab. on leave 1/6-17/65. Captured Bennettsville, S. C. 3/6-8/65 as Pvt., Co. C, 1st S. C. Cav. Sent to New Bern 3/26/65. NFR. Priest, Grace Episcopal Ch., St. Francisville, La. 1867-70. d. Roseville, La. 1/7/72. Bur. there.

JOHNSTON, JAMES H. 1stCpl., Co. D. b. S. C. circa 1835. Farmer, age 25, Sumter, Fla. 1860 census. Enl. Chester CH 9/10/61 age 25. Reduced to Pvt. 10/24/61. Present through 4/62. On picket Townsend's House 5-6/62. Present 7-12/62. Ab. on horse detail 3/4-4/30/63. Present through 8/31/63 & 9-10/63. Ab. recruiting station for horses 11/13-12/31/63. Present 1-10/64. Ab. sick in Brigade hospital 12/26-31/64. Paroled Hillsboro, N. C. 4/65. d. Daytona, Fla. 2/12/06.

JOLLY, MANSON SHERRILL J. 1stSgt., Co. F. b. S. C. 1841. Enl. Colleton Dist. 4/8/62 as Pvt. On picket Jacksonborough through 4/30/62. Promoted 2ndCpl. Present 5-6/62. On detached duty 7-10/62. On scout 11-12/62. Present 3-4/63, promoted 1stSgt. Present through 8/31/63. On scout 9-10/63. Present 11-12/63. Ab. on leave 2/10-29/64. On duty Green Pond 4/25-30/64. Present 5-6/64. Present sick in camp 8/20-31/64. Present 9-12/64. NFR. d. Howard, Milam Co., Texas 7/8/69. Bur. Little River Cem. Mase Jolly Camp, CV, Anderson, S. C. named in his honor.

JOLLY, WILLIAM ENOS. Pvt., Co. F. b. S. C. 12/37. Farmer, age 32, Cherokee, Ala. 1860 census. Enl. Pickens CH 12/4/61. AWOL 12/21/61-4/62. In Capt. Watkins Co. near Columbia 6/62. NFR. Farmer, age 42, De Kalb Co., Ala. 1880 census. Farmer, age 62, Hunt Co., Texas 1900 census. Living Leesburg, Cherokee Co., Ala. 1910 census. d. Mackey, Cherokee Co., Ala. 1918.

JONES, ROBIN ALLEN CADWALLADER. Captain, Co. H. b. Hillsborough, N. C. 1/18/26. Res. York Dist. 1860 census. Enl. Rock Hill 1/9/62 age 35. Present through 4/62. 71 horses present 3/26/62. 65 horses present 4/1/62. 78 horses present 5/18/62. Ab. on leave 5-10/62. Present 11-12/62. Ab. on leave 3-4/63. 61 horses present 5/8/63. Signed for 1 blanket & 4 tent flys 5/25/63. KIA Brandy Station 6/9/63. Body brought home by Pvt. Alexander W. Wise. Bur.St. Matthews Epis. Ch., Orange Co. N. C. beside his wife. Marker on his grave site.

JONES, SAMUEL HENRY. Captain, Co. A. b. S. C. circa 1839. Enl. Abbeville CH 8/13/61 as 3rdLt. Ab. recruiting 11/61-1/62. Present 5-6/62. Ab. sick 7-8/62. Present 9-10/62. On scout 11-12/62, promoted 2ndLt. 20 horses present 1/1/63. Commanding Company 3-4/63, 6/9-8/1/63, however, ab. sick in Richmond hospital 6-15-20/63. 73 horses present 5/1/63. Present through 8/31/63, elected Captain. 26 horses present 7/31/63. Present 9-10/63, commanding sharpshooters. Signed for 2 pr. boots, 19 pr. socks, 3 blankets, 12 pr. drawers & 1 overcoat 9/8/63. Ab. sick with "Feb. Remit" in Charlottesville hospital 11/18-12/4/63. Ab. on leave to Lancaster, S. C. 12/28-31/63. Present 1-2/64. Signed for 21 pr. horse shoes 3/27/64. Ab. on detached service in command of horse recruiting camp 4/18-12/31/64. Signed for 50 lbs. rope & 100 lbs. nails Columbia 5/3/64. Signed for 14 caps, 23 jackets, 30 shirts & 29 pr. socks 9/11/64. NFR. Living Memphis, Tenn. 1915.

JONES, THOMAS PINCKNEY. Pvt., Co. G. b. Abbeville Dist. 10/13/30. Enl. Columbia 5/24/64. Present through 12/64. NFR. d. Jack Co., Texas 11/26/16. Bur. Graceland Cem., Jack Co., Texas.

JONES, WILLIAM W. 3rdSgt., Co. A. b. S. C. circa 1840. Farmer, age 22, Abbeville Dist. 1860 census. Enl. Abbeville CH 8/13/61 as Pvt., age 22. Present through 8/31/63. Detailed a Courier at Brigade Headquarters 9-10/63. Ab. on horse detail 12/23/63-2/64., dismounted since 11/20/63 & remounted 1/1/64. Promoted 1stCpl. On duty Green Pond 3-4/64. Present 5-6/64. Ab. detached 8/24-31/64. Promoted 3rdSgt. Present 9-12/64. Paroled Greensboro, N. C. 5/1/65. Farmer, age 40, Greenwood, Abbeville Co. 1880 census. Retired, age 70, Laurens, S. C. 1910 census.

JORDAN, LUCIUS. 4thSgt., Co. I. b. S. C. 4/14/42. Res. Colleton Dist. 1860 census. Enl. Parker's Ferry 3/2/62 as Pvt. On picket Jacksonborough through 4/30/62. Present 5-10/62. On scout 11-12/62. Present 3-4/63. Detailed recruiting camp for horses through 12/31/63. Promoted 4thCpl. Ab. sick with "Catarrah Fever" in Staunton hospital 1-2/29/64. On detail Green Pond 4/16-30/64. Present 5-6/64. Promoted 3rdCpl. Present 7-8/64, promoted 4thSgt. Present 9-12/64. NFR .Farmer & Merchant, Sheridan, Colleton Co. 1880 census. d. 9/19/96. Bur. Jordan Cem., Colleton Co.

JUSTNOTT, WILLIAM. Asst. Surgeon. On roster but not on muster rolls.

KALSTROM, ------------. Pvt., Co. _. Res. Richland Dist., age 30. KIA in N. C. 1865.

KARICK, ADAM G. Pvt., Co. E. b. S. C. circa 1840. Res. Orangeburg Dist. Enl. Adams Run 7/3/62. Present through 12/62. Ab. sick in Columbia, S. C. hospital 3/28-10/63. Present 11/63-2/64. Ab. on leave 4/26-5/3/64. Present 5-12/64. NFR. Farmhand, age 29, Pine Grove, Orangeburg Co. 1870 census.

KARICK, ARTEMUS G. Pvt., Co. E. b. S. C. 1836. Res. Orangeburg Dist. 1850 census. Enl. Ft. Motte 10/26/61 age 24. Ab. with wagons 11-12/61. Died of disease in Adams Run hospital 3/9/62.

KARICK, EMANUEL D. Pvt., Co. E. b. S. C. 10/36. Res. Orangeburg Dist. 1850 census. Enl. Ft. Motte 10/26/61 age 26. Present through 12/61. Ab. on sick leave 3/20-4/62. Present 5-6/62. Present sick in camp 7-8/62. Present 9-10/62. On picket 11-12/62. Present 3-8/63. Captured on picket Robinson River 9/22/63. Sent to Old Capitol & Pt. Lookout. Exchanged 9/18/64. Ab. sick with "Febris Remit" in Richmond hospital 9/22-26/64. Furloughed for 30 days. Ab. on sick leave through 10/64. Detached Pocotaligo 12/10-31/64. Paroled Hillsboro, N. C. 4/65. Farmer, age 33, Pine Grove, Orangeburg Co. 1870 census & age 63, 1900 census.

KEE, A. MARTIN. 1stSgt., Co. H. b. York Dist. 12/28/32. Farmer, age 28, York Dist. 1860 census. Enl. Rock Hill 1/18/62 as Pvt., age 30. Promoted 1stSgt. Present 3-4/63. Ab. on sick leave 6/9-8/31/62. Present 9-10/62. Ab. on leave 11-12/62. Present 3-8/63. Ab. on horse detail 9-10/63. Present 11-12/63, AWOL 4 days. Present 1-8/64. Present sick in camp 9-10/64. Detached Pocotaligo 12/10-31/64. Paroled Greensboro, N. C. 4/26/65. Farmer, age 38, Ft. Mill, York Co. 1870 census d. Lancaster Co. 6/28/90. Bur. Old Presb. Ch., Lancaster Co.

KEGGS, JACOB PETER. Pvt., Co. I. b. S. C. circa 1838. Res. of S. C. Enl. Camp Abbeville 4/29/62. Present 5/62-6/63. Ab. on horse detail 8/13-31/63. Present 9-10/63. On picket 11-12/63. Present 1-2/64. Detached Green Pond 4/16-30/64. Present 5-6/64. Ab. sick in Charleston hospital 8/27-31/64. Ab. detailed in Charleston Arsenal as Carriage Maker 9-10/64. Detailed as Mechanic in Engineering Dept. 12/1-31/64. Deserted to the enemy Charleston 3/27/65. Age 25, 5'6", light complexion, blue eyes, dark hair. Took oath & released. Wood hauler, age 66, Screven Co., Ga. 1900 census. Retired, age 77, Pickens, Duval Co., Fla. 1910 census. d. Jacksonville, Fla. 11/30/11.

KELLER, JACOB W. (1). Pvt., Co. E. b. S. C. circa 1839. Res. Orangeburg Dist. Enl. Ft. Motte 10/26/61 age 22. Ab. with wagons 11-12/61. On picket at Railroad 3-4/62. Courier, Capt. Walters Battery 5-6/62. Ab. on leave 7-8/62. Ab. sick Summerville hospital 9-10/62. Discharged in Va. 11/19/62.

KELLER, JACOB W. (2). Pvt., Co. E. b. Orangeburg Dist. circa 1836. Res. Orangeburg Dist. 1850 census. Enl. Ft. Motte 10/26/61. Discharged 8/31/62, loss of use of right hand. Age 22, 6', fair complexion, grey eyes, dark hair, Farmer. Re-enl. Co. E, 5th S. C. Cav. 11/24/63. Present through 12/63. Detailed as Teamster 1-10/64, however, ab. sick with "Icterus" in Richmond hospital 9/23/64 & ab. sick with chronic diarrhea in Raleigh, N. C. hospital 10/3-24/64. Issued clothing 11/2/64. NFR.

KELLER, JOHN. Pvt., Co. I. b. S. C. circa 1840. Enl. Parker's Ferry 4/3/62. Ab. on detached service 5-8/62. Present 9/62-8/31/63. WIA (thigh) Culpepper CH 9/13/63. DOW's from gangrene in Richmond hospital 10/24/63.

KELLER, JOHN MICK. Pvt., Co. B. b. S. C. circa 1844. Enl. Ft. Motte 10/26/61 age 22. Ab. with wagons 11-12/61. Present 3-6/62. Present as Courier at Parker's Battery 7-8/62. Present 9-10/62. On scout 11-12/62. Present 3-4/63. Ab. on sick leave Orangeburg Dist. 5-8/31/63. Present 9-12/63. On picket 1-2/64. Detached Green Pond 4/16-30/64. Present 5-8/64. Ab. detailed to guard POW's Florence 9/17-12/31/64. Paroled Hillsboro, N. C. 4/65. Farmer, age 26, Pine Grove, Orangeburg Co. 1870 census.

KELLER, JOHN W. Pvt., Co. B. b. S. C. circa 1840. Res. Orangeburg Dist. 1850 census. Enl. Camp Johnson 10/15/61. Present through 6/62. Ab. on detached service 7-8/62. Ab. driving horses from Summerville, S. C. to Staunton, Va. 9-10/62. Ab. sick with "Bubo syphileticum" in Richmond hospital 11/12/62--1/1/63. Transferred to Farmville hospital with "syphilis" through 1/23/63. Ab. on horse detail 3-4/63. Captured Beverly Ford 6/9/63. Sent to Old Capitol. Exchanged by 8/63. Ab. detached at horse recruiting camp Nelson Co., Va. 8/24-31/63. AWOL 9-10/63. Ab. on detached service with Major Briggs 11-12/63. Present 1-4/64, lost horse 3/7/64. Remounted 4/15/64. Present 5-8/64. Ab. sick in Summersville hospital 10/1-31/64. Present 11-12/64. NFR. d. by 1898.

KELLER, RICHARD. Pvt., Co. I. b. S. C. circa 1841. Res. Colleton Dist. Enl. Parker's Ferry 4/3/62.

Present until ab. on detached service 5-8/62. Ab. on leave 9-10/62. AWOL 11-12/62. Present 3-8/63. d. in Culpepper CH hospital 10/24/63.

KELLER, THOMAS WILLIAM. Pvt., Co. I. b. 1842. Res. Orangeburg Dist. Enl. Co. K, 22nd S. C. Inf. Columbia 1/22/62. Enl. Co. I, 1st S. C. Cav. Parker's Ferry 4/3/62 without authority. AWOL through 4/30/62. Returned to Co. K, 22nd S. C. Inf. Deserted from Savannah, Ga. hospital 8/20/63. Present 11-12/63. AWOL 2/17-12/31/64, however, Issued clothing Kinston, N. C. 4/29/64 & 7/20/64. WIA (right little toe amputated) in Williamsburg hospital 8/10/64. Furloughed for 30 days 8/16/64. NFR. Farmhand, age 28, Amelia, Orangeburg Co. 1870 census. Farmer, age 58, Erata, Jones Co,, Miss. 1900 census & 1910 census. d. 7/22/17. Bur. Zoah Bapt. Ch. Cem., Spartanburg Co., S. C.

KELLETT, JOHN SAMPSON. Pvt., Co. B. b. S. C. circa 1840. Age 21, Laurens Dist. 1860 census. Enl. Clinton 8/22/61 age 21. Present through 12/61. On picket Hallover Bridge 3-4/62. WIA in skirmish at Walpole Gate, John's Island 6/7/62. Ab. wounded through 6/30/62. Ab. on leave 7-8/62. Present 9/62-8/31/63. On detached service 9-10/63. Ab. horse recruiting station 11-12/63, horse died 11/1/63. Present 1-2/64. Detached Green Pond 4/26-30/64. Present 5-8/64. Ab. guarding POW's Florence 9/16-12/31/64. NFR. Farmhand, age 39, Fairview, Greenville Co. 1880 census.

KELLEY, JOSEPH H. Pvt., Co. E. b. 6/13/22. Not on muster rolls. WIA (leg) on pension application, from Walhalla, Oconee Co. 1901. d. 6/19/02. Bur. Neville Cem., Oconee Co.

KELLY, D. S. Pvt., Co. F. b. S. C. circa 1824. Res. Caloosa, Ga. 1850 census. Enl. Pickens CH 12/4/61 age 37. Present through 4/62. Ab. driving ambulance 5-6/62. Ab. on sick leave 7-8/62. d. Pickens Dist. 8/23/62. Left widow.

KELLY, JOSEPH A. Pvt., Co. F. b. S. C. circa 1825. Farmer, age 37, Pickens Dist. 1860 census. Enl. Pickens CH 12/4/61 age 35. Present through 4/62. Ab. on sick leave 6/23-7/31/62. Present 8/62-8/31/63. Ab. sick in hospital 10/18/63. Furloughed for 30 days 10/19/63. Ab. horse recruiting station 12/22/63-2/64. Ab. sick with "gonorrhoea" in Charlottesville hospital 3/1-4/7/64. Present 5-12/64. Paroled 4/65. Farmer, age 45, Pickney, Union Co. 1870 census.

KELSEY, B. On Confederate Veterans list Chester Co. 1904.

KELSEY, JOHN RANDOLPH. Pvt., Co. D. b. Chester Dist. 12/16/44. Enl. Co. A, 6th S. C. Inf. Chester CH 4/11/61 age 16. Discharged 6/62. 5'6", fair complexion, hazel eyes, Farmer. Enl. Co. D, 1st S. C. Cav. Chester CH 4/22/63. Present through 10/63. Ab. horse recruiting station 12/24/63-2/64. Present 3-10/64. Detached Coosawhatchie 12/10-31/64. Paroled Hillsboro, N. C. 4/65. Farmhand, age 25, Four Mile, Barnwell Co. 1870 census. Farmer, age 55, Landsford, Chester Co. 1900 census. d. Ft. Lawn, S. C. 12/13/07. Bur. Elmwood Cem., Chester, S. C.

KENNEDY, D. W. Pvt., Co. D. Enl. by 1862. Discharged 12/28/62. Paid 2/14/63. NFR.

KENNEDY, DAUL J. Pvt., Co. E. Enl. Va. 2/9/64. Detached Green Pond 4/16-30/64. Present 5-6/64. Present sick in camp 7-8/64. Present 9-10/64. Detached Pocotaligo 12/10-31/64. NFR. d. of disease Orangeburg Dist. 4/15/65.

KENNEDY, J. L. Pvt., Co. D. On post war roster.

KEOWN, ROBERT WESTLY. Pvt., Co. G. b. Anderson Dist. 8/9/43. Res. Saluda, Abbeville Dist. 1850 census. Enl. Abbeville CH 3/15/62. Ab. sick in hospital 4/30-7/30/62. Present 8-10/62. Ab. on picket 11-12/62. Present 3-6/63. Lost horse 7/1/63. Detailed to drive wagon. Ab. on horse detail 8/31/63. Present 9-12/63. Ab. detailed at horse recruiting station 11/3/63-2/64. Present 3-12/64. NFR. Married Noxubee Co., Miss. 1866 age 22, Farmer. Farmer, age 66, Edgewood, Fulton Co., Ga. 1910 census.

KEY, HENRY. Pvt., Co. C. b. S. C. circa 1824. Cotton Speculator, age 37, Hamburg, Edgefield Dist. 1860 census. Enl. Hamburg 8/27/61 age 38. Ab. detached on Wadmalaw Island through 12/31/61. Transferred 4th Texas Infantry 2/21/62. NFR. Trader, age 46, Hamburg, Edgefield Co. 1870 census. Alive 1898.

KIMBRELL JAMES. Pvt., Co.'s B & K. b. Spartanburg Dist. 5/8/45. Age 14, Spartanburg Dist. 1860 census. Enl. Co. B Columbia 3/25/64. Transferred Co. K 4/15/64. Detached Green Pond through 4/30/64. Detailed as Courier, Provost Marshal's Office, Charleston 6-/25-30/64. Detailed in Engineering Dept., Charleston 7-12/64, however, Transferred Co. F, 2nd Confederate Engineers 9/17/64 as Carpenter. Paroled Raleigh, N. C. 4/65. Miller, age 25, Cherokee, Spartanburg Co. 1870 census. Farmer, age 35, Cherokee, Spartanburg Co. 1880 census. d. Spartanburg 12/14/19.

KIMBRELL, JOHN S. Pvt., Co. B. b. Ft. Mill, York Dist. 9/24/42. Enl. Rock Hill 1/18/62 age 18. Present through 6/62. Ab. on sick leave 7-8/62. Detailed to drive horses from Summerville, S. C. to Staunton, Va. 9-10/62. On raid 11-12/62. Present 3-6/63. WIA (slightly) Gettysburg 7/3/63. Ab. on horse detail 8/18-10/63. Ab. on leave 12/20-31/63. Present 1-2/64. Detached Green Pond 4/16-30/64. Present 5-10/64. Detached Pocotaligo 12/10-31/64. Paroled Greensboro, N. C. 5/1/65. Farmhand, age 34, Ft. Mill, York Co. 1880 census. Farmer, Ft. Mill, York Co. 1910 census. Receiving pension Ft. Mill, York Co. 4/5/19.

KIMBRELL, SAMUEL J. Pvt., Co. H. b. S. C. 8/30/25. On roster but not on muster rolls. Enl. Co. G, 6th S. C. Reserves (90 days) York Dist. 11/4/62. Never reported on rolls through 2/63. NFR. Farmer, Cherokee, Spartanburg Co. 1880 census. d. 5/22/09. Bur. Unity Cem., Rock Hill, S. C.

KING, A. P. Pvt., Co. A. Not on muster rolls. Enl. by 1864. Deserted to the enemy Morris Island 8/1/64 or 9/28/64. Sent to Hilton Head. Worked in QM Dept. there. NFR.

KING, DAVID. Pvt., Co. F. Enl. Pickens Dist. 4/5/62. Present 5-6/62, mounted 6/27/62. Present 7-8/62. Ab. on detail 9-10/62. On picket 11-12/62. Present 3-6/63. WIA and captured in Md. 7/9/63. DOW's.

KIRBY, BOLEN C. Pvt., Co. C. b. Ga. circa 1848. Served in Co. G, 5th S. C. Troops. Enl. Company C, 1st S. C. Cav. Columbia 2/20/64. Present 3-6/64. Ab. detailed 10-12/64. NFR. Farmer, age 32, Bacon Level, Randolph Co., Ala. 1880 census and age 61, 1910 census.

KIRBY, J. Pvt., Co. K. Enl. by 1863. d. 6/8/63. Bur. Magnolia Cem., Charleston, S. C.

KIRKLAND, WILLIAM LENOX. Pvt., Co. A. b. S. C. 1828. On post war roster. Reenlisted Co. K, 4th S. C. Cav. Pocotaligo 11/20/62. Present until ab. on detailed with Gen. Walker 6/19/63 for 20 days. Present until ab. on sick leave 8/29/63. Present 11/63-2/64. WIA (thigh fractured) Hawes Shop 5/28/64. Leg amputated 6/1/64. DOW's 6/19/64. Bur. Quaker Cem., Camden, S. C.

KIRKPATRICK, HENRY. Pvt., Co. H. b. S. C. circa 1837. Enl. Rock Hill 1/9/62 age 25. Ab. on sick leave 3-4/62. Discharged 6/25/62. NFR. Farmer, age 43, Whiteville, Dorsey Co., Ark. 1880 census.

KITCHENS, CHARNER TERHAMER. 3rdSgt., Co.'s D & K. Enl. Cat Island 12/2/61 as Pvt. Ab. detached guarding baggage through 12/31/61. Present 3-4/62. Transferred Co. K 6/25/62 & promoted 2nd Cpl. Present 7-12/62, promoted 4thSgt. Ab. sick 4/15-30/63. Present 7-8/63, promoted 3rdSgt. WIA James City 10/9/63. DOW's 10/15/63.

KIZER, IRVIN B. Pvt., Co. E. b. Reevesville, Dorchester Co. 2/18/40. Enl. Coosawhatchie 12/13/61. On picket White Point 12/31/61. Ab. sick in hospital 3-4/62. Present sick in camp 5-6/62. On duty Pine Berry 7-8/62. Present 9-10/62. On picket 11-12/62. Present 3-8/63. Ab. on sick leave 9-10/63. Present 11/63-2/64. Detached Green Pond 4/26-30/64. Present 5-7/64. Detached in Commissary Dept. 8/5-31/64. On guard Wappo Bridge 12/29-31/64. Paroled Hillsboro, N. C. 4/65. Farmer, age 40, Cow Castle,

Orangeburg Co. 1880 census. Farmer, age 69, Rogers, Dorchester Co. 1910 census. Receiving pension Reevesville, Dorchester Co. 4/19/19. d. Dorchester Co. 2/18/24. Bur. Reevesville Meth. Ch. Cem.

KIZER, K. P. Pvt., Co. C. b. S. C. cica 1844. Enl. by 1862. Captured and lost horse Pocotaligo 10/22/62. Sent to Ft. Delaware & Ft. Monroe. Exchanged 12/15/62. NFR. Farmer, age 26, Lexington Court House 1870 census.

KNOX, JOHN. 1stSgt. Co. A. b. S. C. circa 1830. Enl. Abbeville CH 8/13/61 as 2ndCpl., age 31. On duty Bear Island 11-12/61. Present 3-6/62, promoted 1stCpl. Present sick in camp 7-8/62. Present 9-12/62, promoted 4thSgt. Present 3-4/63, promoted 3rdSgt. Present 5-6/63. Ab. on horse detail 8/15-31/63, promoted 2ndSgt. Present 9/63-2/64. On duty Green Pond 3-4/64. Present 5-6/64, promoted 1stSgt. Present 9-12/64. NFR. Farmer, age 40, Kershaw Co. 1870 census.

KNOX, SAMUEL BLAIR. Pvt., Co. A. b. Abbeville Dist. 1/9/45. Enl. Abbeville CH 4/1/63. Present through 10/63. Ab. recruiting station for horses 12/21/63-2/64. Present 3-10/64. Detached at C & R Railroad 12/29-31/64. NFR. Farmer, Diamond Hill, Abbeville Co. 1880 & 1900 censuses. d. 11/4/04. Bur. Little Mountain Cem., Abbeville Co.

KOGER, J. H. Pvt., Co. E. Enl. Adams Run 3/17/62. Present through 4/62. Courier at Capt. Walter's Battery 5-6/62. Present 7-10/62. On scout 11-12/62. Present 3-11/63. Ab. on leave 12/2/63-1/1/64. Present through 2/64, dismounted 2/10/64. Detached Green Pond 2/26-4/30/64. Present 5-6/64. Ab. on sick leave 8/21-31/64. On picket 9-10/64. Present 11-12/64. NFR. Bur. Koger Cem., Colleton Co., S. C., no dates.

KRIMMINGER. RUFUS. Pvt., Co. A. b. Cabarrus Co., N. C. circa 1837. Printer, age 23, Abbeville CH, 1860 census. Enl. Co. B, 7th S. C. Inf. 6/4/61. Discharged. Enl. Co. A, 1st S. C. Cav. Abbeville CH 12/15/61. On duty Bear Island through 12/31/61. Present sick in camp 3-4/62. Present 5-6/62. Present sick in camp 7-8/62. Present 9/62-6/63. d. of disease in Gordonsville hospital 8/16/63. Effects: $27.50. Bur. Confederate Cem., Gordonsville, Va.

LACKEY, GEORGE W. Pvt., Co. C. b. S. C. 9/16/39. Enl. 9/6/61 on Mississippi pension application. Discharged for disability 12/8/62. NFR. d. Noxapater, Miss. 3/8/15. Bur. Mt. Carmel Cem., there.

LACKEY, WILLIAM B. Pvt. Co. A. On post war roster. Transferred Co. K. d. 7/28/62.

LACKEY, WILLIAM GEORGE W. Pvt., Co. D. b. York Dist. circa 1836. Age 14, Union, York Dist. 1850 census. Enl. Chester CH 9/10/61 age 24. Present 11/61-4/62. Discharged 5/31/62. NFR.

LANCE, WILLIAM S. Pvt., Co. A. b. S. C. 11/26/33. On post war roster. Reenlisted Co. K, 4th S. C. Cav. Grahamville 5/16/62. Present until ab. detached with Signal Corps as Signal Operator 10/20/62-10/64. Relieved 11/1/64. NFR. d. 10/6/90. Bur. Grahamville Cem., Jasper Co.

LANGSTON, JOHN T. Pvt., Co. B. b. Laurens Dist. 5/10/46. Student, age 14, Tylersville, Laurens Dist. 1860 census. Enl. Co. I, 3rd S. C. Inf. Clinton 4/14/61. WIA Savage Station 6/30/62. Discharged for underage 1862. Enl. Co. B, 1st S. C. Cav. Columbia 4/14/64. Present through 4/30/64. Ab. sick in Charleston hospital 6/25-30/64. Present 7-8/64. Detailed to guard POW's Florence 9/16-12/31/64. Paroled Greensboro, N. C. 5/1/65. Farmer, Scuffletown, Laurens Co. 1870 census & Motts, Florence Co. 1900-1910 censuses. d. Mecklenburg Co., N. C. 9/12/20. Bur. Laurens City Cem., S. C.

LATIMER, PIERCE MATHEW BUTLER. Pvt., Co. G. b. S. C. 8/11/47. Enl. age 21 on postwar roster. Farmer, Woodstock, Ga. 1900 census. d. 3/19/1934. Bur. Enln Cem., Woodstock, Ga.

LAW, JAMES W. Pvt., Co. A. b. S. C. 1839. On postwar roster. Reenlisted Co. H, 2nd S. C. Cav. as Sgt. 7/17/62. WIA (shoulder) Williamsburg, no date. Discharged 11/4/62. Reenlisted Co. K, 4th S. C. Cav. Charleston 12/12/63. Present until captured Cold Harbor 5/30/64. Sent to Pt. Lookout. Exchanged 11/1l5/64. NFR. Farmer, Winnsboro 1870 census. Insurance Agent, Winnsboro 1880 census.

LAWSON, JOSEPH JAMES. Pvt., Co.'s F & K. b. Orangeburg Dist. circa 1816. Enl. Co. F, Pickens Dist. 4/5/62. Transferred Co. K 6/25/62. Present through 10/62. Ab. detailed QM Dept. 11-12/62. Ab. sick with "Icterus" in Charlottesville hospital 2/19-3/9/63 & in Richmond hospital 3/15-4/21/63. Ab. sick in Columbia, S. C. hospital 8/31-10/63. Present 11-12/63. Ab. horse recruiting camp 1-2/64. Ab. on leave 4/23-30/64. Ab. sick in Charleston hospital 5-6/64. Ab. on sick leave 7-8/64. AWOL 10/29-31/64. On picket Chatman's Ford 12/31/64. Rifle fell from stack and discharged killing him, Goldsboro, N. C. 4/12/65 age 54. Probably buried in Confederate Cem., Goldsboro, N. C.

LAYTON, DAVID P. R. Pvt., Co. B. b. Spartanburg Dist. 10/6/40. Farmer, age 19, Spartanburg Dist. 1860 census. Enl. Adams Run 12/16/61. On picket Edisto Island 12/61. On picket Hallover Bridge 3-4/62. WIA in skirmish Wallpole Gate, John's Island 6/7/62. Ab. wounded through 8/62. Present 9/62-8/31/63. KIA Culpepper CH 9/26/63.

LAYTON, JAMES OLIVER. Pvt., Co. B. b. Spartanburg Dist. 1/31/33. Res. Spartanburg Dist. Enl. Clinton 8/22/61 age 25. Ab. sick 10/31/61. Horse died 12/12/61. Ab. sick in Adams Run hospital 1/62. Died of disease there 1/7/62 age 30. Bur. Cedar Shoals Bapt. Ch. Cem., Hobbyville, Spartanburg Co., S. C.

LAYTON, JESSE C. Pvt., Co. B. b. S. C. circa 1839. Farmer, age 22, Spartanburg Dist. 1860 census. Enl. Clinton 8/22/61 age 22. Ab. sick 10/31/61. Detailed as nurse in hospital 11-12/61. Horse died 2/15/62. Present 3-4/62. Ab. nurse in Adams Run hospital 5-6/62. Courier at Railroad Station 7-8/62. Present 9-10/62. On picket Ellis Ford on Rappahannock 11-12/62. Present 3-4/63. Present dismounted 5-6/63. Ab. on horse detail 8/18-31/63. Present 9/63-12/64. NFR. Grocer, age 30, Bartow, Ga. 1870 census.d. in Ga. by 1898.

LAYTON, T. J. Pvt., Co. B. Res. Spartanburg Dist. Not on muster rolls. KIA Rapidan River 9/14/63.

LEAKE, WILLIAM JAMES. Captain, Co. B. b. S. C. 9/26/35. Enl. Clinton 8/22/61 as 1stLt. Ab. sick 10/31/61. Present 11-12/61. Present John's Island 1/62. Present 3-4/62. Commanding Company 5-6/62. Present 7-8/62. Ab. on march with horses from Summerville, S. C. to Staunton, Va. 9-10/62. 120 horses present 10/1/62. Commanding Company 11-12/62. Present 3-4/63. Commanding Company 7-8/63. Signed for 2 pr. shoes 8/2/63. Present 9-10/63. 58 horses present 10/63. Commanding Company 11-12/63. 69 horses present 12/31/63. Present 1-2/64. Detached Green Pond 4/16-5/15/64. On picket 7/1-8/31/64. Elected Captain 9/15/64. Present 9-12/64. NFR. d. 7/4/99. Bur. Clinton City Cem., S. C.

LEAPHART, CHARLTON E. L. Pvt., Co. I. b. Lexington Dist. circa 1845. Res. Lexington Dist. Enl. Co. K, 13th S. C. Inf. 9/3/61 age 16. Present through 10/61. Discharged for typhoid fever 7/22/62. Age 16, 5'6", light complexion, dark eyes, dark hair, Student. Enl. Co. I, 1st S. C. Cav. Columbia 4/1/64. On duty Green Pond 4/16-30/64. Present 5-10/64. Ab. sick in hospital 11/10-12/64. Killed by accidental gun shot James Island 12/15/64 age 18.

LEAPHART, I. G. Pvt., Co. C. On post war roster.

LEE, ANDREW J. Pvt., Co.'s D & K. b. S. C. 1830. Farmer, age 30, Chester Dist. 1860 census. Enl. Chester CH 9/10/61 age 31. Present 11/61-4/62. Transferred Co. K 6/25/62. Present through 8/62. Ab. sick 9-10/62. Died of typhoid fever Chester Dist. 11/2/62. Left widow.

LEE, CORYDON or CARODON FOX B. 1stCpl., Co. D. b. S. C. circa 1844. Age 16, Chester Dist. 1860 census. Enl. Chester CH 9/10/61 as Pvt., age 18. Present through 4/62. On picket Townsend's House 5-6/62. Present 7-10/62, promoted 3rdCpl. Present 11-12/62, promoted 2ndCpl. Present 3-6/63, promoted 1stCpl. Ab. on horse detail 8/18-31/63. Present 9/63-2/64. Detached Green Pond 4/16-30/64. Present 5-10/64. Detached Coosawathchie 10/10-31/64. Paroled Hillsboro, N. C. 4/65. Engineer, age 23, Houston, Texas 1870 census. Farmer, age 35, White, Polk Co., Ark. 1880 census.

LEE, J. E. Pvt., Co. G. On post war roster.

LEE, JAMES Q. Pvt., Co. G. b. S. C. circa 1844. Res. Pickens Dist. 1850 census. Enl. Abbeville CH 1/8/62. Present through 10/62. On extra duty as Teamster 11-12/62. Ab. under arrest 3-4/63. Present 5-6/63, dismounted. Ab. on horse detail 7-8/63. Present 9-10/63. Ab. on detail as Teamster for Provost Guard, Orange CH 11/15-12/31/63. On Provost Guard, Hanover Junction 1-2/64. Present under arrest 4/23-30/64. Transferred Co. A, 20th S. C. Inf. 5/13/64. Present through 6/64. Ab. sick with chronic rheumatism in Richmond hospital 7/24-8/21/64. Furloughed to Batesville, S. C. for 30 days 8/22/64. NFR. Retired, age 75, Pittsburg, Crawford Co., Kansas 1920 census.

LEGARE, EDWARD THOMAS. Pvt., Co. A. b. S. C. 8/29/41. Age 19,Chester, 1860 census. Enl. 11/7/61. Transferred Co. I, 3rd S. C. Cav.2/17/62. Detailed as Signal Sgt., James Island 9/18/62-12/64. NFR. Planter, Charleston 1870 census. d. Charleston 12/4/24. Bur. Magnolia Cem.

LESSLIE, DAVID TAYLOR. Pvt., Co. H. b. Lancaster Dist. 2/8/28. Enl. Co. I, 3rd S. C. Reserves (90 days) Greenville 11/25/63. Present through 12/31/63. NFR. Enl. Co. H, 1st S. C. Cav. Columbia 7/12/64. Present through 12/64. NFR. Farmer, age 50, Catawba, York Co. 1880 census and 1900 census. d. York Co. 10/29/03. Bur. Neeley's Creek Presb. Ch. Cem., York Co.

LEWIS, FRANCIS PORCHER. Pvt., Co. A. b. S. C. 5/24/44. On post war roster. Reenlisted Co. K, 4th S. C. Cav. Charleston 1/1/64. Present until ab. on horse detail 8/11/64 for 40 days. AWOL 9/21-10/31/64. Paroled 4/65. d. 3/25/25. Bur. Magnolia Cem., Charleston.

LEWIS, JAMES M. 3rdCpl., Co. D. b. S. C. 5/40. Enl. Chester CH 10/25/61 as Pvt. Ab. guarding baggage 11-12/61. Present 3-6/62. Ab. on sick leave Chester 8/2-31/62. Ab. driving QM wagon 9-10/62. Ab. attending General Courts Martial, Fredericksburg 12/26-31/62. Present 3-6/63. Ab. on horse detail 8/18-31/63. Ab. sick in Chester Dist. 10/15-31/63. Promoted 3rdCpl. Ab. recruiting horses 12/24/63-2/64. Ab. on horse detail 4/23-30/64. Ab. sick in Charleston hospital 6/27-30/64. Present 7-12/64. Paroled Hillsboro, N. C. 4/65. Farmer, age 62, Lowndesville, Abbeville Co. 1900 census.

LEWIS, JOHN TARELTON. Pvt., Co. F. b. S. C. 6/43. Enl. Pickens CH 1/4/61. Present through 4/62. On picket Camp Means 5-6/62. Ab. on sick leave 7-8/62. Present 9/62-2/64. Ab. on leave 4/27-30/64. Ab. sick in hospital 6/20-30/64. Present 8-12/64, dismounted since 7/16/64 or 9/1/64. NFR. Mechanic, age 57, Gillsville, Hall Co., Ga. 1900 census. d. 1915.

LEWIS, ROBERT S. 1stLt., Co. D. b. S. C. 1/1/33. Farmer, age 27, Chester Dist. 1860 census. Enl. Chester CH 9/10/61 as 3rdLt., age 26. Present through 12/61. Commanding Company 3-4/62. On picket Townsend's House, Stone Road 5-6/62. Present 7-10/62. Signed for 15 pr. boots, 21 pr. socks, 3 pr. shoes, 44 pr. drawers, 15 pr. pants & 1 overcoat 9/10/62. Promoted 2ndLt. 11/62. Present 11-12/62. Commanding Company 3-4/63. Promoted 1stLt. Present 5-6/63. 76 horses present 7/1/63. 77 horses present 7/15/63. 79 horses present 7/22/63. Signed for 10 caps, 19 pr. pants, 31 jackets & 60 shirts 8/24/63. Signed for 7 caps & 9 shirts 8/26/63. WIA Raccoon Ford on Rapidan River 9/13/63. DOW's Georgetown, D. C. 9/14/63. Bur. Fishing Creek Presb. Ch. Cem., Chester Co., S. C.

LEWIS, SAMUEL A. Pvt., Co. D. b. Chester Dist. 9/1/35. Age 25, Honey Path, Anderson Dist. 1860 census. Enl. Chester CH 5/29/62 age 27. Ab. sick Fairfield Dist. 6/13-30/62. Present 7-8/62. AWOL 10/1-31/62. Captured Fredericksburg 12/13/62. Paroled for exchange age 27, 5'10", light complexion, light hair, blue eyes. Ab. paroled POW 12/31/62-6/63. Returned to duty 7/1/63. Present through 8/31/63. Ab. on horse detail 10/23-12/31/63. Present 1-2/64. Detached Green Pond 4/26-30/64. Ab. after stray horses 6/26-30/64. Present sick in camp 7-8/64. Present 9-10/64. Detached Coosawhatchie 12/10-31/64. NFR. Member, Hattiesburg, Miss. Confederate Veterans Camp. d. Hattiesburg, Miss. 6/27/09.

LIBY, F. A. Lt., no company. Not on muster rolls. WIA Upperville 6/21/63 on casualty list. NFR.

LIGON, JOHN HENRY. Commissary Sgt. b. Lebarron, Abbeville Dist. 10/20/38. Clerk, age 21, Abbeville CH 1860 census. Enl. Co. A Abbeville CH 8/13/61 as Pvt., age 23. Ab. on special duty in Commissary Dept. 10/1/61-8/31/63. Appointed Commissary Sgt. 8/1/63. Present through 12/64. Present through 4/15/65 on pension application. Farmer, Cedar Springs, Abbeville Co. 1880 census. & Saluda, Saluda

Co. 1920 census. Receiving pension Saluda 10/20/19. Applied to enter Old Soldier's Home, Columbia 1922. d. Saluda 9/20/22. Bur. Saluda Cem.

LIGON, RICHARD CATER. Pvt., Co. A. b. Abbeville Dist. 6/6/45. Farmhand, age 15, Abbeville Dist. 1860 census. Enl. Abbeville CH 6/2/63. Present through 10/63. Ab. on horse detail 12/20/63-2/64. Present 3-10/64. Detailed Pocotaligo 12/10-31/64. Paroled Greensboro, N. C. 4/26/65. Presb. Minister, Ivy, Abbeville Co. 1876 & Dark Corner, Anderson Co. 1880-1900 censuses. d. Iva, Anderson Co. 9/26/06. Bur. Goodhope Pres. Ch. Wife receiving pension there 5/13/19. Brother of Thomas A. C. Ligon.

LIGON, THOMAS A. CHEVES. 3rdCpl., Co. G. b. S. C. 5/21/40. College Student, age 20, Abbeville Dist. 1860 census. Enl. Co. C, 7th S. C. Inf. 6/4/61. Discharged for disability 12/12/10/61. Enl. Co. G, 1st S. C. Cav. Abbeville CH 5/1/62 as Pvt. Present through 10/62. Ab. detailed in Commissary Dept. 11-12/62. Present 3-6/63. Bay horse KIA Gettysburg 7/3/63. Paid $400.00. Ab. on horse detail 8/63. Present 9-12/63. Ab. detailed in Brigade Commissary Dept. 1/15-2/29/64. Present 3-6/64. Promoted 3rdCpl. Present 7-12/64. Presb. Minister, Floyds, Newberry Co. 1880 census. Presb. Minister, Jefferson, Chesterfield Co. 1900 census. Presb. Minister, Eastatoe, Pickens Co. 1910 census. Res. Saluda Co. 1919. d. Townsville, Anderson Co. 11/13/23. Bur. Townsville Presb. Ch. Cem. Brother of Richard C. Ligon.

LINDSEY, OSCAR L. Pvt., Co. B. b. S. C. 1845. Student, age 15, Spartanburg Dist. 1860 census. Enl. Co. L, Orr's 1st S. C. Rifles 7/30/61 age 17. Transferred Co. L, 26th Ga. 11/6/61. Present through 2/62. Transferred Co. B, 1st S. C. 12/29/62. Joined Co. B, 1st S. C. Cav. Camp Whatley 1/8/63. Present through 10/63. On detached service scouting for Gen. Hampton 11/22/63-6/64. Captured near Petersburg 7/29/64. Sent to City Point, Elmira & Pt. Lookout. Released 6/16/65. 21, 6'2", fair complexion, brown hair, blue eyes, Farmer, res. Hamburg, S. C. d. by 1898.

LINING, ARTHUR PARKER. Pvt., Co. A. b. S. C. 6/22/35. On post war roster. Reenlisted Co. K, 4th S. C. Cav. Grahamsville 3/25/62 as Pvt. Present as Musician through 6/62. Detailed to Commissary until discharged and appointed Captain & Commissary Officer, 4th S. C. Cav. 12/13/62. Present until WIA Trevilllian 6/11/64. Paid 7/9/64. Issued clothing 11/22/64. NFR. d. 3/11/77. Bur. Quaker Cem., Camden, S. C.

LINING, THOMAS, JR. Pvt., Co. A. b. S. C. 12/17/38. Bank Officer, Charleston 1860 census. Enl. 12/11/61 on postwar roster. Reenlisted Co. K, 4th Va. Cav. Grahamville 3/25/62. Promoted Color Sgt. KIA Trevillian 6/11/64.

LINK, EDMOND. Pvt., Co. G. Enl. age 22. Died of disease Adams Run on postwar roster. d. of brain fever Adams Run 3/5/62. Res. of Abbeville Co.

LINK, JAMES THOMAS. Pvt., Co. G. b. S. C. circa 1845. Student, age 15, Abbeville Dist. 1860 census. Enl. Abbeville CH 6/25/63. Present through 10/63. Ab. sick in Richmond hospital with Chills & "Int. Fever" 12/20-31/63. Present 1-2/64. Ab. on horse detail 4/1-30/64. Present 5-10/64. Detached Pocotaligo 12/10-31/64. NFR. Bur. Lebanon Presb. Ch. Cem., Abbeville, S. C., no dates. Brother of Robert W. Link.

LINK, ROBERT W. 1stCpl., Co. G. b. S. C. circa 1840. Farmhand, age 20, Calhoun Mills, Abbeville Dist. 1860 census. Enl. Abbeville CH 1/2/62 as Pvt. On picket Jacksonborough 3-4/62. Present 5-10/62, promoted 4thCpl. 9/15/62. Present 11-12/62, promoted 2ndCpl. 11/27/62. Present through 6/63. Ab. on horse detail 8/63. Present 9-10/63, promoted 1stCpl. Present 1-12/64. NFR. Farmer, age 39, Calhoun Mills, Abbeville Co. 1870 census & age 39, Magnolia, Abbeville Co. 1880 census. d. by 1902. Bur. Lebanon Presb. Ch. Cem., Abbeville Co., no dates. Brother of James T. Link.

LINK, SAMUEL CLARK. Pvt., Co. G. b. S. C. circa 1839. Merchant's Clerk, Mountain View, Abbeville Dist. Enl. Co. C, 7th S. C. Inf. 4/15/61. Discharged for disability 12/7/61. Enl. Co. G, 1st S. C. Cav. Abbeville Dist. 5/12/62. Ab. on picket through 6/62. Present 7-12/62. Ab. sick with "Remit. Fever in Richmond hospital 2/20-3/14/63. Ab. on detached service 3-4/63. Present 5-6/63. Ab. on horse detail 8/31/63. Present 9-10/63. Ab. sick with dyspepsia in Richmond hospital 11/21/63-2/12/64. Ab. as safe guard through 2/29/64. Present 3-8/64. Ab. on detail 9-10/64. Present 11-12/64. NFR. Farmer, age 31, Cedar Springs, Abbeville Co. 1870 census & age 41, Magnolia, Abbeville Co. 1880 census. Farmer, age 71, Newton, Jackson Co., Ga. 1900 census. Bur. Lebanon Presb. Ch. Cem., Abbeville Co., no dates.

LINK, WILLIAM EDWARD. Pvt., Co. G. b. S. C. circa 1844. Farmhand, age 16, Abbeville Dist. 1860 census. Enl. Abbeville CH 1/8/62. d. of disease in Adams Run hospital 3/19/62 age 18.

LINK, WILLIAM EDWIN. Pvt., Co. G. b. Abbeville Dist. 12/19/31. Gd. Medical College, U. of Ga.1859. M. D. Served briefly as Asst. Surgeon, 27th S. C. Inf. Enl. Co. G, 1st S. C. Cav. Columbia 3/7/63. Ab. on detached service through 4/63. Present 5-6/63. Ab. on horse detail 7-8/63. Present 9-10/63. Ab. on horse detail 12/22/63-2/64. Present 3-6/64. Ab. Detailed in Commissary Dept. 8/10-31/64. Present 9-10/64. Ab. sick at home 12/25-31/64. NFR. M. D., Calhoun Mills, Abbeville Co. 1880-1910 censuses. M.D., Mt. Carmel, McCormick Co. 1920 census. Moved to Abbeville 1/11/24. d. there 10/1/25. Bur. Willington Cem. McCormick Co, S. C.

LINSEY, JOHN. Pvt., Co. E. Enl. by 1864. Deserted to the enemy James Island 8/64. In arrest Kinston, N. Y. 10/8-13/64. NFR.

LITTLE, JOHN H. Pvt., Co. A. b. S. C. 8/22/37. Enl. Abbeville CH 7/13/61 age 21. On duty Bears Island 11-12/61. On duty Bears Bluff 3-4/62. Present 5-6/62. Ab. sick leave 7-8/62. Present 9/62-4/63. Ab. sick in Staunton hospital 5-6/63. Present 9-10/63. Ab. on picket 12/29/63-2/2/64. Ab. on safe guard near camp through 2/64. Ab. on leave 4/20-30/64. Present 5-8/64. Ab. on detail guarding POW's Florence 9/17-12/31/64. NFR. d. 2/28/75. Bur. Clinton City Cem.

LITTLE, JOHN P. H. Pvt., Co. A. b. Laurens Dist. circa 1847. Student, age 13, Laurens Dist. 1860 census. Enl. Columbia 3/7/64. Present through 12/64. Paroled Greensboro, N. C. 4/26/65. d. 11/13/74. Wife receiving pension Clinton, Laurens Co. 10/3/19.

LITTLE, JOHN W. Pvt., Co. B. b. Clinton, S. C. 2/10/32. Farmer. Enl. Camp Means 5/19/62. Present through 6/62. Present sick in camp 7-8/62. Ab. on march with horses from Summerville, S. C. to Staunton, Va. 9-10/62. Present 11/62-4/63. Ab. on horse detail 8/18-31/63. Present 9-10/63. Ab. sick with "gonorrhoea" in Richmond hospital 11/30/63-1/26/64. Horse died 12/31/63. Present through 2/64. Detached Green Pond 4/26-30/64, remounted 4/12/64. Ab. on leave 6/22-30/64. Present 7-8/64. Ab. detailed to guard POW's Florence 9/16-12/31/64. Paroled Greensboro, N. C. 5/1/65. Farmer, Jacks, Laurens Co. 1880-1910 censuses. Member, Camp Garlington, CV, Laurens. S. C. d. 11/10/14. Bur. Duncan Creek Presb. Ch. Cem., Laurens Co.

LITTLE, THOMAS JEFFERSON. Pvt., Co.'s B & A. b. S. C. 2/1/40. Age 20, Clinton, Laurens Co. 1860 census. Enl. Co. B Clinton 8/22/61 age 21. Ab. detached Camp Bee through 10/31/61. Courier for Gen. Evans 11-12/61. Present 3-4/62. Transferred Co. A 6/1/62, however, enl. Co. E, 7th S. C. Cav. Camp McKissick 5/31/62. Paroled Appomattox 4/9/65. Farmer, Jacks, Laurens Co. 1870 & 1880 censuses. Farmer, Scuffletown, Laurens Co. 1900 census. d. 12/1/12. Bur. Bethany Presb. Ch. Cem., Laurens Co.

LITTLE, WILLIAM J. Pvt., Co. H. b. Union Dist. 7/43. Farmer, age 17, Union Dist. 1860 census. Enl. Rock Hill 4/14/62. On picket McCord's Battery through 4/30/62. Present 5-6/62. Ab. on sick leave 7-8/62. Detached driving horses from Summerville, S. C. to Staunton, Va. 9-10/62. Present 11-12/62. Ab. sick with acute rheumatism 2/10-3/19/63. Present 4-10/63. Ab. horse recruiting station 11/15/63-2/29/64. Present 3-12/64. NFR. Farmer, Bogensville, Union Co. 1870-1900 censuses. d. Union, S. C. 2/9/04.

LITTLEFIELD, THOMAS C. Pvt., Co. B. b. Spartanburg Dist. 5/38. Enl. Clinton 8/22/61 age 23. Ab. sick 10/31/61. Present 11/61-6/62. Ab. on sick leave 7-8/62. Present 9-10/62. On raid with Gen. Hampton 11-12/62. Present 3-6/63. Ab. on horse detail 8/18-31/63. Present 9-10/63. On picket 11-12/63. Present 1-2/64. Detached Green Pond 4/26-30/64. Present, detailed as Courier Col. Black 5-6/64. Present 7-8/64. Ab. detailed to guard POW's Florence 9/16-12/31/64. NFR. Married Spartanburg Co. 1866. Farmer, age

43, Beech Springs, Spartanburg Co. 1880 census and age 63, 1900 census. Retired, age 72, Chick Springs, Greenville Co. 1910 census.

LIVINGSTON, JOHN FRAZIER. 1stLt., Co. G. b. S. C. circa 1833. Res. Abbeville Dist. 1860 census. Enl. Abbeville CH 1/9/62. Present 3-10/62. 72 horses present 5/2/62. Signed for rations for 24 men 10/2/62. 164 horses present enroute to Staunton 10/22/62. Ab. sick in Richmond hospital 12/12/62. Furloughed for 60 days 12/27/62. Present 3-4/63. 55 horses present 4/1/63 & 5/7/63. Present through 12/31/63. Ab. sick with "tonsilitis" in Richmond hospital 2/7-14/64. Ab. on leave through 2/29/64. Present 3-6/64. Signed for 8 jackets, 8 pr. pants, 5 cotton shirts, 5 pr. drawers, 4 pr. shoes and 4 pr. socks 5/20/64. 69 horses present 6/15/64. Present on James Island 7-8/64. Present 9-12/64. NFR. d. 5/20/04. Wife receiving pension Abbeville Co. 9/19/19.

LIVINGSTON, WILLIAM BARNETT. Pvt., Co. A. b. S. C. 8/26/41. Age 9, Orangeburg Dist. 1850 census. Enl. 11/7/61. Discharged at end of enlistment 2/7/62. NFR. Farmer, Elizabeth, Orangeburg Co. 1880 census. d. 4/20/92. Bur. Branch Bapt. Ch., Orangeburg Co.

LOCKEY, WILLIAM B. Pvt., Co. A. b. England 10/23/33. Plasterer, Abbeville 1860 census. Enl. Abbeville CH 8/13/61. Present through 10/62. Ab. on scout 11-12/62. Present 3-8/63. Ab. on horse detail 9-12/63 & 1-2/64. Ab. recruiting horses 4/20-31/64, dismounted since 7/63. Present 5-12/64. NFR. Plasterer, Memphis, Tenn. 1880 census. Cement Contractor. d. Shelby, Tenn. 4/24/21. Bur. Elmwood Cem.

LOGAN, W. H. Pvt., Co. A. Enl. 8/2/61. Not on muster rolls 10/11/61. NFR.

LOMAX, GEORGE W. Pvt., Co. A. b. S. C. circa 1827. Enl. Abbeville CH 8/13/61 age 34. Present through 9/62. On scout 11-12/62. Present 3-6/63. Ab. on horse detail 8/18-31/63. d. of dysentery in Abbeville Dist. 9/25/63.

LOMAX, WILLIAM JAMES. 1stLt., Co. A. b. S. C. 10/19/18. Farmer, age 43, Abbeville Dist. 1860 census. Enl. Abbeville CH 8/13/61. Ab. on recruiting service 11-12/61. Present Wadmalaw Island 1/62. Present 5-8/62. Present sick in camp 9-10/62. Ab. sick in Richmond hospital 11-12/62, furloughed for 30 days 12/8/62. Elected Lt. Col but resigned for "lateral curvature of spine" 2/22/63. NFR. Huckster, age 52, Augusta, Ga. 1870 census. d. Abbeville Co. 3/90.

LOWNDES, EDWARD RUTLEDGE. Pvt., Co. A. b. Clinton, S. C. 1836. Gd. S, C. College 1859. Enl. 11/15/61. Promoted Lt, Capt. Kirk's Co., S. C. Partisan Ranger 1/5/62. May have served as 1stLt., Co.'s H & K, 1st S. C. Arty. Real Estate Broker, Charleston 1870 census. d. 12/23/79. Bur. Cathedral Ch., Charleston, S. C.

LOWNES, WILLIAM. Pvt., Co. A. b. S. C. 8/1/43. Enl. 11/7/61. Detailed by Gen. Ripley 11/30/61. Reenlisted Capt. Kirk's Co., S. C. Partisan Rangers 1862, but not on muster rolls of that unit. NFR. d. 4/23/79. Bur. St. James Epis. Ch. Cem., Hyde Park, N. Y.

LUSK, ERASTUS CAPEHART. Pvt., Co. F. B. Oconee Dist. 4/4/33. Enl. Pickens CH 7/1/62. Present through 10/62. On scout 11-12/62. Present 3-6/63. Ab. sick in hospital 7/1-8/31/63. Ab. on detached service 9-10/63. AWOL 12/30-31/63. Present 1-2/64. Detached Green Pond 4/25-30/64. Present 5-8/64. Ab. detached guarding POW's Florence 9/17-12/31/64. NFR. d. Marshall Co., Ala. 7/25/01. Bur. Guntersville City Cem., Marshall Co., Ala.

LUSK, LEROY WORTH. 1stLt., Co. F. b. Oconee Dist. 5/10/31. Farmer, age 29, Pickens Dist. 1860 census. Enl. Pickens CH 12/4/61 as Pvt., age 25. Present through 4/62, promoted 2ndCpl. Present 5-10/62, elected 2ndLt. 11/1/62. On duty as scout 11-12/62. Present 3-10/63, elected 1stLt. 11/1/63. On picket U. S. Ford 12/29-31/63. Ab. on leave 2/22-29/64. Present 3/64-1/31/65. NFR. Farmer, Keowee, Oconee Co. 1880 & 1900 censuses. d. 5/9/09. Bur. Lusk Cem., Oconee Co.

LYLES, BELTON ENGLISH. Pvt., Co. K. b. Blair, Fairfield Dist. 7/8/47. Student, age 13, Fairfield Dist. 1860 census. Enl. Columbia 4/12/64. Paroled Greensboro, N. C. 5/1/65. Farmer, Blair, Fairfield Co. 1880-1910 censuses. Receiving pension there 4/1/19. d. Columbia, S. C. 6/18/23. Bur. Winnsboro, S. C. Brother of John W. Lyles.

LYLES, JOHN WOODWARD. Pvt., Co. K. b. Winnsboro, Fairfield Dist. 9/2/45. Student, age 15, Fairfield Dist. 1860 census. Enl. Ridgeway 8/13/63. Present through 12/63. Ab. on leave 2/12-29/64. Present 3-6/64. On duty Signal Station 7-8/64. AWOL 10/25-31/64, however, sent in private physicans certificate of disability. Courier at Tar Bluff 12/31/64. WIA Moccasin's Creek, N. C. 4/65. Paroled Greensboro, N. C. 5/1/65. Att. S. C. College 1866. Served in S. C. House of Reps. Clerk of Court, Fairfield Co. Receiving pension Winnsboro, S. C. 3/25/19. d. Columbia, S. C. 6/28/33. Bur. Winnsboro. Brother of Belton E. Lyles.

LYLES, JOSHUA G. Pvt., Co. F. b. S. C. circa 1842. On postwar roster. Living Pickens Co. circa 1900.

LYLES, JOSHUA Y. 1stCpl., Co. F. b. Pickens Dist. circa 1842. Age 18, Pickens Dist. 1860 census. Enl. Co. D, 1st S. C. Inf. (Butler's) Wahalla 3/2/61. Ab. sick in hospital 6/30/61. Discharged 7-8/61. Enl. Co. F, 1st S. C. Cav. Pickens CH 12/4/61 as 3rdCpl., age 21. Present through 12/31/61. On picket Jacksonborough 3-4/62. Promoted 1stCpl. Present 5-6/62. Ab. on leave 7-8/62. Present 9/62-10/63. Ab. sick with "plurisey" in Richmond hospital 11/163. Furloughed for 30 days 11/21/63. Present 1-12/64. Paroled Greensboro, N. C. 5/1/65. d by 1900 on postwar roster.

LYLES, M. L. Pvt., Co. F. b. S. C. circa 1847. Not on muster rolls. Receiving pension Ft. Madison, Oconee Co., S. C. 1901 age 64.

LYLES, SAMUEL HOLDEN. Pvt., Co. F. b. S. C. 1827. Enl. Pickens CH 12/4/61 age 38. Present through 12/31/61. Ab. on leave 4/22-5/22/62. Present through 8/62. Ab. on detached service 9-10/62. Ab. on scout 11-12/62. Present 3-10/63. WIA (right shoulder) 11/9/63. Ab. wounded through 11/21/63. Furloughed for 30 days. Present 1-4/64, on duty Green Pond 4/25-30/64. Ab. on leave 5/31-6/30/64. Present 7-12/64. NFR. Farmer, age 43, Tugaloo, Oconee Co. 1870 census. d. Oconee Co. 1874.

LYON, W. Pvt., Co. A. Enl. Columbia 4/16/64. Present through 10/64. NFR.

MABRY, SAMUEL W. Pvt., Co. G. b. Edgefield Dist. 1835. Lawyer, age 32, Abbeville 1860 census. Served briefly in 1st S. C. Inf. 1861. Enl. Abbeville Ch 3/15/62. Present through 12/62. Ab. sick with chronic bronchitis in Columbia, S. C. hospital 2/21/63-4/4/63. On detail Richmond as Clerk, C. S. Treasury Dept. 6/14/63-12/32/63. Ab. sick in Richmond hospital 1/28-2/29/64. Detailed in C. S. Treasury Dept., Richmond through12/64. Enl. Co. B, 2nd Texas State Guard & paroled Millican, Texas 7/11/65. d. of "Incepient phithilis" Edgefield Dist. 1869.

MADDEN, DAVID D. Pvt., Co. A. b. S. C. 4/13/28. Enl. Columbia 12/13/63. Present through 6/64. Ab. sick 7/6-8/31/64. Ab. sick in hospital 12/11-31/64. NFR. d. 1/4/77. Bur. New Liberty Bapt. Ch. Cem., Laurens Co.

MADDEN, NATHANIEL MERRITT. 2ndCpl., Co. F. b. Anderson Dist. 9/24/23. Farmer, age 35, Pickens Dist. 1860 census. Enl. Pickens CH 12/4/61 age 40. Present through 4/62. Ab. sick in hospital 5-6/62. Reduced to Pvt. Present 9-10/62. Ab. on leave 9/10/62. Present 11/62-8/63. Captured on picket Robinson River 9/16/63. Sent to Old Capitol & Pt. Lookout. Exchanged 2/24/65. NFR. Farmer, Central, Pickens Co. 1900 census. d. there 1/27/08. Bur. Mt. Zion Cem., Pickens Co.

MAGUIRE, JOHN J. Pvt., Co. I. b. Ireland 1833. Arrived Charleston 1857. Merchant, age 27, Colleton Dist. 1860 census. Enl. John's Island 6/30/62. Present through 2/64. Detached Green Pond 4/22-30/64. Present 5-6/64. Present sick in camp 7-8/64. Discharged Mt. Pleasant, S. C. 9/15/64. NFR.

MAGWOOD, H. MILNE. Pvt., Co. A. b. S. C. 7/10/34. Res. of Clinton 1859. Enl. 11/7/61. Furloughed by

Surgeon 11/24/61. Reenlisted Co. H, 3rd S. C. Cav. Grahamville 10/10/63. Present through 10/64. Ab. sick 12/64. Transferred Co. A, 1st S. C. Inf. (Butler's). NFR. Broker, Charleston 1892. d. there 8/18/94. Bur. Magnolia Cem.

MAHAFFEY, S. B. 2ndLt., Co. B. b. S. C. 11/1/29. Res. Eden, Laurens Co. 1860 census. Enl. Clinton 8/22/61 as 3rdLt. Ab. wounded accidentally by pistol shot 11-12/61. Present John's Island 1/62. Present 3-4/62. Ab. sick 5-/8/62. Promoted 2ndLt. 6/1/62. AWOL 9-10/62. Present 11-12/62. Resigned 1/15/63. NFR. d. 6/22/01. Bur. Raburn Creek Bapt. Ch. Cem., Laurens Co.

MAHON, J. A. Pvt., Co. B. b. S. C. circa 1840. Enl. Clinton 8/22/61 age 21. Ab. sick 10/31/61. On picket Laroches 11-12/61. Present 3-4/62. Killed in skirmish at Wallpole's Gate, John's Island 6/7/62.

MAILHENNY, P. Pvt., Co. C. Enl. Aiken Dist. age 25 on postwar roster. d. by 1898.

MALONE, PETER JERU. Pvt., Co. E. b. S. C. circa 1844. Res. Charleston 1850 census. Enl. Ft. Motte 10/26/61 age 18. On picket White Point through 12/31/61. Present 3-4/62. Courier, Capt. Parker's Battery, Simons Bluff 5-6/62. Present sick in camp 7-8/62. Present 9-10/62. On picket 11-12/62. Present 3-6/64. WIA (right leg & abdomen) Gettysburg 7/3/63 and left in hospital. Captured 7/5/63. Sent to David's Island, N. Y. hospital. Exchanged by 8/31/63. In Williamsburg hospital 9/15-22/63. Furloughed for 30 days 9/25/63. Ab. on sick leave 9-10/63. AWOL 12/1/63-4/64. Present 5-6/64. Retired to Invalid Corps 7/26/64. Detailed in coast Defense to end of war on wife's pension application. Lawyer, age 25, Verdier, Colleton Co. 1870 census. d. 9/18/13. Wife receiving pension Ruffin, Colleton Co. 9/10/19.

MANIGAULT, ALFRED. Pvt. Co. A. b. S. C. 7/15/40. Enl. 11/1/61. Reenlisted Co. K, 4th S. C. Cav. Grahamville 3/25/62. Present until appointed 4thCpl. 1/63. Present until ab. on sick leave 4/23-6/14/63. Promoted 1stCpl. Present until ab. sick 3/20-10/31/64 in Charleston hospital. Issued clothing 12/5/64. d. Winnsboro, S. C. 2/20/65. Bur. St. Phillips Epis. Ch., Charleston.

MANIGAULT, GABRIEL EDWARD. Pvt., Co. A. b. S. C. 1833. On post war roster. Reenlisted Co. K, 4th S. C. Cav. Grahamville 3/25/62. Present until slightly wounded at Pocotaligo 10/22/62. Ab. through 1/4/63. Detailed as Acting Adjutant of regiment 1/25/63. Present through 12/63. Present until ab. on sick leave 2/1/64. Appointed Adjutant 3/64. Present until captured Trevillian 6/11/64. Sent to Pt. Lookout & Ft. Delaware. Exchanged 2/27/65. Paroled 4/65. d. 9/15/99. Bur. St Phillips Epis. Ch., Charleston.

MANN, JOHN THOMAS. Pvt., Co. G. b. Ireland 1845. Enl. Abbeville CH 1/8/62. Present 3-4/62. Ab. on sick leave 5/30-6/30/62. Horse died 5/31/62. Ab. sick in Charleston hospital 6/22-8/31/62, dismounted since 7/20/62. Present 9-12/62. Ab. on detached service 3-4/63. Present 7-8/63. Ab. on detached service 9-12/63, dismounted. Teamster, Brigade Headquarters through 2/64. Ab. on horse detail 4/1-30/64. Present 5-8/64. On detached service guarding POW's Florence 9/17-12/31/64. NFR. Farmer, age 35, Abbeville Co. 1880 census. d. Abbeville 6/23/99. Wife receiving pension MCCormick, McCormick Co. 9/23/19.

MANNING, THOMAS. Pvt., Co. K. Enl. circa 9/64. Ab. sick in Charleston hospital 10/29-31/64. Ab. detached at Dist. Headquarters, Green Pond 12/24-31/64. Paroled Raleigh, N. C. 4/65.

MANNING, WADE HAMPTON. Pvt., Co. A. b. S. C. 10/6/45. On postwar roster. Reenlisted Co. K, 4th S. C. Cav. Charleston 11/19/63. Present until detailed 2/18/64 for 20 days. Present until detailed as Courier, Gen. Hampton 5/27-10/31/64, however, ab. sick with chronic diarrhea in Richmond hospital 7/18/64. Furloughed for 30 days 7/22/64. Recommended for Cadetship, C. S. A. Paroled Greensboro, N. C, 5/5/65. d. 4/5/11. Bur. Elmwood Mem. Gardens, Columbia, S. C.

MARDNER, J. O. Pvt., Co. H. Res. York Dist. Not on muster rolls. d. in Richmond hospital 2/1/65.

MARION, BENJAMIN P. Pvt., Co. A. b. S. C. 10/13/23. On post war roster. Reenlisted Co. K, 4th S. C. Cav. 4l/11/62. Present until ab. on leave 6/18/62. Present until ab. on leave 12/29/62 for 13 days. Promoted 1stCpl. 1/1/63. Detached with Gen. Walker 2/26/63. Ab. on sick leave 4/26-10/31/63. Promoted

1stCpl. Present until ab. sick leave 1/1/64-8/64. Reduced to Pvt., Appointed Assessor in Kind, 2nd Cong. Dist. of S. C. Paroled 7/7/65. d. 3/1/85. Bur. Magnolia Cem., Charleston.

MARION, F. Pvt., Co. C. Res. Edgefield Dist. Enl. by 1862. d. of disease in Richmond hospital 1/11/63.

MARS, WILLIAM D. Pvt., Co. G. b. S. C. circa 1836. Enl. Abbeville CH 3/15/62. Present sick in camp 4-6/62. Hired Henry Howland as substitute and discharged 8/5/62. Reenlisted Co. B, 5th S. C. Cav. 2/1/64. Present 3-4/64. Present in Regimental QM Dept.5-10/64. Issued clothing 12/25/64. NFR. Farmer, Calhoun Mills, Abbeville Co. 1870 census. d. 1893.

MARSHALL, JAMES C. Pvt., Co.'s A & I. b. S. C. 1837. Enl. Co. A. 11/7/61. Detailed Gen. Ripley 11/26/61 NFR. Reenlisted Co. I James Island 10/22/64. Detached in Commissary Dept. Hqtrs. 11/18-12/31/64. NFR. d. of injuries in a train accident 5/13/98. Bur. Magnolia Cem., Charleston.

MARSHALL, JOHN WILSON. 2ndLt., Co. I. b. Charleston, S. C. 7/20/41. Att. Charleston public schools & King's Mountain Academy. Enl. Parker's Ferry 4/3/62 as 3rdLt. Present through 4/30/62. Ab. on sick leave 5-6/62. Present 7-8/62. Promoted 2ndLt. Present in arrest 9-10/62. Ab. on leave 3-4/63. 70 horses present 3/11/63. Present 5-10/63, however, 112 horses present 7/23/63, 32 horses present 8/1/63, signed for 6 pr. shoes Culpepper CH 8/2/63 & ab. sick with chronic "dysenteria" in Richmond hospital 8/10-16/63. 41 horses present 12/8/63. Ab. on horse detail 12/22/63-2/64. 35 horses present 1/22/64. Present 3-4/64. 73 horses present 3/7-8/64. 68 horses present 3/18/64. Signed for 1 one horse wagon, 2 hammers & 2 lbs. horse shoe nails 5/2/64. Ab. on detail East Line, James Island 5/20-12/31/64. Ab. on leave 1/7/65. Present Asheboro, N. C. 4/65. Farmer, Ft. Mill, York Co. 1880 census. Retired from Standard Oil Co., Catawba, York Co. 1920 census. d. Rock Hill 5/22/32. Bur. Laurelwood Cem.

MARSHALL, T. M. 1stLt., Co. I. On roster but not on muster rolls.

MARSHALL, WILLIAM A. J. Pvt., Co. I. b. S. C. 10/20/45. Enl. Co. B, 4th S. C. State Troops Lancaster CH 9/3/63. Present through 2/1/64. Disbanded. Enl. Co. I, 1st S. C. Cav. James Island 9/13/64. Present through 10/64. Detached, Hdqtrs., East Lines, James Island 12/18-31/64. Farmer, Pleasant Hill, Lancaster Co. 1880-1910 censuses. d. 3/29/30. Bur. Salem Cem., Lancaster Co., S. C.

MARTIN, EDWARD HAYNE. Pvt., Co. A. b. S. C. 1842. On post war roster. Reenlisted Co. K, 4th S. C. Cav. Grahamville 3/25/62. Present until detailed with Signal Corps as Signal Operator 11/25/62. Ab. on leaved 1/5/3. Detailed with Signal Corps through 10/31/64. Ordered back to regiment 11/5/64, & 12/20/64. NFR. d. 9/24/25. Bur. Westview Mem. Cem., Wahalla, S. C.

MARTIN, GEORGE M. Pvt., Co. A. b. Scotland 7/2/21. Enl. 11/7/61. Discharged at end of enlistment 2/7/62. Reenlisted Co. K, 4th S. C. Cav. Transferred to Signal Corps. NFR. d. Charleston 2/7/86. Bur. Magnolia Cem.

MARTIN, JAMES H. Pvt., Co. I. b. Laurens Dist. 7/4/49. Age 11, Clarks Mill, Lexington Dist. 1860 census. Enl. Columbia 4/18/64. Detached Green Pond through 4/30/64. Present 5-6/64. Ab. sick in camp 7/15-8/31/64. Present 9-10/64. Detached Pocotaligo 12/10-31/64. NFR. Farmer, Richland Co. 1880 census. d. 9/12/84. Bur. Abbeville City Cem.

MARTIN, JAMES POLLARD. Pvt., Co. A. b. S. C. circa 1840. Res. Abbeville Dist. 1850 census. Enl. Abbeville CH 12/23/61. Present through 8/63. Teamster, 9-10/63. Present 11/63-4/64. Teamster, Regimental QM Dept. 5-12/64. Present 4/65. NFR. d. 7/3/92. Wife receiving pension McCormick, McCormick Co., S. C. 10/20/19.

MARTIN, JAMES T. Pvt., Co. F. b. S. C. 4/65. Enl. Columbia 5/3/64. Present through 6/64. Ab. sick in hospital 8/10-31/64. Present 9-12/64. NFR. Farmer, age 55, Lanes, Morgan Co., Ala. 1900 census.

MARTIN, JAMES W. Pvt., Co. I. b. St. Mathews Parish, S. C. 4/12/32. Timber Cutter, age 28, St. Pauls Parish, Colleton Dist. 1860 census. Enl. John's Island 6/30/62. Ab. on sick leave 8/27-31/62. Present

9-10/62. On scout 11-12/62. Hired a substitute and discharge 1/30/63. Age 32, 5'8", light complexion, blue eyes, light hair, Farmer. NFR. Wood Contractor, St. James, Goose Creek, Charleston 1880 census. d. Colleton Co., S. C. 12/5/98.

MARTIN, JOHN ALVIN. 3rdCpl., Co. E. b. S. C. 1843. Res. Pleasant Mound, Laurens Dist. Enl. Adams Run 4/28/62 as Pvt. On detail Adams Run Railroad Station on Charleston & Savannah Railroad 5-6/62. On duty Pine Berry 7-8/62, promoted 4thCpl. 8/13/62. Present 9-12/62. Promoted 3rd Cpl. Present 3-4/63. Ab. sick with Catarrh in Charlottesville hospital 6/24-7/20/63. WIA (hip) Brandy Station 8/1/63. DOW's in Richmond hospital 8/19/63. Bur. Hollywood Cem. Effects: 1 gold ring, 1 silver watch & $17.25.

MARTIN, JOHN M. 2ndCpl., Co. A. b. S. C. 1/7/40. Enl. Abbeville CH 8/13/61. Ab. on leave 10/31/61. Ab. sick in hospital 11-12/61. Present 3-6/63. Ab. sick in Summerville hospital 7-8/62. Present 9-10/62, reduced to Pvt. Ab. sick 11/62-3/13/63. Present through 6/63, dismounted. Ab. on horse detail 8/18-31/63. Present 9-10/63. Ab. sick in Richmond hospital 11/7/63-2/64. Detached Green Pond 4/7-30/64. Present 5-6/64. Present in arrest 8/11-31/64. Present 11-12/64. NFR. Farmer, Diamond Hill, Abbeville Co. 1880 & 1900 censuses. d. 8/11/05. Bur. Little Mountain Cem., Abbeville Co.

MARTIN, JULIUS C. (1). 1stSgt., Co. G. Enl. Abbeville CH 1/8/62 as 2ndSgt. On picket Jacksonborough 3-4/62. Present 5-19/62, promoted 1stSgt. 9/15/62. Present 11/62-5/63. Ab. on detached service 8/31/63. Present 9/63-12/64. Paroled Greensboro, N. C. 5/1/65. Magistrate & Cotton Buyer, Donalds. Receiving pension Donalds, Abbeville Co. 4/1/19. d. Abbeville Co. 10/10/20. Bur. Donalds Cem.

MARTIN, JULIUS C. (2). Pvt., Co. A. b. Donalds, Abbeville Dist. 10/15/46. Enl. Columbia 4/16/64. Present through 10/64. Detached Pocotoligo 12/10-31/64. NFR. Captialist, Florence, S. C. 1910 census.

MARTIN, NOAH J. 2ndLt., Co. D. b. S. C. circa 1837.Enl. Chester CH 9/10/61 as Pvt., age 24. Present through 4/62. On picket Townsend's House 5-6/62. Present 7-8/62. Ab. on sick leave 8/9-31/62. On scout 11-12/62, elected 2ndLt. 11/62. Present 3-4/63. 49 horses present 3/1/63. Commanding Company 7-8/63. Ab. sick with typhoid fever in Charlottesville hospital 10/19/63 until d. 11/8/63 age 30. Bur. U. of Va. Cem. in unmarked grave. Effects: $213.40.

MARTIN, RICHARD HUTSON. Pvt., Co. A. b. S. C. 3/16/35. On post war roster. Reenlisted Co. K, 4th S. C. Cav. Charleston 1/27/64. Present until Ab. on horse detail 8/11/64 for 40 days. Ab. sick with "Remittens Fever" in Raleigh, N. C. hospital 10/8-30/64. Paroled 4/65. d. Colleton Co. 5/27/17. Bur. Stoney Creek Cem., Beaufort Co., S. C.

MARTIN, ROBERT. Pvt., Co. G. b. Ireland 7/31/43. On roster. d. Charleston 1/1/08. Bur. St. Lawrence Cem.

MARTIN, ROBERT A. Pvt., Co. G. b. Newberry, S. C. circa 1808. Tailor, age 52, Abbeville 1860 census. Enl. Camp Clinton 5/13/62 age 56. Present through 6/62. On picket Gimbell's 7-8/62. Present until ab. sick in Columbia hospital 12/2-31/62. Present 3-10/63. Detailed horse recruiting station 11/20-12/31/63. Present 1-2/64. Detached, Green Pond 4/1-30/64. Present 5-6/64. Discharged for chronic rheumatism 8/5/64. Age 56, 5'8", ruddy complexion, blue eyes, sandy hair, Tailor. In Charlotte, N. C. hospital with "rubeola" 4//18/65. Furloughed 4/21/65. NFR.

MARTIN, SAMUEL STARKE. Pvt., Co. G. b. S. C. 1/45. Student, Abbeville 1860 census. Enl. Columbia 10/15/64. Detached Pocotoligo 12/10-31/64. NFR. Farmer, age 35, Abbeville Co. 1880 census & age 55, Donaldsville, Abbeville Co. 1900 census. d. 1910. Bur. Broadmouth Ch. Cem., Honea Path, Abbeville Co.

MARTIN, THOMAS P. Pvt., Co. A. b. S. C. circa 1843. Res. Abbeville Dist. 1850 census. Enl. Abbeville CH 10/1/61. Ab. on sick leave 10/31/61. On duty Bears Island 11-12/61. Present 3-6/62. Ab. sick in Summerville hospital 7-8/62. Present sick in camp 9-10/62. Ab. sick 11-12/62, in Richmond hospital 12/11/62. Furloughed for 30 days 12/22/62. Present sick in camp 3-4/63. Present 5-6/63 & WIA (slightly) Brandy Station 6/9/63. Ab. on horse detail 8/18-10/31/63. AWOL 11-12/63. Present 1-2/64. On horse detail 4/12-30/64. Present 5-12/64. NFR. Farmer, Burleson Co., Texas 1880 census.

MARTIN, WILLIAM AIKEN. Pvt., Co. A. b. S. C. 4/7/33. On post war roster. Reenlisted Co. K, 4th S. C. Cav. Grahamville 3/25/62. Present until ab. on leave 6/27/62 for 10 days. Transferred to Captain Martin's Co. Arty. 8/5/62. NFR. d. 1/27/04. Bur. Magnolia Cem., Charleston.

MARTIN, WILLIAM W. Pvt., Co. F. b. S. C. circa 1845. Age 16, Muscogie Co., Ga. 1860 census. Enl. Columbia 3/7/64. Present through 12/64. NFR. Farmer, age 35, Warrior Stand, Macon Co., Ala. 1880 census.

MASSEY, STEPHEN BURDETTE. Pvt., Co.'s D & K. b. Lancaster Dist. 6/3/32. Res. Chester Dist. 1860 census. Enl. Co. D Chester CH 2/24/62. Present 3-4/62. Transferred Co. K 6/25/62. Present through 12/62. Deserted 3/16/63. NFR. Grocer, Chester Co. 1880 census. d. 8/6/94. Bur. Evergreen Cem., Chester, S. C.

MASSEY, WILLIAM H. Pvt., Co. D. b. S. C. 2/36. Age 24, Rocky Mount, Chester Co. 1860 census. Enl. Camp Means 4/10/62. Present through 4/30/62. On picket Townsend's House 5-6/62. Present 7-8/62. AWOL in Chester Dist. 10/2-31/62. Attending General CM, Fredericksburg 10/26-31/62. Present 3-4/63. Present or absent not stated 5-6/63. Ab. on horse detail 8/18-31/63. Present 9/63-8/64. Ab. sick Chester Dist. 10/22-31/64. Present 11-12/64. Paroled Hillsboro, N. C. 4/65.
Farmer, Rossville, Chester Co. 1880. Farmer, Navarro, Texas 1900 census.

MATER, JOSEPH. Asst. Surgeon. On roster but not on muster rolls.

MATHENY, ANDERSON P. Pvt., Co. C. b. S. C. circa 1836. Student, age 14, Edgefield Dist. 1850 census. Enl. Camp Butler 9/18/61 age 23. Unfit for duty 10/9-12/31/61. Ab. on leave 1/62. NFR. Probably discharged for disability. Conscript in Camp of Instruction, Columbia 1864. NFR.

MATHEWS, BUD(D). Pvt., Co. A. b. S. C. 8/35. Enl. Abbeville CH 9/12/61. Present through 6/62. Present sick in camp 7-10/62. Ab. sick in Lynchburg hospital 11/62-4/63. Court M'd 3/26/63. Present 5-6/63. Teamster in Brigade QM Dept. 7-10/63. Present 11-12/63. Present as Company Blacksmith 1-2/64. Ab. in arrest 4/27-30/64. Ab. sick in Charleston hospital 5/19-8/29/64. Deserted to the enemy 9/28/64. Not in Federal records. NFR. Age 64, Statesboro, Ga. 1900 census. d. Statesboro, Ga. 3/10/10.

MATHIS, JOHN C. Pvt., Co. G. b. S. C. circa 1836. Enl. Abbeville CH 1/8/62. AWOL 2/14-4/30/62. NFR. Farmer, age 35, Tippah Co., Miss. 1870 census & age 44, 1880 census.

MAULDIN, WILLIAM C. Pvt., Co. F. b. S. C. 10/21/33. Pickens Dist., Militia 1858. Blacksmith, age 30, 6'1", dark complexion, hazel eyes, dark hair. Enl. Pickens CH 12/4/61. Present through 12/31/61. Ab. sick 3/4-4/30/62. Present 5-6/62. Discharged for disability 7/15/62. NFR. d. 6/5/97. Bur. Whitmire Memorial Cem., Salem, S. C.

MAXWELL, PINCKNEY JOHNSTONE. b. S. C. 1/3/34. On post war roster. Reenlisted Co. K, 4th S. C. Cav. Grahamville 3/25/62. Present until ab. on leave 12/19/63. Discharged & appointed Asst. Surgeon 12/30/63. Assigned to 24th S. C. Inf. 5/31/64. Present Atlanta 8/26/64. Ab. sick 10/5/64. Assigned to Opelaka, Ala. hospital by 2/65. Paroled Columbus, Miss. 5/16/65. d. 6/13/93. Bur. 1st Scots Presb. Ch., Charleston.

MAYER, NATHANIEL R. E. Pvt., Co. C. b. Charleston, S. C. circa 1833. Enl. Wadmalaw Island 3/20/62. Discharged for disability (neuralgia) 5/31/62. Age 29, 5'8", dark complexion, hazel eyes, dark hair, Carriage Maker. Reenl. Co. D, 4th S. C. Cav. 1/30/64. Present through 8/64. NFR. Age 45, no occupation, Georgetown, Quitman Co., Ga. 1880 census. d. in Ga. by 1898.

MAYHARUE (MAYHORNN), A. J. A. Pvt., Co. B. b. S. C. circa 1832. Res. Laurens Dist. No on muster rolls. KIA John's Island 1862 age 30.

MAYSON, HENRY LAURENS. Captain & Quartermaster. b. S. C. 8/10/17. Farmer, age 44, Beech Island, Edgefield Dist. 1860 census. Enl. Co. C Hamburg 8/27/61 as Pvt., age 46. Elected Company Commissary, Secretary & Treasurer 9/18/61. Present through 10/31/61. Appoint Asst. QM & Commissary of Bn.11/1/61. Present through 12/31/61. Present Adams Run 1-2/62. Appointed Captain & QM 1st S. C. Cav. 4/17/62. Resigned for rheumatism 1/27/63. Paroled Augusta, Ga. 5/19/65 as Captain, Wilson Rangers, Ga. Local Defense Troops. Had enlisted 2/23/63 and was present, Augusta, Ga. through 3/22/65. Farmer, Hamburg, Edgefield Co. 1870 & 1880 censuses. d. 2/6/00. Bur. Hammond Cem., Beech Island, Aiken Co., S. C.

MC ALLISTER, ANDREW H. Pvt., Co. G. b. S. C. 9/23/27. Farmer, age 33, Abbeville Dist. 1860 census. Enl. Abbeville CH 3/25/62. On picket King's House through 4/62. Ab. on sick leave 6/16-30/62. Ab. sick in Charleston hospital 7/30-8/31/62. Ab. detached service in Surgeon's Dept. as nurse 9-12/62. Present 3-10/63. Detached in Medical Dept. 12/31/63-1/64. Ab. on leave 2/14-29/64. Present 3-4/64. Ab. sick in hospital, James Island 6/20-30/64. On detached service 8/25-31/64. Present 9-12/64. NFR. Farmer, age 76, Calhoun, Abbeville Co. 1900 census. Bur. Willington Cem., McCormick Co., no date.

MC ALLISTER, BENJAMIN FRANKLIN. Pvt., Co. G. b. S. C. 4/13/39. Enl. Abbeville CH 1/8/62. On picket Walter's Battery 3-4/62. Ab. as Teamster 5-6/62. Ab. sick in Charleston hospital 7/30-8/31/62. Ab. as Teamster 9-12/62. Present 3-4/63. Ab. detailed at recruiting station for horses 8/5-12/31/63. Present 1-2/64. Ab. on horse detail 4/1-30/64. Present 5-12/64. NFR. Farmer, age 45, Tippah Co., Miss. 1880 census. d. 5/28/11. Bur. New Hope Cem., Tippah Co., Miss.

MC ALLISTER, CLAUD G. Pvt., Co. G. b. Abbeville Dist. 10/28/45. Enl. Abbeville CH 10/25/63 age 18, 5'5", fair complexion, blue eyes, light hair, Farmer. Present 11-12/63. Ab. with Signal Corps 1/29-2/29/64. Ab. on horse detail 4/4-30/64. Present 5-8/64. Ab. detailed to guard POW's Florence 9/10-12/31/64. NFR. Farmer, Magnolia, Abbeville Co. 1900 & 1910 censuses. d. 8/23/14. Bur. Old Rocky River Presb. Ch. Cem., Abbeville Co.

MC ALLISTER, THADDEUS WARSAW. Pvt., Co. G. b. Abbeville Dist. circa 1841. Farmhand, age 19, Warrenton, Abbeville Dist. 1860 census. Enl. Abbeville CH 7/1/62 age 21. Ab. sick in Charleston hospital 8/22-31/62. Discharged for chronic asthma Summerville, S. C. 9/27/62. Age 21, 6'1", fair complexion, blue eyes, fair hair, Farmer. Conscripted Columbia 7/9/63 & assigned Co. F, 1st S. C. Inf. (Hagood's). Ab. sick with asthma in Williamburg hospital 8/21/63. Transferred Farmville hospital. Returned to duty 9/21/63. WIA (toe amputated) Wilderness 5/6/64. Ab. wounded through 10/25/64. AWOL 10/26/64-2/65. NFR. Laborer, age 40, Due West, Abbeville Co. 1880 census. Wife receiving pension Lowndesville, Abbeville Co. 9/29/19.

MC BRYDE, JOHN MC LAREN. 1stSgt., Co. G. b. Abbeville, S. C. 1/1/41. Att. S. C. College 1858-59. Att. U. VA 1860-61. Farmer, age 20, Abbeville Dist. 1860 census. Enl. Co. A, 1st S. C. Inf. (6 mo.) Morris Island, S. C. 4/22/61 age 24. Discharged for ill health 7/9/61. Enl. Co. G, 1st S. C. Cav. Abbeville CH 1/8/62 as 1stSgt. Ab. sick with typhus in hospital 3-4/62. Ab. on sick leave 5/4-8/31/62. Discharged Legares Point, James Island 9/14/62. Applied for Clerkship in Treasury Dept. 11/15/62. Division Chief, in War Tax Office, Richmond. Enl. Co. K, 3rd Va. Local Defense Troops 10/23/63. On rolls 8/29/64. NFR. Farmer, Buckingham Co. Va. 1865-67. Farmer, Albermarle Co., Va. Farmer, age 40, Mt. Crogan, Chesterfield Co., S. C. 1880 census. Professor of Botany, U. of Tenn. 1879. Chairman of faculty, U. of S. C. 1882-83. President, S. C. College 1883-91. President, U. of Tenn. 1887. President, Va. Agricultural & Mechanical College (now Va. Tech) 1891-1907. d. New Orleans, La. 3/20/23. Bur. Blacksburg, Va. Cem. McBryde Hall at Va. Tech and McBryde Quadrange at U. of S. C. named in his honor.

MC BRYDE, ROBERT JAMES. Pvt., Co. G. b. Abbeville Dist. 6/1/45. Enl. Abbeville CH 3/15/62. On picket Jacksonborough through 4/62. Ab. on sick leave 6/15-30/62. AWOL 7-8/62. Discharged for asthma Camp Myers, Va. 10/11/62. Age 19, 5'11 1/2", fair complexion, blue eyes, light hair, Student. NFR. Presbyterian Minister postwar. Minister, Fredericksburg, Va. 1873. Move to Lexington, Va. Minister, Lexington 1900 census. Member, Robert E. Lee Camp, CV, Lexington. Minister, Fredericksburg 1910 census. College Professor, VPI, Blacksburg, Va. 1920 census. d. Brownsburg, Va. 9/6/16. Bur.

Blacksburg, Va. Cem.

MC CALL, ALFRED P. Pvt., Co. F. b. S. C. 9/27/27. Enl. Pickens CH 12/4/61 age 35. Present through 4/62. Ab. on leave 6/27-30/62. Present 7-12/62. Ab. on detached service 3-4/63. Present through 12/63, dismounted since 11/1/63. Teamster in QM Dept. through 2/64. Ab. on horse detail 4/24-30/64. Teamster for Regimental QM 6/26-30/64. Present sick in camp 7-8/64. Present 9-12/64. NFR. d. Oconee Co. 5/7/90. Bur. Neville Cem.

MC CALLA, JAMES D. 2ndSgt., Co. H. b. S. C. circa 1843. Student, age 17, Spartanburg Dist. 1860 census. Enl. Rock Hill 3/14/62 as Cpl. Present 5-8/62. Ab. in arrest Adams Run 9-10/62. Promoted 4thSgt. Present 11/62-4/63. Promoted 2ndSgt. Present 5-6/63, however, ab. detached service as Scout 4/30-8/31/63. Present 9-12/63. Ab. on recruiting duty 1/2-2/29/64. Ab. detailed to Gen. Hampton as Scout 4/22/64-8/64, reduced to Pvt. KIA on Beefsteak Raid near Petersburg 9/14/64.

MC CANN, RANSOM J. W. Pvt., Co. F. b. Fla. 1844. Enl. Columbia 5/10/64. Present through 6/64. Present sick in camp 8/1-31/64. Ab. sick 9-10/64. Present 11-12/64. NFR. Farmer, age 55, Moutons Cove, Vermillion Co., Fla. 1900 census.

MC CANN, ROBERT JULIUS. Pvt., Co. F. b. S. C. 2/23/45. Tombstone only record of service. d. 8/11/71. Bur. Carmel Presb. Ch. Cem., Pickens Co., S. C.

MC CANTS, NATHANIEL SAMUEL. Pvt., Co. A. b. Abbeville Dist. circa 1834. M. D., age 26, Abbeville Dist. 1860 census. Enl. Abbeville CH 9/19/61 age 27. Ab. recruiting 11-12/61. Present 3-4/62. Ab. on detached service as Hospital Steward, Adams Run Hospital 5-10/62. Discharged for blindness in one eye & imperfect vision in the other at Camp Brown, Staunton, Va. Age 27, 6'1", fair complexion, blue eyes, light hair, M. D. NFR.

MC CASLAN, JOHN F. Pvt., Co. G. b. Abbeville Dist. circa 1840. College Student, age 20, Abbeville Dist. 1860 census. Enl. Abbeville CH 5/26/63. Present through 12/63. Ab. horse recruiting station 1/26-2/29/64. d.of disease in Ladies Hospital, Columbia, S. C. 4/6/64.

MC CAW, C. C. Pvt., no company. d. Richmond, Va. unknown date. Bur. Oakwood Cem.

MC CAW, F. M. Sgt., no company. d. of disease in Richmond hospital 9/17/63.

MC CLANN, RUFUS BRATTON. Pvt., Co. H. b. York Dist. 1/25/45. Student, age 15, York Dist. 1860 census. Enl. Rock Hill 2/10/62 age 17. Ab. with wagons 3-4/62. Present 5/62-4/63. In arrest in camp 5-8/63. Present 9-10/63. Ab. sick with "Remit. Fever" in Richmond hospital 11/27/63. Furloughed for 40 days 12/5/63. Present 2/64. Ab. on leave 3/31-4/29/64. Present 5-6/64. Ab. sick in Charleston hospital 8/22-31/64. Ab. sick 9-10/64. Detached Pocotaligo 12/10-31/64. Paroled Charlotte, N. C. 5/11/65. Mail Carrier, York Co. 1870 census. Wheelwright, Palestine, Ark. 1880 census & Farmer, 1900 census. Town Manager, Washington, Ark. 1910 census & Justice of the Peace 1920 census. d. Hermitage, Bradley Co., Ark. 11/29/29.

MC CLELLAND, SAMUEL L. Pvt., Co. G. b. Abbeville Dist. circa 1838. Res. Abbeville Dist. Enl. Abbeville CH 1/8/62. Ab. sick in hospital 3-4/62. On picket duty 5-6/62. On picket Gimball's 7-8/62. Present sick in camp 9-10/62. Ab. sick in Staunton hospital 11/19-12/31/62, however, ab. sick with "ascities" in Richmond hospital 11/20/62. Furloughed for 50 days 1/4/63. Admitted Petersburg hospital with pneumonia 1/5/63. d. 1/8/63 age 35. Bur. Blandford Ch. Cem., Petersburg. Efects: 30 cents.

MC CLINTON, WILLIAM T. 2ndSgt., Co. A. b. S. C. circa 1825. Farmer, age 25, Abbeville Dist. 1850 census. Enl. Abbeville CH 8/13/61 as Pvt., age 38. Present through 12/61. Promoted 4thCpl. Present 3-4/62. Ab. sick 5-6/62. Present 7-8/62, promoted 3rdCpl. Present 9/62-6/63, promoted 3rdSgt. WIA saber cut to head) Gettysburg 7/3/63. Ab. wounded through 8/31/63. Present 9-10/63. Ab. on horse detail 12/23/63-2/64. Present 3-6/64. Promoted 2ndSgt. Present 9-12/64. NFR. Farmer, Knox Co., Tenn. 1880 census. d. 11/24/91 Bridge, Texas.

MC CLUNEY, SAMUEL ALEXANDER. Pvt., Co. D. b. S. C. 3/28/30. Enl. Chester CH 4/20/62. On picket Townsend's House 5-6/62. Ab. on sick leave 8/4-31/62. AWOL 9/27-10/31/62. On scout 12/25-31/62. Present 3-4/63. Ab. sick in Winchester hospital 7/17/63 until captured there 7/30/63. Exchanged by 11/63. Ab. sick with "Febris" in Staunton hospital 11/8/63. Ab. sick with typhoid fever in Richmond hospital 12/4/63 & furloughed for 40 days. AWOL in Chester Co. 1/25-12/31/64. NFR. d. 7.16.98. Bur. Limestone Cem., Cherokee Co., S. C.

MC COMBE, WILLIAM HENRY. Pvt., Co. G. b. S. C. 3/9/42. Res. Abbeville Dist. 1860 census. Enl. Columbia 3/25/63. Presence or absence not stated through 8/31/63, however, ab. detailed at horse recruiting station 8/5/63-2/64. AWOL 4/27-30/64. Present 5-10/64. Detached Pocotaligo 12/10-31/64. Paroled Greensboro, N. C. 5/11/65. Farmer, Long Cane, Abbeville Co. 1880 census. Farmer, Greenwood, S. C. 1900 & 1910 censuses. d. there 3/11/19. Bur. Edgewood Cem. Wife receiving pension Greenwood, S. C. 11/20/19.

MC CORD, ARCHIE FRANKLIN. 1stCpl., Co. A. b. Abbeville Dist. 3/10/42. Res. Abbeville Dist. 1860 census. Enl. Abbeville CH 8/13/61 as Pvt., age 21. Ab. on leave 10/31/61. On duty Bears Island 3-4/62. Present 3/62-6/63. Horse killed 8/1/63. Paid $600.00. Ab. on horse detail 8/18-31/63. Present 9-12/63, promoted 3rdCpl. Present 1-2/64. Ab. on leave 4/26-30/64. Present 5-10/64, promoted 1stCpl. Present 11-12/64. Paroled Greensboro, N. C. 4/26/65. Farmer, Long Cane, Abbeville Co. 1870 census. Butcher, Liberty, Pickens Co. 1910 census. Receiving pension Liberty, Pickens Co. 9/20/19. d. Pickens Co. 10/29/19. Bur. Enon Bapt. Ch. Cem.

MC CORD, DOCK. Pvt., Co. A. Res. Abbeville Co. Enl. age 19 on postwar roster. Died of wounds, Stevensburg, Va. 1862-63.

MC CORD, FRANCIS MARION "FRANK." 4thSgt., Co. G. b. S. Abbeville Dist. 4/3/37. Student, age 18, Abbeville Dist. 1850 census. Married 1856. Enl. Abbeville CH 1/8/62 as 2ndCpl. Present through 10/62, promoted 1stCpl. 9/15/62. Present 11-12/62, promoted 4thSgt. 11/27/62. On detached service with horses 3-4/63. WIA and in Richmond hospital 6-8/63. d. of dysentery & "erysipelas" in Jackson hospital, Richmond 9/17/63. Effects: Sundries. Left widow. Probably buried in Oakwood Cem., Richmond, Va.

MC CORD, JOHN MILTON. Pvt., Co. G. b. S. C. circa 1845. Enl. Abbeville CH 1/8/62. On picket Jacksonborough 3-4/63. Present 5-10/62. On picket 11-12/62. d. Culpepper Co. 1/31/63 age 17. Bur. Long Cane Cem., Abbeville Co.

MC CORD, THOMAS M. Pvt., Co. G. b. S. C. circa 1835. Overseer, age 23, Abbeville 1860 census. Enl. Abbeville CH 1/8/62. On picket Jacksonborough 3-4/62. Present through 10/62. On picket 11-12/62. NFR. d. by 1902.

MC CORD, WILLIAM L. Pvt., Co. A. b. S. C. circa 1840. Enl. Abbeville CH 3/19/62. Present through 6/62. Present as Farrier 7-12/62. Ab. detached 3-4/63. Present 5-6/63, dismounted 1 month. Ab. on horse detail 8/18-31/63. Present 9-10/63. Ab. on horse detail 11/15/63-2/64. Ab. detached on G. & C Railroad 3-12/64. NFR. d. Richland Co. S. C. 7/21/06. Wife receiving pension Abbeville Co. 11/22/19.

MC CORKLE, WILLIAM L. Pvt., Co. D. b. S. C. 1828. Enl. Camp Means 4/10/62. On picket Townsend's House 5-6/62. Ab. on sick leave Chester Dist. 8/26-31/62. Present 9-10/62. On scout 12/25-31/62. Present 3-8/63. Ab. on horse detail 10/23-12/31/63. Ab. recruiting horses 1/26-2/29/64. Present 3-10/64. Detached Coosawhatchie 12/10-31/64. Paroled Hillsboro, N. C. 4/65. d. 5/4/03 age 75. Bur. Cedar Shoals Presb. Ch. Cem., Chester Co.

MC CORMICK, ROBERT C. 2ndSgt., Co. H. b. S. C. circa 1840. Enl. Co. F, 1st S. C. Inf. 4/11/61 age 21. Present until transferred Co. H, 1st S. C. Cav. 2/15/63. Present 4/63, promoted 3rdSgt. Present 5-6/63, reduced to Pvt. & ab. detached 7-8/63. Horse killed Funktown, Md. 7/8/63. Paid $350.00. Present as 3rdSgt. 9-10/63. Ab. on horse detail 12/22/63-2/64. Detached Green Pond 4/26-30/64. Present 5-6/64, promoted 2ndSgt. Present 7-12/64. NFR.

MC CORMICK, S. B. Pvt., Co. E. Res. Horry Dist. Not on muster rolls. d. Richmond 6/5/64.

MC CULLEY, A. J. Pvt., Co. H. b. S. C. circa 1816. Farmer, age 34, Chester Dist. 1850 census. Res. York Dist. Enl. Rock Hill 1/9/62 age 46. Ab. driving wagon 3-6/62. Present 7-8/62, no horse. Teamster for QM 9-12/62. Ab. on sick leave 3-4/63. Ab. sick in Richmond hospital 5-6/63. Ab. sick with "Fistula in ano" in Richmond hospital 9/23/63. Furloughed for 30 days 9/23/63. Ab. on sick leave through 12/31/63. Present 1-4/64. Ab. detailed as guard for government distillery, Medical Purveyors Dept., Macon, Ga. 5/64-3/22/65, age 52 and unfit for duty. Discharged by habeus corpus. NFR.

MC DANIEL, MATHEW T. Pvt., Co. H. b. S. C. circa 1845. Farmhand, age 15, Pickens Dist. 1860 census. Res. York Co. Not on muster rolls. d. of disease 3/15/62.

MC DOLL, JAMES. Pvt., Co. H. On roster but not on muster rolls. NFR.

MC DONAHUE, --------. Captain. d. Petersburg, Va. 8/19/64.

MC DONALD, DANIEL M. Pvt., Co. D. b. S. C. circa 1847. Age 3, Chester Dist. 1850 census. Enl. Chester CH 11/5/64. Present through 12/31/64. Paroled Greensboro, N. C. 5/1/65 . Laborer, age 24, Chester Co. 1870 census.

MC DONALD, JOHN M. Pvt., Co. H. Enl. Co. G, 19th S. C. Inf. Abbeville CH 12/18/61. Promoted 2ndLt. 6/19/62. Commanding Company 5-6/62. Ab. sick 7-8/62. Present 11-12/62, and wounded (slightly) Murfreesboro, Tenn. 12/31/62. Present 1-2/63. Ab. to get battle flags, Augusta, Ga. 4/25-30/63. Present 7-8/63. Ab. sick in hospital 12/20-31/6363. Furloughed for 30 days. Ab. detailed Conscript Camp, Columbia, S. C. 1-4/64. NFR. Enl. Co. H, 1st S. C. Cav. 1864-65 on roster but not on muster rolls. NFR.

MC DONALD, M. T. Pvt., Co. H. b. S. C. circa 1841. Res. Chester Dist. 1850 census. Enl. Rock Hill 1/9/62 age 21. d. 4/8/62.

MC DOWELL, J. SYLVANIUS. Pvt., Co. D. b. S. C. circa 1840. Enl. Chester CH 9/10/61 age 21. Present through 12/31/61. On picket Townsend's House 3-4/62. Present 5-8/62. Driving wagon for QM 9-10/62. Ab. sick in Staunton hospital 11/19-20/62. Ab. sick in Richmond hospital 11/21/62-1/15/63. Teamster for QM Dept. 3-8/63. Teamster for Brigade QM 9/63-2/64. Present 3-4/64. Ab. on sick leave 5/28-6/30/64. Present 7-10/64. Detached Coosawhatchie 12/10-31/64. Ab. sick with "Febris Int. Tertian" in Charlotte hospital 3/26/65. Paroled Hillsboro, N. C. 4/65.

MC DOWELL, JAMES M. Pvt., Co. D. b. Spartanburg Dist. circa 1840. Res. Spartanburg Co. 1850 census. Enl. Rock Hill 1/9/62 age 23. Present through 4/62. Ab. on horse detail to Adams Run 5-6/62. Ab. on sick leave 7-8/62. Present 9/62-6/63, however, ab. sick with debilitas in Charlottesville hospital 2/19-4/10/63. Ab. on horse detail 8/18-31/63. Present 9/63-2/64. Ab. on leave 4/16-20/64. Present 5-8/64. Present sick in camp 9-10/64. Present 11-12/64. NFR. Retired, age 80, Pennington, Bradley Co., Ark. 1920 census.

MC ELWEE, JEROME MANLIUS. Pvt., Co.'s D & K. b. York Dist. 2/2/44. Enl. Co. D York 4/9/64. Present through 6/64. Detached driving cattle 7/25-8/31/64. Transferred Co. K. 10/1/64. Ab. detached through 10/31/64. Detached at Dist. Hdqtrs., Green Pond 12/24-31/64. NFR. Farmer, Yorkville 1870 census. Farmer, Catawba, York Co. 1880 census. d. Mt. Holley, Berkley Co. 2/3/97.

MC ELWEE, LOUIS J. Pvt., Co.'s D & K. b. S. C. 1846. Res. Charleston 1850 census. Enl. Co. D York 5/30/64. Ab. sick in Charleston hospital 6/27-30/64. Present 7-8/64. Transferred Co. K 10/1/64. Courier, Riverside Post 12/29-31/64. Paroled Thomasville, N. C. hospital 5/1/65.

MC ELWEE, PETER. Pvt., Co.'s D & K. Res. York Dist. Enl. Co. D by 1862. Transferred Co. K 7/62. NFR.

MC FADDEN, THOMAS L. Pvt., Co. D. b. Chester Dist. circa 1820. Farmer, age 30, Chester Dist. 1850 census. Enl. Chester CH 9/10/61 age 40. Present through 4/62. On picket Townsend's House 5-6/62. Present 7-10/62. Sick in camp 11-12/62. Present 3-6/63. Captured near Martinsburg 7/19/63. Sent to Wheeling 7/24/63. Age 45, 5'9 1/2", dark complexion, blue eyes, dark hair, Blacksmith. Transferred Camp Chase & Ft. Delaware. d. of "Inflammation of the brain" 4/4/64. Bur. Finn's Point Nat. Cem., N. J.

MC GAW, JOHN J. Pvt., Co. G. b. Abbeville Dist 1846. Res. Abbeville Dist. Enl. Abbeville CH 4/27/64. Present 5-6/64. Ab. sick in hospital 8/20-31/64. d. in 1st La. Hospital, Charleston of "Congestive Fever" 9/11/624. Effects: Sundries & $45.50. Bur. Magnolia Cem. as. "J. G. McGou."

MC GEE, BENJAMIN MATTISON. Pvt., Co. G. b. S. C. 12/1/47. On roster but not on muster rolls. d. 4/26/25. Bur. Pasadena Mausoleum, Los Angeles Co., Calif.

MC GILL, WILLIAM JACKSON. Pvt., Co. G. b. York Dist. 5/16/43. Age 17, York Dist. 1860 census. Enl. Columbia 8/16/64. Present 11-12/64. NFR. Farmer, age 57, Cherokee Co. 1900 census. d. 12/6/16. Bur. Smyna Cem., York Co.

MC GUINNESS, W. M. Pvt., Co. F. Captured and died of disease Ft. Delaware, dates unknown.

MC INTYRE, ROBERT C. Pvt., Co. K. b. S. C. circa 1843. Res. Marion Dist. 1860 census. Enl. Columbia 4/18/64. Courier, Naval Ordnance Dept., Charleston 6/25-30/64. AWOL 8/30-31/64. Ab. sick leave 10/14-31/64 for 30 days. Retired 12/13/64. Detailed 1/23/65. Paroled Hillsboro, N. C. 4/65. Retired, age 64, Hebron, Marlboro Co., S. C. 1910 census

MC KELVY, PETIS BOOKMAN. Pvt., Co. B. b. Fountain Inn, Laurens Dist. 11/3/46. Enl. by 1865. Paroled Greensboro, N. C. 5/3/65. Farmer, Fairview, Greenville Co. 1900 census & Dials, Laurens Co. 1910 census. Receiving pension Fountain Inn, Laurens Co. 4/1/19. d. Laurens 12/28/25. Bur. Fountain Inn Cem.

MC KENZIE, LADSON. 2ndSgt., Co. E. b. S. C. circa 1844. Age 6, Orangeburg Dist. 1850 census. Enl. Ft. Motte 10/26/61 as Pvt., age 22. Present through 12/31/64, promoted Cpl. Promoted 3rdSgt. & present 3-4/62. AWOL 5-6/62. On scout Edisto Island 7-8/62. Present 9-10/62, promoted 2ndSgt. Ab. sick in private house 11-12/62. Present 3-6/63. Detached with disabled horses 8/18-12/31/63. Present 1-2/64, dismounted since 2/1/64. Detached Green Pond 4/16-30/64. Present 5-10/64. Detached Pocotaligo 12/10-31/64. d. of typhoid fever James Island 1/15/65. Bur. Buck Head Bapt. Ch. Cem., Ft. Motte, S. C.

MC KENZIE, LAWRENCE R. Pvt., Co. E. b. S. C. circa 1843. Res. Orangeburg Dist. Enl. Coosawhatchie 11/14/61. On picket White's Point through 12/31/61. Present 3-4/62. Present sick 5-6/62. Present 7/62-8/63. Ab. on horse detail 9-10/63. Present 11/63-2/64. Ab. on leave 4/3-5/5/64. Present sick 7-8/64. Ab. sick James Island hospital 8/25-31/64. Ab. sick at home 10/1-31/64. Detached as guard Wappoo Bridge 12/29-31/64. Paroled Hillsboro, N. C. 4/65.

MC KERROLL, J. M. Ensign. Enl. Co. C by 1864. Appointed Ensign of regiment. WIA (leg amputated) and DOW's 8/22/64.

MC KEOWN, EDWARD T. Pvt., Co. D. b. S. C. circa 1833. Farmer, age 27, Chester Dist. 1860 census. Enl. Co. A, 6th S. C. Inf. Chester CH 4/11/61 age 30. Present sick 6-8/61. Baggage guard, Manassas 10/11/61-7/1/62. Discharged Richmond 7/17/62. Enl. Co. D, 1st S. C. Cav. Chester CH 8/24/62. Present until AWOL 10/1-31/62. Present 11/62-8/31/63. Ab. on horse detail 10/23-12/31/63. Present 1-2/64. Ab. on leave 4/23-30/64. Present 5-10/64. Detached Coosawhatchie 12/10-31/64. Paroled Hillsboro, N. C. 4/65. Brother of M. T. McKeown.

MC KEOWN, JAMES A. Pvt., Co. D. b. S. C. circa 1838. Student, age 22, Chester Dist. 1860 census. Enl. Chester CH 9/10/61 age 23. Present through 4/62. Ab. on sick leave 5/17-6/30/62. Steward, Regimental hospital 7-8/62. Present 9/62-8/31/63. Ab. on horse detail 10/23-31/63. Present 11/63-2/64.

Detached Green Pond 4/26-30/64. Present sick in camp 5-6/64. Present 7-8/64. Ab. detached a guard for POW's Florence 9/17-12/31/64, however, ab. sick 12/17-31/64. Paroled Hillsboro, N. C. 4/65. Farmer, age 30, Blackstock, Chester Co. 1870 census.

MC KEOWN, MOSES T. Pvt., Co. D. b. S. C. circa 1836. Student, age 14, Chester Dist. 1850 census. Enl. Chester CH 5/16/62. On picket Townsend's House through 6/62. Present sick in camp 7-8/62. Present 9-10/62. Ab. sick in hospital 12/62. Present 3-4/63. Teamster with wagon 8/3-10/31/63. On horse detail 12/22/63-2/64. Present 3-10/64. Detailed Coosawhatchie 12/10/31/64. Paroled Hillsboro, N. C. 4/65. Brother of Edward T. McKeown.

MC KEOWN, SAMUEL THOMPSON. Pvt., Co. K. b. Cornwell, Chester Dist. 8/7/46. Enl. Columbia 4/19/64. Present until on picket Charleston fortifications 6/25-30/64. Present 7-8/64. Ab. detached Ordnance Office, Charleston 9-10/64. On picket Chatman's Ford 12/29-31/64. Paroled Greensboro, N. C. 4/26/65 on pension application. Attended "Survivors of Butler's Brigade" reunion 10/25/05. Receiving pension Chester Co. 4/12/19. d. 11/14/35. Bur. McKeown Cem., Chester Co.

MC KIE, THOMAS JEFFERSON. Pvt., Co. C. M.D., Edgefield Dist. 1860 census. Enl. John's Island 6/24/62. Present through 6/30/62. Discharged by Surgeon 7/26/62. Appointed Asst. Surgeon, 10th S. C. Inf. 4/16/63. Resigned 4/11/64. Contract Surgeon, Augusta, Ga. Relieved & assigned to hospital, Cuthbert, Ga. Contract closed 2/28/65. NFR.

MC LANE, R. Pvt., Co. H. On postwar roster.

MC LAUCHLIN, DONALD L. Captain, Co. G. b. S. C. circa 1836. Druggist, age 25, Abbeville 1860 census. Enl. 1/8/62 as 2ndLt. Ab. detached as QM 3-4/62. Signed for 44 blankets, 1 horse blanket, 4 curry combs, 4 horse brushes, 25 lbs. horse shoe nails, 2 axes, 2 helves, 1 hatchet, 3 lbs. rope, 6 pr. harness & 3 pr. chairs 5/1/62. Signed for 800 blanks, 26 mules, 174 pr. horse shoes, 50 lbs. horse shoe nails, 5 axes, 3 helves, 2 wall tents, 2 army tents & 150 corn bags. Adams Run 5/20/62. Numerous vouchers for forage & supplies in his file. Ab. sick 6/8-30/62. Present sick in camp 7-8/62. Present 9-12/62. Ab. on detached service 3-4/63. Signed for 2 tents & 1 blanket 5/23/63. Present through 12/63. Ab. on leave 2/15-29/64. Present 3-4/64. Present sick in camp 5/64. Ab. sick in hospital 6/27-8/31/64. Promoted Captain by 8/25/64. AWOL at James Island 9/25/64. Present through 12/64. NFR. d. by 1902 on Abbeville County roster.

MC LAUGHLIN, JOHN. Captain & Asst. QM. b. S. C. circa 1830. Captain & QM, 10th Tenn. Inf. 10/61 Ft. Henry, Tenn. Captured there 2/6/62. Sent to Alton, Ill. & Columbus, Ohio. Sent to Vicksburg, Miss. 9/1/62. Exchanged 11/8/62. Assigned 1st S. C. Cav. 12/29/63. Relieved 2/1/64 & assigned to Phillips Ga. Legion. Assigned to 38th Tenn. Inf. 3/8/64 but revoked and appointed QM in charge of transportation, Atlanta, Ga. 9/11/64. Paroled as Division Paymaster, Cheatham's (Bates') Division, Army of Tenn. Greensboro, N. C. 4/26/65.

MC LEES, JOHN FRANKLIN. Pvt., Co. F. b. S. C. 9/21/38. Enl. Co. A, Lucas's 15th Bn. S. C. Arty. Columbia 11/2/64. Transferred Co. F, 1st S. C. Cav. 12/15/64. Present, dismounted through 12/31/64. Paroled Greenville, S. C. 5/23/65. d. 12/6/10. Bur. Roberts Presb. Ch., Anderson Co. Wife receiving pension Anderson, S. C. 5/5/19.

MC LEOD, WILLIAM WALKER. Pvt., Co. A. b. S. C. 1/20. On postwar roster. Reenlisted Co. K, 4th S. C. Cav. Grahamville 3/25/62. age 42. d. 2/65. Bur. St. James Epis. Ch., Charleston, S. C.

MC LURE, R. H. Pvt., Co. K. Res. York Dist. Paroled Raleigh, N. C. 4/65.

MC MILLAN, T. WILLIAM. Pvt., Co. A. Enl. Abbeville CH 3/14/62. Present through 6/62. Ab. on leave 7-8/62. Present 9-10/62. Detailed in Med. Dept. of Regiment 11-12/62. Present 3-8/63. Ab. sick with acute diarrhea in Richmond hospital 10/8-12/8/63. Furloughed for 30 days 12/9/63. Present 1-2/64. Ab. on leave 4/25-30/64. Present 5-6/64. Present as nurse in Med. Dept. of Regiment 7-8/64. Present 9-12/64. NFR.

MC MURRY, J. H. Pvt., Co. D. Enl. Lancaster CH 3/19/64. Present through 10/64. Ab. sick in Brigade hospital 12/23-31/64. Paroled Hillsboro, N. C. 4/65.

MC NAIR, FRANCIS MARION "FRANK." Pvt., Co. A. b. S. C. circa 1830. Laborer, age 30, Long Cane, Abbeville Dist. 1860 census. Enl. Abbeville CH 8/13/61 age 33. Present through 6/63. Ab. on sick leave 7-8/62. Present 9-12/62. Present detailed as Teamster 3-4/63. Present 5-6/63, dismounted 1 month. Ab. on horse detail 8/18-31/63. WIA Culpepper CH 9/18/63 & ab. wounded through 12/31/63. Present 1-2/64. Ab .on horse detail 4/12-30/64. Present 5-10/64. Detached Pocotaligo 12/10-31/64. NFR. Laborer, age 45, Indian Hill, Abbeville Co. 1870 census. d. Abbeville Co. 1879.

MC NAIR, H. F. Pvt., Co. A. On post war roster.

MC NAIR, LEWIS M. Pvt., Co. Enl. age 19 on Abbeville Co. poster war roster.

MC NAIR, WILLIAM E. Pvt., Co. A. b. S. C. 2/13/33. Farmer, age 26, Long Cane, Abbeville Dist. 1860 census. Enl. Abbeville CH 8/13/61 age 29. Present through 6/62. Ab. on sick leave 7-8/62. Present 9/62-6/63. Ab. on horse detail 8/18-31/63. Present 9-10/63. Present dismounted 11-12/63. Present 1-10/64,but ab. on leave 8/27-30/64. Detached Pocotaligo 12/10-31/64. NFR. Farmer, Calhoun, Abbeville Co. 1900 census. d. McCormick, S. C. 3/20/06. Bur. Willington Cem.

MC NEIL, JOHN H. Pvt., Co. G. b. Ireland 1845. Came to U. S. 1848. Enl. age 18 on Abbeville Co. postwar roster. Bookkeeper, Charleston 1870 census. Living Magnolia Dist., Abbeville Co. 1902.

MC PHERSON, D. R. 3rdSgt., Co. D. b. S. C. circa 1836. Peddler, age 24, Chester Dist. 1860 census. Enl. Chester CH 9/10/61 as Pvt., age 25. Present through 4/62. On picket Townsend's House 5-6/62. Present 7-8/62. Present sick in camp 9-10/62, promoted 3rdSgt. Ab. sick in hospital 12/62. Present 3-4/63. Ab. detailed in horse recruiting camp, Botetourt Co., Va. 8/20-12/31/63. Present 1-6/64. Ab. sick in Charleston hospital 8/19-31/64. Ab. sick in Union Dist. 9/12-12/31/64. Detailed as guard at Pontoon Bridge 12/29-31/64. Paroled Hillsboro, N. C. 4/65.

MC PHERSON, JOHN JAMES. Pvt., Co. A. b. S. C. circa 1833. Enl. 11/29/61. Discharged at end of enlistment 2/7/62. Reenlisted Co. K, 4th S. C. Cav. Grahamville 3/25/62 age 25. Present until discharged 5/10/62, appointed Paymaster, C. S. Navy. NFR. d. 11/22/03 age 70. Bur. St. Phillips Epis. Ch., Charleston.

MC TUREOUS, B. WARREN. Sgt., Co. A. b. Charleston 10/37. Clerk, Charleston 1859. Enl. 11/7/61. Detailed to Gen. Ripley 2/7/62. Reenlisted Co. G, 5th S. C. Cav. Charleston 2/20/62. Elected Captain Co. B, 17 Bn. S. C. Cav. Present through 10/64. Sent to Ga. 11/2/5/64. NFR. Salesman, Charleston 1866. Clerk, Charleston 1867. d. 7/12/79. Bur. 1st Bapt. Ch., Charleston.

MC TUREOUS, JOHN C. Pvt., Co. A. b. Beaufort, S. C. 1841. Clerk, Charleston 1859. Enl. 11/7/61. Promoted Sgt., Co. B, 17th Bn. S. C. Cav. 2/20/62. Redesignated Co. G, 5th S. C. Cav. Promoted 1stSgt. Ab. on duty in S. C. for 40 days 10/27/64. NFR. Accountant, Charleston 1867. d. 10/5/82. Bur. Magnolia Cem., Charleston.

MC TUREOUS, JOSEPH COWAN. Pvt., Co. A. b. Beaufort, S. C. 1843. Enl. 11/7/61. Promoted 5thSgt., Co. B, 17th Bn. S. C. Cav. 2/20/62. Became Co. G, 5th S. C. Cav. Acting Commissary & QM 1-4/63. Present 5-6/63. d. of typhoid fever Kingstree, S. C. 7/18/63. Bur. 1st Bapt. Ch. Cem., Charleston.

MC WATERS, JOHN BARNES. Pvt., Co. D. b. Chester Dist. 10/28/43. Enl. Camp Johnson 10/4/61 age 18. Present through 4/62. On picket Townsend's House 5-6/62. Ab. on sick leave in Chester Dist. 8/27-31/62. Present 9/62-6/63. Ab. with horses in the rear 8/3-12/31/63. On picket 2/28-29/64. Present through 12/64. Paroled Greensboro, N. C. 4/26/65 on pension application. Farmer, Lewisville, Chester Co. 1880-1910 censuses. Receiving pension Richburg, Chester Co. 4/3/19. d. 5/9/19. Bur. Union Presb. Ch. Cem., Chester Co. Wife receiving pension Richburg, Chester Co. 9/26/19.

MC WHORTER, SAMUEL T. Pvt., Co. F. b. S. C. circa 1829. Farmer, age 33, Wahalla, Pickens Dist. 1860 census. Enl. Pickens CH 12/4/61 age 30. Present through 4/62. On picket Camp Means 5-6/62. Present 9/62-6/63. KIA Brandy Station 8/1/63.

MEADOWS, J. A. Pvt., Co. B. b. S. C. circa 1844. Enl. Co. D, 9th S. C. Reserves Clinton 11/17/63. Ab. sick at home 2/14/64. NFR. Disbanded. Enl. Co. B, 1st S. C. Cav. Columbia 2/20/64. Detached Green Pond 4/16-30/64. Present 5-6/64. Ab. on horse detail 8/10-31/64. On picket West Lines, James Island 10/1--31/64. Present 11-12/64. NFR. Alive 1898.

MEEK, ROBERT RICHARD. Pvt., Co. D. b. S. C. circa 1839. Enl. Chester CH 10/9/61 age 22. Ab. after baggage at depot 12/31/61. Discharged for disability 3/27/62. NFR. Farmer, Bullock's Creek, York Co. 1880 census. Farmer, Bethel, York Co. 1900 & 1910 censuses. d. 5/16/15 age 76. Bur. Bethel Cem., York Co.

MEETZE, JOHN YOST. Pvt., Co. A. b. S. C. 1836. On post war roster. Reenlisted Co. K, 4th S. C. Cav. Grahamville 3/25/62. Ab. detached with Gen. Pemberton through 8/62. Present 9/62 until detailed on light duty as Wagon Master, Government Stables, Charleston 2/4/63-8/64. Paroled Charlotte, N. C. 5/4/65. d. 1917. Bur. Oakland Cem., Gaffney, S. C.

MELTON, SAMUEL W. Pvt., Co. I. Married York Dist. 1857. Res. York Dist. 1860 census. Res. Colleton Dist. when enlisted by 1863. d. of disease Charleston 1/6/64. Bur. Magnolia Cem., Charleston, no dates.

MERCHANT, DAVID H. Pvt., Co. B. b. Newberry Dist. circa 1838. Enl. Camp Johnson 10/29/61 age 29. Ab. sick 10/31/61. Present 11-12/61. Ab. as Courier, Camp Means 3-4/62. Present 5-6/62. Ab. on sick leave 7-8/62. Present 9-10/62. Ab. sick with "Ascites" in Richmond hospital 12/4/62. Discharged for dropsy from Richmond hospital 12/11/62. Age 29, 5'9", fair complexion, blue eyes, dark hair, Farmer. Enl. Co. K, Palmetto Bn. Arty. 7/1/63. Ab. in hospital 8/63. Present 9-10/63. Ab. sick in hospital 1-12/63. Present 1-4/64. NFR. Served in Cavalry Co., LDT. NFR. Receiving pension Aluchua Co., Fla. 4/9/08 age 64.

MERRITT, JOHN. Pvt., Co. E. Enl. by 1865. Paroled Greenville, S. C. 5/23/65.

MESSER, DAVID H. 3rdSgt., Co. F. b. S. C. 6/21/35. Student, age 16, Cherokee Co., Ga. 1850 census. Enl. Pickens CH 12/4/61 age 24. Present through 12/31/61. On picket Jacksonborough 3-4/62 & Camp Means 5-6/62. Ab. on leave 7-8/62. Present 9-10/62. On scout 11-12/62. Present 3-4/63. Appointed Musician & transferred to Regimental Band 6/1/63. Ab. sick with dyspepsia in Richmond hospital 7/1-8/23/63. Present 9/63-6/64. Ab. sick in hospital 9/1-10/31/64. AWOL 12/2-31/64. NFR. Grocery Salesman, Atlanta, Ga. 1900 census. Receiving pension Ga. 1906. Wrote book of poems called "Grand-Pa's Meditation" in 1908. Retired, Atlanta 1910 census. d. Clark Co., Ga. circa 1910.

MEYER, L. Pvt., Co. A. Enl. 11/7/61. Exempt as M. D. 12/1/61. NFR.

MEYER, THOMAS HAWKINSON. Pvt., Co. C. b. S. C. 11/26/32. Planter, Blacksville, Barnwell Dist. 1860 census. Enl. John's Island 6/28/62. Present 7-10/62. Present sick in camp 11-12/62. Present 3-4/63. Ab. sick with debitas in Williamsburg hospital 5/4/63. Transferred to Farmville hospital 5/8/63. Returned to duty 5/11/63. Ab. on sick leave 6/6-30/63. Ab. on horse detail 8/31-10/31/63. Ab. on detached service 1-2/64. Ab. on horse detail 4/18-5/2/64. Ab. on leave 6/15-30/64. Present 7-12/64. Paroled Greensboro, N. C. 4/26/65. d. 9/26/04. Bur. Meyer-Hawkinson Cem., Aiken Co. Wife receiving pension Beech Island, Aiken Co. 11/25/19.

MIDDLETON, FRANKLIN KINLOCH. Pvt., Co. A. b. S. C. 1835. On post war roster. Reenlisted Co. K, 4th S. C. Cav. Grahamville 3/25/62. Present until detailed with Gen. Ripley 10/23/63. Present 11/63-2/64. WIA (back) & captured Hawes Shop 5/28/64. DOW's 5/30/64. Bur. on farm of Mrs. Newton, on the Pamunkey River. Removed to Magnolia Cem., Charleston.

MIDDLETON, OLIVER HERING, JR. Pvt., Co. A. b. S. C. 7/17/45. Reenlisted Co. K, 4th S. C. Cav. Pocotaligo 1/4/63. Appointed Courier on regimental st rhrough 2/64. DOW's received Cold Harbor 5/30/64 on 5/31/64. Bur. Magnolia Cem., Charleston.

MIKELL, EDWARD W. Pvt., Co. A. b. S. C. circa 1840. On post war roster. Reenlisted Co. K, 4th S. C. Cav. 10/1/63. Detailed as Commissary Agent 10/9-12/31/63. AWOL 1/21/64. Detailed with Brigade Commissary 3-6/64. Transferred Co. G. Detailed with Brigade Commissary through 10/31/64. Paroled Greensboro, N. C. 5/2/65. Accountant, age 40 Charleston 1880 census. Res. of Charleston 1889.

MILES, CHARLES PINCKNEY. Pvt., Co. B. b. Enoree, Spartanburg Dist. 6/18/49. Not on muster rolls. Enl. Newberry 8/1/563 & paroled 4/15/65 on pension application from Enorre, Spartanburg Co. 5/30/15. M .D. d. Spartanburg 11/14/20. Bur. Cedar Shoals Cem., Spartanburg Co.

MILES, JAMES ALLEN. Pvt., Co. A. b. S. C. 1/5/35. On postwar roster. Reenlisted Co. K, 4th S. C. Cav. Grahamville 3/25/62 as 2ndCpl. WIA (slightly) Pocotaligo 10/22/62. Promoted 1stCpl. Promoted 4thSgt. 1/1/63. Promoted 3rd Sgt.. Present until KIA Hawe's Shop 5/28/64.

MILES, JEREMIAH JACKSON Pvt., Co. A. b. S. C. 2/22/36. On post war roster. Reenlisted Co. K, 4th S. C. Cav. Grahamville 3/25/62. Present until ab. on horse detail 4/27/63 for 10 days. Present through 5/64. Ab. sick with "Int. Febris" 6/3-8/64. Present until sent to Stony Creek Station with dismounted men 10/12/64. Captured there 12/1/64. Sent to Pt. Lookout. Released 6/29/65. 5' 3 1/2", fair complexion, brown hair, hazel eyes, res. of Charleston, S. C. Clerk, Charleston, 1870 census. d. 8/5/96. Bur. St. Paul's Churchyard, Summerville, S. C.

MILES, JOHN ALLEN. Pvt., Co. A. b. S. C. 1/5/35. On postwar roster. Reenlisted Grahamville 3/25/62 as 2ndCpl. Promoted 1stCpl. Present until promoted 4thSgt. 1/1/63. Promoted 3rdSgt. Present until KIA Hawes Shop 5/28/64.

MILLAM, GEORGE T. Pvt., Co. B. b. S. C. 1846. Age 5, Laurens Dist. 1850 census. Enl. Columbia 9/1/64. Present through 10/64. On picket West Lines, James Island 11-12/64. NFR. Gunmaker, age 34, Pendleton, Anderson Co. 1880 census.

MILLAM, TURNER R. Pvt., Co. B. b. S. C. circa 1845. Student, age 15, Waterbury, Laurens Co. 1860 census. Enl. Laurens Co. age 18 on post war roster. Alive 1898.

MILLER, B. W. Pvt., Co. A. On roster 10/11/61. NFR.

MILLER, BENJAMIN ROBINSON. Pvt., Co. H. b. S. C. 6/21/32. Enl. Rock Hill 10/1/62. Present through 12/62. Ab. on horse detail 2/21-28/62. Presence or absence not stated 3-6/63. Ab. on horse detail 8/18-31/63. Present 9-10/63. Ab. on horse detail 12/23/63-2/64. Present 5-12/64. Served to 4/10/65 on wife's pension application. d. 1/7/16. Bur. Ebenezer Presb. Ch. Cem., York Co. Wife receiving pension York, S. C. 4/22/19.

MILLER, GEORGE W. (1). Pvt., Co. A. b. S. C. circa 1836. Farmer, age 24, Barnwell Dist. 1860 census. Enl. Abbeville CH 9/12/61. Ab. on sick leave 10/31/61. Ab. on leave 12/30-31/61. Present 3-6/62. Present sick in camp 7-10/62. Present 11/62-4/63. Presence or absence not stated 5-6/63. Transferred to Regimental Band 6/1/63. Ab. on horse detail 8/18-31/63. Ab. on hospital 9-10/63. AWOL 11/17-12/31/63. Present 1-6/64, however, WIA (minnie ball in leg) & in Richmond hospital 1/30/64. Furloughed for 40 days 2/3/64. Ab. on leave 10/15-31/64. Present 11-12/64. NFR. In Georgia Old Soldier's Home, Atlanta, age 74, 1910 census.

MILLER, GEORGE W. (2). Pvt., Co. A. b. Lancaster Co. 1823. On post war roster. Reenlisted Co. K, 4th S. C. Cav. 10/3/63. Present through 2/64. Ab. on horse detail 5/2/64. Ab. sick in Williamsburg hospital 7/4-8/3/64. Ab. sick in Petersburg hospital 8/31/64. Ab. sick in Greensboro, N. C. hospital 10/31/64, age 42. NFR. d. Jefferson, Chesterfield Co. S. C. 1870.

MILLER, J. A. Pvt., Co.'s F & K. b. S. C. circa 1840. Enl. Pickens CH 4/22/62. Present until transferred Co. K 6/25/62. Present 7-8/62. Ab. on detached service 9-10/62. Present 11/62-4/63. Presence or absence not stated 5-7/63. Ab. on horse detail 8/18-31/63. Present 9-12/63. Ab. detached at horse recruiting station 1-2/64. Ab. on horse detail 4/16-30/64. Present 5-6/64. On duty Bee's Ferry 7-8/64. Ab. sick leave from Charleston hospital 10/9/64 for 60 days. AWOL 12/25-31/64. Paroled Raleigh, N. C. 4/65. Farmer, Grant, Neosho Co., Kansas 1905. Retired Farmer, age 74, Ellsmore, Allen Co., Kansas 1920 census.

MILLER, J. B. Pvt., Co. A. Enl. 11/7/61. NFR. Reenlisted Co. B, 17th S. C. Bn. Cav. Charleston 2/28/62. Became Co. K, 4th S. C. Cav. NFR. Reenlisted Co. C, 25th S. C. Inf. James Island 12/7/62. Present through 2/63. Ab. sick with pneumonia in Greensboro hospital 2/23-25/65. In Raleigh, N. C. hospital 3/1-19/65. NFR.

MILLER, JOHN H. Pvt., Co. I. b. Colleton Dist. circa 1838. Res. Charleston. Enl. Parker's Ferry 4/3/62 age 24. Present through 8/62. Ab. on leave 9-10/62. Present sick in camp 11-12/62. Ab. on detached service 3-4/63. Present 7-8/63. Ab. sick in hospital 9-10/63 & with "Febris Remittens" in Richmond hospital 11/30/63-12/15/63. Detached recruiting camp for horses 12/22/63-2/64. Detached Green Pond 4/16-30/64. Present 5-6/64. On Courier duty 7-8/64. Present 9-10/64. AWOL 12/15-31/64. Deserted to the enemy Charleston 2/22/65. Age 24, 6', dark complexion, black hair, black eyes. Took oath & discharged. Paroled Augusta, Ga. 5/18/65 as Pvt., Co. D, 6th S. C. Inf.

MILLER, PINCKNEY B. Pvt., Co. B. b. S. C. circa 1836. Age 14, Spartanburg Dist. 1850 census. Enl. Clinton 8/22/61 age 24. Present through 12/61. On picket Hallover Bridge 3-4/62. Present 5-6/62. d. Spartanburg, S. C. 8/5/62.

MILLER, THOMAS S. (M). 1stLt., Co. C. b. S. C. circa 1833. Farmer, age 27, Edgefield Dist. 1860 census. Enl. Hamburg 8/27/62 as 2ndLt. age 28. Commanding Company 11-12/61. Promoted 1stLt. 6/26/62. Present through 8/62. Ab. sick 9-10/62. On scout 11-12/62. Present 5-8/63. 56 horses present 5/7/63. Ab. on leave 9-10/63. Ab. sick with "gonorrhoea" in Charlottesville hospital 11/10/63-2/11/64. Present sick in camp 2/29/64. Ab. on leave 4/15-30/64. Ab. on leave 6/25-30/64. Ab. sick in hospital 8/18-31/64. Present 9-12/64. NFR Alive 1898.

MILLER, WILLIAM WALLACE. Pvt., Co. C. b. White House, Aiken Co. S. C. 3/29/43. Res. Edgefield Dist. 1850 census. Enl. Hamburg 8/27/61 age 18. Detached Wadmalaw Island 11-12/61. Present 5-6/62. Ab. on sick leave 8/11-31/62. Present 9-10/62. On picket 11-12/62. Ab. on detached service 3-4/63. Present through 12/63. Detailed as scout for Gen. Hampton 4/15-11/64 Captured near Petersburg 11/15/64. Sent to Pt. Lookout. Released 6/29/65. 5'9 1/2", light complexion, hazel eyes, light brown hair. Planter, age 27, Richmond Co., Ga. 1870 census. Farmer, Hammond, Aikens Co. 1880 census. Attended "Survivors of Butler's Brigade Reunion 10/25/05. d. Columbia, S. C. 1/27/10. Bur. Hammond Cem., Beech Island, S. C.

MILLS, G. A. Pvt., Co. K. Paroled Raleigh, N. C. 4/65.

MILLS, JOHN A. Pvt., Co. K. b. S. C. 6/18/21. Enl. Columbia 8/15/64. On duty Signal Station, Charleston 9-10/64. Courier, River Side Post 12/31/64. NFR. d. 4/1/06. Bur. Ora ARP Church Cem., Laurens Co.

MILLS, SIMON MALWOOD D. 4thSgt., Co. H. b. S. C. 2/18/28. Enl. Rock Hill 3/19/62 as 2ndCpl. Present through 8/31/63, promoted 1stCpl. Present 9-10/63. Ab. on horse detail 12/22/63-2/64. Detached Green Pond 4/26-30/64. Present 5-6/64, promoted 4thSgt. Present 9-12/64. NFR. d. Ft. Mill, York Co. 11/23/12. Bur. Unity Cem., Rock Hill. Wife receiving pension Ft. Mill, York Co. 2/20/19.

MILLWEE, JAMES HENRY. 2ndCpl., Co. F. b. S. C. 12/23/40. Stonecutter, Anderson Dist. 1860 census. Enl. Colleton Dist. as Pvt. 4/8/62. On picket Jacksonborough through 4/30/62. Present 5-10/62. On scout 11-12/62. Present 3-4/63, promoted 2ndCpl. Ab. sick with "syphilis" in Richmond hospital 5/16/63, and reduced to 4thCpl. Transferred to Lynchburg hospital 7/8/63. Transferred Charleston hospital with chronic "gonorrhoea" 8/2/63-4/13/64. Requested transfer to 8th Ga. Inf. 4/30/64, however, transferred Co. A, 2nd S. C. Inf. 4/13/64. Not on muster rolls of that unit. Paid 6/22/64. Paroled Greensboro, N. C. 5/2/65. Laborer, Seneca, Oconee Co. post war. d. 1902.

MIMS, JOHN N. Pvt., Co. C. b. Edgefield Dist. 12/31. Enl. Wadmalaw Island 3/23/62 age 31. Present through 6/62. Ab. on sick leave until hired Benjamin Bynum as substitute 8/23/62. Age 31, 5'10", dark complexion, brown eyes, black hair, Merchant. NFR. Farmer, Williston, Barnwell Co. 1880 & 1900 censuses.

MINER, JOHN L. Pvt., No Co. b. S. C. 11/2/35. UDC application only record of service. Miller, Cedar Springs, Abbeville Co. 1880 census.d. 7/29/13. Bur. Keowee Bapt. Ch. Cem., Abbeville Co.

MINOR, JOHN M. Pvt., Co. A. b. S. C. 12/36. Enl. Abbeville CH 8/13/61 age 25. Ab. sick in hospital 11-12/61. On duty Wilson's Point 3-4/62. Present 5-6/62. Ab. on sick leave 7-8/62. Present 9-10/62. Present sick in camp 11-12/62. Present 3-10/63. Ab. on horse detail12/23/63-2/64. Present 3-10/64. Detached Pocotaligo 12/10-31/64. NFR. Farmer, age 63, Diamond Hill, Chester Co.1900 census. Living Abbeville Co. 1902.

MINTER, JOSEPH MONROE E. Pvt., Co.'s D & K. b. S. C. 2/17/30. Enl. Co. D Chester CH 9/10/61 age 31. Present through 4/62. Transferred Co. K 6/25/62. Present through 6/63. Ab. on horse detail 8/16-31/63. Present 9/63-2/64. Detached Green Pond 4/15-30/64. Ab. on leave 6/17-30/64. Ab. sick in Charleston hospital 7-8/64. Furloughed for 60 days 9/5/64. Courier, Stock Causeway 12/29-31/64. Paroled Raleigh, N. C. 4/65. Farmer postwar. d. Chester, S. C. 10/11/08. Bur. Liberty Bapt. Ch. Cem.

MITCHELL, L. LOYD. Pvt., Co. C. b S. C. circa 1819. Farmer, age 31, Edgefield Dist. 1850 census. Enl. Camp Butler 9/1/61 age 42. Detached Wadmalaw Island 11-12/61. Present 5-6/62. Ab. on sick leave 8/19-31/62. Present 9-12/62. d. in Staunton, Va. hospital 3/25/63. Bur. Thornrose Cem., Staunton, Va. Left widow.

MIZZELLE, JOHN. Pvt., Co. D. Not on muster rolls. Bur. Mizzelle Cem., Berkeley Co., S. C., no dates. Tombstone only record of service.

MOBLEY, EDWARD G. Pvt., Co. H. b. S. C. circa 1836. Citadel Cadet 52-53. Att. Mt. Zion College & South Carolina College. Farmer, age 24, Chester Dist. 1860 census. Enl. Rock Hill 8/19/62. On picket Elegaly through 8/31/62. AWOL 9-10/62. Transferred Co. F, 6th S. C. Cav. 11/1/62. Detailed as nurse 8/15/64. NFR. Laborer, age 43, Fairfield, Pickens Co., Ala. 1880 census. Moved to Texas. Farmer. d. Groesbeck, Limestone Co., Texas 4/02.

MOBLEY, WILLIAM DIXON. Pvt., Co.'s D & K. b. Chester Dist. 2/8/41. Age 19, Chester Dist. 1860 census. Enl. Co. D Chester CH 9/10/61 age 19. Present through 12/61. Ab. detailed in Charleston on Company business 4/29-30/62. Transferred Co. K 6/25/62. Ab. detached 7-12/62. Ab. sick 8/15-30/63. Presence or absence not stated through 8/31/63. Ab. on horse detail 9-10/63. Present 11/63-2/64. Detached Green Pond 4/15-30/64. Courier, Naval Ordnance Dept., Charleston 6/25-30/64. Ab. on sick leave 7-8/64. On duty at Entrenchments, James Island 9-10/64. Detailed Dist. Hqtrs., Green Pond 12/24-31/64. Paroled Hillsboro, N. C. 4/65. Farmer, Chester Co. 1880-1910 censuses. d. Chester Co. 12/28/17. Bur. Old Purity Presb. Ch. Cem., Chester.

MONTGOMERY, RICHARD GRANVILLE. Pvt., Co.'s D & K. b. S. C. 1835. M.D., age 24, York Dist. 1860 census. Enl. Co. D Chester CH 9/10/61 age 24. Present through 4/62. Transferred Co. K 6/25/62. Present through 10/62. On scout 11-12/62. Present through 4/63. Ab. sick with chronic diarrhea in Charlottesville hospital 5/24/63. Furloughed for 30 days 6/14/63. Transferred to Band 11-12/63. Musician, present on all rolls through 12/31/64. Paroled Hillsboro, N. C. 4/65. Farmer, age 46, Lewisville, Chester Co. 1880 census. d. by 1890. Bur. Brushy Fork Bapt. Ch. Cem., Chester Co., no dates.

MOODY, WILLIAM F. Pvt., Co. F. b S. C. 12/44. Enl. Co. E, 1st S. C. State Troops Pickens 8/1/63. Served through 12/31/63. Enl. Co. F, 1st S. C. Cav. Columbia 3/7/64. Ab. on leave 4/20-30/64. Present

5-6/64. Present sick in camp 7-8/64. Present 9-12/64. NFR. Age 24, no occupation, Keowee, Oconee Co. 1870 census. Laborer, age 34, Burke Co., Ga. 1880 census. Farmer, age 55, Wageoner, Oconee Co. 1900 census. Retired, age 65, Wageoner, Oconee Co. 1910 census.

MOORE, EDWARD W. 2ndLt., Co. A. b. S. C. 1840. Enl. Abbeville CH 8/13/61 as 4thSgt., age 19. Present through 4/62. Ab. on leave 4-6/62. Present 9-10/62. On picket 11-12/62, promoted 3rdSgt. Present 3-8/63. Ab. on horse detail 9-10/63. Present 11-12/63, promoted 1stSgt. Present 1-2/64. Ab. on leave 4/18-30/64. Present 5-10/64, elected 2ndLt. 8/30/64. Present 11-12/64 & 4/10/65. NFR. Age 29, at home, Abbeville, S. C. 1870 census.

MOORE, THOMAS C. Pvt., Co. I. b. S. C. circa 1842. Enl. Parker's Ferry 4/3/62. On picket Jacksonborough through 4/30/62. Present 5/62-4/63. On guard at Signal Station 8/31/63. Transferred Co. C, 5th S. C. Cav. in exchange for John C. Ackerman 10/2/63. Present 1-2/64. Present as Ambulance Driver 3/1-10/31/64. Deserted to the enemy Charleston 3/12/65. Age 22, 5'2", dark complexion, grey eyes, light hair. Took oath and discharged 3/15/65. NFR.

MOORE, W. B. Pvt., Co. H. On postwar roster but not on muster rolls.

MOORE, W. H. Pvt., Co. K. d. of disease Chester on post war roster.

MOORE, WILLIAM A. Pvt., Co. D. Enl. Lancaster 6/29/64. Detailed with Brigade Commissary 9/10-10/30/64. Present 11-12/64. NFR.

MOORE, WILLIAM AUGUSTUS. Pvt., Co. D. b. S. C. 11/18/21. Served in 4th S. C. State Troops. Enl. Co. D, 1st S. C. Cav. Lancaster 6/29/64. Present until detailed in Brigade Commissary Dept. 9/10-12/30/64. Present 11-12/64. NFR. d. 8/10/78. Bur. Old Presb. Ch., Lancaster Co.

MOORE, WILLIAM C. Pvt., Co. A. b. S. C. 8/11/33. Enl. Abbeville CH 8/13/61. Present as Company Commissary & on duty Wilson's Point 11-12/61. Ab. on leave 3-4/62. Ab. detached 5-6/62. Present 9-10/62. Ab. sick 11-12/62. Ab. detached 3-4/63. Present 5-6/63. Ab. on horse detail 8/18-31/63. Ab. detailed in Brigade QM Dept. 9-10/63. Ab. on horse detail 12/23/63-2/64. Ab detailed in horse infirmary camp 4/18-6/64. Present 9-10/64. Detached Pocotaligo 11-12/64. NFR. At home, Abbeville Co. 1870 census. d. 10/1/00. Bur. Upper Cane Cem., Abbeville Co.

MOORE, WILLIAM H. Pvt., Co. K. b. S. C. 1845. Student, age 14, Abbeville Dist. 1860 census. Enl. Columbia 7/12/64. Present through 10/64. On picket Chatman's Ford 12/29-31/64. Present through 4/65 on wife's pension application. NFR. Farmer, age 54, Walnut Grove, Greenwood Co. 1900 census. Farmer, age 64, Walnut Grove, Greenwood Co. 1910 census. d. 7/16/13. Wife receiving pension Chester, S. C. 11/5/19.

MOORE, WILLIAM R. Pvt., Co. H. b. S. C. circa 1838. Farmer, age 22, York Dist. 1860 census. Enl. Rock Hill 1/18/62 age 26. Present as Bugler 3-4/62. Present 5/62-4/63. Presence or absence not stated 5-6/63, however, transferred to Regimental Band 6/1/63. Ab. on horse detail 8/18-31/63. Present 9-12/64. Ab. sick with "Febris Intermitten" in Charlotte, N. C. hospital 4/23/65. Transferred to another hospital 4/27/65. NFR. Farmer, age 35, Choctaw, Miss. 1870 census. Farmer, age 45, Spring Creek, Phillips Co., Ark. 1880 census. d. 6/17/12. Bur. Hopewell Bapt. Ch., Cherokee Co., S. C.

MORDECAI, J. Pvt., Co. A. Enl. 11/7/61. Discharged for disability 11/27/61. Reenlisted Capt. Kirk's Co. S. C. Partisan Rangers 11/7/61. Discharged 11/27/61. NFR.

MORGAN, GEORGE R. A. Pvt., Co. C. b. S. C. circa 1844. Res. Barnwell Dist. Enl. Wadmalaw Island 3/28/62. Present through 8/62. Ab. sick with typhoid fever in Petersburg & Williamsburg hospitals 10/5-31/62. Furloughed 11/18/62. Horse killed Richard's Ferry 12/30/62. Paid $285.00. Ab. sick at home 3-4/63. Present through 12/63. Ab. on detached service 1/3-2/29/64. Ab. on leave 4/18-5/2/64. Present through 10/64. Detached Pocotaligo 12/10-31/64. NFR. KIA Goldsboro, N. C. 4/15/65 age 21.

MORRAH, DAVID. Pvt., Co. A. b. S. C. circa 1845. Res. Abbeville Dist. 1860 census. Enl. Columbia 2/9/64. AWOL 4/64. Present 5-6/64. Detailed 10/1-31/64. Paroled Augusta, Ga. 5/29/65. Living Calhoun Mills, Abbeville Co. 1870 census. d. 9/6/90. Bur. Lower Long Cane Cem., McCormick Co., S. C.

MORRIS, JOHN W. Pvt., Co.'s D & K. b. S. C. 4/5/33. Res. Darlington Dist. 1850 census. Enl. Co. D, Chester CH 9/10/61 age 29. Ab. on sick leave 10/31/61. Present 11/61-4/62. Transferred Co. K 6/25/62. Present through 6/63. Ab. sick with chronic diarrhea in Richmond hospital 7/29/63. Furloughed for 30 days 9/30/63. Ab. on sick leave from hospital 10/63. Ab. sick Chester, S. C. 11/14-12/31/63. Present 1-2/64. Ab. on horse detail 4/4-30/64. Present 5-6/64. On duty James Island Entrenchments 7-8/64. Ab. on leave for 12 days 10/22-31/64. On picket Chatman's Ford 12/21-31/64. WIA (lost 1 finger on left hand & use of right hand) Combarhee River 2/5/65. Paroled Raleigh, N. C. 4/65. Moved to Ga. 1868. d. 3/7/14. Bur. Union Campground Cem., Carroll Co., Ga.

MORRIS, P. W. Pvt., Co. A. Enl. age 19 on postwar roster. Living Abbeville Co. 1902.

MORROW, L. D. Pvt., Co. F. Served in Co. G, 3rd S. C. Reserves 11/13-12/31/62. Enl. Co. F, 1st S. C. Cav. Columbia 3/22/64. On duty Green Pond 4/25-30/64. Present 5-8/64. Ab. sick 10/24-12/31/64. NFR.

MORTON, JAMES MADISON. Pvt., No Co. b. S. C. 3/28/50. Served as Scout. Paroled Savannah, Ga. 5/10/65. Took oath and released 5/20/65. Married 1866 age 16. Farmer, post war. Cotton Mill Worker, age 61, Spartanburg 1910 census. d. 11/22/18. Bur. Holston Ch. Cem., Spartanburg Co.

MOSELEY, ARTHUR WILLIAM. Pvt., Co.'s C & K. b. S. C. 1828. Res. Barnwell Dist. Enl. Co. C Camp Butler 9/10/61 age 33. Ab. sick in hospital 11-12/61. Presence or absence not stated 1-5/62. Transferred Co. K. 6/27/62. Present through 10/62. On picket 11-12/62. Present 3-6/63. Ab. on horse detail 8/15-31/63. Present 9/63-4/64. Ab. on leave 6/27-30/64. Present 7-12/64. Captured Goldsboro, N. C. 3/31/65. Sent to Hart's Island, N. Y. Released 6/17/65. 5'8", dark complexion, dark hair, blue eyes. d. 1898. Bur. Levels Bapt. Ch. Cem., Aiken Co., no dates.

MOSELEY, JOSEPH. Pvt., Co. C. b. S. C. circa 1830. Farmer, age 30, Barnwell Dist. 1860 census. Enl. Camp Butler 9/10/61 age 30. Ab. on sick leave 12/21-31/61. Presence or absence not stated 1-4/62. Present 5/62-6/63. Ab. on horse detail 8/15-31/63. Present 9/63-10/64. Detached Pocotaligo 12/10-31/64. Captured Cheraw, S. C. 3/3/65. Sent to Hart's Island, N. Y. Released 6/17/65. 5' 11 1/2", light complexion, light hair, blue eyes.

MOSELEY, THOMAS. Pvt., Co. A. b. Troy, McCormick Dist. 12/26/38. Farmer, age 20, Abbeville Dist. 1860 census. Enl. Abbeville CH 12/23/61. On duty Wilson's Point 3-4/62. Present 5/62-12/64. Paroled Greensboro, N. C. 4/28/65. Receiving pension, Troy, McCormick Co. 9/25/19. Farmer, age 81, East Greenwood, McCormick Co. 1920 census.

MULLENHAUER, E. AUGUSTUS. Pvt., Co. C. b. Germany circa 1839. Age 21, Hamburg, Edgefield Dist. 1860 census. Enl. Co. H, 1st S. C. Inf. (6 months) 1/7/61. Discharged 7/9/61. Enl. Co. C, 1st S. C. Cav. Camp Butler 9/10/61 age 21. Detached Wadmalaw Island 11-12/61. Presence or absence not stated 3-4/62. Present sick in camp 5-6/62. Ab. on sick leave 7/12-8/31/62. On detail with Col. Twiggs horses 9-10/62. Present 11/62-4/63. Ab. on detached duty 6/30-10/31/63. On detached service at horse recruiting station 11/25-12/31/63. Present 1-2/64. Ab. on horse detail 4/18-5/2/64. Present through 12/64. Paroled as Courier, Greensboro, N. C. 5/1/65. Policeman, age 41, Charleston 1880 census.

MULLINS, P. S. Pvt., Co. K. Enl. Columbia 4/11/64. Present through 4/30/64. Transferred to Co. B, Lucas' Bn. S. C. Arty. 5/3/64. Present through 12/64. Paroled Raleigh, N. C. 4/65.

MURPHY, DAVID S. Pvt., Co. A. b. Abbeville Dist. circa 1844. Farmhand, age 16, Abbeville Dist. 1860 census. Enl. Abbeville CH 3/19/62. Present through 6/62. Ab. sick in Summerville hospital 7-8/62. Present sick in camp 9-10/62. d. of disease Culpepper CH 12/22/62. Effects: $4l.60.

MURPHY, J. PARSONS. Pvt., Co. E. b. S. C. 12/46. Enl. Columbia 4/15/64. Detached Green Pond 4/16-30/64. Present 5-8/64. On picket 9-10/64. Present 11-12/64. Paroled Hillsboro, N. C. 4/65. Post Master, age 53, Bamberg, S. C. 1900 census & age 62, 1910 census.

MURPHY, JOHN J. B. Pvt., Co. E. b. Branchville, Orangeburg Dist. 3/22/39. Res. Orangeburg Dist. Enl. Branchville 10/21/61 age 20. Present through 12/31/61. Ab. on leave 1/62. On picket at Railroad 3-4/62. died of disease in Adams Run hospital 6/27/62 age 21.

MURRAH, JAMES WILLIAM. 3rdCpl., Co. C. b. S. C. 7/37. Enl. Hamburg 8/27/61 age 24, as Pvt. Detached Wadmalaw Island 11-12/61. Presence or absence not stated 3-4/62. Present 5-6/62. Ab. on sick leave from Richmond hospital 8/6-31/62. Ab. sick in Manchester hospital 9-10/62. On scout 11-12/62. Present 3-4/63, promoted 3rdCpl. Present 5-6/63. Ab. on horse detail 8/15-31/63. Present 9-10/63. Ab. sick with "Ulcurs" in Charlottesville hospital 12/2/63. Ab. sick through 2/64. Detached Green Pond 4/16-30/64. Present 5-6/64. Ab. on leave for 30 days 8/18-31/64. Present 9-10/64. Detached Pocotaligo 12/10-31/64. Paroled Greensboro, N. C. 5/1/65. Farmer, age 62, Hammond, Aiken Co. 1900 census. Retired Farmer, age 72, Gregg, Aiken Co. 1910 census. Applied to enter Old Soldiers' Home, Columbia from Aiken Co. 10/21/16. d. 7/30/19. Bur. Aiken Co.

MURRAY, DAVID N. Pvt., Co. H. b. Va. 1830. Shoemaker, age 30, Blackstock, Chester Co. 1860 census. Enl. Rock Hill 1/22/62 age 31. Present 3-4/62. Ab. driving Ambulance 5-6/62. Ab. sick in Richmond hospital 7-10/62. Present 11/62-8/31/63. Ab. on horse detail 9-10/63. AWOL 10/28-12/31/63. Present 1-2/64.Transferred Co. B, Lucas' Bn., S. C. Arty. 5/13/64. Present through 8/64. Ab. sick 9/8-10/31/64. Present 11-12/64. In Raleigh, N. C. hospital 4/6/65. NFR.

MURRAY, WILLIAM N. Pvt., Co. A. Enl. Abbeville CH 8/2/61, age 21. Not on rolls 10/11/61. on postwar roster. Living Abbeville Co. 1902.

MURRELL, JOHN A. Pvt., Co. A. b. S. C. circa 1834. Enl. Abbeville CH 8/13/61 age 27. On duty Bear Island 11-12/61. Present through 6/62. Ab. sick in Summerville hospital 7-8/62. Present 9-10/62. On scout 11-12/62. Ab. on leave 1-2/63. Present 3-6/63. Ab. detached with Regimental QM 9/63-2/64. Present 3-6/64. Ab. guarding POW's Florence 9/17-12/31/64. Present 4/65. NFR.

MYERS, ALEXANDER. Pvt., Co. E. b. S. C. 9/44. Enl. Columbia 3/29/64. Detached Green Pond 4/16-30/64. Ab. on leave 6/29-30/64. Present 7-10/64. Detached Pocotaligo 12/10-31/64. Paroled Hillsboro, N. C. 4/65. Farmer, age 36, Midway, Barnwell Co. 1880 census. Farmer, age 56, Edisto, Orangeburg Co. 1900 census. Receiving pension Bamberg Co. 1912 age 66.

MYERS, DAVID. D. Pvt., Co. E. b. S. C. circa 1826. Laborer, age 34, Colleton Dist. 1860 census. Enl. Adams Run 2/25/62 age 38. Present 3-4/62. Ab. Courier, Parker's Battery, Simons Bluff 5-6/62. On picket Willtown 7-8/62. Present 9-12/62. Ab. sick with debility in Charlottesville hospital 2/19/63. Transferred to Lynchburg hospital 5/10/63. Present 6-12/63. Ab. on horse detail 2/12-3/3/64. Ab. on leave 4/27-5/12/64. Present 6-8/64. On picket 9-10/64. Detached Pocotaligo 12/10-31/64. Deserted to the enemy Charleston 3/22/65. Age 38, 5'6", light complexion, blue eyes, dark hair. Took oath and released 3/31/65. d. Fairview, N. J. 8/31/04.

MYERS [MIRES], HENRY N. Pvt., Co. A. Enl. Abbeville CH 8/2/61, age 19. Not on roster 10/11/61. NFR.

NAPIER, BENJAMIN C. Pvt., Co. G. b. Abbeville Dist. 10/11/29. Farmer, age 22, Edgefield Dist. 1850 census. Enl. Abbeville CH 3/15/62. On picket King's House through 4/62. Present 5-6/62. Ab. on leave 8/30/62. Present 9-10/62. On scout 11-12/62. On detached service 3-4/63. Present through 8/31/63. Ab. on detached service 9-10/63. AWOL 12/30-31/63. Detailed horse recruiting station 1/13-2/29/64. Detached Green Pond 4/27-30/64. Present 5-8/64. Detached Florence guarding POW's 10/10-12/31/64. Paroled Greensboro, N. C. 5/1/65. d. Abbeville Dist. 3/17/72. Bur. Lower Cane Cem., McCormick Co.

NAPIER, HENRY AUGUSTUS. Pvt., Co. G. b. S. C. circa 1845. Age 5, Edgefield Dist. 1850 census. Not on muster rolls. Paroled Greensboro, N. C. 5/1/65. Married 1/11/66. Collar Maker, age 24, Gregg, Edgefield Co. 1870 census. Farmer, age 33, Gregg, Edgefield Co. 1880 census. d. 3/18/01. Wife receiving pension McCormick Co. 5/20/19.

NAPIER, JOHN F. Pvt., Co. G. b. S. C. 1844. Age 7, Edgefield Dist. 1850 census. Enl. Stevensburg, Va. 9/1/63. Present through 10/31/63. Under arrest & in guard house 11/25-12/31/63. Present 1-2/64. Transferred Co.B, Lucas' Bn. S. C. Arty. 5/13/64. Joined 5/24/64. Present through 12/64, promoted Cpl. 12/1/64. Captured Enfield, N. C. 4/13/65. NFR. Cotton Mill Worker, age 26, Gregg, Edgefield Co. 1870 census. d. 7/4/96. Bur. family cem., Granville, S. C.

NAPIER, NAPOLEON L. Pvt., Co. G. b. S. C. circa 1841. Res. Long Cane, Abbeville Dist. 1860 census. Enl. Abbeville CH 1/8/62, age 21. On picket Jacksonborough 3-4/62. Ab. sick in Columbia hospital 6/62. Present 7-12/62. Ab. detached 3-4/63. Present 5-6/63. Lost saddle and fined $25.00. Ab. on horse detail 8/31/63. Present 9/63-2/64. Transferred Co. B, Lucas' Bn. S. C. Arty. 5/13/64. Joined 6/1/64. Present through 12/64. KIA Averasboro, N. C. 3/16/65.

NEAL, CHARLES. Pvt., Co. F. b. S. C. circa 1832. Blacksmith, age 28, Pickens Dist. 1860 census. Enl. Pickens CH 12/4/61 age 26. Present through 12/31/61. Ab. sick in Brigade hospital 3-4/62. Discharged for disabilty 6/10/62. NFR.

NEAL, D. L. Pvt., Co. B. Served in Co. D, 9th S. C. Reserves 12/28/62-2/14/63. Enl. Co. B, 1st S. C. Cav. Columbia 5/28/64. Ab. on leave 6/30/64. Present 7-8/64. Ab. sick in Summerville hospital 9/20-10/31/64. Ab. on sick leave 11-12/64. NFR. d. by 1898.

NEEL, WILLIAM G. Pvt., Co. A. Enl. Columbia 2/4/64. Ab. on leave 6/21-30/64. Present 9-12/64. Paroled as Teamster, Greensboro, N. C. 5/1/65. d. Abbeville Co. 8/13/90.

NEELEY, THOMAS J. Pvt., Co. H. b. S. C. 1/24/44. Enl. Rock Hill 5/7/62. Present through 6/62. Ab. on leave 7-8/62. Present 9-10/62. On picket 11-12/62. Present 3/63-2/64. Detached Green Pond 4/19-30/64. Present 5-10/64. Detached Pocotaligo 12/10-31/64. NFR. d. 1/30/98. Bur. Gant-Westview-Libert Cem., Pickens Co. Wife receiving pension, Odgen, York Co. 1908.

NEIL, JOHN W. Pvt., Co. G. b. S. C. circa 1839. Enl. Abbeville CH 1/8/62. Ab. attending sick in hospital 3-4/62. Present 5-6/62. On picket Grim Falls 7-8/62. Present 9/62-8/31/63. Ab. sick in hospital 9-11/63. Ab. on leave from hospital 12/12-31/63, for 40 days. Present 3-12/64. Paroled Greensboro, N. C. 5/1/65. Living Marin Co., Fla. 1945. d. Anthony, Marin Co., Fla.

NEIL, W. G. Pvt., Co. G. b. S. C. circa 1838. Enl. Abbeville CH 1/8/62. Present 3-4/62. Present sick in camp 5-6/62. Ab. sick Charleston hospital 7/1-30/62. Hired James Carroll as substitute and discharged 9/10/62. NFR.

NESBITT, JOHN JAMES. Pvt., Co. B. b. S. C. 12/8/43. Enl. Clinton 8/22/61 age 21. Ab. sick 10/31/61. Present 11/61-4/62. Horse killed in skirmish at Hallover Bridge 4/20/62. Present 5-12/62. Ab. sick in hospital 3-6/63. Present 7-8/63. Ab. on horse detail 9-10/63, horse died 10/15/63. Present 11-12/63, however, ab. on horse detail 12/6-31/63. Detached horse recruiting camp 2/4-29/64. Present 3-10/64. Ab. sick McLeod House 12/11/31/64. NFR. d. (stone illegible). Bur. Nanareth Presb. Ch. Cem., Spartanburg Co.

NESBITT, NILES G. Major. b. Spartanburg Dist. 9/28/28. Farmer, age 31, Spartanburg Dist. 1860 census. Enl. Co. B Clinton 8/22/62 as Captain. Present through 12/61. Commanding Regiment 2/15/62. Commanding Company 3-4/62. Horse killed Hallover Bridge 3/20/62. Paid $175.00. 16 horses present in Company 4/6/62. Ab. sick 5-6/62. Ab. on sick leave 9-10/62. On picket Ellis Ford 11-12/62. 45 horses present 2/28/63. Present 3-4/63. 67 horses present 5/7/63. Signed for 3 tent flys 5/25/63. Present Brandy Station & Gettysburg. 28 horses present 7/24/63. Signed for 6 caps, 21 jackets, 14 pr. pants & 34 shirts 8/24/63. and 10 jackets, 2 caps & 53 shirts 8/26/63. Ab. on leave 8/31/63. Present 9-10/63. Signed for 3 pr. boots, 16 pr. socks, 3 blankets, 39 pr. drawers, 13 pr. pants, 4 shirts & 1 over coat 9/8/63. 32 horses present 9/30/63. Ab. on leave 2/24/64. Signed for 18 pr. shoes, 4 blankets, 3 tent flies, 14 pr. pants, 4 shirts, 1 pr. socks, 4 jackets & 2 pr. drawers 3/31/64. 21 horses present 3/31/64. Signed for 1 shovel, 9 Army tents, 3 camp kettles, 9 mess pans, 45 caps, 66 pr. pants, 62 jackets, 3 pr. socks, 79 cotton shirts & 75 pr. drawers 6/30/64. 85 horses present 8/1/64. AWOL 8/3-25/64. 86 horses present 9/1/64. Promoted Major 9/15/64. Signed for 2 jackets, 5 pr. pants, 5 shirts, 6 pr. drawers, 9 pr. shoes, 6 pr. socks & 1 wall tent, James Island 9/30/64. 74 horses present 10/1/64. Present 11-12/64. NFR. Farmer, Woodruff, Spartanburg Co. 1870 census. and Greenville, S. C. 1880 census. d. Woodruff, Spartanburg Co. 2/1/92. Bur. Bethel Bapt. Ch. Cem., Woodruff, S. C.

NETTLES, WILLIAM ALLISTER. Pvt., Co. I. b. S. C. 6/20/42. Student, age 18, Sumter Dist. 1860 census. Enl. Columbia 2/3/64. Present through 8/64. Orderly, 9-12/64. Paroled Greensboro, N. C. 5/1/65. Farmer, age 27, Privateer, Sumter Co. 1870 census. Farmer, age 54, Haywood, Colleton Co. 1900 census. d. 11/8/03. Bur. Wedgewood Presb. Ch., Sumter Co., S. C.

NEVITT, JOSEPH K. Pvt., Co. K. Res. Fairfield Dist. Enl. Columbia 5/10/64. Present through 8/64. Detailed driving cattle for QM Dept. 9-12/64. Paroled Raleigh, N. C. 4/65.

NEWTON, LARKIN L. Pvt., Co. F. b. S. C. 1846. Enl. Co. B, 2nd Bn. S. C. Cav. 1/19/62 as 2ndLt. 12/26/61. Not reelected 5/62. Enl. Co. A, 1st Regiment S. C. Troops (6 months) 8/1/63. Ab. on detached duty through 1/64. Enl. Co. F, 1st S. C. Cav. Columbia 2/6/64. Ab. detached Green Pond 4/25-30/64. Present 5-6/64. Ab. sick in hospital 8/22-10/15/64. Present 11-12/64, dismounted since 10/15/64. NFR.

NEYLE, HENRY MANLY. Pvt., Co. A. b. S. C. 1840. On post war roster. Reenlisted Co. K, 4th S. C. Cav. Grahamville 3/25/62. Present through 2/64. In Richmond hospital 7/6-8/9/64. Ab. on horse detail 8/11/64 for 40 days. Ab. sick at home 9/23-10/31/64. Paroled 4/65. d. 10/21/98. Bur. Live Oak Cem., Walterboro, S. C.

NICHOLSON, EVAN VIRGIL. Pvt., Co. F. b. Oconne Dist. 10/1/44. Enl. Pickens CH 12/4/61. Present through 12/31/61. On picket Jacksonborough 3-4/62. Present 5-6/62. Ab. sick 9-10/62. Present 11/62-2/64. Detached Green Pond 4/16-30/64. Present 5-6/64. Ab. sick in hospital 8/25-31/64. Ab. detached guarding POW's Florence 9/17-12/31/64. NFR. Farmer, White Water, Oconee Co. 1870 census. d. 11/19/02. Bur. Bethlehelm Bapt. Ch. Cem., Mountain Rest, S. C.

NICHOLSON, MILTON. Pvt., Co. F. b. S. C. 2/23/30. Enl. Pickens CH 12/4/61 age 30. Present through 4/42. On picket Camp Means 5-6/62. Present 7/624/63. Ab. detached recruiting horses 5-12/63. Present 1-2/64. On detail Pickens CH 4/30/64. Present 5-12/64. NFR. d. Oconee Co. 5/25/16. Bur. Whitmire Meth. Ch. Cem., Oconee Co.

NIMMO, N. Pvt., Co. C. Enl. Aiken Dist. age 25 on post war roster.

NIVENS, WILLIAM HARVEY. Pvt., Co. H. b. S. C. 1828. Enl. Rock Hill 1/28/62 age 31. On picket Rasatole Bridge 3-4/62. Present 5-8/62. Detailed to drive wagon from Summerville, S. C. to Staunton, Va. 9-10/62. Present 11/62-6/63. Horse killed Funktown, Md. 7/8/63. Paid $300.00. Ab. on horse detail 8/18-31/63. AWOL 9-10/63. Present 11/63-6/64. AWOL 8/30-31/64. Ab. detached in QM Dept. 9-10/64. Present 11-12/64. Paroled Charlotte, N. C. 5/20/645. d. 4/10/02, age 74. Bur. Adnah Meth. Ch. Cem., York Co. Wife receiving pension Tirazah, York Co. 1903.

NOBLE, EZEKIEL PICKENS. Pvt., Co. G. b. S. C. 1840. Student, Age 20, Willington, Abbeville Dist. 1860 census. Enl. Co. C., 7th S. C. Inf. 4/15/61. Present through 10/61. Ab. sick with fever "Interm. qust." 3/9-4/62. Discharged 4/12/62. Enl. Co. G, 1st S. C. Cav. Camp Clinton 6/1/62 age 22. On picket 6/30/62. On picket Gimball's 7-8/62. Present, sick in camp 7-12/62. Ab. sick with debility in Richmond hospital 1/20-2/8/63. Present 3-4/63. Ab. wounded in left leg with "secondary hemmorage" in Richmond hospital 5-6/63 on muster rolls. However, not admitted to Richmond hospital until 8/11/63. Furloughed for 30 days 9/3/63. Ab. on sick leave through 10/63. Ab. in hospital 11/63-8/64. AWOL 9-12/64, since 12/31/63. NFR.

NOWELL, EDWARD WIGFALL. Pvt., Co. A. b. Charleston 1834. Planter, Charleston. On post war roster. Reenlisted Co. K, 4th S. C. Cav. Grahamville 3/25/62. Present until detailed in Signal Corps by Gen. Beauregard 10/9/62. Ab. sick leave 1/24-4/30/63. Present 5/63 until ab. sick in hospital 9-10/63. Present 11/63-2/64. Present until WIA (hand) Hawes Shop 5/28/64. Ab. wounded in Richmond hospital until furloughed for 30 days 6/15/64. Transferred to Invalid Corps 12/30/64. Assigned to Army of Northern Va. 3/2/65. Paroled Greensboro, N. C. 5/2/65. d. Charleston 3/22/79. Bur. St. Paul's Epis. Ch., Charleston.

NOWELL, LIONEL CHAMBERS. Pvt., Co. A. b. S. C. 2/14/36. Merchant, Charleston. Enl. 11/7/61. Reenlisted Co. K, 4th S. C. Cav. Grahamville 3/25/62 as 2ndLt. Present until promoted 1stLt. 12/12/62. Detached with Gen. Walker 1/5-6/30/63. Present, commanding Co. 7-12/63. Ab. on leave 2/26/64 for 20 days. Present until captured Cold Harbor 5/30/64. Sent to Pt. Lookout & Ft. Delaware. Released 6/12/65. 6' 1", dark complexion, dark hair, gray eyes. d. 4/3/96. Bur. Prince George Winijah Cem., Georgetown, S. C.

NORWOOD, WESLEY R. Pvt., Co. A. b. Abbeville Dist. 1839. Enl. Abbeville CH 8/13/61 age 21. Detailed as Courier, Gen. Evans 11-12/61. Present 3-12/62. Ab. sick with secondary "syphilis" in Richmond hospital 2/20-3/14/63. Ab. sick in Staunton hospital with "primary syphilis" 3-4/63. d. 5/1/63. Probably buried in Thornrose Cem., Staunton, Va.

O'BRIEN, TIMOTHY. Pvt., Co. A. b. Tipperary, Ireland 1832. Came to U. S. 1834. Enl. 11/7/61. Reenlisted Co. K, 4th S. C. Cav. Grahamville 3/25/62. WIA (thigh) Hawes Shop 5/28/64. WIA (shoulder) 10/27/64. RTD 1111/1/64. Transferred to Charlotte, N. C. Ordnance Works. Paroled Charlotte, N. C. 5/3/65. Butcher, Charleston 1870 & 1880 censuses. Became U. S. citizen 7/29/71. d. Charleston 4/9/88. Bur. St. Lawrence Cem.

O'HEAR, JAMES W. Pvt., Co. A. b. S. C. 1836. Bookkeeper, Charleston 1860 census. Enl. 11/7/61. Ab. sick Charleston 2/7/62. Reenlisted Co. K, 4th S. C. Cav. Grahamville 3/25/62, as 3rdLt. Present until elected 2ndLt. 12/16/62. Ab. on leave 1/27/63 for 10 days. Detailed with Gen. Walker 2/2763. Present until detached 10/25/63. Present 11/63-2/64. KIA Hawes Shop 5/28/64.

O'HEAR, LAWRENCE WITSELL. Pvt., Co. A. b. S. C. 12/24/44. Enl. 11/7/61. Exempt for underage 2/7/62. Enl. Co. B, S. C. Cadet Bn., L. D. T. Reenlisted Co. K, 4th S. C. Cav. 11/17/63. Ab. on horse detail 8/11/64. AWOL 9/21/64. Paroled Greensboro, N. C. 4/16/65. Bookkeeper, Charleston 1870 census. d. 6/17/77. Bur. Magnolia Cem.

O'NEAL, JOHN FERRIS. Pvt., Co. H. b. York Dist. 8/13/29. Farmer, age 31, Tailor's Creek, York Dist. 1860 census. Enl. Rock Hill 1/9/62 age 33. Ab. sick in hospital 3-4/62. Ab. on sick leave 6/25-10/31/62. Ab. sick in Columbia hospital 11-12/62. Discharged 1863. Reenlisted Columbia 6/28/64. Present sick in camp 8/12-31/64. Ab. detailed in hospital as nurse 9-10/64. Detached Pocotaligo 12/10/31/64. Paroled Charlotte, N. C. 5/3/65. d. 3/5/94. Bur. Laurelwood Cem., Rock Hill, S. C.

O' NEILLE, J. J. A. Pvt., Co. A. b. S. C. circa 1838. Enl. 11/7/61. Reenlisted Co. K, 4th S. C. Cav. Grahamville 3/25/62. 5'1", fair complexion, light hair, grey eyes. Present until WIA (leg fractured) Yamasee 10/22/62. Ab. wounded through 10/64. However, detailed in Ordnance Works, Charlotte, N. C. Paroled Charlotte 5/3/65. At. home, Charleston 1870 census. d. Charleston 12/4/75.

ORR, JOHN M. Pvt., Co. F. b. S. C. 1844. Enl. Pickens CH 10/14/62. Ab. on leave 10/31/62. Ab. sick in hospital 11-12/62. Ab. sick 1/18-10/63. Discharged for disabilty Camden, S. C. 10/29/63. NFR. Farmer, age 25, Garvin, Anderson Co. 1870 census and age 36, 1880 census.

OTTS, WILLIAM J. Pvt., Co. B. b. S. C. 2/7/45. Enl. Clarksville, Va. 5/8/63. Present through 11/63. Ab. recruiting camp for horses 12/18-2/64. Ab. on horse detail 4/28-30/64. Present 5-6/64. Ab. sick McLeod hospital 8/20-31/64. Ab. on sick leave 9/25-10/31/64. Present 11-12/64. NFR. Farmer, Forest, Spartanburg Co. 1880 census. d. 5/25/99. Bur. Nazareth Presb. Ch. Cem., Spartanburg Co.

OWEN, JOHN THOMAS. Pvt., Co. A. b. S. C. 12/9/35. Jeweler, age 24, Abbeville CH 1860 census. Enl. Co. D, 7th S. C. Inf. 4/15/61 age 25, ast 1stLt. Present through 6/30/61, but retired 6/21/61. Enl. Co. A, 1st S. C. Cav. Abbeville CH 3/14/62. Present through 4/62. Discharged 6/22/62. NFR. Jeweler, Cartersville, Ga. 1870 & 1880 censuses. d. 10/17/13. Bur. Oak Hill Cem., Cartersville, Ga.

OWENS, ALEXANDER. Pvt., Co. A. b. S. C. circa 1848. Enl. 11/8/61. Probably discharged for under age. Served in 1st S. C. Militia in Charleston. Reenlisted Co. F, 6th S. C. Cav. Charleston 6/20/62. Present until ab. sick in Charleston hospital 7/12/63-4/64. Present 7-8/64. Ab. with dismounted Bn. 9-10/64. Ab. sick with "Febris" in Richmond hospital12/24/64. Transferred to another hospital 12/26/64. NFR. Farmhand, age 22, Gerald, Ala. 1870 census.

OWENS, MOSES TAGGART. Captain, Co. A. b. S. C. 1825. Silversmith & Jeweler, Abbeville CH 1860 census. Enl. Abbeville CH 8/13/61, age 35. Present through 12/31/61. Present Wadmalaw Island 1/62. Present 3-6/62. Appointed Major at organization of Regiment but declined 6/62. Ab. on detached service.7-8/62. Present 9-12/62. 148 horses present Columbia, S. C. 10/2/62. Presence or absence not stated 1-6/63. 39 horses present 2/28/63. 74 horses present 5/8/63. Signed for 2 tent flies 5/25/63. WIA (heel, left foot) Funktown, Md. 7/8/63. In Richmond hospital 7/15/63. Furloughed for 30 days 7/20/63. DOW's (tetnas) Abbeville CH 8/6/63. Bur. Upper Long Cane Cem., Abbeville Dist.

OWENS, YOUNG J. PERRY. Pvt., Co. B. b. Laurens Dist. 11/20/28. Res. Charleston. Enl. Columbia 3/28/64. Present through 6/64. Ab. on leave 10/10/64. Deserted to the enemy Morris Island 10/12/64. Age 25, 5'9", fair complexion, blue eyes, dark hair. Took oath. Working in QM Dept., Hilton Head 10/31/64. NFR. d. Laurens Co. 9/5/77.

PACE, WILLIAM THOMAS. Pvt., Co. A. b. S. C. 9/4/37. Enl. Abbeville CH 8/2/61. Not on muster rolls 10/11/61. NFR. Farmer, Red River, Tillman Co., Okla. 1910 census. d. Tillman, Okla. 10/24/18.

PACK, THOMAS ALEXANDER. Pvt., Co. I. b. Paxville, Clarendon Dist. 6/27/36. Enl. Co. K, 23rd S. C. Inf. 11/15/61. Discharged. Enl. Co. I, 1st S. C. Cav. Columbia 4/13/64. Present through 4/30/64. Detailed Hqtrs., East Lines, James Island 6/8-30/64. Present 7-12/64. NFR. Store Clerk, Sumter, S. C. 1870 census. Bookkeeper, Greenville, S. C. 1896. Store Clerk, Atlanta, Ga. 1900 census. d. 1/5/01.

PAGE, B. T. Pvt., Co. B. b. S. C. 1845. On roster but not on muster rolls. Farmer, age 25, Hammond, Edgefield Co. 1870 census.

PAGE, WILLIAM B. Pvt., Co. B. b. S. C. 10/44. Enl. Co. D, 16th S. C. Inf. 12/12/61. Discharged for underage 7/3/62. Enl. Co. B, 1st S. C. Cav. 10/7/62. Present through 12/62. Detailed as Teamster, Brigade Hqtrs. 3-12/63. Present 1-2/64. Detached Green Pond 4/15-30/64. Present 5-8/64. Ab. sick in McLeod House hospital 10/15-31/64. Present 11-12/64. Paroled Greensboro, N. C. 5/3/65. Farmer, Woodruff, Spartanburg Co. 1880 & 1900 censuses. d. 9/30/10. Wife receiving pension Woodruff, Spartanburg Co. 9/16/19.

PAGETT, WILLIAM HENRY. 3rdLt., Co.'s C. & K. b. S. C. 11/4/43. Enl. Co. C, Adams Run 12/15/61 as Pvt. Present through 12/3/61. Presence or absence not stated 1-4/62. Transferred Co. K 6/26/62. Present through 9/1/62. Ab. sick 9-10/62. On scout 11-12/62. Present 3-4/63. Ab. recruiting horses 8/10/63-2/64. Present 3-4/64, promoted 4thCpl. Present 5-6/64. On duty Bridge of Ashley River 7-8/64. On duty Telegraph Office, Charleston 9-10/64. Detailed Dist. Hdqtrs., Green Pond 12/22-31/64, promoted 3rdLt. Paroled Greensboro, N. C. 5/26/65. Farmer, Fairfield Co. 1870 & 1900 censuses. d. Richland, S. C. 1/25/15. Bur. Mill Creek Ch. Cem.

PALFREY, ALFRED CONRAD. Pvt., Co. A. b. S. C. 3/20/39. Gd. Yale 1860. Accountant, Charleston. On post war roster. Reenlisted Co. K, 4th S. C. Cav. Grahamville 3/25/62. Present until detailed as Clerk for Gen. Pemberton & Gen. Beauregard 7/16/62-10/63. Present 11/63-2/64. Ab. in hospitals 8/5/64. Kept in Columbia, S. C. by Gov. Bonham 9-10/64. Retired for ill health. Paroled 4/65. d. 6/79. Bur. St. Michaels

Cem., St. Martinsville, La.

PALMER, JAMES J. Pvt., Co. A. b. Union Dist. circa 1841. Student, age 20, South Santee Ferry, Charleston 1860 census. Enl. Co. I, Palmetto Sharpshooters Charleston 5/9/61 age 21. NFR. Enl. Co. A, 1st S. C. Cav. Columbia 2/1/64. Present through 6/64. WIA (slightly) and ab. wounded at home 7/2-12/31/64. NFR. Receiving pension Abbevlle Co. 1888.

PANKEY, JOHN BAGBY. Pvt., Co. H. b. Appomattox Co., Va. 10/4/18. Teacher, Rossville, Chester Dist. 1860 census. Enl. Rock Hill 1/9/62 age 43. Ab. sick in hospital 3-4/62. Horse killed James Island 6/8/62. Ab. guarding baggage Church Flats through 6/30/62. Present dismounted 7-10/62. Present 11-12/62. Discharged for overage 3/1/63. Paid 3/24/63. NFR. Teacher, Marshall Co., Miss. 1870 census & Union Co., Miss. 1880 census. d. 1895 age 77.

PARDUE, JAMES H. (1) Pvt., Co. C. b. S. C. circa 1843. Enl. Camp Butler 9/10/61 age 18. Ab. on sick leave 12/4-31/61 & 1/62. Presence or absence not stated 3-4/62. Ab. sick in Church Flats hospital 5-6/62. Discharged for disability 7/1/62. Enl. Co. A, 22nd S. C. Inf. and KIA 8/23/62.

PARDUE, JAMES H. (2) Pvt., Co. D. b. S. C. circa 1843. Farmhand, age 17, Edgefield Dist. 1860 census. Enl. Chester CH 5/12/62. Present through 4/63. Present sick in camp 8/31/63. Ab. on horse detail 10/23-31/63. Present 11/63-4/64. Transferred Co. B, Lucas's Bn. S. C. Arty. 5/13/64. Present through 12/64. KIA Averasboro, N. C. 3/16/65.

PARKHAM, FRANK. Pvt., Co. B. b. S. C. circa 1840. Res. Spartanburg Dist. Enl. Clinton 8/22/61 age 21. Ab. sick 10/31/61. Present 11/61-4/62. Ab. on sick leave 5-6/62. d. of disease in camp, Adams Run 9/1/62 age 25.

PARKS, ROBERT D. Pvt., Co. K. b. S. C. 5/9/46. Enl. Columbia 8/2/64. Ab. sick in Summerville hospital 10/4-31/64. Present 11-12/64. Ab. sick with "Febris Int. Tertian" in Charlotte hospital 2/13-25/65. Paroled Raleigh, N. C. 4/65. Farmer, age 23, Bullocks Creek, York Co. 1870 census and age 33, 1880 census. d. 12/24/17. Res. Winston, Co., Miss. Bur. Beauvoir Old Soldiers' Home Cem., Biloxi, Miss.

PARLER, SHADRAH B. Pvt., Co. B. b. S. C. 5/11/12. Farmer, age 48, Poplar, Orangeburg Dist. 1860 census. Enl. Ft. Motte 10/26/61 age 49. Present through 12/31/61. Acting Commissary Sgt. for Company 1/62. Discharged for disability 4/4/62. NFR. Farmer, Poplar, Orangeburg Co. 1870 & 1880 censuses. d. Elloree, Orangeburg Co. 3/15/87.

PARRISH, JAMES ADKINS. Pvt., Co. H. b. S. C. 5/28/28. Farmer, age 30, Fur Coats Tavern, York Dist. 1860 census. Enl. Rock Hill 1/9/62 age 32. Present through 6/62. Present sick in camp 7-8/62. Present 9-10/62. Driving wagon for QM 11-12/62. Present 3-6/63. Ab. on horse detail 8/18-31/63. Present 9/63-8/64. Driving wagon for QM 9-12/64. NFR. Farmer, Yorkville 1870 census & Catawba, York Co. 1880 censuses. d. 3/10/00. Bur. Harmony Bapt. Ch. Cem., Chester Co. Wife receiving pension Rock Hill 1903.

PARRISH, JAMES M. Pvt., Co. H. b. York Dist. 1849. On postwar roster. Served in Capt. Senn's Co., Post Guard. NFR.

PARRISH, JOHN B. Pvt., Co. H. b. S. C. 1832. Farmhand, age 26, Chester Dist. 1860 census. Enl. Rock Hill 1/20/61 age 30. Present through 4/62. Ab. on sick leave 6/14-8/31/62. Present 9-10/62. Ab. as guard 11-12/62. Ab. sick in hospital 3-8/31/63. Ab. sick with ulcers from vaccine 3/2/63. Transferred Farmville hospital 4/30/63. Returned to duty 5/21/63. Ab. sick with chronic diarrhea in Richmond hospital 9/3/63. Furloughed for 30 days 10/14/63. Ab. on sick leave 10/31/63. Detached horse recruiting station 12/15/63-2/64. Present 3-4/64. Transferred Co. C, 20th S. C. Inf. 5/13/64. d. Chester Dist. 1864 age 32.

PARSONS, F. C. Pvt., Co. F. b. S. C. 9/5/28. Enl. Columbia 6/13/64. AWOL through 7/31/64. Present 8-12/64. NFR. d. 1/1/05. Bur. Mill Creek Cem., Pickens Co.

PARTLOW, D. R. Pvt., Co. H. b. S. C. circa 1841. Enl. Co. E, 17th S. C. Inf. Hampton 12/13/62. Present until transferred Co. H, 1st S. C. Cav. 12/17/63 by order of Sec. of War. in exchange for another man. Not yet reported Rock Hill 1-2/64, age 22. Detailed in QM Dept. Returned to duty 9/15/64. NFR. Constable, Ft. Mill, York Co. 1870 census.

PATTERSON, A. J. Pvt., Co. H. b. S. C. circa 1843. Enl. Rock Hill 3/19/62. Present through 6/62. Ab. on sick leave 7-8/62. Present 9/62-6/63. Ab. on horse detail 8/18-31/63. Present 9-10/63. Ab. sick in hospital 12/25-31/63. Present 1-8/64. Present sick in camp 9-10/64. Detached Pocotaligo 12/10/31/64. NFR. Age 27, at home, Seneca, Oconee Co. 1870 census. Receving pension Jackson & Fort Mill, S. C. 1905-13.

PATTERSON, ANGUS ALEXANDER. Pvt., Co. C. b. Barnwell, S. C. 1/17/42. Student, age 18, Barnwell Dist. 1860 census. Enl. Wadmalaw Island 4/15/62. Present through 10/62. Ab. on leave Culpeper CH 11-12/62. Ab. on detached service 3-4/63. Present 5-6/63. Ab. on horse detail 8/15-31/63. Present 9-10/63. Ab. on horse detail 12/22/63-4/64. Present 5-6/64. Ab. detailed a Wagoner for Lt. England 7-8/64. AWOL 10/4-31/64. Present 11-12/64. NFR. Farmer, Blakee, Colleton Co. 1880 census, Walterboro, Colleton Co. 1900 census. and Verdier, Colleton Co. 1910 census. d. Waterboro, S. C. 5/4/29. Bur. Live Oak Cem.

PATTERSON, ANGUS Q. Pvt., Co. I. b. S. C. 1836. Age 23, Barnwell Dist. 1860 census. Enl. Co. A, Hampton Legion Columbia 6/26/61 age 25. Present through 10/61. Ab. sick with convulsions from typhoid fever in Charlottesville hospital 12/20/61-2/13/62. Discharged. Enl. Co. I, Wadmalaw Island 4/15/62. Present through 12/64, except for remounts. NFR. Farmer, Three Mile, Barnwell Co. 1870 census & Fish Pond, Barnwell Co. 1880 census. d. Barnwell 3/19/11. Bur. Old Bethelehm Bapt. Ch. Cem., Barnwell, S. C.

PATTERSON, GEORGE S. Pvt., Co. A. b. S. C. 8/44. Res. Abbeville Dist. 1850 census. Served in 5th S. C. Reserves 1862-63 & 1st S. C. State Troops 1863-1864. Enl. Columbia 2/14/64. AWOL through 4/64. Present 5-12/64. NFR. Farmer, Pendleton, Anderson Co. 1900 census.

PATTERSON, WILLIAM F. Pvt., Co. H. b. S. C. 4/3/46. Res. Pickens Dist. 1860 census. Enl. Columbia 2/18/64. Present until detached Green Pond 4/16-30/64. Present 5-10/64. Detached Pocotaligo 12/10-31/64. NFR. Farmer, Ft. Mill, York Co. 1880-1910 censuses. Receiving pension 1907. d. 4/1/12. Bur. Unity Cem., Rock Hill.

PAUL, ANDREW. Pvt., Co.'s A & G. b. circa 1826.Enl. Co. A, Abbeville CH 3/19/62, age 36. Present through 4/62. Transferred Co. G 5/1/62 as Company Physician. Present sick in camp 7-8/62. Ab. detached in Surgeon's Dept. 11-12/62. Present 3-4/63. Present Gettysburg. In Richmond hospital with "Fistula in Ano" 7/18/63-3/28/64. Discharged Columbia, S. C. 4/25/64. Appointed Asst. Surgeon on postwar roster. d. by 1902 on Abbeville roster.

PEARCE, JOHN HAMPTON. Pvt., Co. K. b. S. C. 9/29/48. On post war roster. d. 7/31/04. Bur. Calvary Bapt. Ch., Dorchester Co.

PEARSON, CHARLES M. Pvt., Co. H. b. N. C. 7/14/31. Enl. Rock Hill 1/27/62 age 27. Present through 4/62. Ab. guarding baggage Church Flats 5-6/62. Present 7-8/62, horse died 7/13/62. Detailed to drive wagon from Summerville, S. C. to Staunton, Va. 9-10/62. Driving wagon for QM 11/62-4/63. Ab. sick in hospital 8/31/63. Present 9-10/63. Ab. sick with rheumatism in Richmond hospital 12/13-23/63. Ab. on horse detail 1-2/64. Ab. on leave 3-4/64. Transferred Co. B, Lucas's Bn. S. C. Arty. 5/13/64. Joined 5/20/64. Present through 8/64. Ab. detailed as Mechanic in Engineer Dept. 10/14-12/31/64. NFR. Farmer, age 38, Yorkville, York Co. 1870 census. Farmer, Hazelwood, Chester Co. 1880 census. d. 3/92. Bur. Mt. Holly Meth. Ch. Cem., York Co. Wife receiving pension Odgen, York Co. 1901.

PEARSON, WILLIAM PERRY. Pvt., Co. B. b. S. C. circa 1835. Enl. Co. B, 13th S. C. Inf. Spartanburg CH 8/31/61 age 26. Discharged for "scrufulous constitution" of neck & body. Enl. Co. B, 1st S. C. Cav.

Columbia 2/10/64. Detached Green Pond 4/26-30/64. Detached as Commissary 5-6/64. Ab. sick in Charleston hospital 7/25-8/31/64. Present 9-10/64. Detached Pocotaligo 12/7-31/64. NFR. Farmer, age 47, Grove, Greenville Co. 1880 census. Alive 1898.

PEAY, AUSTIN FORD. Pvt., Co.'s C & K. b. Fairfield Dist. 9/10/44. Age 14, Oktibbelia, Miss. 1860 census. Enl. Hamburg 8/27/61. On duty White Point 11-12/61. Presence or absence not stated 3-4/62. Transferred Co. K 6/27/62. Transferred back to Co. C 7/8/62. Ab. on sick leave 7/23-8/31/62. Present 9-12/62. On detached service 3-4/63. Present 5-6/63. Ab. on horse detail 8/15-31/63. Present 9/63-2/64. Detached Green Pond 4/16-30/64. Present 5-8/64. Ab. Courier for Col. Campbell 9-10/64. Present 11-12/64. Paroled Greensboro, N. C. 4/26/65. Planter, Winnsboro, Fairfield Co. 1870-1900 censuses. d. Kershaw, S. C. 1/10/10. Wife receiving pension Longtown, Fairfield Co. 10/2/19.

PEAY, THOMAS L. Pvt., Co. K. b. Fairfield Dist. 7/21/46. Student, age 13, Edgefield CH,1860 census. Enl. Columbia 4/13/64. Detached Green Pond 4/15-30/64. On picket Charleston fortifications 6/25-30/64. On duty at Bridge over Ashley River 7-8/64. On duty Military Telegraph Office, Charleston 9-10/64. Detached Dist. Hqtrs., Green Pond 12/24-31/64. Paroled Augusta, Ga. 5/20/65. Leaf Tobacco Dealer, Durham N. C. 1880 & 1900 censuses. d. Center, Chatham Co., N. C. 8/13/16.

PEDEN, DAVID MARTIN. Pvt., Co. B. b. Laurens Dist. 2/18/41. Res. Greenville Dist. 1860 census. Enl. Clinton 8/22/61 age 20. Ab. sick 10/31/61. Ab. sick in hospital 12/31/61-1/62. Present 2-8/62. Ab. bringing horses from Summerville, S. C. to Staunton, Va. 9-10/62. Present 11/62-8/31/63. Ab. on horse detail 10/23-12/31/63. Present sick in camp 1-2/64. Ab. sick with debilitas & bronchitis in Richmond hospital 3/4-4/14/64. Ab. on horse detail 4/26-30/64. Present 5-6/64, however, on duty with Commissary 6/15-12/31/64. NFR. Artist-Traveling Photographer, Spring Place, Murray Co., Ga. 1870 census. Farmer, Red Clay, Whitfield Co. Ga. 1900 & 1910 censuses. d. Whitfield Co., Ga. 7/16/16. Bur. Hopewell Cem., Whitfield Co., Ga.

PEDEN, JAMES BOYD. Pvt., Co. B. b. Laurens Dist. 8/2/37. Res. Laurens Dist. Enl. Camp Means 6/21/62. Present through 10/62. On picket 11-12/62. Present 3-4/63. Ab. sick with typhoid fever in Gordonsville hospital 9/3/63. d. 9/5/63 age 26. Bur. Gordonsville Conf. Cem. Effects: $23.00. Son of Thomas A. Peden.

PEDEN, JOHN BALENTINE. Pvt., Co. B. b. Laurens Dist. 9/29/44. Enl. Camp Means 10/8/62. Present through 10/31/62. On picket Ellis Ford 11-12/62. Present 3-4/63. Ab. sick with chronic diarrohea in Richmond hospital 5/31-7/20/63. Ab. on sick leave through 8/31/63. Present 9-12/63. Present sick in camp 1-2/64. Ab. on horse detail 4/15-30/64. Present 5-10/64. Detached Pocotaligo 12/2-31/64. NFR. Farmer, Dial, Laurens Co. 1870-1910 censuses. d. 2/26/18. Bur. Fairlawn Cem., Fountain Inn, S. C.

PEDEN, ROBERT MILTON. Pvt., Co. B. b. Laurens Dist. 4/5/36. Res. Laurens Dist. Enl. Clinton 8/25/61 age 25. Ab. sick 10/31/61. Present 11/61-4/63. Detached horse recruiting camp, Nelson Co., Va. 8/3/63-2/10/64. Detached Green Pond 4/26-30/64. Present 5-10/64. Detached Pocotaligo 12/2-31/64. NFR. d. Laurens Dist. 1865.

PEDEN, THOMAS ALEXANDER. Pvt., Co. B. b. Greenville, S. C. 9/27/08. Enl. Clinton 8/22/61 age 52. Ab. sick 10/31/61. Present 11-12/61. Nurse, Adams Run hospital 1/62. Present 3-4/62. Discharged for over age and disability 6/20/62. Age 54, 5' 10", red complexion, gray hair, blue eyes, Farmer. Paid 10/21/62. NFR. Farmer, Laurens 1870 census. d. Laurens 11/75. Father of James Boyd Peden.

PEDEN, THOMAS M. Pvt., Co. D. b. Chester Dist. 6/10/41. Res. Chester Dist. Enl. Co. F, 6th S. C. Inf. Summerville 4/11/61 age 19. WIA 5/31/62. Ab. wounded through 8/62. Discharged by Medical Board 12/23/62. Enl. Co. D, 1st S. C. Cav. Chester CH 4/17/64. Ab. on horse detail 4-23-30/64. Present 5-8/64. Detached as guard for POW's Florence 9/18-12/31/64. NFR. Farmer, Landsford, Chester Co. 1870 & 1880 censuses. Res. of Chester 1900 census. d. 3/28/11. Bur. Cedar Shoals Presb. Ch., Chester Co.

PEDEN, WILLIAM A. Captain & Asst. Commissary of Subsistence. b. S. C. 10/11/24. Farmer, age 34, Tishomingo, Miss. 1860 census. Enl. Co. D Chester CH 9/10/61 as Pvt., age 36. Present through 12/61.

Ab. Clerk in Commissary Dept., Adams Run 3-6/62. Appointed Captain & Asst. Commissary of Subsistence 8/18/62. Present through 12/62 & 3/28-4/30/63. Dropped 8/1/63. Ab. on leave in S. C. 9-10/63. Reinstated 9/17/63. Ab. on leave through 12/31/63. Present 1-2/64. Ab. sick 4/26-30/64. Ab. detached as Asst. Commissary of Subsistence, East Lines, James Island 5/3-12/31/64. NFR. d. 11/7/--. Bur. Old City Presb. Ch. Cem., Chester, S. C.

PENDERGRASS, WATERS or WAITES D. Pvt., Co. D. b. Chester Dist. 4/41. Enl. Chester CH 9/10/61 age 20. Present 11/61-6/62. Ab. on sick leave 7-8/62. Present 9-10/62. On scout 12/25-31/62. Present 3-6/63. Ab. on horse detail 8/18-31/63. Present 9/63-2/64. Ab. on horse detail 4/19-30/64. Present 5-6/64. Present sick in camp 7-8/64. Present 9-12/64. Paroled Greensboro, N. C. 4/26/65. Farmer, age 25, Baton Rouge, Chester Co. 1870 census. Farmer, age 59, Collins Co., Texas 1900 census and age 68, 1910 census. d. 1912. Bur. Pecan Grove Cem., Collins Co., Texas.

PENNEL, THOMAS R. Pvt., Co. G. b. S. C. circa 1844. Student, age 16, Abbeville Dist. 1860 census. Enl. Abbeville CH 4/1/63. Present through 12/63. Ab. detached recruiting station for horse 1/1-2/29/64. Present 3-10/64. Detached Pocotaligo 12/10-31/64. NFR.

PENNEL, WILLIAM M. Pvt., Co. G. b. S. C. circa 1844. Farmhand, age 16, Abbeville Dist. 1860 census. Enl. Abbeville CH 4/4/63. On detached service 4/30/63. Ab. sick with "Parotitis & debilitas" in Farmville hospital 5/6-13/63. Ab. in Richmond hospital 8/31/63. Ab. on sick leave 9/63-2/64. Ab. on horse detail 4/2-30/64. Present 5-10/64. Detached Pocotaligo 12/10-31/64. Captured and paroled Cheraw, S. C. 3/5/65. Farmer, age 64, Prentiss, Miss. 1910 census.

PENNY, JOHN H. Pvt., Co. A. b. Abbeville CH 9/2/44. Enl. Abbeville CH 9/22/62. Present through 10/62, dismounted. Present 11/62-8/31/63. Captured Raccoon Ford 9/13/63. Sent to Old Capitol & Pt. Lookout. Exchanged 4/27/64. Ab. sick with debilitias in Richmond hospital 5/1-8/64. Furloughed for 30 days. Present sick in camp 7-8/64. Detached Pocotaligo 10-12/64. Paroled Greensboro, N. C. 5/1/65. Farmer, Abbeville Co. 1880-1900 censuses. Receiving pension Abbeville 4/1/19. d. 12/27/26. Bur. Sharon Meth. Ch., Abbeville Co.

PENNY, WILLIAM THOMPSON. Pvt., Co. A. b. Laurens Dist. 1/11/38. Merchant's Clerk, age 22, Abbeville CH 1860 census. Enl. Abbeville CH 3/19/62. Present through 6/62. Ab. on leave 7-8/62. Ab. on detached service 9-10/62. Ab. detailed in Medical Dept. 11-12/62. Present 3-6/63. Ab. on horse detail 8/31/63. Ab. detailed in Medical Dept. 9-10/63. Ab. on horse detail 12/23/63-2/14/64. Present through 6/64. Present as Hospital Steward, Regimental Med. Dept. 7-12/64. NFR. Druggist, Abbeville 1870 & 1880 censuses. d. 1908. Bur. Upper Long Cane Cem., Abbeville Co.

PERRY, BENJAMIN JOSIAH. 2ndSgt., Co. H. b. Kershaw Dist. 7/20/36. Farmer, age 24, Hinds Ford, Chester Dist. 1860 census. Enl. Rock Hill 3/14/62 age 25 as Pvt. Promoted 2ndSgt. On picket Jacksonborough through 4/62. Present 5-6/62. Ab. on sick leave 7-10/62. AWOL 11-12/62. Ab. sick with chronic diarrhea in Richmond hospital 2/19/63. Furloughed for 30 days 3/16/63. Ab. on sick leave 3-4/63. Discharged 5/9/63. Reenlisted Co. K, 7th S. C. Cav. 2/3/64. Present sick 10-11/64. NFR. d, 11/25/68. Bur. Liberty Hall Presb. Ch. Cem., Kershaw Co.

PETERSON, JAMES WOFFORD. Pvt., Co. B. b. S. C. 12/25/46. Enl. Laurens Dist. age 17 on post war roster. Clerk, Auditor's Office, Laurens, S. C. 1880. d. Laurens 11/8/99. Bur. Laurens City Cem.

PHILLIPS, OTIS B. Pvt., Co. A. On post war roster. Reenlisted Co. K, 4th S. C. Cav. Green Pond 7/27/73. Present until mashed by a horse 6/2/64. Ab. in Richmond hospital with "leg stitus" Transferred to Charlottesville hospital 6/15-8/31/64. Ab. sick with chronic dysentery i & "paralysis of the bladder" Charlottesville 10/28-11/9/64. Paroled Charlotte, N. C. 5/4/65.

PHILLIPS, THOMAS J. Pvt., Co. G. b. Brogdon, Sumter Dist. 1845. Enl. Columbia 6/16/64. Present through 10/64. Detached Pocotaligo 12/10-31/64. NFR. Farmer, Providence, Sumter Co. 1880 census. Retired, Andrews, Georgetown Co. 1920 census. d. Kingstree, Williamsburg Co. 1920. Bur. Ader Swamp Cem., Kingstree, S. C.

PHILSON, SAMUEL A. 2ndCpl. Co. B. b. S. C. 8/25/40. Age 17, no occupation, Laurens Dist. 1860 census. Enl. Clinton 8/21/61 age 21 as Pvt. Detached at Camp Bee through 10/31/61. Present 11/61-4/62. Ab. on leave 5-6/62. Ab. on sick leave & in hospital 7-8/62. Present 9-12/62, promoted 3rdCpl. Present 3-8/63. Ab. on horse detail 10/26-12/31/63. Present 1-6/64. Ab. detailed with Commissary 7/13-8/31/64. Promoted 2ndCpl. Present 9-12/64. NFR. Farmer, Jacks, Laurens Co. 1880 & 1900 censuses. Retired, Clinton, S. C. 1910 census. d. 7/10/14. Bur. Clinton City Cem.

PIERCE, JOHN HENRY. Pvt., Co.'s C & K. b. S. C. 11/6/42. Farmer, Newton, Barnes Co., Ga. 1860 census. Enl. Co. C Wadmalaw Island 3/12/62. Presence or absence not stated 3-4/62. Present 5-6/62. Transferred Co. K 6/27/62. Present 7-10/62. On picket Richards Ford 11-12/62. Present 3-6/63. WIA Upperville 6/26/62. Ab. on horse detail 8/16-31/63. Ab. on detached service 9-10/63. On scout over Rappahannock River 11-12/63. Present 1-2/64. Captured Morrisville 3/2/64. Sent to Old Capitol. Sick with Variola & eruptive fever Kalorama hospital, Washington, D. C. 3/24-4/2/64 & with erysipelias 5/5-16/64. Transferred to Ft. Delaware. Released 6/10/65. 5'10", light complexion, light hair, blue eyes. Farmer, age 27, Miller, Ga. 1870 census and Richmond, Ga. 1900 census. d. Columbia Co., Ga. 8/12/13. Bur. Abilene Bapt., Ch., Martinez, Columbia Co., Ga.

PINCKNEY, CHARLES W. Surgeon. b. S. C. 10/30/33. M.D., age 26, Walterboro, Colleton Dist. 1860 census. Contract Surgeon 9/28/61. Appointed Surgeon from S. C. 9/19/62. Discharged for chronic bronchitis 10/11/62. However, on scout with Gen. J. E. B. Stuart 11-12/62. Apparently reappointed Surgeon. Ab. sick 2/20-4/30/63 & 6/11-12/31/63. Present 1-2/64. Ab. sick with chronic "hepatitus" in Richmond hospital 3/18/64. Transferred to Macon, Ga. hospital 3/26/64. Retired 3/28/64. Paid Augusta, Ga. 12/10/64. NFR. M.D., Atlanta, Ga. 1870 & 1880 censuses. Possibly the Charles Pinckney b. Charleston 1828 who died Charleston 1919 age 91.

PITTS, CHARLES. Pvt., Co. G. b. S. C. circa 1841. Enl. Abbeville CH 7/1/62. On picket Gimball's through 8/62. Ab. sick in Richmond hospital 10/20-12/31/62. Present 3-10/63. On picket Chancellorsville 12/31/63. Present 1-2/64. Transferred Co. B, Lucas' Bn. S. C. Arty. 5/13/64. Joined 5/20/64. Present through 12/64. NFR. Bur. Mountain Grove Cem., Raburn Co. Ga., no dates.

PITTS, ROBERT. Pvt., Co. I. b. Laurens Dist. 9/5/19. Res. Laurens Dist. 1860 census. Not on muster rolls. d. Columbia 2/15/64 age 45.

PITTS, T. A. Pvt., Co. I. Res. Colleton Dist. Enl. Columbia 2/1/64. Detached Green Pond 4/16-30/64. Present 5-10/64. Detached Pocotaligo 12/10-31/64. d. of chronic diarrhoea Columbia hospital 1/25/65. Bur. Elmwood Cem., Columbia.

PLATT, WILLIAM W. 2ndSgt., Co. I. b. S. C. circa 1838. Farmhand, age 22, Walterboro, Colleton Dist. 1860 census. Enl. Co. K, 2nd S. C. Inf. Charleston 5/12/61. Discharged for medical reasons 11/19/61. Enl. Co. I, 1st S. C. Cav. Parker's Ferry 4/3/62 as 4thSgt. On picket Jacksonborough through 4/30/62. Present 5-6/62. Ab. on sick leave 8/27-31/62. Present 9-10/62. Promoted 2ndSgt. Present 11-12/62. d. of typhoid pneumonia Lynchburg hospital 3/15 or 22/63. Bur. Lynchburg Old City Cem. Effects: Sundries & $73.50.

PLAXICO, JAMES EDWARD. Pvt., Co. C. b. York Dist. 1846. Student, age 12, York Dist. 1860 census. On postwar roster. Farmer, age 23, Blainsville, York Co. 1870 census & age 33, Broad River, York Co. 1880 census. Receiving pension Sharon, York Co. 1917.

POINCETT, PAUL. Musician, Co. A. (Colored). Enl. 11/17/61. NFR.

POOL, LEWIS. Pvt., Co. B. b. S. C. 1836. Enl. Spartanburg age 25 on post war roster. Farmer, age 62, Bull Swamp, Lexington Co. 1900 census.

POOL, SETH M. Pvt., Co. B. b. Greenville, S. C. 2/10/40. Enl. Columbia 9/8/64. On picket 10/31/64. AWOL 12/31/64. NFR. Carpenter, age 29, Torn Grove, Knox Co., Tenn. 1870 census. House Carpenter, age 40, Davidson Co., N. C. 1880 census. d. Greenville, S. C. 12/26/91. Bur. Enoree Bapt. Ch. Cem., Greenville.

POOLE, MARTIN BOBO P. Pvt., Co. B. b. Laurens Dist. 8/22/41. Age 18, Scuffletown, Laurens Dist. 1860 census. Enl. Camp Johnson 10/29/61. Present through 8/31/63. Ab. sick in hospital 9-10/63. AWOL 12/20-31/63. Present 1-2/64. Ab. on horse detail 4/15-30/64. Present 5-12/64. NFR. Farmer, Scuffletown, Laurens Co. 1870 & 1880 censuses. d. 3/22/19. Bur. Langston Bapt. Ch. Cem., Laurens Co.

POOLE, JOHN TERRY. Pvt., Co. B. b. S. C. 8/25/36. Res. Laurens Dist. 1850 census. Enl. James Island 6/29/64. Present 7-8/64. Ab. on sick leave from Summerville hospital 10/14-12/31/64. NFR. d. 12/14/09. Bur. Laurens City Cem.

PORCHER, PERCIVAL RAVENEL. Pvt., Co. A. b. S. C. 7/17/29. Planter, Charleston 1860 census. On post war roster. Reenlisted Co. K, 4th S. C. Cav. McPhersonville 7/26/62. Present until detailed with Gen. Beauregard l2/27/63-2/64. Present until WIA (bayonet wound - left thigh) Hawes' Shop 5/28/64. DOW's and gangrene 6/3/64. Bur. Hollywood Cem., Richmond.

PORTER, JOSEPH M. 3rdCpl., Co.'s F & K. b. S. C. circa 1843. Farmhand, age 18, Cleveland Co., N. C. 1860 census. Enl. Co. F Pickens CH 12/4/61 as Pvt. Present through 12/31/61. Ab. on sick leave 4/11-5/1/62. Transferred Co. K 6/27/62. Present through 10/62. Ab. detached in QM Dept. 11/62-4/63. Captured Upperville 6/21/63. Sent to Old Capitol. Exchanged by 7/9/63, when issued clothing Richmond. Ab. on horse detail 8/24-31/63. Present 9-12/63. Ab. on leave 2/12-29/64. Ab. on leave 4/25-30/64. Promoted 3rdCpl. Present 5-6/64. On duty Signal Station, Charleston 7-8/64. Ab. on leave from Charleston hospital for 60 days 10/9/64. Present 11-12/64. Paroled Raleigh, N. C. 4/65. Retired, age 72, Rocky Grove, Aiken Co. 1920 census. Bur. Aiken Co.

POSEY, HAMILTON. Pvt., Co. B. b. S. C. 1824. Enl. Spartanburg 1/25/64. Detached Green Pond 4/24-30/64. Present 5-6/64. Ab. sick in Charleston hospital 8/20-31/64. Ab. on sick leave 9/20-10/31/64. Present 11-12/64. NFR. d. 1867. Bur. Nazareth Presb. Ch. Cem., Spartanburg Co.

POSEY, WILLIAM HAMILTON. Pvt., Co. B. b. S. C. circa 1846. Enl. Spartanburg 1/25/64. Detached Green Pond 4/26-30/64. Present 5-10/64. Detached Pocotaligo 12/2-31/64. NFR. No occupation, age 24, Fair Forest, Spartanburg Co. 1870 census. Dry Goods Merchant, age 34, Spartanburg 1880 census. d. by 1898.

POWE, JOHN HENRY. Captain, Co. D. Not on muster rolls. Probably the J. H. Powe listed as serving in the 36th Ala. Inf. & records misfiled. Retired to Invalid Corps 10/19/64. d. 4/18/18. Bur. Old Davids Epis. Ch. Cem., Chesterfield Co.

POWER, J. HENRY. Pvt., Co. A. b. S. C. circa 1837. Enl. Abbeville CH 8/13/61 age 34. Ab. on sick leave 10/31/61. Ab. sick in hospital 12/31/61. Present 3-6/62. Ab. on sick leave 7-12/62. Present 3-8/63. Ab. on horse detail 9-10/63. AWOL 12/20-31/63. Present 1-2/64. Detached Green Pond 4/26-30/64. Present 5-6/64. Ab. on sick leave 9/16-12/31/64. Paroled Greensboro, N. C. 5/1/65. Living Abbeville Co. 1902.

POWER, JOHN WILLIAM. Pvt., Co. G. b. S. C. 2/25/34. Farmer, Abbeville Dist. 1860 census. Enl. Co. D, 7th S. C. Inf. 7/15/61. Discharged for disability 8/20/62. Enl. Co. G, 1st S. C. Cav. 1/28/64. Present through 12/31/64. NFR. d. 12/7/04. Bur. Little Mountain Presb. Ch. Cem., Antreville, S. C.

POWER, JOSEPH WILLIAM. Pvt., Co. G. b. Greenville Dist. 1/21/46. Enl. Abbeville CH 1/28/64. Present through 6/64. Ab. detached 8/9-31/64. Present 9-12/64. NFR. Merchant, New Market, Ala. postwar. d. Franklin Co., Tenn. 12/4/98 age 52. Bur. Rice Cem., New Market, Ala.

POWER, L. S. Pvt., Co. B. Enl. Laurens by 1862 age 35. Discharged 1862. Alive 1898.

POWER, LEWIS D. Pvt., Co. B. Pvt., Co. B. b. S. C. 7/21/47. Enl. Florence 3/1/65. Paroled Greensboro, N. C. 5/1/65. d. 8/22/93. Bur. New Herring Presb. Ch., Laurens Co.

POWER, SAMUEL L. Pvt., Co. B. b. Laurens Dist. circa 1827. Enl. Clinton 8/22/61 age 34. Ab. sick 10/31/61. Ab. detailed as nurse in hospital 12/31/61. Present sick in camp 3-4/62. Discharged for disability Adams Run 6/20/62. Age 39, 6' red complexion, dark hair, blue eyes, Blacksmith. NFR. d. 7/26/86. Wife receiving pension Laurens, S. C. 12/2/19.

POWERS, NATHANIEL. Pvt., Co. I. b. S. C. circa 1830. Res. Charleston. Enl. Parker's Ferry 4/3/62. Present through 6/62. Ab. on sick leave 8/27-31/62. Ab. sick in Summerville hospital 9-10/62. AWOL 11/62-8/63. Ab. sick with debilitas in Richmond hospital 10/11/63. Furloughed for 30 days 10/21/63. Ab. on sick leave through 10/63. AWOL 11/21/63-4/64. Present 5-8/64. On picket 9-10/64. AWOL 11-12/64. Deserted to the enemy Charleston 3/6/65. Age 36, 5'10", dark complexion, blue eyes. Took oath & discharged.

PRATT, JAMES C. 2ndCpl., Co.'s D & K. b. Abbeville Dist. 1845. Res. Chester Dist. 1860 census. Enl. Co. D Chester CH 5/16/62 as Pvt. Transferred Co. K 6/25/62. Promoted 3rdCpl. Present through 8/62. Ab. on detached service 9/2-10/31/62. Present 11-12/62, promoted 2ndCpl. Ab. sick 4/15-30/63. Reduced to Pvt. Present 5-8/63, however, ab. sick with acute diarrhea in Charlottesville hospital 8/10-25/63. Ab. on horse detail 9-10/63. AWOL 11-12/63. Ab. sick Chester until died of disease 1/28/64.

PRATT, JOHNATHAN. Pvt., Co. D. b. S. C. circa 1829. Res. Chester Dist. Enl. by 1862. d. Richmond 10/62 on post war roster.

PRESLEY, JOHN SHELTON. Bugler, Co. D. b. S. C. 3/23/27. Farmer, Chester Dist. 1850 census. Enl. Chester CH 9/10/61 age 32. Present 11/61-6/62. Ab. on sick leave 8/4-31/62. Transferred Captain Barber's Co., 6th S. C. Cav. 9/3/63. Present 11/63-10/64. NFR. Farmer, Taylorsville, Miss. 1880 census. d. Miss. 2/13/05. Bur. Magnolia Cem., Coldwater, Miss.

PRESSLEY, WILLIAM BASWELL. Pvt., Co.'s F & K. b. Pickens Dist. 1835. Farmer, age 26, Cherokee, Pickens Dist. 1860 census. Enl. Pickens CH 12/4/61. Present through 4/62. Transferred Co. K 6/25/62. Present 7-8/62. Ab. on detached service 9-10/62. Present 11-12/62. Ab. sick 2/22-4/30/63. Ab. sick in hospital 8/31/63. Present 9/63-2/64. Ab. on horse detail 4/1-30/64. On picket, Charleston fortifications 6/25-30/64. Ab. on leave 7-8/64. Ab. on leave from Charleston hospital for 60 days 10/9-31/64. Ab. on sick leave for 30 days 12/9-31/64. NFR. Farmer, age 45, Eastatoe, Pickens Co. 1880 census. Farmer, age 80, McCurtain, Okla. 1910 census. d. Manasa, Colo. 12/8/07. Bur. McCurtin, Okla.

PRIOLEAU, CHARLES E. Pvt., Co. A. b. Charleston circa 1840. Res. Charleston 1860 census. On post war roster. Reenlisted Co. K, 4th S. C. Cav. Grahamville 3/25/62. age 22. Present until ab. on leave 8/29/62 for 10 days. Present until ab. on sick leave 9/22-10/3l1/63. Present 11/63-2/64. Present until KIA Hawes Shop 5/28/64.

PRICE, R. S. 3rdSgt., Co. I. b. S. C. circa 1838. Res. Colleton Dist. Enl. Captain Sheridan's Co., 11th S. C. Inf. 10/15/61. Present through 12/61. Company disbanded. Enl. Co. I, 1st S. C. Cav. Parker's Ferry 4/3/62 as 2ndCpl. Present through 10/62, promoted 4thSgt. Present 11-12/62, promoted 3rdSgt. d. Columbia, S. C. 3/23/63 age 35.

PRICE, SAMUEL J. Pvt., Co. K, b. S. C. circa 1836. Age 24, Chester Dist. 1860 census. Enl. Wadmalaw Island 7/10/62. Deserted 7/12/62. NFR. Farmer, Winnsboro, Fairfield Co. 1870 census. Farmer, age 47, Halselville, Chester Co. 1880 census.

PRIDE, RICHARD J. Pvt., Co. C. Enl. Aikens Dist. age 20, on post war roster. d. by 1898.

PRINCE, JESSEY. Pvt., Co. F. b. Ga. circa 1836. Enl. Pickens CH 4/5/62. Present 5-6/62. Ab. sick in hospital 7-8/62. Ab. on leave 9-12/62. AWOL 3-4/63. AWOL 7/1-12/31/63. Dropped as a deserter 4/1/64. NFR. Farmer, age 34, Collins, Edgefield Co. 1870 census and age 44 1880 census. d. 11/1/12 age 76. Bur. Red. Oak Grove Bapt. Ch. Cem., Edgefield Co., S. C.

PRINGLE, DOMINIC LYNCH. Pvt., Co. A. b. S. C. 6/12/46. On post war roster. Reenlisted Co. K, 4th S. C. Cav. Columbia 5/1/64. Present through 8/64. AWOL 10/22/64. Paroled Greensboro, N. C. 5/2/65. Farmer, Pee Dee, Georgetown, S. C. 1880 census. Living San Francisco, Cal. 1900. d. before 1919.

PRINGLE, JOEL ROBERT POINSETT. Pvt., Co. A. b. S. C. 1845. On post war roster. Reenlisted Co. K, 4th S. C. Cav. Pocotaligo 2/19/63. Present until ab. on leave 6/25/63 for 15 days. Present through 2/64. WIA & DOW's in the hands of the enemy 5/31/64 or 6/2/64. Bur. Holywood Cem., Richmond 8/26/64. Brother of John Julius Pringle.

PRINGLE, JOHN JULIUS. Pvt., Co. A. b. S. C. 6/21/42. On postwar roster. Reenlisted Co. K, 4th S. C. Cav. Mc Phersonville 9/14/62. Detailed with Gen. Walker 12/16/62 for 15 days. Present 1/63 until detailed by Gen. Ripley 10/22/ for 20 days. Present 11/63-2/64. Present until ab. on horse detail 8/11/64 for 40 days. Present 9-10/64. Paroled 4/65. d. 8/21/76. Bur. Magnolia Cem., Charleston. Brother of Joseph R. P. Pringle.

PRIOLEAU, CHARLES EDWARD. Pvt., Co. A. b. Charleston 2/6/40. Att. U. Va. 1858-60. Res. Charleston. Enl. 11/7/61. Reenlisted Co. K, 4th S. C. Cav. Grahamville 3/25/62. Present until ab. on leave 8/29/62 for 10 days. Present until ab. sick leave 9/22/63-10/63. Present 11/63-2/64. KIA Hawes Shop 5/28 or 30/64.

PRIOLEAU, JAMES M. Pvt., Co. A. b. S. C. circa 1843. Att. Union College, N. Y. 1856-58. Enl. 3/15/62. Reenlisted Co. K, 4th S. C. Cav. 9/23/62. WIA (right thigh) Pocotaligo 10/22/62. Detailed as Medical Purveyor, Macon, Ga. 11/13/62-3/63. Paroled 4/65.

PRIOR, RICHARD DUNCAN, JR. Pvt., Co. C. b. S. C. circa 1839. Age 21, Beech Island, Edgefield Dist. 1860 census. Enl. Hamburg 8/27/61 age 22. Detached Wadmalaw Island through 12/31/61. Present 5-6/62. Ab. on sick leave 8/2-31/62. Present 9-10/62. On scout 11-12/62. Present 3-8/63. Ab. on horse detail 10/23-12/31/63. Present 1-10/64. Detached Pocotoligo 12/10-31/64. NFR. Wheelwright, age 30, Silverton, Barnwell Co. 1870 census. Grave stone stolen from Ebenezer Bapt. Ch. Cem.

PRITCHETT, GEORGE E. Pvt., Co. A. b. S. C. 1/11/36. Merchant, Charleston 1860 census. On post war roster. Reenlisted Co. K, 4th S. C. Cav. Grahamville 5/19/62. Present until ab. on sick leave 9/30/63-2/64. Detailed for light duty as Hospital Steward, Williamsburg hospital through 8/64. NFR. d. 6/11/92. Bur. Fairview Cem., Frankford, Mo.

PRYOR, SETH THORNTON, SR. Pvt., Co. B. b. Charleston, S. C. circa 1818. Shoe Dealer, Charleston 1850 census. Enl. Adams Run 9/9/62 as substitute for R. L. Cleveland. Ab. on march with horses from Summerville, S. C. to Staunton, Va. 9-10/62. Present 11/62-4/63. Ab. sick in hospital 8/27-31/63. Ab. sick with Lumbago in Richmond hospital 9/21/63. Furloughed for 30 days 12/12/63. Present 1-6/64. Present as Saddler 7-8/64. Ab. nurse in McLeod's hospital 10/15-12/31/64. NFR. Living Charleston 1884. d. by 1898. Wife applied for pension from Darcusville, Pickens Co. 4/1/01.

PRYOR, SETH THORNTON, JR. Pvt., Co. B. b. Laurens Dist. 10/1/40. Age 21, Pleasant Mound, Laurens Dist. 1860 census. Enl. Clinton 8/22/61 age 21. Present through 6/62. Horse killed in skirmish Walpole Bridge 6/7/62. Ab. on sick leave 7-8/62. Present 9/62-4/63. Horse killed Fallings Waters 7/9/63. Paid $325.00. Ab. on horse detail 8/18-31/63. Present 9/63-12/64. Paroled near Durham, N. C. 4/18/65. Farmer, Easley, Pickens Co. 1920 census. d. Darcusville, Pickens Co. 12/26/27. Bur. Darcusville Ch. Cem.

PUCELL, JAMES. Pvt., Co. A. b. S. C. 7/24/28. Enl. 11/7/61. Ab. on leave 2/6/62. Reenlisted Co. K, 4th S. C. Cav. Grahamville 3/25/62. Present until ab. on sick leave 12/14/62-1/63. Present 2/63 until detailed with Gen. Beauregard 7-8/63 Present until discharged 10/1/63. Appointed Asst. Surgeon 10/6/63. Assigned to Light Arty., 6th Military Dist of S. C., Ga. & Fla. through 3/2/64. Assigned to Chestnut Arty. 4/27/64. Paroled Greensboro, N. C. 5/1/65. d. 7/28/78. Bur. St. Lawrence Cem., Charleston.

PURDY, WILLIAM NORRELL. Pvt., Co. G. b. S. C. 1811. Enl. Abbeville CH 1/8/62. Present through 6/62. Present sick in camp 7-10/62. Discharged Camp Brown, Staunton, Va. 11/3/63. NFR. d. 1879.

RADFORD, THOMAS W. Pvt., Co. I. b. S. C. circa 1835. Res. Colleton Dist. Enl. Parker's Ferry 4/3/62. Present as Teamster 5-12/62. Ab. sick in Farmville hospital with "Febris Intermittens" 5/2/63. d. of "Erysipelas" 6/11/63 age 29. Probably buried with other other unknown Confederate soldiers in Farmville Confederate Cem.

RADFORD, W. O. Pvt., Co. I. Enl. Parker's Ferry 4/3/62. Presence or absence not stated through 4/30/62. NFR.

RAGSDALE, BURR AUGUSTUS. Pvt., Co. D. b. Rossville, Chester Dist. 11/9/49. Not on muster rolls. Postmaster, Richburg, Chester Co. 1894. d. 4/21/19. Bur. Old Catholic/Presb. Ch. Cem., Chester Co.

RAGSDALE, CHARLES HENRY. Adjutant. b. Chester Dist. 10/28/39. Student, age 21, Citadel, Charleston 1860 census. Enl. Chester CH 5/1/62 as Pvt. Present 5-6/62. Ab. on sick leave 8/27-31/62. Ab. on detached service 9-10/62. On scout 12/25-31/62. Present 1/63, appointed Adjutant of Regiment 1/5/63. Present through 6/63. Signed for 2 tent flies 5/23/63. Present Brandy Station 6/9/63. WIA Gettysburg 7/3/63. Present 9/63-12/64. Signed for 4 caps, 6 jackets, 4 pr. pants, 7 pr. drawers, 15 shirts, 3 pr. socks, 2 pr. boots, 2 blankets & 3 overcoats for staff 9/3/63. Signed for forage for 10 horses of staff 12/31/63. Signed for 4 Army tents, 3 camp kettles, 6 mess pans, 1 wall tent, 13 pr. pants, 12 jackets, 4 pr. shoes, 18 shirts, 15 pr. drawers & 11 caps for staff James Island 6/30/64. NFR. Farmer, Rossville, Chester Co. 1870 & 1880 censuses. Surveyor, Rossville, Chester Co. 1900 census. d. 10/3/03. Bur. Evergreen Cem., Chester S. C.

RAGSDALE, JAMES PARKER. Pvt., Co. K. b. Rossville, Chester Dist. 6/20/46. Age 14, Rossville, Chester Dist. 1860 census. Enl. Columbia 4/26/64. On picket Charleston fortifications 6/25-30/64. Detailed to drive cattle for QM 7-12/64. Paroled Greensboro, N. C. 4/26/65. Farmer, Hazelwood, Chester Co. 1880-1900 censuses. d. 1/9/04. Bur. Old Catholic/Presb. Ch. Cem., Chester Co. Wife receiving pension Blackstock, Chester Co. 10/17/19. Brother of William H. Ragsdale.

RAGSDALE, WILLIAM H. 2ndLt., Co.'s D & K. b. S. C. 1838. Student, age 21, Rossville, Chester Dist. 1860 census. Enl. Chester CH 9/25/62 as Pvt. Present through 6/63. Ab. on horse detail 8/18-31/63. Present 9-10/63. Ab. recruiting horses in the Valley 11/13/63-2/64. Present 3-4/64. Transferred Co. K 5/16/64. Age 26, 6', fair complexion, blue eyes, dark hair. Present 5-6/64. Detached at Bridge over Ashley River 7-8/64. Ab. drawing rations 9-10/64. Courier, Tarr Bluff 12/29-31/64. WIA Goldsboro, N. C. 4/65. Commanding detachment of Couriers as 2ndLt. &. paroled Greensboro, N. C. 5/1/65. Laborer, Belton, Anderson Co. 1870 census and Farmer, there 1880 census. d. Anderson, S. C. 2/19/15 age 77. Bur. Big Creek Cem. Brother of James P. Ragsdale.

RAINEY, SAMUEL, JR. Pvt., Co. H. b. S. C. 1845. Res. York Dist. 1860 census. Enl. Columbia 1/27/64. Present through 12/64. NFR.

RAMAGE, JAMES E. Pvt., Co. B. b. S. C. 9/20/27. Farmer, Laurens Dist. 1860 census. Enl. Clinton 8/22/61 age 34. NFR. Enl.Co. D, 3rd Bn. S. C. Inf. 1/15/62. Discharged for disability. NFR. d. 12/3/73. Bur. Holly Grove Bapt. Ch. Cem., Laurens Co.

RAMEY, A. Pvt., Co. I. b. Oconee Dist. 1848. Enl. 1864. Paid 5/17/64 for 4/3-6/30/64. Captured and paroled. NFR. d. Oconee Co. 5/4/23 age 75 years, 6 months & 20 days. Bur. Hally Stone Beth. Ch. Cem., no marker.

RAMPLEY, JAMES M. Pvt., Co. B. b. S. C. circa 1840. Enl. Clinton 8/22/61 age 21. Ab. sick 10/31/61. Present 11/61-6/62. Present sick in camp 7-10/62. Present 11-12/62. Ab. sick with diarrhea in Lynchburg hospital 4/13-5/9/63. Ab. on horse detail 8/18-31/63. Present 9-10/63. On picket 11-12/63. Present 1-2/64. Detached Green Pond 4/26-30/64, remounted 4/15/64. Present 5-12/64. NFR. Farmer, Pendleton, Anderson Co.1880 census. d. Cartersville, Ga. 10/19/26. Bur. Mt. Olive Ch. Cem., there.

RATCHFORD, JAMES ALBERTUS. 2ndLt., Co. H. b. S. C. 2/25/25. Enl. Rock Hill 1/15/62 age 37 as Pvt. Present as Company Commissary 3-19/62. Promoted 3rdLt. 11-12/62 & ab. on CM duty. Present 3-4/63. Promoted 2ndLt. 6/13/63. Ab.on detached service Richmond 8/31/63. Ab. with "Fractura" in Charlottesville hospital 9/10/63. Furloughed for 30 days 9/29/63. Ab. sick through 12/31/63. Present 1-2/64. Detached Green Pond 4/16-30/64. Present 5-8/64, commanding company 5/21/64.Signed for 7 caps, 17 pr. pants, 17 jackets, 7 shirts, 9 pr. drawers & 6 pr. socks Green Pond 5/21/64. Ab. on leave 9/16-10/31/64. Present 11-12/64. Paroled Greensboro, N. C. 5/1/65. Farmer, York Co. 1880 census. d. 7/9/97. Bur. Rose Hill Cem., York, S. C.

RATCHFORD, R. H. Pvt., Co. H. Enl. Columbia 11/1/64. Present through 12/31/64. NFR.

RATCHFORD, ROBERT WALKER. Pvt., Co. H. b. S. C. 1/14/20. Not on muster rolls. Staff officer in obit. d. 4/29/65. Bur. Rose Hill Cem., York, S. C.

RAWLINSON, JOHN MORGAN., Pvt., Co. H. b. S. C. 9/23/31. Enl. Co. E, 17th S. C. Inf. 11/26/61. NFR on muster rolls. Enl. Co. H, 1st S. C. Cav. on postwar roster. d. 5/26/21. Bur. Rock Hill Cem., York Co., S. C.

RAY, WILLIAM HARDY. Pvt., Co. B. b. S. C. circa 1845. Res. Spartanburg Dist. Enl. Columbia 1/30/64. Present 3-8/64. Ab. on leave 10/27-31/64. Present 11-12/64. NFR. Minister, age 65, Marlow, Okla. 1910 census & age 75, 1920 census. d. Eva, Cooke Co., Texas 8/18/25. Bur. Marlow, Okla.

REDMAN, MALACHI B. 1stCpl., Co. I. b. S. C. 3/30/30. Enl. Parker's Ferry 4/3/62 as Pvt. On picket Jacksonborough through 4/30/62. On picket John's Island Ferry 5-6/62. Present 7-1/62, promoted 3rdCpl. Present 11-12/62, promoted 2ndCpl. Present 3-6/63, however, Ab. sick with chills & fever in Richmond hospital 3/18/63. Transferred to Farmville hospital 4/18/63. Returned to duty 4/25/63. Ab. sick with "Intermittent Fever" in Farmville hospital 5/8-12/63. Present through 8/63. Ab. on horse detail 9-10/63. AWOL 12/29-31/63. Present 1-2/64. Ab. on leave 4/26-30/64. Present 5-6/64, promoted 1stCpl. Present sick in camp 7-8/64. Present 9-12/64. d. 3/23/72. Bur. Ackerman Cem., Colleton Co.

REDWOOD, ALBERT. Pvt., Co. I. Enl. Parker's Ferry 4/3/62. Present through 4/30/62. Ab. sick with "Int. Fever" in Richmond hospital 8/2/63. The rest of his records are missing.

REEDY, WILLIAM MC LURE. Pvt., Co. D. b. Chester Dist. 1/28/46. Enl. Chester CH 1/29/64. Detached Green Pond 4/26-30/64. Present 5-6/64. Ab. detached Chester CH 8/23-31/64. Present 9-10/64. Detailed as Courier, Col. Campbell at Legare's Point 11/24-12/31/64. Paroled Greensboro, N. C. 5/1/65. Farmer, Chester CH 1870 census. M. D., Clio, Marlboro Co. 1920 & 1930 censuses. d. Marlboro, S. C. 8/21/35. Bur. Hebron Cem., Clio, S. C.

REEVES, FRANKLIN E. Pvt., Co. E. b. S. C. circa 1833. Enl. Coosawahatchie 12/13/61. Detailed as nurse in hospital 5-6/62. On scout Edisto Island 7-8/62. Present 9-10/62. On picket 11-12/62. Present 3-4/63. Ab. detailed horse recruiting station 8/31/63. Present 9/63-2/64. Ab. on leave 4/21-5/1/64. Present 5-8/64. Present sick in camp 9-10/64. Present 11-12/64. Paroled Hillsboro, N. C. 4/65. Farmer, age 40, Putnam Co., Fla. 1880 census. Retired, age 85, Emporia, Volusia Co., Fla. 1920 census. d. 1923. Bur. Purdom Cem., Volusia Co., Fla.

REEVES, GEORGE W. Pvt., Co. I. b. Colleton Dist. 10/31. Farmhand age 28, Morgan Co., Ala. 1860 census. Enl. Parker's Ferry 4/3/62 age 31. Ab. sick in hospital 4/30/62. Present 5-6/62. Ab. on sick leave

8/27-31/62. Discharged for loss of night vision & limited use of right arm Staunton, Va. 11/17/62. Age 31, 5'11", dark complexion, dark hair, grey eyes, Farmer. NFR. Blacksmith, Somersville, Morgan Co., Ala. 1870 census. Blacksmith, Cherokee, Colbert Co., Ala. 1880 census. Retired, age 68, Pike Co., Ark. 1900 census.

REEVES, HENRY. Pvt., Co. E. b. Colleton Dist. 11/24/38. Res. Orangeburg Dist. Enl. Coosawhatchie 12/10/61. Present through 12/31/61. Died in hospital Adams Run 3/24/62.

REID, JACKSON C. Pvt., Co. H. b. York Dist. 4/43. Enl. Rock Hill 2/14/62. On picket Rantslee Bridge 3-4/62. Present 5-6/62. On picket 7-8/62. Present 9/62-4/63. WIA (left hip) and in Richmond hospital 7/17/63. Returned to duty 8/10/63. Ab. sick through 8/31/63. AWOL 9-10/63. Present 11-12/63. Ab. detailed horse recruiting station 2/4-29/64. AWOL 3-4/64. Present 5-8/64. Ab. detailed in Charleston Arsenal 10/24-1/15/65. Paroled Greensboro, N. C. 5/1/65.

REID, JOHN B. Pvt., Co. F. b. Easley, Pickens Dist. 10/14/46. Res. York Co. 1860 census. Enl. 9/63. Captured 1/64. Sent to Ft. Delaware. Released 6/16/65, all on pension record. Mill worker, Easley, S. C. postwar. d. Pickens Co. 10/3/26. Bur. Easley Cem.

REID, M. Pvt., Co. K. d. 8/24/61. Bur. Magnolia Cem., Charleston, S. C.

REID, JOHN B. Pvt., Co. F. b. Easley, Pickens Dist. 10/14/46. Res. York Dist. 1860 census. Enl. 9/63. Captured 1/64. Sent to Ft. Delaware. Released 6/15/65, all on pension application. No on muster rolls. Mill worker, Easley, Pickens Co. postwar. Receiving pension 9/22/19. d. Pickens Co. 10/23/26. Bur. Easley Cem.

REID, JOSEPH S. Pvt., Co. F. b. S. C. 1831. Res. Abbeville Dist. 1860 census. Enl. Pickens CH 12/4/61 age 35. Present through 12/31/61. On picket Jacksonborough 3-4/62. AWOL 6/17-8/31/62. Ab. sick 9-12/62. AWOL 3-6/63. Present 9-10/63. Ab. on horse detail 12/22/63-2/64. Ab. sick with "bronchitis & hepatitis" in Richmond hospital 3/25-4/4/64. Ab. sick in hospital 5-6/64. Present sick in camp 7-8/64, dismounted since 5/1/64. Present 9-12/64. NFR. Farmer, Keowee, Oconee Co. 1870 census. d. 1871 age 40.

RHODE, HIRAM C. Pvt., Co. I. b. S. C. 4/27/40. Farmhand, age 20, Colleton Dist. 1860 census. Enl. Parker's Ferry 4/3/62. Ab. on leave 4/30/62. On picket John's Island Ferry 5-6/62. Ab. on leave 8/30/62. Present 9/62-4/63. Ab. detached horse recruiting station 5-6/63. Ab. sick in Scottsville hospital 8/5/63. Ab. sick with typhoid fever in Richmond hospital 10/1-31/63. Ab. sick in hospital 11-12/63. Ab. sick 2/16-29/64 & 3-4/64. Present 5-8/64. Ab. on sick leave 9/21-10/31/64. Present 11-12/64. NFR. Farmer, Sheridan, Colleton Co. 1900 census. d. 1/17/04. Bur. Jordan Cem., Colleton Co.

RHODE, JAMES P. Pvt., Co. I. b. S. C. circa 1837. Farmhand, age 23, Colleton Dist. 1860 census. Enl. Parker's Frerry 4/3/62. d. of disease Adams Run 4/27/62 age 27.

RHODE, JOHN M. Pvt., Co. I. b. 12/45. Enl. 4/61 on wife's pension application. Paroled Hillsboro, N. C. 4/65. Farmer, Sheridan, Colleton Co. 1870 & 1880 censuses. Farmer, Koger, Dorchester Co. 1900 census. d. 9/27/00. Bur. Ackerman Cem., Red Oak, Colleton Co. Wife receiving pension Cottageville, Colleton Co. 10/22/19.

RHODE, WILLIAM JOHN. 3rdSgt., Co. I. b. Colleton Dist. 8/8/42. Farmhand, age 18, Colleton Dist. 1860 census. Enl. Parker's Ferry 4/3/62 as Pvt. Ab. on leave 4/27-30/62. Ab. on sick leave 5-6/62. AWOL 8/30-31/62. Promoted 3rdCpl. Present 9-10/62. Promoted 1stCpl. On scout 11-12/62. Promoted 3rdSgt. Present 3-6/63, however, captured Beverly Ford 6/9/63. Sent to Old Capitol. Exchanged 6/29/63. Issued clothing Richmond 6/30/63. Ab. on horse detail 8/18-31/63. Present 9/63-6/64. Ab. sick in hospital 8/28-31/64. Ab. on sick leave 9/25-10/31/64. Present 11-12/64. NFR. Farmer, Sheridan, Colleton Co. 1870-1910 censuses. d. 3/17/15. Bur. Fox Cem., Colleton Co.

REVELL, MITCHELL H. Pvt., Co. C. b. S. C. circa 1837. Res. Edgefield Dist. Enl. Camp Butler 9/12/61 age 24. Present 11-12/61. Presence or absence not stated 3-4/62. Present 5-6/62. Ab. on sick leave 7/12-31/62. Present 9/62-6/63. Ab. on horse detail 8/15-31/63. Present 9-10/63, however, ab. sick with bronchitis in Richmond hospital 10/3-12/18/63. Present 1-2/64. Detached Green Pond 4/16-30/64. Present 5-12/64. Captured Goldsboro, N. C. 3/31/65. Sent to Hart's Island, N. Y. Released 6/17/65. 6'1 1/2", light complexion, light hair, gray eyes. Alive 1898.

RHETT, BENJAMIN S., JR. Pvt., Co. A. b. S. C. 6/15/32. Clerk, Charleston 1860 census. Enl. 11/7/61. Discharged at end of enlistment 2/7/62. Reenlisted Co. K, 4th S. C. Cav. Grahamville 3/27/62. Promoted 4thCpl. 6/7/62. Detailed Gen. Beauregard 12/14/62-5/26/64. Paroled Burn's House, N. C. 4/20/65. Merchant, Charleston 1870 census & Clerk, Charleston 1880 census. d. 10/12/93. Bur. St. Peter's Epis. Ch., Charleston.

RHETT, ROLAND R. Pvt., Co. A. b. S. C. 1840. Res. Charleston. Enl. 11/7/61. Detailed Gen. Ripley 11/28/61. Appointed Major & QM from S. C. 11/16/61. Post QM Charleston 12/6/64. NFR. A3 30, Williamsburg, Mo. 1870 census.

RHODE, HIRAM C. Pvt., Co. I. b. S. C. 4/27/40. Farmhand, Colleton Dist. 1860 census. Enl. Parker's Ferry 4/3/62. Ab. on leave 4/30/62. On picket John's Island Ferry 5-6/62. Ab. on leave 8/30/62. Present 9/62-4/63. Detailed horse recruiting station 5-6/63. Ab. on sick leave 7-10/63. Ab. sick in hospital 11-12/63. Ab. sick 2/16-4/64. Present 5-8/64. Ab. on sick leave 9/21-10/31/64. Present 11-12/64. NFR. Farmer, Sheridan, Colleton Co. 1900 census. d. 1/17/04. Bur. Jordan Cem.

RHODE, JAMES P. Pvt., Co. I, b. S. C. 1837. Farmhand, age 23, Colleton Dist. 1860 census. Enl. Parker's Ferry 4/3/62. d. Adams Run 4/15 or 27/62 age 27.

RHODE, JOHN M. Pvt., Co. I, b. S. C. 12/45. Paroled Hillsboro, N. C. 4/65. d. 9/27/00. Bur. Ackerman Cem., Red Oak, Colleton Co.

RHODE, WILLIAM JOHN. 3rdSgt. Co. I. b. Colleton Dist. 8/5/42. Farmhand, age 18, Colleton Dist. 1860 census. Enl. Parker's Ferry 4/3/62 as Pvt. Ab. on leave 4/22-30/62. Ab. sick 5-6/62. AWOL 8/30-31/62. Promoted 3rdCpl. Present as 1stCpl. 9-10/62. On scout 11-12/62. Promoted 3rdSgt. Present 3-6/63, however, captured Beverly Ford 6/9/63. Sent to Old Capital. Exchanged and issued clothing Richmond 6/30/63. Ab. on horse detail 8/18-31/63. Present 9/63-6/64. Ab. sick in hospital 8/28-31/64. Ab. on sick leave 9/25-10/31/64. Present 11-12/64. NFR. Farmer, Sheridan, Colleton Co. 187-1910 censuses. d. 3/17/15. Bur. Fox Cem., Colleton Co.

RICHARDSON, HENRY WARREN. Pvt., Co. A. b. S. C. 8/21/44. On post war roster. Reenlisted Co. K, 4th S. C. Cav. McPhersonville 7/21/62. Present until detailed with Gen. Walker 8/28/62 for 23 days. Present 9/62 until detailed to buy forage 3-/8/63. Ab. on sick leave 10-12/63, but in Scottsville, Va. hospital with typhoid fever 8/5/63 and transferred to Richmond hospital 10/1/63. Present 1-2/64. Present until captured Cold Harbor 5/30/64. Sent to Pt. Lookout. Exchanged 3/14/65. Paroled 4/65. d. 1/4/16. Bur. Black Swamp Meth. Ch., Hampton Co.

RICHARDSON, JAMES BURCHELL. Pvt., Co. A. b. S. C. 1/2/32. On post war roster. Reenlisted Co. K, 4th S. C. Cav. Grahamville 3/25/62. Present until ab. on leave 6/28/62 for 10 days. Present until detailed with Gen. Walker 9/20/62. Present 11/62 until ab. on sick leave 8/28/63. Present 9/63 until ab. on leave 12/28/63 for 30 days. Present through 10/64. Ab. sick with "nephritis" in Danville hospital 11/18/64. Furloughed for 60 days 12/8/64. Paroled 4/65. d. 8/29/09. Bur. St. Marks Epis. Ch., Sumter Co.

RICHARDSON, JOHN. Pvt., Co. A. b. S. C. 1/4/45. On post war roster. Reenlisted Co. B, 2nd S. C. Cav. Grahamville 6/19/61. Present through 10/63. NFR. d. Asheville, N. C. 8/6/20. Bur. Prince Ford Cem., Georgetown Co.

RICHEY, JAMES ALEXANDER. Pvt., Co. A. b. S. C. 1844. Res. Saluda, Abbeville Dist. 1850 census. Enl. Abbeville CH 3/14/62. On duty Willson's Point through 4/62. Present 5-10/62. On scout 11-12/62. Present 3-4/63. WIA (slightly) Brandy Station 6/9/63. Ab. detached horse recruiting station 8-10/63. Ab. sick 11-12/63. Ab. on horse detail 1-2/64. Present 3-10/64. Detached Pocotaligo 12/10-31/64. Paroled Greensboro, N. C. 5/1/65. Farmer, age 26, Bartow, Ga. 1870 census. Farmhand, age 36, Anderson, Texas 1880 census. Retired Laborer, Palestine, Anderson Co., Texas 1900 & 1910 censuses, however, listed as d. 12/3/02. Bur. Lebanon Presb. Ch., Abbeville, S. C.

RICHEY, JOHN C. Pvt., Co. F. b. S. C. circa 1832. Enl. Walhalla 4/25/63. Present through 10/63. Ab. on horse detail 12/22/63-2/64. Ab. on horse detail 3/29-4/30/64. Ab. sick in hospital 5/25-12/31/64, however, recommended for duty in Commissary Dept., Anderson, S. C. for deafness 10/20/64. NFR. Receiving pension Denver, Anderson Co. 1901 age 69.

RICHEY, JOHN SAMUEL P. 2ndSgt., Co.'s F & K. b. S. C. 9/12/33. Enl. Pickens CH 12/4/61 age 28 as Pvt. Present through 4/62. Transferred Co. K 6/25/62. Promoted 3rdSgt. Ab. on sick leave 9/1/62. Present through 10/62, promoted 2ndSgt. Present 11/62-6/63. Ab. on horse detail 8/16-31/63. Present 9-10/63. On detached service with Signal Corps, Port Royal 12/20/63-2/64. Present 3-4/64. On picket Bee's Ferry 6/20-30/64. Present 7-8/64. On picket Bee's Ferry 9-10/64. On picket Chatman's Ford 12/29-31/64. Paroled Raleigh, N. C. 4/65. Farmer, Harmony Grove, Jackson Co., Ga. 1880 census. & Minish, Jackson Co., Ga. 1900 census. d. 12/3//03. Bur. Commerce Cem., Jackson Co., Ga.

RICHEY, OLIVER. Pvt., Co.'s G & A. b. S. C. circa 1838. Enl. Co. G Abbeville CH 1/8/62. Present through 6/62. Ab. on leave for 7 days 8/29/62. Present 9-10/62. On picket 11-12/62. Ab. on detached service 3-4/63. Present 5-10/63. Ab. horse recruiting station 11/20/63-2/64. Ab. on horse detail 4/11-30/64. Present 5-6/64. Transferred Co. A 8/8/64. Present 9-12/64. Paroled Greensboro, N. C. 5/1/65. Farmer, Fox Mills, Wilcox Co., Ala. 1880 census. d. 9/25/99 on wife's pension application.

RICHEY, WILLIAM ADDISON. Pvt., Co. G. b. S. C. 6/4/32. On postwar roster. d. 2/6/85. Bur. Greenville Presb. Ch. Cem., Abbeville Co.

RICHEY, WILLIAM A. (2). Pvt., Co. G. b. Pickens Dist. 11/25/46. Enl. Columbia 8/2/64. Present through 12/64. Paroled Greensboro, N. C. 5/1/65. Farmer, Grimlog, Franklin Co., Ga. 1880 census. d. Pickens Co. 4/11/25. Bur. Liberty, S. C.

RIGBY, CHARLES P. Pvt., Co. E. b. S. C. 9/4/30. Enl. Adams Run 1/8/62. Ab. on sick leave 4/23-8/31/62. Ab. sick Summerville hospital 9-10/62. Present 11/62-4/63. Ab. with disabled horse 8/2-31/63. Ab. horse recruiting station 9/63-2/64. Detached Green Pond 4/26-30/64. Ab. on sick leave 6/1-7/31/64. Present 8/64. On picket Western line, James Island 9-12/64. NFR. Farmer, age 47, Cow Castle, Orangeburg Co. 1880 census. d. 6/8/08. Bur. Reevesville Meth. Ch. Cem., Dorchester Co.

RIGBY, CHARLES SIMON. Pvt., Co. E. b. near Reevesville, Colleton Dist. 11/30/40. Farmer, St. George, Colleton Dist. 1860 census. Enl. Coosawahatchie 12/13/61. Present through 6/62. Ab. on detached service 7-8/62. Present 9/62-2/64, however, ab. sick with bronchitis & itch in Richmond hospital 11/20-12/22/63. Ab. on horse detail 4/20-5/15/64. Present 6-7/64. Ab. sick in Charleston hospital 8/18-31/64. Ab. on sick leave 9/4-20/64. Present 11-12/64. Paroled Hillsboro, N. C. 4/65. Farmer, St. George's, Colleton Co. 1870 census. Canvasser, Charleston 1900 census. d. Charleston 7/5/04. Bur. Reevesville, Bapt. Ch. Cem., Reevesville, S. C.

RISER, ROBERT W. Pvt., Co. E. Enl. Adams Run 7/2/62. On scout Edisto Island through 8/62. Present 9/62-6/63. Presence or absence not stated 7-8/63. Present 9-10/63. On picket 12/29-31/63. Present 1-2/64. Ab. on leave 4/21-5/1/64. Present through 8/64. On picket Western lines, James Island 9-10/64. Present 11-12/64. NFR. Married Jefferson Co., Ga. 1900.

RISHER, JOHN SILAS "JACK." Pvt., Co. I. b. Colleton Dist. circa 1842. Age 18, St. Bartholemew, Colleton Dist. 1860 census. Enl. Parker's Ferry 4/3/62. Ab. on leave 4/25-30/62. Present 5-6/62. Ab. on sick leave 8/27-31/62. Present 9/62-8/31/63. Ab. sick with chronic diarrhea in Lynchburg hospital 9/3/63. d. there 9/7/63. Bur. Old Lynchburg City Cem.

RISHER, SILAS. Pvt., Co. I. b. Colleton Dist. circa 1840. Farmhand, age 20, Colleton Dist. 1860 census. Enl. Parker's Ferry 4/3/62. Present through 8/62. d. at private house near Staunton, Va. 11/20/62, age 24. Probably buried in Thornrose Cem., Staunton, Va. with unknowns.

RITCH, J. T. Pvt., Co. K. b. circa 1845. Enl. Charlotte, N. C. 10/10/62 under 18 years of age. Detached with horses through 10/31/62. On picket Richards Ford 11-12/62. Present 3/63-6/64. Ab. on leave 7-8/64. Ab. on leave from Charleston hospital for 60 days 10/9-31/64. Detailed Dist. Hqtrs., Green Pond 12/24-31/64. Paroled Raleigh, N. C. 4/65.

RIVERS, GEORGE W. OSWALD. Pvt., Co. I. b. S. C. 3/5/44. Student, Colleton Dist. 1860 census. Enl. Co. G, 10th S. C. Inf. 9/4/61. Discharged Mfor disability 2/21/62. Age 17, Farmer, 5' 7", sallow complexion, hazel eyes, dark hair. On post war roster Co. I, 1st S. C. Cav. Reenlisted Co. I, 3rd S. C. Cav. Adams Run 3/16/62. Present until ab. on sick leave 9-10/62. Present 11/62-4/64. Ab. sick 5-8/64. Present 9-10/64. NFR. Census Taker, Verdier, Colleton Co. 1880 census. d. 3/10/15. Bur. Live Oak Cem., Walterboro, S. C.

RIVERS, PRESTON H. Pvt., Co. I. b. S. C. circa 1833. Farmer, age 27, St. Paul's, Colleton Dist. 1860 census. Enl. Parker's Ferry 4/3/62. On picket Jacksonborough through 4/30/62. Present 3-12/62. Ab. sick with "catarrhus" & debility in Staunton hospital 3/11-4/28/63 & with "Paratia." Abs. in Charlottesville hospital 5/15-6/12/63. Present 6-8/63. Ab. detached horse recruiting station 9/11/63-2/64. Detached Green Pond 4/16-30/64. Present 5-10/64. AWOL 12/20-31/64. Deserted to the enemy Charleston 3/6/65. Age 31, 5'11", dark complexion, blue eyes, dark hair, Took oath & discharged. Enl. Co. F, 2nd Confederate Engineers & paroled Chester, S. C. 5/5/65. Farmer, Burns, Colleton Co. 1880 census.

ROACH, JOHN J. Pvt., Co. H. b. S. C. 11/7/39. Store Clerk, age 20, York Dist. 1860 census. Enl. Rock Hill 7/16/62. Present through 8/31/63. Ab. on horse detail 9-10/63, however, AWOL 10/23-12/31/63. Present 1-2/64. Detached Green Pond 4/64. Present 5-6/64. Present detailed QM Dept. 8/20-31/64. Present 9-10/64. Present detailed QM Dept. 11-12/64. Paroled Salisbury, N. C. 5/1/65. Store Clerk, Yorkville 1870 census. & Rock Hill 1880 census. Retired, Rock Hill 1900 census. d. 9/31/05. Bur. Laurelwood Cem., York Co.

ROACH, THOMAS JEFFERSON. Pvt., Co. H. b. Rock Hill, York Dist. 9/21/41. Farmer, age 19, York Dist. 1860 census. On roster but not on muster rolls. Enl. Co. B, Hampton Legion Infantry Columbia 4/1/64. In Richmond hospital with powder burns 7/31-8/30/64. Present 9/64-2/65. NFR. Farmer, Yorkville 1870 census & Bethesda, York Co. 1900 census. d. Rock Hill 1/30/16. Bur. Ebenezer ARP Presb. Ch. Cem., York Co.

ROBBINS, WILLIAM J. Pvt., Co. F. b. Chester Dist. 9/41. Age 18, Carmel Hill, Chester Dist. 1860 census. En. Pickens CH 12/4/61. Ab. on leave 12/31/61. Present as Teamster in QM Dept. 3-4/62. Present as Brigade Wagon Master 5-6/62. Ab. on detached service through 12/31/62. Present until ab. on horse detail 12/22/63-2/64. Present 3-4/64. Present sick in camp 6/27-30/64. Ab. on sick leave from Charleston hospital 8/22-10/31/64. Present 11-12/65. NFR. House Carpenter, age 58, Pinckney, Union Co. 1900 census. d. 1/5/07. Bur. Mt. Pleasant Presb. Ch., Chester Co.

ROBBINS, WILLIAM M. Pvt., Co. A. b. S. C. 5/14/24. Mexican War. Vet. Farmer, Sandersville, Chester Dist. 1860 census. Enl. Camp Mears 4/20/62. Present through 4/30/62. Ab. on sick leave 5/13-6/30/62. Present sick in camp 7-8/62. AWOL 9/28-12/31/62. Horse died 1/17/63. Present until ab. sick with "Affection Kidney" in Farmville hospital 5/63. Returned to duty 5/21/63. Present until ab. sick with "int. Fever" in Richmond hospital 9/13/63. Furloughed for 30 days 10/3/63. Ab. on sick leave through 12/31/63. Present 1-2/64. AWOL 4/27-30/64. Transferred Co. A, Lucas' Bn. S. C. Arty. 5/16/64. Present through

12/64. NFR. Farmer, Baton Rouge, Chester Co. 1870 census. d. 1/6/88. Bur. Bullock's Creek Presb. Ch., York Co.

ROBERDS, WILLIAM GEORGE. 2ndLt., Co. A. b. Beaufort Dist., S. C. 1844. Enl. Abbeville Dist. 1/3/62 as 3rdCpl. On duty Wilson's Point through 4/30/62. Present 5-10/62. On scout 11-12/62. Ab. on leave 1-2/63. Present 3-4/63. Promoted 2ndLt. Ab. sick in Culpepper hospital 8/31/63. Ab. sick with debilitas in Charlottesville hospital 9/17-23/63 and Richmond hospital 10/10/63. Furloughed for 30 days 10/11/63. Present 11/63-12/64. 14 horses present 1/1/54. 11 horses present 3/30/64. 11 horses present 4/3/64, 4/30/64 & 5/4/64. Signed for 9 caps, 12 jackets, 12 pr. pants, 9 shirts, 20 pr. socks, 6 pr. shoes & 10 pr. drawers 5/20/64. 50 horses present Green Pond 6/30/64. Present through 12/64. NFR. Farmer, Hardeesville, Beaufort Co. 1870 census. d. 1/14/93.

ROBERTS, ALFRED R. Pvt., Co. G. b. Abbeville 2/25/26. M. D. On postwar roster. Appointed Asst. Surgeon. NFR. d. 6/6/76.

ROBERTS, EMANUEL M. Pvt., Co. A. b. Ga. 8/30/33. M.D., Mt. Carmel, Abbeville Dist. 1860 census. Enl. Abbeville CH 3/7/63. On detached service a nurse & dispensing medicine through 4/63. Present through 8/31/63. On detached service 9-10/63. AWOL 12/30-31/63. Ab. sick in hospital 1/8-2/29/64. Transferred 20th S. C. Inf. 5/13/64. Appointed Asst. Surgeon 6/24/64. Asst. Surgeon, Jackson hospital, Richmond 8/30-9/28/64. Assigned to Invalid Bn., Belle Island 9/64. Appointed Asst. Surgeon, PACS 3/17/65 to rank from 6/25/64. Relieved from hospital duty Richmond 3/8/65 and ordered to Gen. Hampton's command in S. C. NFR. M.D., Utica, Hinds Co., Miss. 1870 census. d. 3/7/84. Bur. Green Hill Cem., Mt. Pleasant, Texas.

ROBERTS, W. G. Pvt., Co. A. b. S. C. 10/36. On post war roster. Reenlisted Co. G, 7th S. C. Cav. 3/1/63. Surrendered Appomattox CH 4/9/65. Mail Contractor, Hill, Texas 1900 census. Receiving pension 1911.

ROBERTSON, ALEXANDER, JR. Pvt., Co. A. b. S. C. 8/18/40. Enl. 11/7/61. Reenlisted Co. K, 4th S. C. Cav. Grahamville 3/25/62. Present until detailed with Gen. Walker 12/10/62 for 30 days. Present until detailed with Gen. Ripley 10/12/63 for 15 days. Present 11/63 until KIA Hawes' Shop 5/28/64. Bur. Magnolia Cem., Charleston.

ROBERTSON, EBER RABB. Pvt., Co. A. b. S. C. 4/3/47. On post war roster. Reenlisted Co. K, 4th S. C. Cav. Charleston 11/5/63. Present through 2/64. Present until detailed as Courier for Gen. Butler 8/31/64. Wounded accidentally 9/25/64. DOW's Winnsboro 9/27/64. Bur. St. Johns Cem., Winnsboro.

ROBINSON, ARTHUR. Pvt., Co. A. b. S. C. 6/11/42. On post war roster. Reenlisted Co. K, 4th S. C. Cav. Grahamville 3/25/62. Present until promoted 1stCpl. 11-12/63. Present 1-2/64. Present until KIA Hawes Shop 5/28/64. Bur. 2nd Presb. Ch., Charleston.

ROBINSON, AUGUSTUS C. Pvt., Co. G. b. S. C. circa 1846. Enl. Camp Pines, Stevensburg, Va. 1/16/63 age 17. Ab. sick in hospital 3-4/63. Present 5/63-2/64. Transferred Co. G, Lucas's Bn. S. C. Arty. 5/13/64. Present through 12/31/64. Transferred Co. G, 1st Ga. Heavy Arty. 1/14/65. NFR. d. Catawba, N. C. 2/26/27.

ROBINSON, JOHN. Pvt., Co. A. b. S. C. circa 1831. Enl. 11/7/61. Reenlisted Co. K, 4th S. C. Cav. Grahamville 3/25/62 age 29. Present until detailed with Gen. Walker 10/28/62. Present until detailed as Courier for Col. Rutledge 3/26/63. Transferred Co. C 11/29/63. Transferred back to Co. K 10/22/64. Paroled Greensboro, N. C. 5/2/65.

ROBINSON, JOHN W. Pvt., Co. A. b. S. C. 1841. Enl. Abbeville CH 10/1/61. Ab. sick in hospital 11-12/61. Present sick in camp 3-4/62. Ab. on leave 5-10/62. Present 11/62-8/31/63. Captured Culpepper Co. 9/13/63. Sent to Old Capitol & Point Lookout. Exchanged 9/18/64. Ab. sick with chronic diarrhea in Richmond hospital 9/21/64. Furloughed for 40 days 9/25/64. Present 11-12/64. NFR. Farmer, Louisville, Clay Co., Ill. 1870 census. Living Gregg, Aiken Co., S. C. 1930 age 89.

ROBINSON, MOSES G. 1stCpl., Co. D. b. S. C. circa 1830. Res. Chester Dist. 1860 census. Enl. Chester CH 9/10/61 as Pvt., age 31. Present 11-12/61. Promoted 1stCpl. Present 3-4/62, horse killed 4/28/62. On picket Townsend's House 5-6/62. Present 7-8/62. Ab. sick with "Re. Febris" in Williamsburg, Va. hospital 10/6-11/20/62. Present sick in camp 11-12/62. d. of disease in camp Stevensburg, Va. 1/15/63 age 32.

ROBINSON, SAMUEL W. Pvt., Co. G. b. Barnwell Dist. 10/7/44. Enl. Abbeville CH 1/8/62, age 18. Present 3-4/62. Present sick in camp 5-6/62. Presence or absence not stated 7-8/62. Present 9/62-10/63. Detached horse recruiting camp 12/22/63-2/64. Present 3-6/64. Ab. under arrest 7-10/64. Detached Pocotaligo 12/10-31/64. NFR. d. Washington, Fla. 1/4/13.

ROBINSON, WILLIAM BANKS. 1stSgt., Co. D. b. S. C. 2/4/38. Res. Chester Dist. Enl. Chester CH 9/10/61 age 23 as 3rdCpl. Ab. guarding baggage 11-12/61. Present 3-6/62. Ab. on sick leave 8/4-31/62. Promoted 2ndSgt. Ab. sick Chester Dist. 9-10/62. Ab. sick 11/26-12/31/62. Present 3-6/63. Ab. on horse detail 8/18-31/63. Promoted 1stSgt. Present 9/63-12/64. Paroled Hillsboro, N. C. 4/65. Farmer, Lewisville, Chester Co. 1870 census. Retired, Blackstock, Chester Co. 1910 census. d. 3/8/17. Bur. Pleasant Grove Presb. Ch. Cem., Chester Co.

ROBINSON, WILLIAM S. Pvt., Co. G. b. S. C. 1839. Enl. Abbeville CH 1/8/62. Present 3-4/62. Ab. sick in hospital 6/62. Present a Wagoner 7-8/62. Present 9-10/62. On scout 11-12/62. Present 3-4/63. Present dismounted 5-6/63. Ab. sick with debilitas in Charlottesville hospital 7/2763. Transferred Lynchburg hospital 7/28/63. Ab. on horse detail 8/31/63. Ab. on leave 9/63. AWOL 10/15-12/31/63. Present 1-2/64. Retired to Invalid Corps 6/19/64. NFR. Farmer, Beech Springs, Spartanburg Co. 1880 census.

RODDY, JOHN BARBER. Pvt., Co. H. b. York Dist. 3/16/46. Enl. Frederick Co., Va. 3/6/64. Present through 8/64. Ab. guarding POW's Florence 9-10/64. Ab. on sick leave 11-12/64. NFR. Farmer, York Co. 1870 census. d. 6/3/76 age 30.

RODDY, WILLIAM L. Pvt., Co. D. b. Chester Dist. 1834. Merchant, age 27, Chester Dist. 1860 census. Enl. Chester CH 9/10/61 age 25. Present 11/61-6/62. Ab. on sick leave 8/27-31/62. Present 9/62-4/63. Present sick in camp 5-6/63. Ab. on horse detail 10/23-12/31/63, horse killed near Culpepper CH 9/13/63. Paid $175.00. Present 1-2/64. AWOL 4/29-30/64. Present 5-6/64. On picket 7-8/64. AWOL 10/28-31/64. Present 11-12/64. NFR. Dry Goods Merchant, Yorkville, S. C. 1870 census. Capitalist, Rock Hill, York Co. 1900 census. d. Chester 4/12/02. Bur. Evergreen Cem., Chester Co. Wife receiving pension Whitmire, Newberry Co. 10/16/19.

RODGERS, WILLIAM N. Pvt., Co. F. b. S. C. circa 1838. Enl. Columbia 8/3/64. Present through 12/64. NFR. Farmer, age 42, Lee Co., Miss. 1880 census.

ROGERS, N. C. Pvt., Co. F. b. S. C. 3/16/38. Enl. 3/7/63 on Mississippi pension record. Served to 5/4/65. d. Tupelo, Miss. 11/22/19--. Bur. Old Union Ch. Cem., near Shannon, Miss.

ROLAND, M. O. Pvt., Co. K. Enl. Columbia 5/10/64 age 26. Ab. on leave from Charleston hospital 10/9/64 for 60 days. On picket Chatman's Ford 12/31/64. d. at home 1865 on post war roster.

ROLAND, THOMAS JEFFERSON. Pvt., Co.'s F & K. b. S. C. 12/27/38. Enl. Co. F Pickens CH 4/22/62. Present through 4/30/62. Transferred Co. K 6/25/62. Present 7-10/62. On picket Richards Ford 11-12/62. Present 3-6/63. Horse killed Upperville 6/21/63. Paid $200.00. Ab. on horse detail 8/16-31/63. AWOL 9-10/63. Present 11/63-4/64. On picket Charleston fortifications 6/25-30/64. Ab. sick in Charleston hospital 7-8/64. Furloughed for 60 days 9/5/64. On picket Chatman's Ford 12/31/64. Paroled Raleigh, N. C. 4/65. Farmer, Mountain Dist. Walton Co., Ga. 1900 census. d. Penton, Chambers Co., Ala. 12/25/06. Bur. Mt. Hickory Bapt. Ch. Cem., near Penton, Ala.

ROMANISKI, TADOR. Pvt., Co. E. Enl. Adams Run 2/6/62 age 42. Present sick in camp 3-4/62. Present 5-8/62. Present as Farrier 9-10/62. Present 11/62-10/63. Discharged by order of Sec. of War 12/13/63.

NFR.

RONEY, WILLIAM THOMAS. Pvt., Co. D. b. Surry Co., N. C. 8/10/37. Married in S. C. 1859. Landlord, Chester Dist., 1860 census. Enl. Columbia 6/8/63. Discharged for disability 8/26/63. Enl. Co. A, 17th S. C. Inf. Columbia 5/1/64. Present 7-8/64. Present sick 9/6-2/65. Surrendered Appomattox CH 4/9/65. d. Richland, Navarro Co., Texas 10/15/27.

ROPER, NEWMAN. Pvt., Co. F. Enl. Pickens CH 12/4/61. Present through 12/31/61. d. at home in Pickens Dist. 3/1/62.

ROSE, --------. Pvt., Co. C. Present 8/64. NFR.

ROSE, ALEXANDER. Pvt., Co. A. b. S. C. 1/14/39. Res. Spartanburg Dist. Enl. 11/7/61. Reenlisted Co. K, 4th S. C. Cav. Grahamville 3/25/62 as 4thCpl. Promoted 3rdCpl. 4/25/62. Present until detailed with Gen. Walker 9/22/62 & with Gen. Beauregard 11/7/62. Discharged and appointed 1stLt. & Aide-De-Camp, Gen. W. G. Gardner 12/3/62. Promoted Captain. d. of typhoid fever in Jackson, Miss. hospital 3/24/63. Bur. Unitarian Ch. Cem., Charleston.

ROSE, EDWARD M. Pvt., Co. K. b. Spartanburg Dist. 5/6/45. Enl. Co. I, 5th S. C. Inf. 10/8/61. Present through 12/31/61. Ab. sick with typhoid fever in Danville hospital 1/25/62. Transferred Charlottesville hospital. NFR. Probably discharged for underage. Enl. Co. K, 1st S. C. Cav. Columbia 4/11/64. Ab. on horse detail through 4/30/64. Present 5-6/64. Detailed to drive cattle for QM Dept. 7-8/64. Ab. on sick leave for 30 days 10/24/64. Detailed driving cattle to Florence, S. C. 11-12/64. Paroled Raleigh, N. C. 4/65. Hotel Clerk, Columbia 1880 census. County Treasurer, York Co. d. Columbia 3/17/90. Bur. Rose Hill Cem., York, S. C.

ROSENDORFF, MICHAEL J. Pvt., Co.'s K & C. b. Prussia 10/39. Enl. Hamburg 8/27/61 age 18. Courier for Gen. Evans 11-12/61. Transferred Co. C 7/8/62. Ab. sick in Richmond hospital 8/21-10/31/62. AWOL 11/63-4/63. NFR. Storekeeper, age 41, Richmond, Va. 1880 census. Cigar Dealer, Boston Mass., 1900 census. Died Boston, Mass. 9/6/15 Buried St. Joseph's Cemetery.

ROSS, ALLEN A. Pvt., Co. H. b. S. C. circa 1841. Enl. Rock Hill 4/13/63. Ab. on leave through 4/63. WIA (left lung by minnie ball) & captured Gettysburg 7/3/63 age 22. Sent to General hospital 7/24/63. Transferred to Provost Marshal 9/14/63. In West buildings hospital, Baltimore 9/25/63. Exchanged. In Richmond hospital 9/29/63. Furloughed for 30 days. NFR. May have served later in 5th S. C. Cav. but not on muster rolls of that unit.

ROSS, DAVID. Pvt., Co. E. b. Orangeburg Dist. 2/19/31. Farmer, age 28, Poplar, Orangeburg Dist. 1860 census. Enl. Fort Motte 10/26/61 age 30. On picket White Point through 12/31/61. Present 3-6/62. On picket at Railroad 7-8/62. Present 9-10/62. On picket 11-12/62. Present 3-4/63. Ab. on horse detail 8/31/63. Present 9-12/63. Ab. on leave 2/12-3/3/64. Ab. on leave 4/20-5/20/64. Present 5-8/64. Detailed with QM 9-10/64. Present 11-12/64. Paroled Hillsboro, N. C. 4/65. Farmer, Pinckney, Union Co. 1870 census & Pine Grove, Orangeburg Co. 1880 census. d. Atmore, Ala. 10/30/09.

ROSS, JEFFERSON JACOB. Pvt., Co. E. b. S. C. 11/17/45. Res. Orangburg Dist. Enl. by 1865. Paroled Hillsboro, N. C. 4/65. Blacksmith, Pine Grove, Orangeburg Co. 1870 census. Farmer, Friendship, Claredon Co. 1880 & 1900 censuses. d. 12/6/19. Bur. Andrews Chapel, Clarendon Co., S. C.

ROSWELL, W. D. Pvt., Co. A. WIA & captured Gettysburg 7/3/63 on Record of Events for Co. A. Not on muster rolls.

ROTHROCK, DAVID BENJAMIN. Pvt., Co. K. b. Stokes Co., N. C. 9/14/25. Master Blacksmith, age 35, Chester Co. 1860 census. Enl. Columbia 4/1/63 age 38. AWOL 7/63-2/64. Present 3-4/64. Retired to Invalid Corps 5/3/64. Age 38, 5'10", fair complexion, blue eyes, light hair, Blacksmith. Detailed in Subsistence Dept., Cheraw, S. C. 5/29/65. NFR. d. Washington, D. C. 4/6/89.

ROTHROCK, SOLOMON JACOB J. Pvt., Co. K. b. N. C. 2/12/34. Blacksmith, age 26, Chester Dist. 1860 census. Enl. Columbia 5/20/64. Present through 8/64. d. in Charleston hospital 9/27/64 age 32. Bur. Magnolia Cem., Charleston, S. C.

ROWELL, GEORGE R. Pvt., Co. C. b. S. C. 8/23/31. Enl. Wadmalaw Island 5/21/62. Present sick in camp through 6/62. Ab. on sick leave from hospital 8/21-31/62. Present 9-10/62. On picket 11-12/62. Present 3-4/63. Ab. on detached service 8/24-31/63. Present 9-12/63 & 1-2/64. Detached Green Pond 4/16-30/64. Present 5-8/64. Ab. on leave for 12 days 10/27/64. Present 11-12/64. NFR. Farmer, Hammond, Aiken Co. 1880 & 1900 censuses. d. 4/12/06. Wife receiving pension Aiken, S. C. 10/9/19.

ROWELL, M. H. Pvt., Co. C. b. S. C. circa 1839. Res. Lancaster Co. 1850 census. On post war roster.

ROWLING, M. O. Pvt., Co. K. Res. Spartanburg Dist. Paroled Raleigh, N. C. 4/65.

ROY, F. S. Pvt., Co. H. Enl. Rock Hill 1/9/63. Ab. sick with debilitas in Charlottesville hospital 1/12-13/63. Present 3-4/63. Captured date & place unknown and on paroled 8/31/63. Present 9-12/63. Ab. detached 1/12-2/29/64. Ab. on leave 3-4/64. Transferred C. S. Navy 5/13/64. NFR.

RUMBLEY, JOHN DAVID. Pvt., Co. C. b. Bath, Aiken Dist. 2/22/41. Farmer, age 18, Greenville, Edgefield Dist. 1860 census. Enl. Camp Butler 915/61. Present 11-12/61 & 5-10/62. Ab. sick in Richmond hospital 11/21-12/31/62. Ab. on leave 1/10/63 for 40 days. Present 3-4/63. Ab. on detached service 7/15-8/31/63. Present 11/63-6/64. Ab. sick in hospital 8/19-31/64. Ab. on sick leave for 60 days 9/5/64. Present 11-12/64. Paroled Augusta, Ga. 5/29/65. Farmer, Langley, Aiken Co. 1880-1920 censuses. Retired, Langley, Aiken Co. 1930 census. d. Aiken 5/17/31. Bur. Beth Armsby Cem.

RUMBLEY, THOMAS R. Pvt., Co. C. b. S. C. 3/17/39. Farmer, Graniteville, Edgfield Dist. 1860 census. Enl. Adams Run 12/23/61. Ab. detached Church Flats 5-6/62. Present 7-10/62. Ab. sick in Richmond hospital 11/21/62-1/15/63. Present 3/63-2/64. Ab. on leave 4/20-5/3/64. Present 6-10/64. Detached Pocotaligo 12/10-31/64. Paroled Augusta, Ga. 5/20/65. Farmer, Gregg, Aiken Co. 1880 census. Farmer & Carpenter, Hammond, Aiken Co. 1900 census. d. 11/9/09. Bur. Old Pine Grove Cem., North Augusta, S. C.

RUMPH, ARCHIE WASHINGTON. 1stSgt., Co. I. b. Dorchester Dist. 8/24/39. Enl. as 1stLt. in Smith's Co. (Beech Hill Rangers) 1st (Martin's) S. C. Mounted Militia 9/61. Served through 2/62. Disbanded. Enl. Co. I, 1st S. C. Cav. Parker's Ferry 4/3/62 as 1stSgt. Present through 10/62. Ab. Fredericksburg as witness for CM 11-12/62. Present 3-8/63. Ab. on horse detail 9-10/63. AWOL 12/30-31/63. Present 1-2/64. Detached Green Pond 3-4/64. Present 5-12/64. NFR. Farmer, Kozer, Dorchester Co. 1880-1900 censuses. Farmer, Rogers, Dorchester Co. 1910 census. d. Burns, Dorchester Co. 7/28/16. Bur. Murray Cem., Grover, Dorchester Co.

RUSH, BENJAMIN. Pvt., Co. E. b. S. C. 7/20/41. On post war roster. d. 8/23/73. Bur. Trinity Luth. Ch., Orangeburg Co.

RUSH, JOHN W. Pvt., Co. I. b. S. C. 1836. Age 25, Orangeburg Dist. 1860 census. Enl. Abbeville CH 4/22/62. Ab. on leave through 6/62. Present sick in camp 7-8/62. Ab. sick in Summerville hospital 9-10/62. d. of "Febris Congest" in Gordonsville hospital 11/29/62. Effects: Sundries & #3.25. Bur. Confederate Cem., Gordonsville, Va. as unknown.

RUSH, LEWIS BENJAMIN. Pvt., Co. E. b. S. C. 7/20/41. Age 19, Poplar, Orangeburg Dist. 1860 census. Enl. Ft. Motte 10/26/61 age 25? On picket White Point 11-12/61. Present sick in camp 3-4/62. Present 5-6/62. On scout Edisto Island 7-8/62. Present 9/62-6/63. Ab. on horse detail 8/18-31/63. Present 9-10/63. On picket 12/29-31/63. Present 1-2/64. Ab. on horse detail 3/31/64. Present 5-6/64. Detailed horse infirmary camp 8/23-32/64. Present 9-10/64. Detached Pocotaligo 12/10-31/64. Paroled Hillsboro, N. C. 4/65. Farmer, Poplar, Orangeburg Co. 1870 census. d. Goodby's, Orangeburg Co. 8/23/73. Bur.

Trinity Lutheran Ch. Cem., Orangeburg Co. Wife receiving pension Elloree, Orangeburg Co. 10/11/19.

RUSSELL, DAVID HAMILTON. Pvt., Co. F. b. Anderson Dist. 1/4/41. Student, age 19, Anderson Dist. 1860 census. Att. U. of Va. Served in 4th S. C. Inf. 1861. Enl. Co. F, 1st S. C. Cav. Colleton Dist. 5/10/62. Present through 10/62. On detached service with Signal Corps 12/18/62-10/63. Ab. on horse detail 12/22-31/63. On detached service with Signal Corps 1-2/64. Age 23, 5'11", light complexion, dark hair, blue eyes, Lawyer. Ab. on leave 4/27-30/64. Present 5-6/64. Courier, Col. Black 7/28-12/31/64. NFR. Att. U. of Va. Farmer, Hopewell, Anderson Co. 1880 census. Publisher, "Daily Mail," Anderson, S. C. Journalist & Editor, Anderson, S. C. 1900 census. Insurance Agent, Anderson, S. C. 1910 census. d. Anderson, S. C. 11/9/15. Bur. Anderson. Brother of Thomas W. & William W. Russell.

RUSSELL, HENRY D. Pvt., Co. A. b. S. C. circa 1843. Farmhand, age 17, Abbeville Dist. 1860 census. Enl. Abbeville CH 8/13/61 age19. On duty Bears Island 11-12/61. Present 3-6/62. Ab. on sick leave 7-8/62. Present 9/62-6/63. WIA (abdomen, severely & right hand) Gettysburg 7/3/63. Captured 7/4/63. Sent to David's Island, N. Y. Exchanged 9/23/63. Ab. wounded in Richmond hospital 9/28/63. Furloughed for 30 days 10/7/63. Ab. on leave through 4/30/64. Present 5-6/64. Retired to Invalid Corps 8/18/64. NFR. Laborer, age 28, Mobile, Ala. 1870 census.

RUSSELL, LOUIS HENRY. 1stLt., Co. A. b. Abbeville Dist. 5/12/36. Livery Stable Keeper, age 27, Abbeville, 1860 census. Served briefly in Capt. James M. Perrin's Co. Inf. from Abbeville Dist. in 1861. Enl. Co. A, 1st S. C. Cav. Abbeville CH 8/13/61 age 25 as 2ndLt. On duty Bears Island 11-12/61. Accidently wounded (foot) by discharged of a pistol 1/3/62. Ab. wounded through 6/62. On detached duty 11/62-2/63 Abbeville Dist. Transferred to Enrolling Dept. 5-6/63. Requested leave 9/23/63. Present Lexington, S. C. 4/9/65. NFR. Grocer, Abbeville 1880 census. Farmer, Abbeville 1900 census. Liveryman, Abbeville 1910 census. d. Abbeville 4/19/10. Bur. Upper Cane Cem., Abbeville Co. Wife receiving pension Abbeville 1919.

RUSSELL, M. A. H. Pvt., Co. D. b. S. C. circa 1841. Age 9, Chester Dist. 1850 census. Enl. Chester CH 9/10/61 age 20. Ab. on sick leave 10/31/61. Present 11/61-6/62. Ab. on sick leave 8/28-31/62. Present 9-10/62. On scout 12/25-31/62. Present 3/63-8/64, however, ab. sick with abscess of jaw & chronic diarrhea in Richmond hospital 12/13-22/63. Ab. detailed guarding POW's Florence 9/18-12/31/64. Paroled Hillsboro, N. C. 4/65. Farmer, age 28, Toccopola, Lafayette Co., Miss. 1870 census.

RUSSELL, SAMUEL LEE. Pvt., Co. A. b. Abbeville Dist. circa 1839. Farmhand, Due West, Abbeville Dist. 1860 census. Enl. Abbeville CH 8/13/61 age 22. On duty Bears Island 11-12/61. Present 3-6/62. Ab. sick in Summerville hospital 7-8/62. Present 9/62-6/63. Captured Harrisburg, Pa. 7/7/63. Sent to Ft. Delaware. Released 6/10/65. 6'4", light complexion, dark hair, grey eyes. d. Orangeburg, S. C. 10/26/18 age 84. Bur. Sunny Side Cem., Orangeburg.

RUSSELL, SAMUEL OLIVER. Pvt., Co. A. b. Abbeville Dist. circa 1842. Enl. Abbeville CH 3/19/62. Present through 12/62. Ab. detached 3-4/63. Present until WIA Brandy Station 8/1/63. DOW's in Culpepper CH hospital 8/15/63. Probably buried in Fairview Cem., Culpepper in unknown grave site.

RUSSELL, THOMAS HOGAN. 2ndLt., Co. F. b. Cross Creek, Pickens Dist. 9/10/20. Farmer, Equality, Anderson Dist. 1860 census. On postwar roster. Landloard, Williamston, S. C. 1900 census. d. by 1902. Bur. Carmel Presb. Ch., Pickens Co.   Father of Thomas W. Russell.

RUSSELL, THOMAS WALLACE. 2ndLt., Co. F. b. Anderson Dist. 3/8/43. Student, age 16, Equality, Anderson Dist. 1860 census. Enl. Colleton Dist. 5/10/62 as Pvt. Present through 10/62. On picket 11-12/62. Present 3-4/63, promoted 4thSgt. Present 5/63-6/64. Promoted 2ndLt. 8/27/64. Present sick in camp 9-10/64. Present 11-12/64. NFR. Farmer, Brushy Creek, Anderson Co. 1870 census. d. Slabtown, Anderson Co. 3/9/79. Bur. Carmel Presb. Ch. Cem., Pickens Co. Brother of David H. & William W. Russell.

RUSSELL, WILLIAM WALKER. Pvt., Co. F. b. Slabtown, Anderson Dist. 3/6/45. Student, age 14, Equality, Anderson Dist. 1860 census. Enl. Pickens CH 12/4/61 age 18. On picket Jacksonborough

3-4/62. On picket Camp Means 5-6/62. Ab. on leave 7-8/62. Present 9-10/62. Ab. detailed with Signal Corps 3-8/31/63. Ab. on scout 9-10/63. Ab. on horse detail 12/22/63-2/64. Ab. on detail as scout 3/25-10/64, with Gen. Hampton's Division. WIA (shoulders, thigh & stomach) & captured Proctor's Cross Roads near Petersburg 10/11/64. Sent to Pt. Lookout. Exchanged 3/15/65. Paroled Greensboro, N. C. 4/26/65. Farmer, Pendleton, Anderson Co. 1880 census. Contractor, Pendleton, Anderson Co. 1900 & 1910 censuses. Receiving pension Anderson 9/23/19. Retired, Anderson 1920 census. d. Minneapolis, Minn. 3/18/24. Brother of David H. & Thomas W. Russell.

RUTLEDGE, BENJAMIN HUGER. Captain, Co. A. b. S. C. 6/4/29. Gd. Yale U. 1848. Lawyer. Enl. 11/7/61. Reenlisted Co. K, 4th S. C. Cav. Grahamville 3/25/62. Present until ab. on leave 8/21/61 for 10 days. Present until appointed Colonel, 4th S. C. Cav. 12/17/62. Present through 1/64. Commanding Brigade Hertford, N. C. 11/17/64 & Belfield, Va. 12/31/64 & 3/8/65. NFR. d. 4/30/93. Bur. Magnolia Cem., Charleston. .

RYAN, MICHAEL. Pvt., Co.'s C & K. b. Ireland circa 1839. Enl. Co. C Hamburg 8/27/61, age 22. Detailed Wadmalaw Island 11-12/61. Transferred Co. K 6/27/62. Present 7-10/62. Ab. on scout 11-12/62. Present 3-4/63. Detached recruiting horses 8/11-31/63. Present 9/63-4/64. Wagon Master for Regiment, James Island 5-12/64. Paroled Augusta, Ga. 5/25/65. Possibly the Michael Ryan who died Charleston 2/3/68 age 26. Bur. St. Lawrence Cem.

SAMMONS, DANIEL M. Pvt., Co. D. Enl. Chester CH 7/28/62. Present sick in camp through 8/62. Ab. sick in Summerville hospital 9/28-10/31/62. AWOL 11-12/63. Present 3-4/63. d. of disease Lynchburg, Va. 6/63. Bur. Old City Cem., Lynchburg.

SAMS, CALHOUN. Surgeon. b. Beaufort, S. C. 1/12/28. Gd. S. C. Medical Collage 1861 age 33. Enl. 9th (11th) S. C. Inf. Hardeesville 11/1/61 as Pvt. & Acting Asst. Surgeon. Present until discharged 4/12/62. Assigned to Winder hospital, Richmond. Appointed Asst. Surgeon PACS from S. C. at Columbia 5/14/62. Assigned to 1st N. C. Cav. 12/10/62-3/15/63. Assigned to 1st S. C. Cav. 4/63 as Surgeon Yates was a POW. Present through 2/64. Ab. on leave 3-4/64. Promoted Surgeon 4/16/64. Assigned to Phillips Ga. Legion 6/25/64. Relieved 7/18/64 & assigned to 1st S. C. Arty. Assigned to Battery Dantzler, Charleston, S. C. 11/64. Ab. on leave 3/7/65. Ab. sick in Stuart hospital, Richmond 4/2/65. Ab. sick in Jackson hospital, Richmond 4/27/65, however, paroled 4/18/65 as Surgeon, 1st S. C. Arty. Deserted from Jackson hospital, Richmond 5/2/65. d. Taylor, Texas 11/08. Bur. City Cem., Taylor, Texas.

SAMS, WILLIAM BERNERS. Pvt., Co. K. b. Charleston 2/12/27. Enl. Co. I, 3rd S. C. Cav. 2/17/62. NFR. Enl. Co. K, 1st S. C. Cav. 4/11/62. Surrendered Greensboro, N. C. 4/65 on pension application. Receiving pension Brevard Co., Fla. 8/17/07. d. Contrey, Brevard Co., Fla. 1/3/10. Bur. St. Luke's Cem., Brevard Co., Fla.

SAMUELS, GEORGE W. 1stCpl., Co. C. b. S. C. circa 1839. Res. Edgefield Dist. 1850 census. Enl. Hamburg 8/27/61 age 21 as 4thCpl. Present 11-12/61. On picket Bears Bluff 1-2/62. Present sick in camp 5-6/62. Promoted 3rdCpl. Ab. on sick leave 8/6-31/62. Promoted 1stCpl. Present 9/62-6/63. Ab. on horse detail 8/15-31/63. Present 9-10/63. Ab. on detached service 12/19/63-2/64. Ab. on leave 4/20-5/3/64. Present through 8/64. d. of typhoid fever in 7th Dist. hospital, James Island 10/5/64.

SAMUELS, WILLIAM R. Pvt., Co. C. b. S. C. 9/41. Enl. Wadmalaw Island 3/28/62. Present sick in camp 5-6/62. Ab. on sick leave 7/12-8/31/62. Present 9-10/62. On picket 11-12/62. Present 3-10/63. On detached service recruiting horses 12/22/63-2/64. Present 3-6/64. Ab. on leave for 30 days 8/22/64. Present through 12/64. NFR. Farmer, Schultz, Aiken Co. 1880-1910 censuses. d. 4/10/15. Bur. Sweetwater Cem., Aiken Co., but no marker.

SANDERS, JOHN. Pvt. Co. A. Enl. Abbeville CH 8/2/61, age 20, M. D. Not on muster rolls 10/11/61. Living Abbeville Co. 1902. Possibly John Josiah Sander 7/14/44-1/15/12. Bur. Friendship Bapt. Ch., Barnwell Co.

SANDERS, WILLIAM H. Pvt., Co. F. b. 7/3/43. Not on muster rolls. Reenlisted Co. K, 7th S. C. Cav. d.

5/26/14. Bur. Sardis Cem., Comanche Co. Texas. Tombstone only record of service.

SARGENT, HENRY WARREN. Pvt., Co.'s F & K. b. Pickens Dist. 1838. Enl. Co. F, 4th S. C. Inf. Columbia 6/2/61 age 19, as Sgt. WIA (knee) Bull Run 7/21/61. Discharge 11/2/61. Reenlisted Co. F, 1st S. C. Cav. Pickens CH 12/4/61. Ab. on sick leave 12/31/61. Ab. sick in Brigade hospital 3-4/62. Transferred Co. K 6/27/62. Discharged for gunshot wound in knee 7/2/1/61 & total loss of vision in one eye & partial loss in the other 9/18/62. Age 23, 5'10", dark complexion, black hair, grey eyes, Farmer. Enl. Ferguson's Beauregard Battery 4/8/63. Present through 12/63. NFR. d. DeKalb, Ala. 10/23/96. Bur. Piney Grove Cem.

SARGENT, W. H. Pvt., Co. K. d. in S. C. during the war on post war roster.

SAUNDERS, THOMAS MABRY. 2ndCpl., Co. D. b. S. C. 2/3/44. Enl. Chester CH 9/10/61 as Pvt. Ab. on sick leave 11/22/61-1/62. Present 3-6/62. Captured & ab. on parole 7-8/62. Present 9-10/62, promoted 4thCpl. On scout 11-12/62. Promoted 3rdCpl. Present 3-4/63, promoted 2ndCpl. Ab. detailed as Courier on Gen. Stuart's staff 4/13-10/31/63. Ab. on horse detail 12/22/63-2/64. Detached Green Pond 4/26-30/64. Present 5-12/64. Paroled Greensboro, N. C. 5/1/65. Farmer, Baton Rouge, Chester Co. 1870 census & Chester City 1880 census. d. 10/9/11. Bur. Evergreen Cem., Chester, S. C.

SCHILLER, LOUIS. Pvt., Co.'s C & K. b. Sachen, Saxony, Germany circa 1841. Age 19, Hamburg, Edgefield Dist. 1860 census. Enl. Co. C Hamburg 8/27/61 age 20. Detached on Wadmalaw Island 11-12/61. Transferred Co. K 6/27/62. Transferred back to Co. C 7/8/62. Ab. on sick leave 8/7/62. Present 9-10/62. On scout 11-12/62. Present 3-10/63. Ab. sick with "Haemorrhois" in Richmond hospital 11/29/63. Furloughed for 40 days 12/9/63. Ab. on sick leave 12/12/63-2/64. Ab. on horse detail 4/10-5/10/64. Present 5-8/64. Detailed as Clerk for Surgeon Leby, Dist. hospital, James Island 9/6-12/31/64. NFR. Trial Justice, age 29, Hamburg, Edgefield Co. 1870 census. Post Master, Schultz, Aiken Co. 1900 census. d. Milam, Texas. 6/12/18.

SCHNEIRLE, JOHN M. Pvt., Co. I. Gd. Yorkville, S. C. Mil. Academy & served as Principal of the school. Enl. 23rd S. C. Inf. 9/23/61 and appointed Adjutant. Present through 12/31/61. Appointed Drillmaster, 1st Mil. Dist. of S. C. 9/12/62.1stLt., & Enrolling Officer, Charleston, Beaufort & Colleton Districts 10/30/62-11/24/62. 1stLt. & Drillmaster, Camp of Instruction, Columbia 4/4/63-2/20/64, however, served as Aide de Camp Gen. Ripley 7/8-18/63., same for Gen. Colquitt 7/16/63 & Col. Keitt 8/2-4/63, & Gen. Ripley 1/27/64. Enl. Co. I, 1st S. C. Cav. Columbia 4/27/64. Ab. on leave through 4/30/64. Present 5-6/64. Detailed Headquarters, Legares Point 7/10-12/31/64. NFR.

SCOTT, JOHN C. Pvt., Co. G. b. S. C. circa 1844. Served in Co. B, 4th S. C. Bn. Cav. 1861. NFR. Enl. Abbeville CH 3/15/62. Present through 4/62. Ab. on sick leave 6/27-30/62. AWOL 7-8/62. Present 9-10/62. Present sick in camp 11-12/62. NFR. Farmhand, age 26, Calhoun Hill, Abbeville Co. 1870 census. Farmer, age 36, Desoto Co., Miss. 1880 census.

SCOTT, THOMAS B. 2ndSgt., Co. G. Enl. Abbeville CH 1/8/62 as 3rdSgt. Present through 4/62. Present sick in camp 5-8/62. Promoted 2ndSgt. 9/15/62. Present through 10/62. Ab. sick in Richmond hospital 12/9-31/62. Present sick in camp 3-4/63. Ab. sick in Farmville hospital 5/3-6/1/63. Ab. sick with chronic bronchitis in Richmond hospital 7/3/63. Transferred Staunton hospital 8/11/63. Furloughed to Calhoun Mills, S. C. for 40 days 9/7/63. Ab. on sick leave through 4/64. Present 3-4/64. Detailed to horse infirmary camp Lancaster, S. C. 5/2-12/64. NFR. d. by 1902 on Abbeville Co. postwar roster.

SEABRIGHT, JOHN H. Pvt. Co. I. On post war roster. d. 2/28/10. Bur. Stonewall Confederate Cem., Winchester, Va.

SEABROOK, HENRY. Pvt., Co. A. b. S. C. 9/13/36. On post war roster. Reenlisted Co. K, 4th S. C. Cav. Grahamville 3/25/62. Present until discharged 7/19/63. Apt. Lt., Smith's S. C. Bn. Appointed 2ndLt. PACS from S. C. 10/4/62. Assigned to 2nd B., S. C. Arty. Appointed Captain 10/6/63. Served as Ordnance Officer 1st. Mil. Dist. of S. C. & Charleston Armory. Paroled Greensboro, N. C. 5/2/65. d. 7/18/72. Bur. Magnolia Cem., Charleston.

SEABROOK, JOSEPH WHALEY. Pvt., Co. A. b. S. C. 1835. On post war roster. Reenlisted Co. K, 4th S. C. Cav. Grahamville 3/21/62. Present until transferred to Marion, S. C. Arty. 9/13/62, in exchange for J. M. Prioleau. Present until detached to Signal Corps 3/3/63-12/31/64, as Signal Operator. NFR. d. 1/24/06. Bur. Trinity Epis. Ch., Charleston.

SEABROOK, PERONNEAU FINLEY, Pvt., Co. G. b. Edisto Island, S. C. 2/2/47. Student, age 13, St. Paul's, Colleton Dist., S. C. 1860 census. Enl. Co. A, 25th S. C. Inf. Charleston 11/1/62. WIA (foot & leg) 5/16/64. Ab. wounded through 10/64. NFR. Enl. Co. G, 1st S. C. Cav. by 1865. Captured Bacon's Bridge, S. C. 4/10/65. Age 18, 5'4", fair complexion, light hair, blue eyes. Confined Charleston 5/27/65. NFR. d. 12/13/02. Bur. Magnolia Cem., Charleston.

SEAGO LAWRENCE T. Pvt., Co. C. b. Edgefield Co. circa 1839. Farmhand, age 21, Hamburg, Edgefield Dist. 1860 census. Enl. Camp Butler 9/17/61. Ab. sick in hospital 11-12/61. Present 5-6/62. Ab. on sick leave 8/26/62-4/63. Detached recruiting horses 11/20/63-2/64. Ab, on horse detail 4/12-5/2/64. Present 5-8/64. On picket 9-10/64. Present 11-12/64. NFR. Farmer, Butler, Edgefield Co. 1870 census. d. in Ga. 1/07. Bur. Raytown Cem., Sharon, Ga.

SEALS, JOHN R. Pvt., Co. G. b. S. C. circa 1830. Enl. Capt. Perryman's Co., Co. F, 2nd S. C. Inf. (Palmetto), Greenwood, S. C. 4/17/61. On detached service through 4/63. Enl. Co. G, 1st S. C. Cav. Abbeville CH 3/7/63, age 33. Present 5-6/63. Ab. on horse detail 8/31/63. Present 9/63-2/64. Transferred Co. G, 2nd S. C. Cav. 3/20/64. Present through 6/30/64. Ab. sick in Georgetown, S. C. hospital 8/12-31/64. Present 9-12/64. NFR. Living Abbeville 1902.

SEAWRIGHT, JOHN H. Pvt., Co. I. b. S. C. 1/15/48. Res. Anderson Dist. 1860 census. Served as 1stLt., Co. B, 4th Bn. S. C. Reserves 1863-64. Enl. Co. I, 1st S. C. Cav. Coosawahatchie 2/2/64. AWOL 4/3/64. Transferred Co. E, 20th S. C. Inf. 5/23/64. Present through 6/64. Paroled Greensboro, N. C. 5/2/65. Clerk, Greenville, Butler Co., Ala. 1870 census. Clerk, Bath, Anderson Co., S. C. 1880 census. d. 8/24/11. Bur. Greenville Presb. Ch. Cem.

SEGARS, LOVE D. Pvt., Co. A. Enl. Columbia 2/1/64. Detached Green Pond 3-4/64. Ab. sick 6/78/31/64. AWOL 11-12/64. Ab. sick with chronic rheumatism in Charlotte, N. C. hospital 2/6/65. Returned to duty same day.

SHANDLEY, JOHN ARTHUR. Pvt., Co. E. b. S. C. circa 1846. Enl. before 1/2/62 when discharged being a Student & underage. Reenlisted 2 /6/62 Adams Run. Present 3-10/62. On scout 11-12/62. Present 3-10/63, however, ab. injured by horse kick in Farmville hospital 5/6-6/11/63 & ab. sick with bronchitis in Charlottesville hospital 6/18-7/2/63. On picket 12/29-31/63. Detached at horse recruiting station 1/20-2/29/64. Detached Green Pond 4/26-30/64. Present 5-12/64. Paroled Hillsboro, N. C. 4/65. Farmer, age 54, Throckmorton Co., Texas 1900 census & age 64, 1910 census. d. Throckmorton, Texas 1/1/11 age 65.

SHANKLIN, JULIUS LEWIS. Pvt., Co. F. b. Oconee Co. S. C. 9/30/28. Enl. Co. K, 4th S. C. Inf. as Captain 6/7/61. Disbanded 62. Reenlisted Co. A, 12th Bn. Ga. Arty. Augusta, Ga. 4/10/62. Transferred Co. A, 63rd Ga. Inf. 5/26/62. NFR. Enl. Co. F, 1st S. C. Cav. Pickens CH 4/1/63. Present 5-6/63. Ab. on detached service 8/18-31/63. AWOL 9/30-12/31/63. Present 1-2/64. Ab. detached Signal Corps, Gen. Hampton's Division 3/26-12/64. NFR. Receiving pension Columbus, Ga. 1906. Farmer, Wagner, Oconee Co. 1900 census. d. 12/6/06. Bur. Richland Presb. Ch., Oconee Co.

SHARPE, EDWIN ALPHONSO. Ordnance Sgt. b. S. C. circa 1831. Lawyer, age 19, Mateer, Anderson Dist. 1850 census. Enl. Co. I, 4th S. C. Inf. Fair Play, S. C. 6/7/61 age 30. Ab. detailed as Nurse by order of Surgeon 7-8/61. Discharged 8/12/61. Reenlisted Co. F, 1st S. C. Cav. Pickens CH 12/4/61 age 30 as Pvt. Present through 12/31/61 as Company Commissary. Present 3-6/62. Promoted Ordnance Sgt. Present 7/62-2/64. Detached Green Pond 4/24/-30/64. Present 5-6/64. Ab. on leave for 30 days 10/22/64. Present 11-12/64. Captured Hartwell, Ga. 5/3/65. NFR. Paid Taxes Pendleton, S. C. 1866 as Auctioneer.

Farmer, Youngs, Laurens Co. 1880 census. Bur. Old Stone Ch. Cem., Clemson, S. C., no dates.

SHARPE, ELAM. Captain, Co. F. b. Pendleton, Anderson Dist. circa 1821. Farmer, age 37, Pickens Dist. 1860 census. Enl. Pickens CH 12/4/61. Present through 12/31/61. Present Camp Ripley 1/62. On picket Jacksonborough 3-4/62. 86 horses present Adams Run 4/1/62. 28 horses present Adams Run 4/30/62. Signed for 5 Army tents, 3 camp kettles, 2 axes & heaves & 5 mess pans 5/7/62. Signed for 12 lbs. horse shoe nails 5/24/62. Ab. on sick leave 6/29-7/17/62 & 8/62. Present 9-12/62, however, ab. sick with "Intermittent fever" in Richmond hospital 12/10/62. 40 horses present 2/1/63. 24 horses present in Nelson Co., Va. 1/6/63. 256 horses present Rockfish Valley, Recruiting Depot 2/28/63 & 3/31/63. In Columbia hospital 4/15/63. Returned to duty 6/23/63. Present Gettysburg. 22 horses present 7/31/63. Present Brandy Station 8/2/63. Signed for 2 pair horse shoes Culpepper CH 8/4/63. Ab. sick with "Feb. Int. Quo."in Charlottesville hospital 8/20/63. Transferred Lynchburg hospital 9/22/63. 32 horses present 9/30/63. Resigned for "spirnal irritation, great anacenia & debility" 11/2/63. NFR. Depot Agent, age 47, Limestone, Ala. 1870 census. At home, age 58, Dallas Co., Texas 1880 census. d. French, Dallas Co., Texas 11/24/88. Father of Elam H. Sharpe.

SHARPE, ELAM HAYNIE. 4thCpl., Co. F. b. Charleston, S. C. 9/2/45. Student, age 14, Cherokee, Pickens Dist.,1860 census. Enl. Pickens CH 10/1/62 as Pvt., under age 18. Present through 6/63. On detached service 8/18-31/63. Present 9-12/63. Detached horse recruiting station 1/20-2/29/64. Detached Green Pond 4/16-30/64. Promoted 4thCpl. Present through 12/64. NFR. Asst. Depot Agent, age 23, Athens, Ala. 1870 census. Married Denton, Texas 4/21/76. Farmer, age 35, Clay Co., Texas 1880 census. d. East Dallas, Texas 2/10/97 age 51. Son of Elam Sharpe.

SHARPE, HUGH T. Pvt., Co. E. b. Laurens Dist. 11/25/45. Enl. Co. E, Williams Bn., S. C. Cav. 7/14/64. Transferred Co. E, 1st S. C. Cav. no date. Not on muster rolls. Farmer, age 23, Youngs, Laurens Co. 1870 census. Carpenter. d. Spartanburg Co. 5/11/15. Bur. Laurel Run Cem.

SHARPE, WILLIAM W. Pvt., Co. G. b. Lexington, S. C. 8/44. Enl. 10/19/64. Ab. on horse detail through 10/31/64. Present 11-12/64. Ab. sick with Variola in Richmond hospital 4/19/65. Paroled 4/21/65. Farmer, Lone Cane, Abbeville Co. 1880 & 1900 censuses. d. Lexington, S. C. 9/10.

SHAW, DAVID LAMAR. Pvt., Co. C. b. S. C. circa 1822. Farmer, Edgefield Dist. 1850 census. Served in Co. I, 2nd S. C. Troops 1863-64 on roster but not on muster rolls. Enl. Co. C, 1st S. C. Cav. Columbia 2/20/64 age 50, res. Aiken Dist. Detached Green Pond 4/16-30/64. Present 5-6/64. Ab. on horse detail 8/9-31/64. Detailed to guard POW's Florence 9/17-12/31/64. NFR. Farmer, Butler, Edgefileld Co. 1870 census. Alive 1898.

SHAW, ROBERT W. Pvt., Co. H. b. Lancaster, S. C. 1826. Laborer, Clay Hill, York Dist. 1860 census. Enl. Rock Hill 2/1/62 age 36. Present 3-4/62. Present on extra duty as Blacksmith 5-6/62. Present sick in camp 7-8/62. Present as Blacksmith 9-12/62. Present 3-6/63. Ab. sick with debilitas in Charlottesville hospital 8/31-11/21/63. Ab. sick with typhoid fever in Richmond hospital 11/26-12/23/63. Ab. on horse detail 12/24-31/63. Present 1-2/64. Ab. on leave 3-4/64. Transferred Co. A, 20th S. C. Inf. 5/20/64. Ab. sick with typhoid fever in Richmond hospital 7/31-9/17/64. Furloughed for 30 days to Yorkville, S. C. Transferred to 5th S. C. Inf. 12/31/64. Present 1-2/65. Paroled Appomattox CH 4/9/65. Miller, Yorkville, York Co. 1870 census. Farmer, York Co. 1880 census. d. in Arkansas date unknown.

SHEBLIN, C. R. Pvt., Co. A. Enl. by 1865. Captured Anderson, S. C. 5/3/65 and paroled. NFR.

SHELOR, G. S. Pvt., Co. F. Enl. Columbia 2/27/64. Present 3-6/64. Ab. sick in hospital 7/9-10/31/64. Present 11-12/64. Captured Anderson, S. C. 5/3/65 and paroled. Living Pickens Co. 1900. Living Atlanta, Ga. 1924.

SHEPPARD, JAMES A. Pvt., Co. F. b. Greenville, S. C. 3/10/42. Cadet, age 18, Arsenal Academy (Citadel), Charleston, 1860 census. Enl. CSA Decatur, Ga. 5/11/61 age 19. NFR. Enl. Co. F, 1st S. C. Cav. Colleton Dist. 5/13/62. Present through 6/62. Ab. on leave 7-8/62. Present 9-10/62. Ab. as wagon guard 11-12/62. Ab. detached service 3-4/63. Captured Uppperville 6/21/63. Exchanged by 8/63. Present through 8/31/63. AWOL 9/20-12/31/63. Present 1-2/64. On picket Rapidan River 2/20-29/64. Ab. sick in Richmond hospital 4/20-6/30/64. Present 7-12/64. NFR. Farmer, Stone Mountain, Ga. 1870-1900 censuses. d. there 10/18/19. Bur. Mountain View Bapt. Ch. Cem.

SHEPPARD, JOHN A. Pvt., Co. F. b. Greenville Dist., S. C. 8/2/32. Res. Newberry Dist. Enl. Pickens CH 12/4/61 age 30. Present through 12/31/61. On picket Walters Battery 3-4/62. On picket Camp Means 5-6/62. Present 7-10/62. Ab. sick in hospital 11-12/62. Present 3-4/63. Ab. on detached service 8/31/63. Present 9-10/63. Ab. on horse detail 12/22/63-2/64. Present 3-12/64. NFR. Farmer, Butler, Greenville Co. 1880 census. d. Greenville Co. 6/8/04. Bur. Rocky Creek Bapt. Ch. Cem.

SHEPPARD, THOMAS H. Pvt., Co. F. b. S. C. 8/22/45. Enl. Columbia 9/16/64. Present dismounted through 12/31/64. NFR. Farmer, Butler, Greenville Co. 1870-1910 censuses. d. Greenville Co. 11/11/14. Bur. Rocky Creek Bapt. Ch. Cem.

SHEPPHERD, WILLIAM GARRISON. Pvt., Co. F. b. S. C. 3/9/38. Enl. Pickens CH 12/4/61 age 20. Present through 12/31/61. On picket Simmons House 3-4/62. On picket Camp Means 5-6/62. Ab. on leave 7-8/62. Ab. sick 9-12/62. Ab. on detached service 3-4/63. Present 5-6/63, however, ab. sick with acute diarrhea in Richmond hospital 6/24-10/31/63. AWOL 11/1-12/31/63. On picket Gold Mine Ford, Rappahannock River 2/27-29/64 & captured. Sent to Old Capitol. Transferred Ft. Delaware & Pt. Lookout. Exchanged 10/31/64. NFR. Farmer, Butler, Greenville Co. 1880-1920 censuses. d. Butler, Greenville Co. 10/5/23. Bur. Rocky Creek Bapt. Ch. Cem.

SHERLOCK, THOMAS. Pvt., Co. E. b. Ireland circa 1820. Came to U. S. 1841. Ditcher, Orangeburg Dist. 1850 census. Enl. Ft. Motte 10/26/61 age 40. Present through 12/31/61. Detailed as nurse in hospital 3-4/62. Present 5-10/62. On picket 11-12/62. Present 3-10/63. Detached as wagon guard 10/1/63-2/64. Present 3-10/64. Present sick in camp 11-12/64. Paroled Hillsboro, N. C. 4/65. Res. Philadelphia, Pa. 1870 census.

SHILLITO, HENRY HUGH. Pvt., Co. A. b. S. C. 1847. Student, age 13, Abbeville Dist. 1860 census. Enl. Abbeville CH 9/1/63. Present through 10/63. Ab. sick with pneumonia in Richmond hospital 12/18-29/63. Present 1-6/64. On picket 9-10/64. Ab. on sick leave 11/25-12/31/64. NFR.

SHILLITO, JAMES ANDREW. Pvt., Co. A. b. Abbeville Dist. 11/9/40. Enl. Abbeville CH 8/13/61 age 21. Ab. Courier for Gen. Evans 11-12/61. Present 3-6/62. Present sick in camp 7-8/62. Present 9-12/62. Ab. sick in Staunton hospital 3/25-4/30/63. Present 5-6/63. Ab. on horse detail 8/18-31/63. Present 9/63-10/64. Detached Pocotaligo 12/10-31/64. Paroled Greensboro, N. C. 5/1/65. Tinner, Abbeville, S. C. 1870 & 1880 censuses. d. Abbeville 3/27/97. Bur. Sharon Meth. Ch. Cem., Abbeville. Wife receiving pension Abbeville 12/20/19.

SHILLITO, WILLIAM W. "BILL." Pvt., Co. A. b. Abbeville Dist. circa 1832. Enl. Abbeville CH 2/6/62. Present 3-10/62. Ab. detached 11/62-4/63. Presence or absence not stated 5-6/63. Ab. on horse detail 8/18-31/63. Ab. detached at Brigade Hqtrs. 9-10/64. Detailed to Enrolling Dept. in S. C. 11/1/63-12/64. NFR. Farmer, Little River, Johnson Co., Ark. 1870 census. Farmer, Caddo, Montgomery Co., Ark. 1880 census. d. Montgomery, Ark. 3/81.

SHIRER, JAMES H. 4thSgt., Co. E. b. S. C. 9/41. Res. Orangeburg Dist. Enl. Coosawatchie 11/22/61. Promoted 4thCpl. Present 3-6/62. On scout Edisto Island 7-8/62, promoted 5thSgt. 8/13/62. Present 9/62-6/63, promoted 4thSgt. Ab.on horse detail 8/18-31/63. Present 9/63-2/64. Ab. on leave 4/21-5/1/64. Present 5-12/64. Paroled Hillsboro, N. C. 4/65. Farmer, age 58, St. Stephens, Berkeley Co. 1900 census. Living Pine Grove, Calhoun Co. 1910 census. Barber, Columbia, S. C. 1918.

SHIRLEY, JOHN F. Pvt., Co. G. b. S. C. 12/42 or 10/22/40. Enl. Abbeville CH 3/15/62. Present through 10/62. On scout 11-12/62. Present 3-6/63. Ab. on horse detail 8/31/63. AWOL 9-10/63. On picket Fredericksburg 12/31/63. Ab. detailed horse recruiting station 2/4-29/64. Present 3-10/64. Detached Pocotaligo 12/10-31/64. NFR. Farmer, Anderson Co. 1870 census. Farmer, Parker, Texas. 1900 census. d. 1/15/06. Bur.Rose Hill Cem., Greenville, S. C. Brother of Thomas H. Shirley.

SHIRLEY, THOMAS H. Pvt., Co. H. b. Anderson Dist. circa 1850. Student, age 11, Anderson Dist. 1860 census. Enl. Columbia 4/15/64. Detached Green Pond 4/16-30/64. Present 5-8/64. Ab. detailed as guard for POW's Florence 9/18-12/31/64. Paroled Greensboro, N. C. 5/1/65. Farmer, age 20, Williamston, Anderson Co. 1870 census. Farmer, age 32, Rusk, Texas 1880 census. Brother of John F. Shirley.

SHOEMAKER, HENRY M. Pvt., Co. G. b. S. C. circa 1845. Student, age 15, Abbeville Dist. 1860 census. Enl. Abbeville CH 1/8/62. Present through 6/62. Courier, Ft. Lamar 7-8/62. Present 9-12/62. Ab. sick in Lynchburg hospital 3-4/63. Present 5-10/63. Detached with Signal Corps 12/31/63. Present 1-2/64. Ab. on leave 4/25-30/64. Present 5-12/64. NFR.

SHOEMAKER, HENRY W. Pvt., Co. G. Enl. age 39 on postwar roster Abbevlle Co. 1902.

SHORT, H. J. or J. H. Pvt., Co. K. On post war roster.

SHORT, ISHAM HARLIN. Pvt., Co. K. b. S. C. circa 1845. Enl. Chester CH 2/28/63, underage. Present through 6/63. WIA Culpepper CH 1863. Present dismounted 8/31/63. Present 9-12/63. On picket Rapidan River 1-2/64. Present 3-4/64. On picket Charleston fortifications 6/25-30/64. Detailed as Courier, Provost Marshal's Office 7-8/64. Ab. sick in hospital & furloughed for 60 days 10/24/64. Present 11-12/64. Paroled Raleigh, N. C. 4/65. d. Tipton, Tenn. 1906.

SHOTWELL, GEORGE. Pvt., no company. b. Va. Not on muster rolls. Res. Weld Co., Colorado postwar.

SHRIRER, WILLIAM DAVID. b. S. C. circa 1842. Age 18, Orangeburg Dist. 1860 census. Enl. Ft. Motte 10/26/61 age 20. Present through 12/31/61 & 3-6/62. On duty Pine Berry 7-8/62. Present 9/62-4/63. WIA Chancellorsville 5/3/63. WIA (right elbow shattered by shell) & captured Gettysburg 5/3/63. Admitted Union hospital 7/6/63. Arm amputated 7/31/63 and pyaemia developed. DOW's 8/14/63.

SHUTTLESWORTH, IRA A. Pvt., Co. F. b. S. C. 1837. Enl. Pickens CH 12/4/61 age 25. Present through 12/31/61. Discharged on Surgeon's Certificate 4/2/62. Enl. Ferguson's Beauregard S. C. Arty. Pickens CH 1/12/63. Present through 5/63. Ab. sick in hospital 6/11-12/31/63. Discharged 3/21/64. Paroled Meridian, Miss. 5/12/65. Married 1871. Farmer, age 43, Big Smiths, Franklin Co., Ga. 1880 census. Res. Goodwill, Franklin Co., Ga. 2887. d. by 1900.

SIBERT, JAMES ISAAC. Pvt., Co. A. b. Abbeville Dist. 11/12/32. Age 28, Dora Gold Mine, Abbeville Dist. 1860 census. Enl. Abbeville CH 5/6/63. Present through 8/31/63. Ab. on horse detail 9-10/63. Present 11-12/63. Ab. detailed with Brigade QM 2/26-29/64. Ab. on horse detail 4/25-30/64. Present 5-12/64. NFR. Age 38, Bradley Mills, Abbeville Co. 1870. d. Valdosta, Brooks Co., Ga. 5/12/13.

SIBLEY, RICHARD. (Colored). Blacksmith, Co. C. Enl. Hamburg 8/27/61. Presence or absence not stated through 4/62. Present 5/62-6/63. Ab. on leave 8/15-31/63. Present 9/63-2/64. Ab. on leave 4/10-5/1/64. Present 5-6/64. Discharged by Medical Examining Board 7/2/64. NFR.

SIFFORD, MICHAEL LAWRENCE. Pvt., Co. H. b. Lincoln Co., N. C. 6/28/28. Enl. Rock Hill 1/15/62 age 33. Present sick in camp 3-4/62. Ab. on sick leave 6/3-30/62. Present 7-8/62. Detailed to drive horses from Summerville, S. C. to Staunton, Va. 9-10/62. On raid 11-12/62. Present 3-6/63. Ab. on horse detail 8/18-31/63. Present 9/63-2/64. Ab. on leave 3/31-4/29/64. Present 5-10/64. Present detailed as Blacksmith 11-12/64. NFR. Farmer, Bethel, York Co. 1880 census. d. York, S. C. 3/4/90. Bur. Bethel Presb. Ch. Cem.

SIMPKINS, CHARLES WHIPPLE. 1stCpl., Co. C. b. S. C. 3/42. Student, age 18, Barnswell Dist. 1860 census. Enl. Wadmalaw Island 4/2/62 as Pvt. Present 5-6/62. Ab. on sick leave 8/27-30/62. Promoted 3rdCpl. Present 9-10/62. On scout 11-12/62. Present 3-4/63. Promoted 1stCpl. Present through 10/31/63. Ab. on horse detail 12/22/63-2/64. Ab. on horse detail 4/23-5/4/64. Present 5-8/64. Detailed to guard POW's Florence 9/27-12/28/64. AWOL 12/29-31/64. NFR. Overseer, age 27, Sleepy Hollow, Barnwell

Co. 1870 census. Farmer, Hammond, Aiken Co. 1880-1900 censuses. d. 4/23/14. Bur. Hammond Cem., Beech Island, Aiken Co. Brother of M. P. Simpkins.

SIMPKINS, MARSHALL P. 2ndSgt., Co. C. b. S. C. circa 1839. Student, age 21, Barnwell Dist. 1860 census. Enl. Hamburg 8/27/61 age 21, as 2ndCpl. Detached Wadmalaw Island 11-12/61. Present 5-6/62, promoted 1stCpl. Ab. sick 6/26-30/62. Present 9-10/63, promoted 3rdSgt. Present 11/62-6/63, promoted 2ndSgt. Ab. on horse detail 8/15-31/63. Present 9-12/63. Ab. on detached service 1/3-2/29/64. Detached Green Pond 4/16-30/64. Ab. on horse detail 6/15-30/64. Present 7-12/64. NFR. d. by 1898.

SIMMONS, CHARLES Y. L. Pvt., Co. A. b. S. C. circa 1840. Farmer, age 20, Laurens CH 1860 census. Enl. Abbeville CH 9/25/61. On duty Bears Island 11-12/61. Present 3-6/62. Present sick in camp 7-8/62. Present 9-10/62. Present in arrest 11/62-4/63, however, CM'd 3/26/63 & ab. sick with debilitas in Richmond hospital 2/22-4/6/63. Present dismounted 5-6/63. Ab. on horse detail 8/18-31/63. Present 9-10/63. Detailed with Signal Corps 12/17/63-2/64. On horse detail 4/25-30/64. Present 5-6/64. Ab. sick 7-8/64. Present 9-12/64. NFR. Merchant, age 40, Hodges, Abbeville Co. 1880 census.

SIMMONS, D. Pvt., Co. D. Not on muster rolls. d. Lynchburg hospital 6/5/63 and buried in Lynchburg City Cem.

SIMMONS, JOHN S. (1). Pvt., Co. K. b. S. C. circa 1839. Farmer age 21, Walterboro, Colleton Co. 1860 census. d. Mackey's Point on postwar roster.

SIMMONS, JOHN S. (2). Pvt., Co. K. Enl. Anderson CH 5/16/63. Present through 6/63. Ab. on horse detail 8/16-31/63. Present 9-10/63. Ab. on horse detail 11/1/63-2/64. Detached Green Pond 4/15-30/64. On duty Charleston entrenchments 7-8/64. Detailed Engineering Dept., Charleston 8-12/64. NFR. d. Charleston 11/2/79 age 38. Bur. Charleston, S. C.

SIMONS, SAMUEL WAGG. 1stCpl. Co. A. b. S. C. circa 1840. Enl. 11/7/61, age 21. Ab. on leave 2/7/62. Reenlisted Co. K, 4th S. C. Cav. Grahamville 3/25/62. Promoted Sgt. 6/7/62. Detailed with Gen. Beauregard 6/22/62. WIA (neck) Hawes' Shop 5/26/64. Detailed in QM Dept., Columbia through 8/31/64. NFR. Cotton Buyer, Charleston 1880 & 1900 censuses. d. Charleston 10/26/17. Bur. Magnolia Cem.

SIMONS, WILLIAM. Pvt., Co. I. Enl. Parker's Ferry 4/3/62. NFR. May have served in S. C. Arty. unit.

SIMPSON, JOHN BASKIN. Asst. Surgeon. b. S. C. 12/34. Appointed from S. C. 6/1/64 to rank from 12/24/63. Had passed exams 1/16/64. Had served as Hospital Steward, 1st S. C. Cav. 1/1-22/64. Paid 4/24/64. Ab. sick with chronic diarrohea in Richmond hospital 6/23-16/64. Assigned as Asst. Surgeon 17th Ga. Inf. 6/16/64. Paid 6/30/64. Assigned to Medical Director, Cavalry Corps, A. of N. Va. 10/24/64. Captured near Petersburg 10/27/64. Sent to Old Capitol. Transferred Ft. Delaware. Transferred to Ft. Monroe 12/5/64. Exchanged 1/6/65. Paid 1/11/65. NFR. M. D., age 45, Stony Battery, Newberry Co. 1880 census & age 65, 1900 census. d. 9/20/06. Colony Luth. Ch., Newberry Co.

SIMS, CHARLES W. 1stSgt., Co. B. b. S. C. 1/28/31. Farmer, age 27, Newberry Dist. 1860 census. Enl. Clinton 8/22/61 as 27, as 1stSgt. Present through 10/31/61. In arrest 11-12/61 & reduced to Pvt. Present 3/62-2/64. Ab. on leave 4/26-30/64. Present 5-10/64. Detached Pocotaligo 12/2-31/64. NFR. Farmer, Jacks, Laurens Co. 1880 census. d. there 3/19/84. Bur. Odell Chapel Cem., Laurens Co.

SIMS, LAURENS VANCE. Pvt., Co. G. b. Abbeville Dist. 10/19/45. Student, age 14, Abbeville Dist. 1860 census. Served in 5th S. C. Reserves 90 days 1862-63. Enl. Co. G, 1st S. C. Cav. Columbia 9/27/64. Present through 11/64. Ab. on sick leave for 30 days 12/6-31/64. Paroled Greensboro, N. C. 5/1/65. M. D., age 35, Atlanta, Ga. 1880 census. Bur. Sims Family Cem., Milner, Ga., no date of death.

SIMS, RICHARD A. Pvt., Co. I. b. S. C. 12/39. Enl. Parker's Ferry 4/3/62. Present through 10/62. Present as Blacksmith 11-12/62. Ab. sick ulcer on left leg in Staunton hospital 3/12-4/63. Present as Blacksmith 5-10/63. Ab. sick with ulcer left leg in Richmond hospital 1/6-2/29/64 & 3/23-8/27/64. Present 9-10/64. Detached Pocotaligo 12/10-31/64. NFR. Farmer, White Store, Anson Co., N. C. 1900 census. d.

Marshville, Union Co., N. C. 6/10/21. Bur. Anson Co., N. C.

SITGREAVES, FREDERICK ALLEN. 2ndLt., Co.'s D & K. b. Rock Hill, York Dist. 7/19/34. Stable Operator, age 27, Chester Dist. 1860 census. Enl. Co. D Chester CH 9/10/61 age 26 as Pvt. Ab. on sick leave through 10/31/61. Ab. on recruiting service 12/25/61-1/5/62. Appointed Sgt. Major and transferred to staff 3-4/62. Present 5-6/62. Appointed 2ndLt., Co. K 6/25/62. Present through 8/62. Presence or absence not stated 9-10/62. Commanding Company 11-12/62. Ab. on deta ched service 3-4/63. WIA(right leg) Upperville 6/20/63. Ab. wounded in Richmond hospital 6/28/63. 55 horses present 6/30/63. Furloughed for 30 days 7/20/63. Ab. on leave through 8/31/63. Leave extended 11/3/63, however, signed for 10 mules Columbia, S. C. 10/4/63. Ab. detailed as Brigade Ordnance Officer 1-2/64. 12 horses present 3/31/64, & signed for 7 pr. shoes, 4 jackets, 3 fly tents, 13 pr. pants, 2 overcoats, 1 axe, 5 shirts, 1 pr. drawers. Ab. on leave 4/15-30/64. Commanding Company Green Pond 5/20/64, and signed for 6 caps, 14 jackets, 16 pr. trousers, 8 shirts, 9 pr. drawers, 5 pr. shoes & 14 pr. socks. Present 5-6/64. 38 horses present 6/30/64. Signed for 1 shovel, 7 Army tents, 2 camp kettles, 7 mess pans & 1 wall tent 6/30/64, commanding Company. Present sick in camp 7-8/64. 30 horses present 7/30/64. Ab. on sick leave for 30 days 10/5/64. Present 11-12/64. Paroled Charlotte, N. C. 5/22/65. Farmer, Lansford, Chester Co. 1870 census. Stock Dealer, Winnsboro, Fairfield Co. 1880 census. d. 4/26/92. Wife receiving pension Winnsboro, S. C. 10/2/19.

SITGREAVES, JOHN M. Pvt., Co. K. Served in Co. L, 5th S. C. Troops (6 months -1863-64). Enl. Co. K, 1st S. C. Cav. Columbia 1/16/64. Detached Green Pond 4/15-30/64. On picket Mooreland's Warf, Charleston 6/25-30/64. Ab. on leave from Charleston hospital 7-8/64. AWOL 10/29-31/64. Present 11-12/64. Paroled Charlotte, N. C. 5/22/65.

SKELTON, SAMUEL AUGUSTUS. Pvt., Co. K. b. Anderson Dist. 4/12/46. Enl. Columbia 4/14/64. Ab. on horse detail through 4/30/64. On picket Bee's Ferry 6/25-30/64. Present 7-8/64. Detailed to QM Dept. to drive cattle to Florence 9-12/64. Paroled Greensboro, N. C. 4/26/65. Farmer, Fork, Anderson Co. 1880 & 1900 censuses. Retired, Rock Mills, Anderson Co. 1920 census. d. 5/9/26. Bur. Roberts Presb. Ch. Cem.

SKINNER, RICHARD M. Pvt., Co. I. Res. Clarendon Dist.b. S. C. 1/26/33. Enl. 2/17/61 on postwar roster. Enl. Co. A, 14th Bn., S. C. Cav. as 2ndLt. 1/2/62. Promoted 1stLt. of Arty. 5/21/62. Present until WIA Pocotaligo 10/22/62. Promoted Captain 11/24/62. Company reassigned as Co. H. 5th S. C. Cav. 1/22/63. Present until KIA 6/24/64. Bur. Skinner Cem., Clarendon Co.

SMALL, ANDREW. Chief Musician. b. Scotland 1815. Bookkeeper, age 46, Abbeville 1860 census. Enl. Co. A. Abbeville CH as Trumpeter 8/23/61 age 48. Present Bugler for Company A through 10/31/61. Appointed Chief Musician & Bugler for Battalion 11-12/61. No further record until present 9/62-6/64. AWOL 2 days since 6/28/64. Retired for overage 7/18/64 over 50 and wife and 8 children. Paid Columbia 9/15/64. NFR. Auctioneer, age 55, Abbeville 1870 census.

SMALL, JOHN A., SR. Captain & Quartermaster. b. S. C. 1841. Farmer, age 19, Flint Ridge, Lancaster Co. 1860 census. Enl. Co. A Abbeville CH 8/23/61 age 19 as Pvt. Appointed QM Sgt. on rolls 11-12/61. Present through 12/62. Appointed Captain & QM of Regiment 2/1/63. Present through 10/63. Resigned 11/18/63. NFR. Machinist, age 78, Ocala, Marion Co., Fla. 1900 census.

SMALL, JOHN A., JR. Pvt., Co. A. b. S. C. 4/45. Enl. Caroline Co., Va. 11/24/63. Detailed in Regimental QM Dept. through 12/31/63. Present 1-2/64. Ab. on leave 4/24-30/64. Present 9-12/64. NFR. Farmer, Floyds, Horry Co. 1900 census. d. 6/26/00. Bur. Douglas Presb. Ch. Cem., Lancaster Co. S. C.

SMITH, ADAM L. Pvt., Co. E. b. S. C. circa 1825. Blacksmith, age 35, Orangeburg Dist. 1860 census. Enl. Ft. Motte 10/26/61 age 33. Killed by accidental discharge of a pistol Fripp's Bluff 12/20/61.

SMITH, ALEXANDER KOHASH. 4thCpl., Co. I. b. S. C. 2/1/37. Enl. Parker's Ferry 4/3/62 as Pvt. Present through 12/62, however, ab. sick with typhoid fever in Richmond hospital 11/21-12/7/62. Promoted 4thCpl. Present 3-8/31/63, reduced to Pvt. Ab. on horse detail 9-10/63. AWOL 12/30-31/63. Present 1-2/64. Detached Green Pond 4/16-30/64. Present 5-10/64. Present in arrest 11-12/64. Transferred

Lucas's 15th Bn. S. C. Arty. but not on muster rolls of that unit. Deserted to the enemy Charleston 3/22/65. Age 27, 5'10", light complexion, black hair, blue eyes. Took oath and discharged. d. 3/17/19. Bur. Laureland Cem., York Co.

SMITH, ALEXIUS J. Pvt., Co. K. b. S. C. 11/20/43. Enl. by 1864. Transferred Co. B, Lucas's 15th Bn. S. C. Arty. 5/64. NFR. Farmer, age 27, Statesburg, Sumter Co. 1870 census. d. 3/8/13. Bur. Beaver Dam Bapt. Ch. Cem., Laurens Co.

SMITH, ARCHIBALD K. Pvt., Co. I. b. S. C. 10/38. Not on muster rolls. Present 4/65 on wife's pension application. Farmer, age 42, Burns, Colleton Dist. 1880 census. Furniture Collector, Charleston 1900 census. d. 3/30/07. Wife receiving pension Charleston 10/31/19.

SMITH, CHARLES BOYD. Pvt., Co. D. b. S. C. circa 1831. Enl. Chester CH 9/10/61 age 31. Present 11/61-6/62. Ab. on sick leave 8/19-31/62. Present 9-10/62. On scout 12/5-31/62. Ab. sick with neuralgia in Charlottesville hospital 2/19-3/9/63. Present 3-6/63. Ab. on horse detail 8/18-31/63. Present 9-10/63. Ab. on horse detail 12/22-2/29/64. Present 3-12/64. Paroled Hillsboro, N. C. 4/65. Farmer, age 48, Chester, S. C. 1880 census. Receiving pension Chester 1903. Retired, age 79, York, S. C. 1910 census.

SMITH, DAVID BUFFUM. Pvt., Co. G. b. Providence, Rhode Island 7/23/30. Enl. Abbeville CH 1/8/62. On picket Jacksonborough 3-6/62. Horse died Secessionville. Paid $150.00. Ab. on detail Charleston Arsenal 8/16/62-2/64. Relieved of detail 5/1/64. Present 5-6/64. Transferred Co. B, Lucas's 15th S. C. Bn. Arty. 7/1/64. Paroled 1865. Carpenter, Abbeville 1880 census. d. Abbeville 6/22/99. Bur. Trinity Episc. Ch., Abbeville.

SMITH, EBENEZER T. Pvt., Co.'s F & K. b. S. C. circa 1838. Enl. Co. F Pickens CH 12/4/61 age 22. Present through 12/31/61. On picket Simmons House 3-4/62. Transferred Co. K 6/25/62. Present 7-10/62. On picket Richards Ford 11-12/62. Present 3-4/63. WIA (right thigh) & captured Hanover, Pa. 7/2/63. Sent to David's Island, N. Y. Exchanged 8/63. In Williamsburg hospital wounded 8/28/63. Ab. on leave from hospital 9-10/63. Present 11-12/63. On picket Rapidan River 1-2/64. Captured 2/29/64. Sent to Old Capitol. Transferred to Ft. Delaware. Released 6/7/65. 5'9", dark complexion, brown hair, grey eyes. Farmer, age 32, Bogansville, Union Co. 1870 census.

SMITH, EDWARD C. Pvt., Co. A. b. Ga. 9/49. Enl. Abbeville CH 1865. Paroled Greensboro, N. C. 5/2/65. Married 1867. Res. Talluhah Falls, Ga. 1900. Res. Pulaski, Oconee Co. 1910 census.

SMITH, J. T. Pvt., Co.'s F & K. On post war roster.

SMITH, JAMES HARVEY. Pvt., Co. H. b. S. C. 9/46. Age 14, Little River, Cherokee Co., Ga. 1860 census. Enl. Columbia 4/14/64. Present through 4/30/64. Detached, Dist. Hqtrs., Green Pond 5-6/64. Detached Medical Dept., Adams Run 5/18-12/31/64 as Forage Master. NFR. Farmer, Little River, Cherokee Co., Ga. 1880 & 1900 censuses. d. 9/6/07. Bur. Holbrook-Campground Cem., Cherokee Co., Ga.

SMITH, JOHN HARVEY. Pvt., Co. A. b. S. C. circa 1845. On roster but not on muster rolls. Receiving pension 9/20/19. Farmer, age 75, Beech Springs, Spartanburg Co. 1920 census.

SMITH, LEVI T. Pvt., Co. K. Res. Pickens Dist. Enl. by 1865. Paroled Raleigh, N. C. 4/65. d. Williston, S. C. 6/30/13. Bur. Big Creek Cem., Anderson Co., S. C.

SMITH, LELAND P. Pvt., Co. B. b. Laurens Dist. 2/16/32. Res. Laurens Dist. Enl. Clinton 8/22/61 age 30. Ab. sick 10/31/61. Present 11/61-4/62. Wounded by premature discharge of his own gun 6/5/62. Ab. on leave 7-8/62. Present 9-10/62. d. of pneumonia in Staunton, Va. hospital 11/18/62. Bur. Thornrose Cem., Staunton.

SMITH, LEVI T. Pvt., Co. K. b. Pickens Dist. 5/20/45.   Res. Pickens Dist. Enl. by 1865. Paroled Raleigh, N. C. 4/65. d. Williston, S. C. 6/30/13. Bur. Big Creek Cem., Anderson Co., S. C.

SMITH, M. R. Pvt., Co. H. b. S. C. 12/5/45. Student, age 13, Choctaw, Ala. 1860 census. Enl. Columbia 10/26/64. Present 11-12/64. Paroled Greensboro, N. C. 5/1/65. Clerk & Bookkeeper, age 23, Butler, Choctaw Co., Ala. 1870 census. d. 10/13/07. Bur. Arbor Springs Cem., Reform, Pickens Co., Ala.

SMITH, R. A. Pvt., Co. H. b. York Dist. circa 1824. Stone Mason. Enl. Rock Hill 1/8/62. Ab. on picket Rentoules through 2/62. Present 3-4/62. Ab. on sick leave 6/30/62. Present 9/62-4/63. Ab. in arrest 4/30-8/31/63. WIA (right hand) 9/63. Ab. wounded in Richmond hospital 9/15-10/23/63. Ab. on leave for 30 days 10/24/63-20/64. Transferred Co. A, Lucas's Bn. S. C. Arty. 5/16/64. Present through 6/64. Discharged for disability 8/15/64. Age 40, 6', dark complexion, blue eyes, dark hair. NFR.

SMITH, ROBERT A. Pvt., Co. H. b. N. C. 3/14/49. Enl. by 9/63 but not on muster rolls. Ab. sick in Richmond hospital 9/24-12/2/63. NFR. Farmer, age 22, York Co. 1870 census. d. Ebenezer, York Co. 4/24/09. Bur. Ebenezer Cem.

SMITH, SAMUEL ROBERT. Pvt., Co. K. b. S. C. 11/20/44. Res. Pickens Dist. 1850 census. Enl. Plank Road 1/14/63, under 18 years. Present 3-10/63. On picket Rapidan River 11-12/63. Present 1-4/64. Ab. sick in Charleston hospital 5-6/64. Present 7-8/64. On duty Bee's Ferry 9-10/64. On picket Chatman's Ford 12/29-31/64. NFR. Retired, age 85, La Plata, Durango Co., Colo. 1930 census. Bur. Greenmont Cem., La Plata, Colo., no dates.

SMITH, SIDNEY L. Pvt., Co. H. b. S. C. circa 1817. Overseer, age 43, York Dist. 1860 census. Enl. Rock Hill 1/15/62 age 43. Present 3-6/62. AWOL 9-10/62. Present 11/62-6/63. Ab. on horse detail 8/18-31/63. Present 9/63-2/64. Ab. on leave 4/25-30/64. Present 5-8/64. Detailed as Overseer on Government Land, James Island 9-10/64. Present 11-12/64. d. York Co. 7/2/98. Bur. Flint Hill Bapt. Ch. Cem., York Co.

SMITH, SILAS V. Pvt., Co.'s F & K. b. S. C. circa 1841. Enl. Co. B, 1st Bn. (Orr's) S. C. Rifles Sullivan's Island 11/1/61. Transferred Co. F, 1st S. C. Cav. 4/9/62. Transferred Co. K 6/25/62. Ab. on sick leave in Pickens Dist. through 12/62. Ab. sick with debilitas in Charlottesville hospital 2/19-3/9/63. Present through 4/63. Ab. sick in hospital 6/20-7/31/63. Ab. sick with fever in Staunton hospital 8/9-31/63, furloughed for 20 days. Present 9-10/63. On picket Rapidan River 11-12/63. Present 1-2/64. Ab. on horse detail 4/10-30/64. Ab. as Courier, Charleston Telegraph Office 6/25-30/64. Present 7-8/64. Ab. on sick leave for 20 days 10/29/64. Present 11-12/64. Paroled Raleigh, N. C. 4/65. Farmer, Boone, Crawford Co., Mo. 1880 census. Carpenter, Boone, Crawford Co., Mo. 1910 census. Applied for pension in Mo. 11/10/15. d. 2/12/29. Bur. Mt. Hope Cem., Webb City, Mo.

SMITH, THEODORE A. Sgt., Co. A. b. S. C. 2/13/34. Enl. 11/7/61. Transferred to S. C. Rangers 1/6/62. Enl. Co. D, 5th S. C. Cav. Charleston 2/17/62. Present until ab. on horse detail 9/15/64. Ab. sick in hospital 9/2010/31/64. NFR. Clerk, Charleston 1870 census. d. 1/15/77. Bur. St. John's Luth. Ch., Charleston.

SMITH, THOMAS W. 1stSgt., Co. A. b. Abbeville Dist. 10/7/29. Enl. Abbeville CH 8/13/61 as 2ndSgt., age 32. Ab. recruiting duty 11-12/61 & 1-2/62. Present sick in camp 3-4/62. Present 5-8/62. Ab. on detached service 9-10/62. Present 11-12/62. Ab. detached 3-4/63. 16 horses present Farmville 5/1/63, commanding detachment. Ab. sick through 8/31/63. Promoted 1stSgt. Ab. sick with typhoid fever in Charlottesville hospital 9/4/63. d. there 9/19/63 age 35. Effects: Sundries. Bur. Confederate Cem., U. of Va., Charlottesville.

SMITH, ZADOCK D. Pvt., Co. D. b. S. C. circa 1822. Farmer, age 38, Clay Hill, York Dist. 1860 census. Enl. Chester CH 9/10/61 age 35. Present 11/61-8/62. d. in Summerville hospital 10/4/62.

SMOKE, GEORGE WASHINGTON. Pvt., Co. E. b. S. C. circa 7/39. Overseer, age 25, Orangeburg Dist. 1860 census. Enl. Ft. Motte 10/26/61 age 30. Present through 10/63. Present 11-12/63, dismounted since 12/4/63. Present 1-4/64, dismounted through 5/1/64. Transferred Co. B, 20th S. C. Inf. 5/13/64. Transferred Hammond's Co., 25th S. C. Inf. 6/64. NFR. Farmer, age 40, Orangeburg Co. 1870 census, age 50 1880 census and age 60, 1900 census. d. St. Matthews, Calhoun Co. 1/10/03.

SMOKE, SAMUEL P. Pvt., Co. E. b. S. C. circa 1844. Age 16, Macon Co., Ala. 1860 census. Enl. Columbia 4/12/64, conscript assigned for the war. Detached Green Pond 4/16-30/64. Ab. sick James Island hospital 6/22/64. Ab. on sick leave 6/23-10/31/64. Ab. sick Summerville hospital 11-12/64. NFR.

SOFGE, THEODORE A. Musician, Co. C. b. circa 1836. Served in Ga. Militia 1861-62. Enl. Wadmalaw Island 3/19/62 as Pvt. age 25. Appointed Bugler & present 5-6/62. Ab. on sick leave 8/31/62. AWOL 9-10/62. Present 11/62-4/63. Transferred to Band 6/1/63. Present as Musician through 2/64. Ab. on horse detail 4/1-5/1/64. Present 5-12/64. NFR. Res. Augusta, Ga. 1866-73. d. 1/4/76. age 47. Bur. Magnolia Cem., Augusta, Ga..

SPELL, BENJAMIN E. T. Pvt., Co. I. b. Colleton Dist. 1819. Farmer, Colleton Dist. 1860 census. Enl. Parker's Ferry 4/3/62. AWOL through 4/30/62. Probably discharged. Enl. Co. C, 5th S. C. Cav. Cheraw 11/10/62. Present 1-3/63. Ab. on sick leave 10 days 4/27/63. Present 5-9/63. Detached as Sutler 10/9/63-2/64. Sent to Lynchburg with unserviceable horses 3/1-8/31/64. AWOL 10/24-31/64. Issued clothing 12/25/64. NFR. Farmer, age 60, Sheridan, Colleton Co. 1880 census. d. Colleton Dist. 11/26/95 age 76. Father of Henry W. & John R. Spell.

SPELL, HENRY W. Pvt., Co. E. b. S. C. 1841. Farmhand, age 19, Walterboro, Colleton Dist. 1860 census. Enl. Adams Run 2/25/62. Present through 6/62. On scout Edisto Island 7-8/62. Present 9/62-6/63. Ab. on horse detail 8/31/63. Present 9-12/63. Detached with Signal Corps 1/1-2/29/64. Ab. on leave 4/26-5/6/64. Present through 8/64. Detailed to drive cattle for QM Dept. 9/18-12/31/64. Paroled Hillsboro, N. C. 4/65. Farmer, age 38, Burns, Colleton Co. 1880 census, age age 58, Sheridan, Colleton Co. 1900 census. d. 1927. Son of Benjamin E. T. Spell & brother of John R. Spell.

SPELL, JOHN R. Pvt., Co. E. b. S. C. 1844. Student, age 16, Colleton Dist. 1860 census. Enl. Columbia 7/20/64. Present through 8/64. Ab. on sick leave 10/24-31/64. Present 11-12/64. Paroled Hillsboro, N. C. 4/65. Farmer, Colleton Co. 1870-1900 census. d, Leavenworth, Darlington Co. 9/4/1942 age 94. d. 1927. Son of Benjamin E. T. & brother of Henry W. Spell.

SPENCER, JOHN WASHINGTON. Pvt., Co. D. b. S. C. 1840. Enl. Chester CH 9/10/61 age 20. Present through 6/62. Ab. on sick leave 8/11/62. Ab. sick with debility in Manchester hospital, Richmond 10/12-30/62. Ab. sick with debility in Charlottesville hospital 12/5-1/1/63. Present 3-12/63. Ab. on leave 2/12-29/64. Ab. on leave 4/26-30/64. Present 5-6/64. Present sick in camp 7-8/64. Present 9-12/64. Paroled Hillsboro, N. C. 4/65. Farmer, Lewisville, Chester Co. 1870-1880 censuses. Confederate Veteran, Chester Co. 1904. Retired, Catawba, York Co. 1910 census.

SPIGENER, JOEL. Pvt., Co. E. Res. Orangeburg Dist. Enl. as volunteer Columbia 7/15/64. Present 7-8/64. Present 9-10/64, dismounted. Detached Pocotaligo 12/10-31/64. Captured Orangeburg, S. C. 2/13/65. Sent to New Bern & Pt. Lookout. Released 6/19/65. 5'5 1/2", dark complexion, black hair, blue eyes. NFR.

SPIGENER, JOHN PAULING. Pvt., Co. E. b. Orangeburg Dist. 12/22/31. Res. Orangeburg Dist. 1860 census. Enl. Ft. Motte 10/26/61 age 28. Present through 6/62. Ab. on leave 8/30/62. Present 9-12/62. Ab. on horse detail 3-4/63. Present through 8/31/63. Ab. on horse detail 10/21-11/24/63. AWOL through 12/31/63. Present 1-2/64. Detached Green Pond 4/16-30/64. Present 5-10/64. Detached Pocotaligo 12/10-31/64, dismounted. Paroled Hillsboro, N. C. 4/65. Farmer, Caw Caw, Orangeburg Co. 1870-1900 censuses. d. Swansea, Lexington Co., S. C. 11/2/09.

SPROUSE, WILLIAM W. Pvt., Co. A. b. Spartanburg Dist. 2/22/40. Married Abbeville Dist. 1860. Enl. Abbeville CH 3/19/62. On duty Bears Bluff through 4/62. Present 5-8/62, horse died 7/26/62. Present 9-10/62, dismounted. Present 11-12/62. Ab. detached service as Teamster for Brigade QM 3-10/63. Ab. on horse detail 12/23/63-2/64. Present 3-4/64. Ab. sick in hospital 6/26-30/64. Ab. sick 9/3-10/31/64. Present 11-12/64. NFR. Farmer, Abbeville Co. 1880 census. Res. Spartanburg 1900 census. d. Cherokee Co. 5/18/18. Bur. Sardis Meth. Ch., Sardis, S. C.

STALNAKER, JOSEPH WILLIAM. Pvt., Co. A. b. Orangeburg Dist. 1837. Overseer, age 23, Saluda, Edgefield Dist. 1860 census. Enl. Abbeville CH 8/13/61 age 23. On duty Bear's Island 11-12/61. Present 3-6/62. Ab. sick in Summerville hospital 7-8/62. Ab. sick 9/62-12/63. Present 1-4/64. Ab. sick 6/26-30/64. Present 9-10/64. Detached Pocotaligo 12/10-31/64. Paroled Greensboro, N. C. 5/1/65. Farmer, age 30, Saluda, Edgefield Co. 1870 census. d. Autauga, Ala., no dates.

STARKE, S. J. H. Pvt., Co. G. b. S. C. circa 1839. Enl. Abbeville CH 1/6/62, age 23.. Present through 6/62. On duty as Courier, Ft. Lamar 7-8/62. Present 9-10/62. On scout 11-12/62. d. of disease in camp 3/25/63.

STEADMAN, EVERETT JOHN. Pvt., Co. C. b. S. C. circa 1844. Enl. Aiken Dist. age 20 on post war roster. Merchant, Bamburg, Barnwell Co. 1870 census. d. by 1898.

STEADMAN, GEORGE E. Pvt., Co. C. b. S. C. 5/38. Merchant, age 24, Barnwell Dist. 1860 census. Enl. Wadmalaw Island 3/18/62. Discharged by Surgeon 6/22/62. Enl. Mathewes Co., S. C. Arty. 3/20/63 and promoted 2ndLt. Resigned 6/14/64. NFR. Merchant, Blackstock, Barnwell Co. 1880 census. Salesman, age 62, Columbia, S. C. 1900 census. d. Rockton, Richland Co. 12/31/12. Bur. Greenbrier Meth. Ch., Winnsboro, S. C.

STEADMAN, WILLIAM MUNROE. Quartermaster Sgt. b. Aiken Dist. 1840.   Merchant, age 21, Barnwell Dist. 1860 census. Enl. Co. C Wadmalaw Island 3/18/62 as Pvt. Present 5-6/62. Ab. sick 8/2-31/62. Present 9/62-6/63. Ab. on horse detail 8/15-31/63. Present 9/63-4/64. Promoted QM Sgt. of Regiment 3/31/64. Present 5-12/64. NFR. Dry Goods Merchant, age 30, Athens, Ga. 1870 census. Farmer, age 39, Aiken Co. S. C. 1880 census. d. 1880. Bur. 1st Bapt. Ch., Aiken.

STEELE, GEORGE JAMES, JR. Pvt., Co. H. b. Yorkville, York Dist. 10/14/46. Enl. Columbia 10/26/64. Present through 12/31/64. Paroled Greensboro, N. C. 4/26/65. Farmer, Yorkville, York Co. 1870 census. Farmer, Chester Co. 1910 census. Receiving pension Lowrysville, Chester Co. 10/20/19. d. Lowrysville, Chester Co. 6/7/20. Bur. Zion Cem.

STEELE, JOHN GILLIAM. Pvt., Co. K. b. S. C. 12/11/41. Farmer, age 18, Ebenezer, York Dist. 1860 census. Enl. President's Guard. Transferred Co. K, 1st S. C. Cav. 8/7/62. Ab. on raid 11-12/62. Ab. as wagon guard 3-4/63. Captured Beverley Ford and bay horse killed 6/9/63. Sent to Old Capitol. Exchanged 6/25/63. Paid $450.00 for horse. Present 9/63-2/64. Detached Green Pond 4/16-30/64. Present 5-10/64. Detailed as Courier, Gen. Elliott 11-12/64. Paroled Salisbury, N. C. 5/1/65. Farmer, Catawba, York Co. 1870 & 1900 census. d. 7/10/05. Bur. York Co.

STEELE, JOSEPH NEWTON. Pvt., Co. H. b. S. C. 9/18/38. Merchant, age 25, Rock Hill, York Dist. 1860 census. Enl. Co. H, 1st S. C. Cav. Columbia 5/17/64. Present through 12/64. NFR. Farmer, age 36, York Co. 1870 census. d. 10/9/17. Bur. Ebenezer Presb. Ch. Cem., York Co.

STEELE, ROBERT ALEXANDER. 2ndCpl. Co. H. b. S. C. 1/6/32. Enl. Rock Hill 1/9/62 age 28 as Pvt. Present through 10/62. Ab. on raid 11-12/62. Promoted 3rdCpl. Ab. on horse detail 3/20-4/30/63. Presence or absence not stated 5-6/63. Ab. on horse detail 8/18-31/63. Promoted 2ndCpl. Present 9-10/63. Ab. on horse detail 12/22/63-2/64. Present 3-12/64. Paroled Salisbury, N. C. 5/1/65. Farmer, York Co. 1870 census. d. York Co. 11/24/05. Bur. Ebenezer Pres. Ch. Cem., York Co.

STEELE, WILLIAM DANIEL. Pvt., Co. H. b. S. C. 1/1/46. Res. York Dist. Enl. Rock Hill 1/9/62 age 16. Ab. on sick leave 3-4/62. Present 5-6/62. Ab. on sick leave 7-8/62. Detailed to drive horses from Summersville, S. C. to Staunton, Va. 9-10/62. Present 11/62-4/63. WIA Brandy Station 8/1/63. Ab. wounded in Richmond hospital 8/3-24/63. Furloughed for 30 days. Ab. wounded through 10/31/63. Present 11-12/63. d. of disease 1/25/64 age 19. Bur. Ebenezer Presb. Ch. Cem., York Co.

STEINNIKER, J. W. Pvt., Co. A. On roster but not on muster rolls.

STEWART. GEORGE W. Pvt., Co. G. Not on muster rolls. d.4/84. Wife's pension application is the only record of service. She was living in Iva, Anderson Co. 9/12/19.

STEWART, JAMES H. Pvt., Co. H. b. S. C. 11/20/44. Enl. Rock Hill 10/6/62. Present 11/63-4/63. Present until WIA Brandy Station 8/1/63. Ab. wounded in Richmond hospital 8/2-15/63. Furloughed for 30 days. Present 9/63-2/64. Detached Green Pond 4/16-30/64. Present 5-8/64. Ab. detailed guarding POW's Florence 9/15-12/31/64. Paroled Greensboro, N. C. 5/1/65. d. 3/25/96. Bur. Unity Cem., Rock Hill, S. C.

STEWART, JOAB F. Pvt., Co. K. b. S. C. 3/13/46. Res. Tugaloo, Oconne Dist. Enl. Hamilton's Crossing 3/7/64. Present through 4/64. On picket Bee's Ferry 5-6/64. Present 7-8/64. Ab. on leave for 13 days 10/20/64. Present 11-12/64. Shot in knee on pension application. Paroled Greensboro, N. C. 4/26/65. Received pension Tugaloo, Oconee Co. Att. reunion Walhalla, S. C. 1922. d. Oconee Co. 10/28/27. Bur. Toxaway Bapt. Ch. Cem.

STOKES, BENJAMIN J. Pvt., Co. E. b. S. C. 3/18/25. Res. Colleton Dist. Enl. Adams Run 3/27/62. Present through 8/62. Present sick in camp 9-10/62. Discharged 1862, elected to S. C. Legislature. NFR. Farmer, Verdier, Colleton Co. 1870 census and Landlord, age 75, 1900 census. d. Colleton Co. 6/6/03. Bur. Live Oak Cem., Walterboro, S. C.

STONE, WILLIAM FRANKLIN. Pvt., Co.'s D & K. b. S. C. 5/29/29. Farmer, age 31, Chalksville, S. C. 1860 census. Enl. Co. D Chester CH 9/10/61 age 32. Present 11/61-4/62. Transferred Co. K 6/25/62. Present through 4/63. WIA (fracture upper right thigh) & captured Upperville 6/17/63. Sent to U. S. hospital, Washington, D. C. DOW's Washington, D. C. 7/18/63 age 34. Bur. Arlington Nat. Cem.

STOUDEMIRE, DANIEL CHASTAIN. Pvt., Co. E. b. S. C. 11/10/37. Enl. Ft. Motte 10/26/61 age 22. Ab. sick in hospital 12/31/61. Ab. on sick leave 1/62. Present 3/62-4/63. Ab. on horse detail 8/18-31/63. Present 9/63-2/64. Detached Green Pond 4/22-30/64. Present through 10/64. Detached Pocotaligo 12/10-31/64. Paroled Hillsboro, N. C. 4/65. Farmer, Pine Grove, Orangeburg Co. 1900 census. d. d. Calhoun, S. C. 9/5/16. Bur. Pine Grove Cem., Orangeburg Co.

STOUDEMIRE, DAVID H. Pvt., Co. E. b. S. C. circa 1841. Age 19, Ft. Motte, Orangeburg Dist. 1860 census. Enl. Ft. Motte 10/26/61 age 19. Present through 6/62. On scout Edisto Island 7-7/62. Present 9-10/62. d. of disease Orange CH, Va. 11/19/62. Probably buried there.

STOWE, ROBERT S. Pvt., Co. K. b. Banks Co., Ga. circa 1839. Res. Banks Co., Ga. Enl. Culpepper CH 6/12/63. Ab. on horse detail 8/16-31/63. AWOL 9-12/63. Ab. horse recruiting camp 1-2/64. Detached Green Pond 4/15-30/64. Present 5-6/64. On duty James Island entrenchments 7-8/64. Ab. sick Summerville hospital 9-10/64. WIA & captured Pocotaligo 12/16/64. Sent to Hilton Head hospital. DOW's 12/30/64 age 25. Bur. Hilton Head Military Cem. Removed to Beaufort Nat. Cem.

STOWE, ROBERT W. Pvt., Co. K. b. S. C. 1849. Enl. Anderson, S. C. 5/16/63. Present until captured Upperville 6/21/63. Sent to Old Capitol. POW 6/24/63. NFR. Loom Fixer, age 81, Rock Hill, S. C. 1920.

STREELY, JOHN G. Pvt., Co D. b. circa 1843. Enl. 10/2/61, age 18. WIA (face) South Island 12/20/61. Discharged 5/18/62, Reenl. Co. I, Hampton Legion Infantry Clinton 3/15/64. AWOL10-12/64. NFR. d. before 1902.

STREETER (STREATER), EDWARD G. Pvt., Co. D. b. S. C. circa 1845. Enl. Chester CH 9/10/61 age 16. Accidentally wounded by pistol shot South Island 12/26/61. On leave Chester Dist. 1/62. Present 3-4/62. Discharged 5/18/62. Reenlisted. Co. I, Hampton Legion Infantry Chester CH 3/15/64. AWOL 9-12/64. NFR. d. before 1902.

STRIBLING, THOMAS JABEZ. Pvt., Co. F. b. S. C. 7/13/39. Student, age 21, Snow Creek, Pickens Dist. 1860 census. Enl. Co. E, Orr's S. C. Rifles, Anderson CH 7/20/61 age 22. Present until WIA

Chancellorsville 5/3/63. Ab. wounded through 10/63. Transferred Co. F, 1st S. C. Cav. 12/16/63. Detached Green Pond 4/24-30/64. Present 5-6/64. Present sick in camp 8/29-31/64. Present 9-12/64. NFR. d. Broadwater, Montana 4/25/65 age 25.

STRINGFELLOW, EDWARD H. 4thSgt., Co. D. b. S. C. circa 1836. Res. Chester Dist. 1850 census. Enl. Chester CH 9/10/61 age 26 as 4thSgt. Present through 4/62. Discharged for defective eyesight 5/7/62. Reenl. Co. H, 2nd S. C. Cav. Columbia 4/21/64. Detailed Major Melton, QM, Charleston 4/26/64-10/64. NFR. d. Macon Co., Ala. 6/11/22.

STROBLE, JONAS DURANT. Pvt., Co. E. b. S. C. 7/28/42. Enl. 1/6/62. On picket on Railroad 3-4/62. Present 5-6/62. On duty Pine Berry 7-8/62. Ab. sick in Summerville hospital 9-10/62. On scout 11-12/62. Present 3/63-8/64. On picket 9-10/63. Present 11-12/64. Paroled Hillsboro, N. C. 4/65. Laborer, Gregg, Aiken Co. 1900 census. Receiving pension Graniteville. Aiken Co. 9/19/19. d. Graniteville 2/15/24. Bur. Graniteville Cem. with full uniform and saber.

STROUD, JAMES H. Pvt., Co.'s D & K. b. Chester Dist. 10/14/14. Farmer, age 45, Chester Dist. 1860 census. Enl. Co. D Chester CH 9/10/61 age 47. Present through 12/61. Ab. on sick leave 4/25-30/62. Transferred Co. K 6/25/62. Ab. on sick leave through 9/1/62. Present 9-12/62. Discharged for overage & chronic rheumatism 12/31/62. Age 51, 5'10", light complexion, light hair, light eyes, Farmer. NFR. Farmer, Rossville, Chester Co. 1870 and Lewisville, Chester Co. 1880 census. d. 4/28/03. Bur. Union ARP Ch., Chester Co.

STROUD, PINCKNEY. Pvt., Co. B. b. S. C. 1836. Farmer, age 24, Spartanburg Dist. 1860 census. Enl. Camp Fripp 5/26/62. Ab. sick in Adams Run hospital 5/30/62. d. of disease in Adams Run hospital 7/21/62 age 25. Left widow. 5'10", fair complexion, black hair, Farmer. Bur. Cedar Shoals Bapt. Ch. Cem., Spartanburg Co. "He was a good soldier and an upright man."

STUART, W. J. K., Pvt., Co. K. Enl. by 10/64. Present 11-12/64. Paroled Raleigh, N. C. 4/65. d. Baltimore, Md. 6/3/66.

STURKEY, MILLEDGE L. B. Pvt., Co. A. b. S. C. circa 1836. Enl. Abbeville CH 8/2/61, age 25. Not on muster rolls 10/11/61. NFR. Living Abbeville Co. 1902.

SUMERALL, PLEASANT MATTISON. Pvt., Co. F. b. S. C. 8/1/34. Student, age 25, Pickens Dist. 1860 census. Enl. Pickens CH 12/4/61. Ab. sick 12/61-1/62. On picket Walters Battery 3-4/62. Present 5-6/62. Ab. sick Charleston hospital 7-8/62. Present 9/62-12/63. Detailed horse recruiting camp 1/20-2/29/64. Present 3-6/64. Detached to drive cattle for QM Dept. 8/20-12/31/64. NFR. Receiving pension Roberts, Anderson Co. 1901 age 65. Farmer, age 76, Forks, Anderson Co. 1910 census. d. 4/14/20. Bur. New Hope Cem., Senaca, Oconee Co.

SWEENY, WILLIAM. Pvt., Co. D. b. S. C. circa 1843. Res. S. C. Enl. by 1865. Deserted to the enemy Charleston 3/1/65. Age 22, 5'8", dark complexion, dark hair, dark eyes. Took oath and discharged.

SYPHRETT, RICHARD A. Pvt., Co. E. b. Orangeburg Dist. circa 1847. Res. Orangeburg Dist. Enl. as Volunteer Columbia 5/10/64. Present through 6/64. d. of fever in Charleston hospital 8/29/64.

TABER, ALBERT RHETT. 2ndLt., Co. E. b. S. C. 1833. M. D. Enl. Ft. Motte 10/26/61 as Pvt. 1/62. Present through 12/31/61. Ab. on leave 1/62. Elected 2ndLt. Present 2-4/62. Signed for forage for 83 horses 3/27/62. Ab. on leave 5/26-10/31/62. Resigned 11/26/62. Appointed Asst. Surgeon from S. C. 4/4/63. Assigned to Coosawhatchie, S. C. 5-1/63-4/24/64. Served with Jefford's 4th Bn. S. C. Cav. & captured Sandy Run, S. C. 2/15/65 while on leave. Released 2/25/65. NFR. d. 182/21/80. Bur. St. Matthews Epis. Ch., Ft. Motte, S. C. Wife receiving pension Ft. Motte, Calhoun Co. 11/3/19.

TAGGART, WILLIAM MOSES. Pvt., Co.'s G & A. b. S. C. 8/10/34. M.D., Preston, Webster Co., Ga. 1860 census. Enl. Co. A. Abbeville CH 8/2/61. Not on muster rolls 10/11/61. Renlisted Co. G Abbeville CH 3/7/63. On detached service 3-4/63. Transferred Co. A. Ab. sick with dysentery in Richmond hospital

7/27-9/15/63 & 10/9/63. Furloughed for 30 days 10/11/63. Present 11-2/63. Ab. recruiting horses 1/4-2/29/64. Detached Green Pond 4/64. Present 5-6/64. Detailed in Med. Dept. 7-8/64. Appointed Asst. Surgeon from S. C. 11/3/64 to rank from 9/24/64. Assigned to 1st La. Cav. 11/22/64. NFR. M.D. & Farmer, Magnolia, Abbeville Co. 1880 census. d. 5/8/90. Bur. Old Rocky River Presb. Ch. Cem., Abbeville Co.

TALLEY, WALTER T. Pvt., Co. F. b. S. C. circa 1829. Enl. Pickens CH 12/4/61 age 32. Present through 4/62. On picket Camp Means 5-6/62. Present 7/62-8/31/63. Ab. detached 9-10/63. Ab.on leave 10/22-12/31/63. AWOL 1-2/64. Transferred Co. A, Lucas's 15th Bn. S. C. Arty. 5/13/64. Joined 5/16/64. Present through 12/64. Paroled Greensboro, N. C. 5/1/65 .

TAYLOR, ALEXANDER ROSS, JR. Pvt., Co. A. b. S. C. 8/9/45. Res. Richland Dist. On post war roster. Reenlisted Co. K, 4th S. C. Cav. 8/24/63. WIA Battery Wagner on post war roster. Present through 2/64. Present until ab. on horse detail 8/11/64 for 40 days. Detailed as Orderly, Gen. Butler through 10/64. Paroled 4/65. d. 7/22/65. Bur. Taylor Meth. Ch., Columbia, S. C.

TAYLOR, E. J. Pvt., Co. G. b. Hampshire Co., Mass. circa 1825. Carriage Maker, age 25, Edgefield Dist. 1850 census. Enl. Abbeville CH 1/8/62. AWOL 4/28-5/30/62. Ab. sick in hospital 6/62. Present 7-8/62, as Commissary of Company. Discharged for "Phithisis Pulmonalis" Camp Myers, Va. 10/14/62. Age 36, 5'7", light complexion, blue eyes, dark hair, Carriage Maker. Paid 10/20/62. Served in Co. I, 5th S. C. Reserves 90 days 1862-63. NFR.

TAYLOR, EDWARD. Pvt., Co. G. b. S. C. 2/37. Enl. age 28 on postwar roster. Farmer, Fairfield, S. C. 1900. Living Abbeville Co. 1902. Living Fairfield Co. 1910 census.

TAYLOR, GEORGE W. 3rdSgt., Co.'s C & K. b. S. C. 5/24/40. Enl. Camp Butler 9/17/61 age 21 as Pvt. Detached Wadmalaw Island 11-12/61. Transferred Co. K 6/27/62 and promoted 4thSgt. Present 8-10/62, promoted 3rdSgt. Present 11/62-6/63. Captured Upperville 6/21/63. Sent to Old Capitol. Exchanged by 7/1/63. Present through 10/63. Ab. on horse detail 12/28/63-2/64. Present 3-4/64. Flag Bearer, James Island, 5-12/64. Ab. sick with "diaorrhia" in Charlotte, N. C. hospital 1/19-24/65. Paroled Augusta, Ga. 5/27/65. Farmer, Aiken Co. 1900 census. d. Aiken Co. 10/12/08. Bur. Levels Bapt. Ch. Cem. Wife receiving pension Aiken, S. C. 9/5/19.

TAYLOR, GEORGE WASHINGTON. Pvt., Co. E. b. S. C. 7/2/42. Res. Orangeburg Dist. Enl. by 1865. Paroled Hillsboro, N. C. 4/65. Married Aiken Co. 9/15/68. d. Thehane, Oklahoma 5/21/12.

TEBLER, SEBASTIAN E. Sgt., Co. E. Res. Orangeburg Dist. Enl. by 1862. Transferred to Artillery 1862 on postwar roster.

TENNANT, GILBERT C. Pvt., Co. G. b. S. C. 6/40. Res. Abbeville Co. Enl. Co. C, 7th S. C. Inf. 4/15/61. WIA Sharpsburg 9/17/62. WIA Fredericksburg 12/13/62. Discharged for disability 4/23/63. Enl. Co. G, 1st S. C. Cav. age 23, on postwar roster. Farmer, Richmond Co., Ga. 1880 census. d. Augusta , Ga. 9/3/15.

TENNENT, J. B. 3rdCpl., Co. I. b. Warrenton, Ga. circa 1843. Enl. Parker's Ferry 4/3/62 as Pvt. On picket Jacksonborough 4/30/62. Ab. on sick leave 6/62. Present 7-10/62, promoted 4thCpl. Present sick in camp 11-12/62. Promoted 3rdCpl. Ab. sick with catarrah fever in Staunton hospital 12/31/62-2/28/63. Ab. sick through 4/63. Paid 6/16/63. AWOL 6/25-10/31/63 and reduced to Pvt. Present 11/63-2/64. Detailed 4/16-30/64. Transferred Co. F, 1st S. C. Inf. (Butler's) Charleston 5/18/64. Present through 6/30/64. Ab. sick in Charleston hospital 7/64. Ab. sick in Summerville hospital 7/8-10/31/64. d. 3/8/65 age 22. Bur. Blandford Cem., Petersburg, Va.

TENNENT, WILLIAM. H. Pvt., Co. G. b. S. C. 1837. Gd. Citadel. Lawyer, age 23, Charleston 1860 census. Enl. Abbeville CH 3/15/62. Present through 4/62. Present sick in camp 5-6/62. Hired Lewis Howland as substitute and discharged 8/5/62. NFR. Civil Engineer, age 32, Charleston 1870 census. d. Spartanburg, S. C. 1893. Bur. Elmwood Memorial Gardens, Columbia, S. C.

THOMAS, WALTER J. JR. Pvt., Co. A. b. S. C. circa 1830. Overseer, age 20, Richland Co. 1850 census. Res. Abbeville Dist. Enl. Abbeville CH 8/13/61 age 28. On duty Bears Island 11-12/61. Nurse for Lt. Russell, Adams Run hospital 1/62. Present 3-4/62. KIA John's Island 6/8/62.

THOMASSON, JAMES B. Pvt., Co. H. b. S. C. circa 1849. Res. York Dist. 1850 census. Enl. Columbia 3/12/64. Present 11-12/64. NFR. Farmer, age 30, York Co. 1880 census. Retired, age 60, Gastonia, N. C. 1910 census. d. Catawba, N. C. 7/18/13.

THOMPSON, CHARLES ROBERT. Cpl., Co. E. b. S. C. 1826. M. D. Enl. Ft. Motte 10/26/61 age 35 as Pvt. Promoted Cpl. & present 11-12/61. NFR. Promoted Asst. Surgeon from S. C. 5/6/62. Assigned to College hospital, Columbia, S. C. 5/12/62. Assigned to 1st Ga. Inf. (Olmstead's) & promoted Surgeon 7/1/62. Assigned Savannah, Ga. hospital 11/25/63. Ab. on leave 12/29/63-1/28/64. Present 2-5/64. Assigned East Paul, Ga. 8/25/64, Tuscumbia, Ala. 11/7/64, & West Point, Miss. 1/13/65. NFR. M.D., age 44, Poplar, Orangeburg Co. 1870 census.

THOMPSON, JOHN A. Pvt., Co. D. b. S. C. circa 1841. Age 20, Oden, Effingham Co., Ga. 1860 census. Enl. Co. A, 2nd Bn. S. C. Volunteers Chester CH 1/16/62. Transferred Co. D, 1st S. C. Cav. 9/1/62. Present through 6/63. Ab. on horse detail 8/18-31/63. Ab. carrying Regimental Mail 9-10/63. Present 11-12/63. Ab. recruiting horses 1/19-2/29/64. Detached Green Pond 4/26-30/64. Present 5-6/64. Ab. on leave 8/21-31/64. Present 9-10/64. Detached Coosawhatchie 12/10-31/64. Paroled Hillsboro, N. C. 4/65. M.D., & Farmer, Red River, Texas 1880 census. d. Bogata, Red River Co., Tex. 8/17/22. Bur. Bogata Cem.

THOMSON, JOHN ANDREW. Pvt., Co. G. b. S. C. 3/29/47. Enl. Abbeville Ch 3/24/63. Ab. on detached service through 4/63. Present 5-6/63, detached in Regimental Band 6/1/63. Ab. on horse detail 8/18-31/63. Present as Musician 9/63-12/64. NFR. Farmer, age 63, Pacolet, Spartanburg Co. 1910 census. d. Union Co., S. C. 12/4/34. Bur. Rosemont Cem.

THOMPSON, JAMES S. Pvt., Co. E. b. St. Matthews Parish, S. C. circa 1827. Enl. Ft. Motte 10/26/61 age 32. On picket White Point through 12/31/61. Present sick in camp 3-4/62. Present 5-6/62. Present sick in camp 7-8/62. Present 9-10/62. Ab. Fredericksburg as witness in CM 11-12/62. Discharged for disability 2/15/63. Age 33, 5'11", dark complexion, dark eyes, black hair, Planter. Reenlisted Co. K, 4th S. C. Cav. 11/10/63. Dropped 4/9/64. Farmer, Lansford, Chester Co. 1880 census.

THOMPSON, JOHN CALDWELL CALHOUN. Pvt., Co. F. b. Anderson Dist. 3/12/46. Att. Citadel. Res. Student, age 14, Anderson Dist. 1860 census. Enl. Columbia 4/1/64. Ab. on leave 4/20-30/64. Ab. sick in hospital 5/2-6/64. d. of diptheria in Ladies Hospital, Columbia 6/29/64. Bur. Roberts Presb. Ch. Cem., Anderson Co.

THOMPSON, SAMUEL ROBINSON. 4thSgt., Co. D. b. S. C. circa 1840. Res. Chester Dist. Enl. Chester CH 9/10/61 age 21 as 2ndCpl. Present through 6/62. Ab. on leave 8/27-31/62. Present 9-10/62. Promoted 4thSgt. Present 11/62-6/63. Ab. sick with typhoid fever in Lynchburg hospital 8/3/63. d. there 8/13/63 age 23. Bur. Lynchburg Confederate Cem., grave no. 7. Effects: $28.25.

THOMPSON, WILLIAM G. Pvt., Co. H. b. S. C. circa 1830. Enl. Rock Hill 1/30/62 age 30. On picket Routole Bridge 3-4/62. Present 5/62-6/63. Ab. on horse detail 8/18-31/63. Present 9-10/63. Ab. sick in Richmond hospital with acute diarrhoea 12/13/63-2/12/64 & 3/20-28/64. Present 4-8/64. Ab. on leave 9-10/64. Present 11-12/64. NFR. Farmer, Mallorys, Wilkes Co., Ga. 1900 census.

THURSTON, JOHN GOUGH. Pvt., Co. A. b. S. C. circa 1837. Clerk, Charleston 1860 census. Enl. 1/1/62. Reenlisted Co. K, 4th S. C. Cav. Grahamville 3/25/62 age 23. Present until ab. sick 6/63. Present until ab. on sick leave 12l17/63-2/64. Present 3-10/64. Paroled 4/65. Cotton Factor, Summerville 1880 census. Res. of Charleston 1896. Wife living Taylors, N. C. 1/17/30.

225

TILLEY, LEWIS M. Pvt., Co. K. b. Anderson Dist. 11/19/18. Res. Pickens Dist. Enl. Columbia 5/25/64. Ab. sick in Columbia hospital 10/24-31/64. On picket Chatman's Ford 12/29-31/64. Transferred Co. H, 2nd S. C. Arty. in exchanged for B. Foreman 2/11/65. NFR. d. Anderson Co. 11/14/97. Bur. Providence Cem.

TILLOTSON, JAMES WILSON. Pvt., Co. B. b. Spartanburg Dist. 8/15/41. Enl. 5th S. C. Inf. Spartanburg 4/13/61. Present until ab. sick with pneumonia in Richmond hospital 2/6/62-3/24/62. Transferred to hospital in S. C. NFR. Reenlisted Co. B, 1st S. C. Cav. Spartanburg 1/26/64 age 23. Present through 12/64. NFR. d. Spartanburg 7/4/09. Bur. Cannons Camp Meth. Ch., Spartanburg Co.

TIMMONS, WILLIAM MARION. Pvt., no company. b. S. C. 9/18/37. Not on muster rolls. d. 4/19/28. Bur. Waverly East Cem., Cook Co., Ill.Tombstone only record of service.

TINDALL, DAVID D. Pvt., Co. E. b. S. C. 10/8/38. Enl. Ft. Motte 10/26/61 age 35. Present through 12/61. Ab. on leave 6 days 1/62. Present on all rolls through 12/64. Paroled Hillsboro, N. C. 4/65. d. 3/28/97. Bur. Denmark Cem., Bamberg Co., S. C.

TOLBERT (TALBERT), T. B. Pvt. Co. A. Enl. Abbeville CH 8/2/61. Not on muster rolls 10/11/61. NFR. Possibly Thomas M. Tolbert b. 8/19/41. d. 6/9/15. Bur. Tolbert Cem., Abbeville Co.

TOLBERT, WILLIAM K. 3rdSgt., Co. G. b. S. C. 1840. Enl. Abbeville CH 1/8/62 as 4thSgt. Ab. sick in hospital 3-4/62. Ab. on sick leave 6/3-30/62. On picket Grimball's 7-8/62. Promoted 3rdSgt. 9/15/62. Ab. sick in Summerville hospital 10/1-12/31/62. Present 3-4/63. Present 5-6/63, dismounted. Ab. on horse detail 8/31/63. AWOL 10-12/63 & reduced to Pvt. Present 1-4/64. Transferred Co. C, Lucas's 15th Bn. S. C. Arty. 5/13/64. NFR. d. 12/1/69. Bur. Tranquil Meth. Ch., Greenwood Co., S. C.

TOOLE, BENJAMIN F. 3rdCpl., Co. C. b. Barnwell Dist. 4/1/40. Enl. Adams Run 9/1/62 as Pvt. On detached service through 10/62. Ab. on leave Culpepper CH 11-12/62. Ab. sick with "Exzema" in Charlottesville hospital 1/29-3/10/63. Present 3-4/63. Promoted 3rdCpl. Present 5-8/63. Ab. sick with gonorrhea in Richmond hospital 9/22/63. Furloughed for 30 days 10/21/63. Ab. on leave through 12/31/63. Present 1-2/64. Ab. on horse detail 4/12-5/12/64. Present through 12/64. NFR. Teacher, Homer, Banks Co., Ga. 1880 census. d. 3/16/89 age 48. Bur. Williston Cem., Barnwell Co., S. C.

TOOLE, EVERETT G. Pvt., Co. C. b. S. C. 1830. M.D., age 30, Barnwell Dist. 1860 census. Enl. Hamburg 8/27/61 age 31. On recruiting duty 12/20-31/62. Present 5-6/62. Ab. on sick leave 8/31/62. Discharged 9/4/62. NFR. Alive 1898.

TOOLE, LUCIUS B. 4thSgt., Co.'s C & K. b. Barnwell Dist. 11/43. Enl. Co. C Wadmalaw Island 1/1/62 as Pvt. Presence or absence not stated through 4/62. Transferred Co. K 6/27/62. Ab. sick leave 8/31/62. Promoted 2ndCpl. Present 9/62-4/63. Promoted 4thSgt. Present 5-8/63. Ab. on horse detail 9/22-12/31/63, however, ab. sick with "Int. Fever" in Richmond hospital 9/22-29/63. Ab. on detached service 2/1-29/64. Detached Green Pond 4/16-30/64. Present 5-10/64. Detached Pocotaligo 12/10-31/64. Paroled Augusta, Ga. 5/19/65. Farmer, age 38, Williston, Barnwell Co. 1880 census. Farmer, Bainbridge, Ga. 1900 & 1910 censuses. d. Bainbridge, Ga. by 1920.

TOOLE, WALKER WARREN N. Pvt., Co. C. b. Aiken Dist. circa 1828. School Teacher, age 32, Barnwell Dist. 1860 census. Enl. Aiken Dist. age 35 on post war roster. Discharged or transferred 2nd S. C. Arty. by 1864. AWOL 3/24/64-8/30/64. NFR. At. home, age 42, Millbrook, Barnwell Co. 1870 census. d. 1873.

TOOMER, FRANCIS STALL. Pvt., Co. A. b. Charleston circa 1841. Merchant, age 19, Charleston 1860 census. Enl. 11/7/61. Ab. on leave 2/62. Reenlisted Co. C, 5th S. C. Cav. Cheraw 3/12/62. Present until detailed in Medical Dept. 8/8/63. Present 11-12/63. Ab. on leave 2/26/64 for 20 days. In Richmond hospital 6/22-30/64. Furloughed for 30 days 8/6/64. AWOL 10/2/64. Ab. sick with ulcus in Charlotte, N. C. hospital 2/21/22/65. NFR. d. in England 1866 age 25.

TRENHOLM, EDWARD LEONARD. Pvt., Co. A. b. S. C. 1844. On post war roster. Reenlisted Co. K, 4th S. C. Cav. Charleston 1/17/64. Present until WIA (contusion) & in Richmond hospital 6/3/64 & with "Febris Int." 6/23/64. Present until ab. on horse detail 8/11/64 for 40 days. Ab. sick in Columbia, S. C. hospital 9/22-10/1/64. Deserted from hospital 10/16/64. NFR. d. Greenville, S. C. 8/7/73. Bur. Epis. Ch. of Advent, Spartanburg.

TREZEVANT, JAMES DAVIS. Captain, Co. E. b. Columbia, Orangeburg Dist. 11/25/22. Planter, 1860 census. Enl. Ft. Motte 10/26/61. Present Adams Run 1/62. Present through 6/62. 85 horses present 4/1/62. Signed for 60 pr. horse shoes & 25 lbs. horse shoe nails Adams Run 4/15/62. Signed for 336 heads of salt beef 5/13/62. On scout Edisto Island 7-8/62. Present 9-10/62. Ab. on CM Fredericksburg 11-12/62. Paid $75.00 to have 75 horses shoed in Caroline Co., Va. 12/17/62. Present 3-10/63. 44 horses present Stevenburg 2/1/63. 69 horses present 5/1/63. 62 horses present 5/9/63. Signed for 20 lbs. horseshoe nails 6/10/63. 30 horses present 7/31/63. 32 horses present 8/1/63. Signed for 8 jackets, 5 caps & 4 shirts at Stevensburg 8/6/63. Signed for 12 caps, 17 jackets, 12 pr. pants, & 9 shirts Stevensburg 8/24/63. Signed for 13 pr. boots, 17 pr. socks, 36 blankets, 30 pr. drawers, 15 pr. pants, 1 cap, 1 overcoat & 3 shirts 9/8/63. 42 horses present in Madison Co., Va. 10/1/63. Ab. procuring forage for Brigade 11-12/63. Ab. on leave 2/4-3/3/64. Present 4-8/64. 26 horses present, Graham, N. C. 4/10/64. 1 officer and 37 men present James Island 4/64. Signed for 8 pr. pants, 19 jackets, 1 pr. shoes, 6 pr. socks, 18 shirts, 15 pr. drawers, 10 caps, 3 Army tents, 5 camp kettles, 13 mess pans, 1 spade & 2 axes James Island 6/30/64. 51 horses present 8/1/64. 52 horses present 9/1/64 & 48 horses present 10/1/64. Gen. Hampton wrote the Sec. of War wanting Trezevant to command a Bn. of Sharpshooter 8/64. Commanding Regiment 9-10/64. Signed for 1 jacket, 4 pr. pants, 5 shirts, 5 pr. drawers, 7 pr. shoes, 1 pr. socks & 1 wall tent James Island 9/30/64. Detached Pocotaligo 12/10-31/64. Commanding Regiment Branchville & Barnwell 2/65. Ordered to Columbia by Gen. Hampton 2/9/65. Colonel Black was still on James Island. Major Nesbitt was at Columbia with half the men dismounted who wanted to go home for horses. Reported the Yankees were taking everything. "They are a lawless set. There are horses to be gotten if they don't get them first." NFR. Paid taxes Richland Co. 1865. Living Rayeville, La. 6/27/74. d. Columbia, Orangeburg Co. 1892.

TREZEVANT, WILLOUGHBY FARQUHAR. Pvt., Co. E. b. Columbia, Richland Dist. 1846. Student, age 13, Columbia 1860 census. Served briefly in Co. E. Appointed Aide de camp Gen. Evans. DOW's received Sharpsburg 9/17/62 at Shepherdstown, Va. 9/24/62 age 16. Bur. Trinity Ch. Cem., Columbia, S. C.

TROTTER, HENRY ODUS. Pvt., Co. F. b. Pendleton, Anderson Dist. 12/22/13. Farmer, age 42, Pickens Dist. 1860 census. Enl. Pickens CH 12/4/61 age 48. Present through 12/31/61. On picket Simmons House 3-4/62. d. of typhoid fever in Adams Run hospital 6/2/62. Left widow. Bur. Oonleney Bapt. Ch. Cem., Pickens Co.

TROTTIN, FRANK B. Pvt., Co.'s C & K. b. Augusta, Ga. 9/46. Age 5, Edgefield Dist. 1850 census. Enl. Co. C Wadmalaw Island 5/12/62. Transferred Co. K 6/27/62. Ab. on sick leave 7-8/62. Transferred back to Co. C 7/8/62. Present 9-12/62. Ab. on sick leave 3-4/63. Present 5-6/63. Ab. on horse detail 8/15-31/63. Present 9-10/63. Ab. sick with chronic diarrhoea in Richmond hospital 11/29/63. Furloughed for 30 days 12/22/63. Present 2/64. Ab. on horse detail 4/20-5/10/64. Present 5-6/64. Transferred Co. C(3rd), Manigault's Bn. S. C. Arty. in exchange for Thomas Turner 7/15/64. Present through 12/64. Paroled Augusta, Ga. 5/18/65. Farmer, Saluda, Edgefield Co. 1870-1910 censuses. Applied for pension in Ga. 1915. d. Saluda, Edgefield Co. 5/17/16. Bur. Leesville, S. C.

TRUESDELL, JOHN R. Pvt., Co. H. b. Kershaw Dist. circa 1844. Enl. Rock Hill 1/9/62 age 27? Ab. sick in hospital 3-4/62. Present 5-8/62. Ab. sick with rheumatism in Richmond hospital 10/11-17/62. Furloughed for 30 days. Ab. sick 11-12/62. Present 3-8/63. Ab. on sick leave 9-10/63. Present 11/63-2/64. Detached Green Pond 4/16-30/64. Present 5-6/64. Ab. on sick leave 8/2-31/64. Ab. detailed 9-10/64. Present 11-12/64. NFR. Farmer, Kershaw Co. postwar. d. Buffalo, Kershaw Co. 9/15/20 age 76.

TUCKER, BENJAMIN F. Pvt., Co. G. b. Greenville Dist. 1836. Farmer, age 24, West's Store, Spartanburg Dist. 1860 census. Enl. Abbeville CH 3/15/62. Present through 4/62. Present sick in camp 5-6/62. Detached in Commissary Dept., Dills Bluff 7-8/62. AWOL 9-10/62. Present 11/62-4/63. Present 5-6/63, dismounted. Ab. on horse detail 8/31/63. Present 9-10/63. Ab. on horse detail 12/22/63-2/64. Ab. detached 4/14-30/64. Present 5-6/64. Present sick in camp 7-8/64. Detached guarding POW's Florence 9/10-12/31/64. NFR. Living Abbeville Co. 1902. d. 4/28/12. Bur. Plum Cem/. Lakeview, Ohio.

TUCKER, J. A. Pvt., Co. I. b. S. C. circa 1841. Enl. Columbia 4/15/64. Detached Green Pond through 4/30/64. Present 5-8/64. Ab. sick in hospital 10/1-12/31/64. NFR. Farmer, age 39, Oktibbeha Co., Miss. 1880 census. d. Barbour Co., Ala. 6/6/08.

TUMBLESTON, CHARLES A. Pvt., Co. A. b. S. C. 1837. Farmer, age 24, Colleton Dist. 1860 census. Enl. Parker's Ferry 4/3/62. AWOL through 4/30/62. NFR. Farmer, age 43, Burns, Colleton Co. 1880 census. d. 1880 age 43. Brother of Josiah J., Nathaniel T. & Richard P. Tumbleston.

TUMBLESTON, JOSIAH J. Pvt., Co. I. b. S. C. 1840. Age 10, Colleton Dist. 1850 census. Enl. Parker's Ferry 4/3/62. Present through 6/62. Ab. sick in hospital 7-8/62. Present 9/62-6/63. Ab. on horse detail 8/18-31/63. Present, detached with wagon after corn 9-10/63. Present 11/63-2/64. Detached Green Pond 4/16-30/64. Present 5-12/64. Detached in Commissary Dept. by 3/65. Deserted to the enemy Charleston 3/22/65. Age 22, 5'7", dark complexion, dark hair, dark eyes, Took oath and discharged. Farmer, Verdir, Colleton Co. 1870 census. d. 6/26/94 age 54. Wife receiving pension Charleston 10/14/19. Brother of Charles A., Nathaniel T. & Richard P. Tumbleston.

TUMBLESTON, NATHANIEL T. Pvt., Co. I. b. S. C. 1839. Res. Colleton Dist. Enl. Columbia 5/4/64. Present through 10/64. On picket 11-12/64. Deserted to the enemy Charleston 3/22/65. Age 26, 5'9", dark complexion, dark hair, blue eyes. Took oath and discharged. Farmer, age 45, Colleton Co. 1880 census. d. circa 1888. Bur. Shepherd Grove Ch., Sugar Hill, Dorchester Co. Brother of Charles A., Josiah J. & Richard P. Tumbleston.

TUMBLESTON, RICHARD POSTEL. Pvt., Co. I. b. Colleton Dist. circa 1838. Farmer, Colleton Dist. Enl. Parker's Ferry 4/3/62. Present through 10/62. Present sick in camp 11-12/62. Ab. sick in Staunton hospital 2/28-3/15/63. Present 5-6/63. Detached horse recruiting station 8/2-12/31/63. Present 1-2/64. Ab. on leave 4/22-5/2/64. Present through 10/64. AWOL 12/23-31/64. Deserted to the enemy Charleston 3/22/65. Age 27, 6', dark complexion, black hair, dark eyes. Took oath and released. Farmer, Colleton Co. 1870 census. d. 1895. Wife receiving pension Round O, Colleton Co. 10/1/19. Brother of Charles A., Josiah J. & Nathaniel T. Tumbleston.

TUMBLESTON, WILLIAM MILES. Pvt., Co. C. b. Colleton Dist. 4/22/42. Enl. Wadmalaw Island 1862. NFR. Enl. Co. C, 5th S. C. Cav. Cheraw 3/12/62. Present 1-5/63. Ab. on leave 12 days 6/26/63. Present 7/63-2/64. Ab. in hospital 3-4/64. WIA (left thigh by minnie ball) and in Richmond hospital 7/5/64. Furloughed for 50 days 7/12/64. AWOL 10/6-31/64. NFR. Deserted to the enemy Charleston 4/22/65. Age 33, 5'10", dark complexion, black hair, grey eyes. Took oath and discharged. d. Dorchester Co. 6/16/20. Bur. Shepherd Grove Ch., Dorchester Co.

TURNER, A. T. Pvt., Co. F. b. S. C. circa 1842. Enl. Pickens CH 12/4/61 age 19. Present through 4/62. AWOL 6/29-30/62. Present 7/62-10/63. Detached recruiting horses 11/16/63-2/64. Present 3-6/64. Present driving wagon 8/26-31/64. Transferred Co. A, Lucas's 15th Bn. S. C. Arty. in exchange for J. F. McLees 9/15/64. Present through 12/64. Paroled Greensboro, N. C. 5/1/65.

TURNER, C. Pvt., No Co. d. Richmond 10/15/64. Bur. Hollywood Cem.

TURNER, JOSEPH ROBERT. Pvt., Co. H. b. Chester Dist. 1834. Farmer, age 26, York Dist. 1860 census. Enl. Rock Hill 1/2/62 age 34. Present sick in camp 3-4/62. Present 5-10/62. On raid 11-12/62. Ab. on sick leave 3-4/63. Ab. sick in hospital 5-6/63. Present 9-10/63. Ab. on horse detail 12/22/63-2/64.

Present 3-8/64. Ab. sick 10/6-31/64. NFR. Res. Flat Rock, York Co. 1870 & 1880 censuses. Res. Lanford, S. C. 1900-1920 censuses. d. 10/12/27. Bur. Chester Co.

TURNER, RICHARD A. Pvt., Co. C. b. S. C. 5/37. Enl. Camp Butler 9/17/61. NFR. Enl. Co. D, 23rd S. C. Inf. Charleston 9/23/61 age 32. Present through 10/61. Served to 5/65 on wife's pension application, but not muster rolls after 10/61. d. Pickens, S. C. 4/5/12. Wife receiving pension Pickens Co. 10/2/19.

TURNER, THOMAS. Pvt., Co. C. b. Marion Dist. 1846. Age 14, Sawyers Mill, Lexington Dist. 1860 census. Enl. James Island 8/1/64. Present through 10/64. Detached Pocotaligo 12/10-31/64. NFR. Farmhand, age 37, Mobley, Edgefield Co. 1880 census. Alive 1898.

TURNER, W. HOLLY. Sgt., Co's. D & K. Enl. Chester Dist. age 41. on post war roster. Transferred Co. K 7/62. NFR.

TUSTIN, HIRAM HARRISON TILMAN. Pvt., Co. A. b. Abbeville Dist. 12/18/28. Enl. Abbeville CH 3/19/62. Discharged for disabilty 4/24/62. 5' 8", dark complexion, black eyes, black hair, Watchmaker. NFR. Jeweler, Abbeville 1870 & 1880 censuses. & 1902. Silversmith. d. 2/7/15. Bur. Upper Cane Cem., Abbeville Co.

TWIGGS, JOHN DAVID. Lieutenant Colonel. b. Richmond Co., Ga. 4/6/26. Planter, Edgefield Dist. Enl. Co. C, 1st S. C. Bn. Cav. Hamburg 8/27/61 as Captain, age 34. Present through 10/61. Signed for 5 camp kettles, 2 lanterns, & 1 wheelbarrow Camp Butler 10/15/61. Present Wadmalaw Island 1/62. Appointed Major 5/26/62 & Lt. Colonel 6/25/62. Present 9-10/62. Ab. attending CM Fredericksburg 11-12/62. Ab. on CM 4/30/63. Present through 10/63. Signed for forage for 6 horses of staff Winchester 7/6/63. Injured in riding accident 11/63. Ab. sick at home in Hamburg with rheumatism 11/19-12/31/63. Present 1-2/64. Detached Green Pond 4/15-5/18/64. Signed for 10 Army tents Green Pond 5/29/64. Ab. on leave 6/16/64 for 20 days. Signed for 1 wall tent, 1 office desk, 1 camp kettle & 2 men pans 6/30/64. Commanding East Lines, James Island 8/6-25/64. Ab. on leave 9/2/64. Killed in street fight by a man named Butler Hamburg 9/15/64. Bur. Magnolia Cem., Augusta, Ga. Father of Joseph A. Twiggs.

TWIGGS, JOSEPH A. Pvt., Co.'s C & K. b. Augusta, Ga. 1844. Student, age 16, Richmond Co., Ga. 1860 census. Enl. Co. C Hamburg 8/17/61 age 17. Detached Wadmalaw Island 11-12/61. Present 5-6/62. Transferred Co. K 7/1/62. Ab. on sick leave 7-8/62. Present 9-10/62. On scout 11-12/62. Present 3-6/63. Black horse killed Brandy Station 8/1/63. Paid $450.00. Ab. on horse detail 8/16-31/63. On scout 11-12/63, however, ab. on horse detail 12/23/63-2/64. Ab. on horse detail 4/20-30/64. Orderly for Lt. Col. Twiggs 5-8/64. Ab. sick Charleston hospital 12/20-31/64. Paroled Raleigh, N. C. 4/65. Policeman, age 36, Augusta, Ga. 1880 census. Member, Ga. Camp No. 435, Confederate Veterans. d. Augusta, Ga. 5/31/25. Son of John David Twiggs.

TYLER, BENJAMIN F. Pvt., Co. F. Enl. Adams Run 1/1/62. Present 5/62-10/64. Detached Pocotaligo 12/10-31/64. KIA N. C 1865 on post war roster.

TYLER, CHARLES. Pvt., Co. C. Not on muster rolls. KIA Moccasin Creek, N. C. 4/10/65.

TYLER, ELMON. Pvt., Co. C. Not on muster rolls. KIA 4/65 on postwar roster.

TYLER, MARTIN VAN BUREN. Pvt., Co. C. b. S. C. 12/11/39. Enl. Wadmalaw Island 3/18/62. Present 5-6/62. Ab. on sick leave 7/27-831/62. Present 9-10/62. On wagon guard 11-12/62. Ab. sick with chronic diahorrea in Richmond hospital 1/16-19/63. Present 3-4/63. Ab. sick in hospital 8/16-31/63. Present 9-10/63. Ab. sick with "Int. Fever" in Richmond hospital 11/27-12/8/63. Ab. horse recruiting camp 12/22-31/63. Present 1-2/64. Detached Green Pond 4/16-30/64. Present 5-10/64. Detached Pocotaligo 12/10-31/64. NFR. d. Aiken Co. 5/30/93. Bur. Bethany Cem., Aiken Co.

TYLER, R. ELMORE. Pvt., Co. C. b. S. C. circa 1840. Enl. Wadmalaw Island 3/18/62. Present 5-6/62. Ab. on sick leave 7/27-8/31/62. Present 9-10/62. On scout 11-12/62. Present 3/63-2/64. Detached Green Pond 4/16-30/64. Ab. on leave 6/16-30/64. Present 7-12/64. Present Fayetteville, N. C. 4/65. Farmer, age

30, Silverton, Barnwell Co. 1870 census and age 39, Sleepy Hollow, Aiken Co. 1880 census. d. 6/16/90. Bur. Rhett-Green Cem., which has been moved. Wife receiving pension North Augusta, Aiken Co. 10/16/19.

VANDERHORST, LEWIS. Pvt., Co. A. b. S. C. 1826. Enl. 11/7/61. Reenlisted Co. K, 4th S. C. Cav. 3/25/62 age 35. Present until ab. on leave 12/27/63 for 5 days. Present 1/64 until KIA Hawes' Shop 5/28/64. Bur. Magnolia Cem., Charleston.

VAUGHN, DAVID. Pvt., Co. G. b. Greenville Dist. circa 1838. Enl. Abbeville CH 1/8/62, age 20. Present sick in camp 3-4/62. On picket 5-6/62. Ab. sick in Charleston hospital 7/30/62. Present 9/62-4/63. NFR. Enl. Co. F, 1st S. C. Inf. (Butler's) Ft. Moultrie 1/27/64. Issued clothing 3/64. NFR. d. 3/13/64. Bur. Magnolia Cem., Charleston, S. C.

VERRELL, BELTON O. Pvt., Co. A. b. S. C. 9/39. Farmhand, age 20, Abbeville Dist. 1860 census. Enl. Abbeville CH 8/13/61 age 22. Ab. on sick leave 10/31/61. Present 11/61-6/62. Present sick in camp 7-8/62. Present 9/62-4/63. Ab. in charge of recruiting horses 5-6/63. Ab. on horse detail 9-10/63. Present 11/63-2/64. Ab. on leave 4/23-30/64. Present 5-12/64. NFR. Farmer, Smithville, Abbeville Co. 1870 census. Farmer, Greenwood, Abbeville Co. 1900 & 1910 censuses.

VERRELL, JAMES. F. Pvt., Co. A. b. S. C. 1833. Painter, age 27, Abbeville, S. C. 1860 census. Enl. Abbeville CH 8/13/61. On duty Bears Island 11-12/61. Present 3/62-8/31/63. Ab. in charge of recruiting horses 9/63-2/64. Detached Green Pond 4/64. Present 5-12/64. NFR. Res. Beauregard, Ark. 1870 census. Res. Spring Hill, Ark. 1880 census. Res. Wilmar Town, Drew Co., Ark. 1900 census. d. Drew, Ark. 1910.

VINCENT, WILLIAM E. Pvt., Co. A. b. S. C. 5/23/44. On post war roster. Reenlisted Co. K, 4th S. C. Cav. Grahamville 3/25/62. Present until detailed with Asst. Engineer by Gen. Pemberton 7/3-/62-11/26/63. Present 11/63-2/64. Ab. on horse detail 8/11/64 for 40 days. Ab. sick in Columbia, S. C. hospital 10/22-10/31/64. Paroled 4/65. d. 8/23/88. Bur. Magnolia Cem., Charleston.

WADE, AMBROSE WILLIAM. Pvt., Co. D. b. Chester Dist. 1844. Age 16, Chester Dist. 1860 census. Enl. Chester CH 9/10/61 age 17. Present 11/61-6/62. Ab. on sick leave 8/4-31/62. Present 9-10/62. On scout 12/25-31/62. Present sick in camp 3-4/63. WIA (hand severley) Brandy Station 6/9/63. Ab. wounded in Charlottesville hospital 6/10-7/24/63. Furloughed for 60 days. Ab. wounded through 2/64. Detached to Enrolling Officer, Chester Dist. 4/15-12/31/64. NFR. d. S. C. 1880. Bur. Brushy Creek Bapt. Ch., Chester Co.

WADE, EARL THOMPSON. Pvt., Co. D. b. S. C. 1838. Farmer, age 22, Chester Dist. 1860 census. Enl. Chester CH 9/10/61 age 24. Ab. sick 10/28/61-1/62. Present 3-4/62. Captured on picket John's Island 6/7/62. Sent to Ft. Columbus, N. Y. & Ft. Delaware. Paroled for exchange 10/6/62. Exchanged 11/10/62. Present 3-6/63. Ab. on horse detail 8/18-31/63. Present 9-10/63. Ab. on horse detail 12/22/63-2/64. Present 3-6/64. Ab. sick in hospital 7/26-31/64. AWOL 10/1-12/31/64. Paroled Greensboro, N. C. 4/26/65. Farmer, age 42, Baton Rouge, Chester Co. 1880 census. d. 9/20/98. Bur. Brushy Fork Bapt. Ch. Cem., Chester Co., S. C.

WADE, J. T. Pvt., Co. D. b. S. C. circa 1840. On roster but not on muster rolls. NFR.

WADE, WILLIAM D. Pvt., Co. C. b. S. C. 10/34. Farmer, age 27, Chester Dist. 1860 census. Enl. Summerville 10/1/62. Present through 10/31/62. On scout 11-12/62. On detached service 3-6/63. Ab. sick in hospital 8/10-31/63. Present 9/63-10/64. Detached Pocotaligo 12/10-31/64. NFR. Laborer, age 65, Bates, Greenville Co. 1900 census. Probably William Dickson Wade 1833-9/4/11. Bur. Sowell Cem., Akubutta, Miss.

WAGNER, A. C. Pvt., Co. A. b. S. C. circa 1827. On post war roster. Reenlisted Co.G, 4th S. C. Cav. Grahamville 3/25/62 age 35. Present until ab. sick 6/62 until discharged for disability 7/1/63. Served later in Co. C, 3rd S. C. State Troops 1863-64. NFR. Clerk, State Cotton Press, Charleston 1875.

WALKER, JERRY T. Pvt., Co. D. b. Chester Dist. 1841. Farmer, Chester Dist. 1860 census. Enl. Co. F, 6th S. C. Inf. Summerville 6/11/61 age 19. Transferred Co. A 7/8/61. Present through 10/61. Discharged for hernia 12/4/61. Age 19, 5'11", sandy complexion, grey eyes, auburn hair, Planter. Enl. Co. D, 1st S. C. Cav. Camp Ripley 2/16/62. Present through 6/62. Ab. on sick leave in Chester Dist. 8/1-31/62. Present 9-10/62. On scout 12/25-31/62. Present 3-6/63. Ab. on horse detail 8/18-31/63. Present 9-10/63. Ab. on horse detail 12/22/63-2/64. Detached Green Pond 4/26-30/64. Present 5-10/64, however, ab. on sick leave 10/21-31/64. Present 11-12/64. Paroled Greensboro, N. C. 5/1/65. Farmer, Chester Co. 1870 & 1880 censuses. d. Sheep Island, Fla. 3/11/89 age 55 on wife's pension application.

WALKER, JOHN GADDEN. Pvt., Co. G. b. Abbeville CH 11/22/43. Enl. Abbeville CH 1/8/62. Present through 10/62. Muster rolls missing until present 9-10/64. Detached Pocotaligo 12/10/31/64. NFR. Farmer, Abbeville Co. 1880-1910 censuses. d. Abbeville 6/1/20. Bur. Sharon Meth. Ch. Cem.

WALKER, JOSEPH A. Pvt., Co. H. b. S. C. 5/10/46. Enl. Columbia 4/12/64. Present through 6/64. Detailed in Commissary Dept. 7-10/64. Detached Pocotaligo 12/10-31/64. NFR. Sawyer, age 34, Barnwell Co. 1880 census. Mill Hand, age 52, Bamburg, S. C. 1900 census. Applied to enter Old Soldiers' Home, Columbia, S. C. from Lancaster Co. 11/3/13. d. 5/29/15. Bur. Poplar Springs Bapt. Ch. Cem., Laurens Co.

WALKER, WILLIAM ALEXANDER. Lieutenant Colonel. b. Chester, S. C. 6/14/19. Gd. S. C. College 1840. Lawyer & Banker, Chester, S. C. Elected Captain, Co. D, 1st S. C. Bn. Cav. 9/10/61. Appointed Major, 1st S. C. Cav. 6/62. Commanding Regiment Christmas Raid 1862. WIA Gettysburg 7/3/63 while commanding regiment. Promoted Lt. Col. 9/15/64. Present Asheboro and Moccassin Swamp, N. C. 4/65. Member, S. C. Legislature 1865-67 & 1877-1882. d. Chester, S. C. 4/21/82. Bur. Purity Presb. Ch. Cem., Chester Co. Most muster rolls on him are missing.

WALKER, WILLIAM JAMES. Pvt., Co. C. b. S. C. 10/15/53. Enl. Columbia 2/21/64. Ab. on detail 4/1-5/1/64. Present 5-6/64. In Charleston hospital 8/13/64. Detailed to Edgefield Dist. to collect supplies for Wayside Home, Charleston 8/21-31/64. On picket 9-10/64. Present 11-12/64. Courier, Gen. Hardee 2-4/26/65. Paroled Augusta, Ga. 5/19/65. d. by 1898.

WALL, JAMES RANDALL PICKETT. Pvt., Co. D. b. Chester Dist. 8/24/35. Farmer, age 25, Sumter Co., Fla. 1860 census. Enl. Chester CH 4/10/62. Ab. on leave 4/28-30/62. Captured John's Island 6/7/62. Sent to Ft. Columbus, N. Y. & Ft. Delaware. Exchanged 10/4/62. Ab. sick through 12/62. Present 3-6/63. Ab. on horse detail to Chester, S. C. 8/18-31/63. Present 9-10/63. Ab. on horse detail 12/22/63-2/64. Detached Green Pond 4/26-30/64. Present 5-6/64. Paroled Hillsboro, N. C. 4/65. Farmer, Rutland, Sumter Co., Fla. 1870-1910 censuses. d. Sumter Co., Fla. 10/14/13 on wife's pension application.

WALLACE, ROBERT. Saddler, Co. D. b. Chester Dist. circa 1835. Enl. Co. G, 6th S. C. Inf. Chester CH 4/11/61 age 30. Ab. sick Manassas hospital 10/12/61. Transferred to Richmond hospital with chronic dysentery 10/13-11/5/61. Furloughed for 30 days. In Richmond hospital with chronic diarrhea 12/4-6/61. Discharged for disability 12/17/61. Age 30, 5'7", dark complexion, black eyes, black hair, Farmer. Enl. Co. D, 1st S. C. Cav. Chester CH 5/12/62. On picket 12/29-31/62. Present sick in camp 3-4/63. Present 5-6/63. Ab. on horse detail 8/18-31/63. Detailed to Regiment QM as Teamster 10/29/63-2/64. Present 3-6/64. Present as Teamster in QM Dept. 7/8-12/31/64. Paroled Hillsboro, N. C. 4/65. Farmer, age 35, Rossville, Chester Co. 1870 census. d. S. C. circa 1890.

WALLACE, WILLIAM L. Pvt., Co. D. b. York Dist. 9/11/40. Enl. Chester CH 8/20/62. Present through 10/62. On scout 12/25-31/62. Present 3-8/63. Black horse killed Robinson River 9/18/63. Paid $200.00. Ab. on horse detail 10/23-12/31/63. Present 1-6/64. Ab. sick in hospital 8/18-31/64. Present 9-12/64. Paroled Hillsboro, N. C. 4/65. Farmer, York Co. 1870-1900 censuses. d. 5/13/01. Bur. Smyrna Ch. Cem., York Co.

WARD, JAMES H. Pvt., Co. I. b. Colleton Dist. 1813. Enl. Parker's Ferry 4/3/62 age 45. Present through 6/62. Ab. sick in hospital 7-8/62. Discharged Richmond 10/9/62. Age 45, 5'8", dark complexion, brown eyes, light hair, Farmer. NFR. Farmer, age 57, Flat Rock, Henderson Co., N. C. 1870 census. d. Brunswick, Ga. 7/24/71 on wife's pension application from Marion Co., Fla.

WARD, JOHN W. 2ndSgt., Co. B. b. near Spartanburg, S. C. 12/25/34. Farmer, age 25, Spartanburg Co. 1860 census. Enl. Clinton 8/22/61 age 25 as Pvt. Ab. sick 10/31/61. Present 11/61-4/62. Promoted 4thSgt. Present 5-10/62. On raid with Gen. Hampton 11-12/62. Promoted 3rdSgt. Present 3-8/63. Horse KIA 9/12/63. Ab. on horse detail 9-10/63. Present 11-12/63, remounted 12/6/63. Present 1-2/64. Detached Green Pond 4/26-30/64. Present 5-6/64. Promoted 2ndSgt. Ab. driving cattle for QM Dept. 8/15-12/31/64. WIA near Goldsboro, N. 4/16/65. NFR. Farmer, Fair Forest, Spartanburg Co. 1880 census. Farmer, Suttons, Williamburg Co. 1900 census. Retired Farmer, Walnut Grove, Spartanburg Co. 1910 census. d. 3/4/14. Bur. Bethlehem Bapt. Ch. Cem., Spartanburg Co.

WARD, RUFUS J. P. 1stCpl., Co. F. b. S. C. circa 1834. Sawyer, age 36, Wahalla, Pickens Dist. 1860 census. Enl. Pickens CH 12/4/61 age 36 as Pvt. Present through 12/61. On picket King's House 3-4/62. AWOL 6/19-30/62. Present 7-10/62. On picket 11-12/62. Present 3-4/63. Promoted 3rdCpl. Present 5-8/63. On detached service 9-10/63. Promoted 2ndCpl. Present 11/636/64. Present on extra duty as Ordnance Sgt. 8/10-31/64. Promoted 1stCpl. & on extra duty as Ordnance Sgt. 9-10/64. Present 11-12/64. NFR. Living Pickens Co. circa 1900 on postwar roster.

WARD, THOMAS JEFFERSON. Pvt., Co. B. b. S. C. 1844. Enl. Adams Run 9/26/62. Present through 10/63. Detached at horse recruiting station 12/20/63-2/64. Detached Green Pond 4/26-30/64. Present 5-6/64. Detached to drive cattle for QM Dept. 8/15-12/31/64. NFR. Retired, age 66, Spartanburg Co. 1910 census.

WARDLAW, DAVID J. Pvt., Co. G. b. S. C. circa 1841. Enl. Co. C, 7th S. C. Inf. 4/15/61, but not on muster rolls. Discharged circa 4/12/62. Enl. Co. G, 1st S. C. Cav. Columbia 2/9/64. AWOL 2/29/64. Present 5-12/64. Paroled Augusta, Ga. 5/29/65. At home, age 29, Abbeville 1870 census. d. 4/25/24. Wife receiving pension McCormick, S. C.

WARE, WILLIAM HARRISON. Pvt., Co. H. b. S. C. 8/9/44. Enl. Columbia 6/28/64. Present through 12/64. NFR. Railroad Clerk, age 55, Washington, D. C. 1900 census. d. 5/31/19. Bur. Phantom Cem., Jones Co., Texas.

WARING, JOSEPH HALL. Pvt., Co. C, 1st. S. C. Inf. & on postwar roster Co. A, 1st. S. C. Cav. Reenlisted Co. K, 4th S. C. Cav. Green Pond 8/20/63. Present through 10/64. Discharged 12/22/64 elected to Civil Office. Paroled 4/65. Bur. Waring Cem., Pine Hill, Dorchester Co., no dates.

WARING, MORTON NATHANIEL. Pvt., Co. A. b. S. C. 6/18/35. M. D. On post war roster. Reenlisted Co. K, 4th S. C. Cav. McPhersonville 8/12/62. Present until detailed by Gen. Beauregard to Black Oak Dist., St. Joh's Parish, Berkeley Co. as M. D. 5/22/63-2/64. Present until sent to Gordonsville with disabled horses 8/64. Detailed with disabled horses Lynchburg 8/18-10/31/64. Paroled 4/65. d. 2/16/20. Bur. Black Oak Cem., Berkeley Co.

WATSON, WILLIAM. Pvt., Co. I. b. Darlington Dist. 1839. Student, S. C. College, Columbia 1860 census. Enl. Spring 1863 and served to 1865 on wife's pension application. Not on muster rolls. Farmer, age 29, Saluda, Edgefield Co. 1870 census. d. 2/13/88. Wife receiving pension Sellers, Marion Co. 4/23/19..

WATT, JOHN L. Pvt., Co. I. b. S. C. circa 1835. Enl. Ft. Motte 10/26/61 age 26. On picket White Point 12/31/61. Present sick in camp 3-4/62. Present 5-6/62. Ab. on sick leave 8/30-31/62. Present 9/62-6/63. Ab. on horse detail 8/18-31/63, dismounted since 7/1/63. Present 11-12/63, dismounted. Ab. on leave 4/23-5/5/64. Present sick in camp 5-6/64. Ab. sick in hospital 8/25-31/64. Ab. on sick leave 9/23-10/31/64, dismounted. Present 11-12/64. Paroled Hillsboro, N. C. 4/65.

WATT. WILLIAM R. Pvt., Co. E. b. S. C. circa 1833. Enl. Ft. Motte 9/26/61 age 28. Present through 6/62. On scout Edisto Island 7-8/62. Present 9-10/62. Present, shoeing horses for Company 11-12/62. Present 3-6/63. Ab. on horse detail 8/18-31/63. Present 9-12/63, dismounted since 7/30/63. Present 1-2/64, dismounted. Present 3-10/64. On picket Wappo Bridge 12/29-31/64. Paroled Hillsboro, N. C. 4/65.

WATTERS, WILLIAM J. Pvt., Co. H. b. S. C. 10/44. On roster but not on muster rolls. Farmer, age 25, Yorkville, S. C. 1870 census. Farmer, age 36, Catawba, York Co. 1880 census. Insurance Agent, age 55, Chester, 1900 census.

WAY, JOSEPH. Pvt., Co. C. b. S. C. circa 1825. Farmer, age 22, Edgefield Dist. 1850 census. Enl. 1862. Present 9-10/62. Ab.detached in QM Dept. 11-12/62. Present 3-6/63. Ab. on horse detail 8/16-31/63. Present 9-10/63. Ab. on horse detail 12/23/63-2/64. Ab. detailed horse recruiting camp 8/25-31/64. Wagon Master, Green Pond 6/10-12/31/64. Paroled Raleigh, N. C. 4/65. Farmer, age 45, Hammond, Edgefield Co. 1870 census. Farmer, age 66, Council, Washington Co., Idaho 1910 census.

WAY, MADISON PARLER. Pvt., Co. E. b. Orangeburg Dist. 12/24/26. Age 36, Orangeburg Dist. 1860 census. Enl. Ft. Motte 9/26/61. Present through 12/31/61. Ab. sick in hospital 3-4/62. Ab. on sick leave 5/12-12/31/62. Transferred Co. A, 5th S. C. Cav. 3/26/63. Ab. sick most of war. Farmer, Poplar, Orangeburg Co. 1870-1880 censuses. d. 5/10/96. Bur. Jerusalem Meth. Ch. Cem., Elloree, S. C.

WEATHERSBEE, JUDSON D. Pvt., Co. K. b. S. C. 1837. Planter, age 27, Barnwell Dist. 1860 census. Enl. Co. F, 1st S. C. Inf. (Hagood's) Ft. Johnson 9/18/61 for 6 months. NFR. Reenlisted Co. K, 1st S. C. Cav. John's Island 6/21/62. Ab. on horse detail 8/16-31/62. Ab. sick 9-10/62. Present 11/62-10/63. Ab. on horse detail 12/23/63-2/64. Present as Blacksmith 3-4/64. Present as Bugler 5-6/64. Transferred Co. A, 2nd S. C. Arty. 6/7/64 in exchange for J. G. Bush. Present 7-8/64. Present in arrest 9-10/64. Present sick in camp 11-12/64. d. Charleston on post war roster.

WEATHERSBY, CHARLES. Pvt., Co. A. Enl. date unknown and not on muster rolls. d. Richmond 3/15/63.

WEATHERSBY, DAVID H. JUDSON. Pvt., Co. E. On roster but not on muster rolls. d. 1/16/65. Bur. Weathersbee Cem., Aiken Co.

WEBB, DUDLEY HAMMOND. Saddler, Co. K. b. S. C. 1831. Farmer, age 27, Anderson Dist. 1860 census. Enl. Columbia 5/6/64. Present through 6/64. Ab. on sick leave 7-8/64. Ab. sick in Charleston hospital 10/5-12/31/64. NFR. Farmer, age 39, Hopewell, Anderson Co. 1870 census.

WEED, ANDREW WILLIAM WILSON. Pvt., Co. G. b. Abbeville Co. 3/24/44. Enl. Columbia 1/5/64. Ab. on horse detail 1/25-2/29/64. Present 3-4/64. Ab. on leave 6/22-30/64. Retired to Invalid Corps 8/16 or 20/64. NFR. Farmer, age 37, Indian Hill, Abbeville Co. 1880 census. Farmer, age 57, Plum Branch, Edgefield Co. 1900 census. d. Abbeville 4/28/14.

WEEKS, JAMES. Pvt., Co. I. b. S. C. 1841. Farmhand, age 19, West Springs, Union Dist. 1860 census. Enl. John's Island 6/30/62. Present 7-10/62. On scout 11-12/62. On detached service 3-4/63. Present 5-6/63. Detached recruiting station for horses 9/63-2/64. Ab. on leave 4/26-5/6/64. Present 5-10/64. On picket 11-12/64. NFR.

WEEKS, LEWIS W. Pvt., Co. E. b. Orangeburg Dist. circa 1845. Age 15, Orangeburg Dist. 1860 census. Enl. Ft. Motte 10/26/61. Present 11-12/61. Discharged for underage. Reenl. Co. E as substitute for S. F. Felder under 16 years, now 18 years old Stevensburg, Va. 8/22/63. Present 11-12/63, lost Springfield Rifle & fined $42.00. Detached with disabled horses 2/10-29/64. Ab. on horse detail 3/31-4/30/64. Present 5-12/64. Paroled Greensboro, N. C. 5/1/65.

WEEKS, WARREN. Pvt., Co. I. Res. Colleton Dist. 1860 census. Enl. Parker's Ferry 4/3/62. Present through 8/31/63. Ab. on horse detail 9-10/63. AWOL 12/30-31/63. Present 1-2/64. Detached Green Pond 4/11-30/64. Present 5-10/64. Detached Pocotaligo 12/10-31/64. NFR.

WEEMS, J. Pvt., Co. A. Captured and died of disease Pt. Lookout 1/27/64, no other information

WEEMS, THOMAS CORNELIUS. Pvt., Co. K. b. S. C. circa 1834. Res. Pickens Dist. Enl. by 1863 but not on muster rolls. d. Stevenburg, Va. 1863 age 28.

WEEMS, THOMAS H. Pvt., Co.'s F & K. b. S. C. circa 1840. Res. Abbeville Dist. Enl. Co. F Pickens CH 12/4/61 age 21. Present through 12/31/61. On picket King's House 3-4/62. Transferred Co. K 6/25/62. Present 7-10/62. On picket Richards Ford 11-12/62. d. of chronic diarrhea in Lynchburg hospital 3/28/63. Left widow. Bur. Lynchburg City Cem. Effects: $32.50 & sundries.

WEEMS, W. C. Pvt., Co. K. d. in Va. on post war roster.

WELLS, JOHN F. Pvt., Co. B. b. S. C. circa 1840. Res. Spartanburg Dist. Enl. Clinton 8/22/61 age 21. Ab. sick 10/31/61. On picket White Point 11-12/61. Present 3-4/62. Ab. on sick leave 5-6/62. Nurse in Adams Run hospital 7-8/62. Ab. sick 9-10/62. On raid with Gen. Hampton 11-12/62. Present 3-6/63. WIA & captured Boonsboro, Md. 7/7/63. DOW's 7/8/63 age 24.

WELLS, WILLIAM H. Pvt., Co. G. b. S. C. 1837. Res. Abbeville Dist. Enl. Co. C, 7th S. C. Inf. 4/15/61 age 22. Farm Hand. Discharged for disability 8/30/61. Enl. Co. G, 1st S. C. Cav. Abbeville CH 1/8/62. Present 3-4/62. Present sick in camp 5-6/62. Ab. sick in Charleston hospital 7/30-8/31/62. Present 9/62-6/63. Ab. sick with debilitas in Richmond hospital 7/1-8/26/63. Ab. on horse detail 8/31/63. Ab. on sick leave 9-10/63, however, returned to duty 10/12/63. AWOL 10/15-12/31/63. Present 1-2/64. Ab. sick 3-4/64. Present 5-12/64. Paroled Hillsboro, N. C. 4/65. Farmer, age 44, Lynchburg, Sumter Co. 1880 census. Living Abbeville Co. 1902.

WESTBROOK, JACOB HARVEY. 4thSgt., Co. D. b. S. C. circa 1843. Student, age 17, Chester Dist. 1860 census. Enl. Chester CH 9/10/61 age 19 as Pvt. Present 11/61-6/62. Ab. on patrol 7-8/62. Present 9/62-6/63. Promoted 4thCpl. Ab. on horse detail 8/18-31/63 & 10/23-31/63. Promoted 4thSgt. 11/1/63. On horse detail through 12/31/63. Present 1-2/64. Ab. sick in Columbia hospital 4/28-30/64. Ab. sick 6/8-8/31/64. Present 9-10/64. Ab. detailed light duty Lancaster Dist. 12/3-31/64. NFR. Farmer, age 37, Lewisville, Chester Co.1880 census. Brother of Robert H. Westbrook.

WESTBROOK, ROBERT H. Pvt., Co. D. b. S. C. circa 1845. Student, age 15, Chester Dist. 1860 census. Enl. Chester CH 5/6/63. Ab. sick with debilitas in Charlottesville hospital 7/31/63. Transferred Lynchburg hospital 9/21/63. Ab. on sick leave through 10/31/63. Present 11-12/63. On picket 2/28-29/64. Present 3-10/64. Ab. sick Summerville hospital 12/11-31/64. NFR. Farmer, age 25, Lewisville, Chester Co. 1870 census and age 35, Hazelwood, Chester Co. 1880 census. Brother of Jacob H. Westbrook.

WESTMORELAND, JAMES WHITE. Pvt., Co. A. b. S. C. 8/4/45. Enl. Abbeville CH 8/2/61, age 19. On muster rolls 10/11/61. NFR. Merchant, Woodruff, S. C. 1900 & 1910 censuses. d. Spartanburg 1/17/1941. Bur. Bethel Cem.

WESTON, R. A. Pvt., Co. A. On post war roster. Reenlisted Co.K, 4th S. C. Cav. Charleston 11/21/63. Present until ab. sick 1/18-2/28/64. Present detailed in Division QM Dept. 3=8/31/64. & on light duty with Brigade 9/16-10/31/64. Paroled Augusta, Ga. 5/22/65.

WHALEY, M. Pvt., Co. A. Enl. 11/15/61. Promoyed 1/14/62. WIA Seccessionville 6/16/62. Served later in Co. G, Charleston Bn.

WHALEY, WILLIAM SMITH, JR. Asst. Surgeon. b. Charleston 9/23/30. M. D., St. John's Island Colleton Dist. 1860 census. Appointed Asst. Surgeon from S. C. 8/5/64. Assigned 1st S. C. Cav. 9/12/64. Present through 11/64. Ab. at Pocotaligo 12/10-31/64 by order Gen.(sic) Black. NFR. d. Erwin, Tenn. 1902. Bur. Ocone Hill Cem., Athens, Ga.

WHARTON, H. Pvt., Co. F. Res. Pickens Co. d. of disease Adams Run on postwar roster.

WHARTON, SAMUEL. Pvt., Co. G. Enl. age 32 on postwar roster. Farmer, Anderson Co. 1870 & 1880 censuses. Living Abbeville Co. 1902.

WHATLEY, JOSEPH L. 3rdSgt., Co. C. b. S. C. circa 1843. Student, age 17, Edgefield Dist. 1860 census. Enl. Hamburg 8/27/61 age 18 as 3rdCpl. Present through 6/62. Promoted 2ndCpl. Present 7-8/62. Promoted 4thSgt. Present 9-10/62. Present sick in camp 11-12/62. Present 3-4/63. Promoted 3rdSgt. Present 5-6/63. Ab. on horse detail 8/15-31/63. Present 9-12/63, however, ab. sick with "Cardetis" in Richmond hospital 12/4-15/63. Present 1-2/64. Ab.on horse detail 4/12-5/15/64. Present through 8/64. Ab. sick in hospital 10/22-31/64. Present 11-12/64. NFR.

WHATLEY, LAWTON J. 3rdSgt., Co. C. Enl. Aiken Dist. age 20 on post war roster.

WHATLEY, THOMAS W. Captain, Co. C. b. S. C. 9/10/26. Res. Edgefield Dist. 1860 census. Enl. Hamburg 8/27/61 age 34 as 1stLt. Commanding Company 10/31/61. Ab. on leave 12/24-31/61. On picket Bears Bluff, Wadmalaw Island 1/62. Present 3-4/62. Elected Captain 5/26/62. Present 6-12/62. 38 horses present 2/28/63. 31 horses present 3/30/63. Present 3-10/63. 56 horses present 5/8/63. Signed for 14 caps, 20 jackets, 10 pr. pants, 35 shirts, 1 blank book, 2 tent flies, 1 lb. horse shoe nails & 20 pr. horse shoes 5/23-23/63. 24 horses present 7/25/63. Dark bay horse Killed Brandy Station 8/1/63. Paid $600.00. Signed for 6 jackets & 6 shirts 8/26/63. Signed for 1 pr. boots, 16 pr. socks, 3 blankets, 25 pr. drawers, 13 pr. pants & 2 overcoats 9/8/63. 27 horses present 1010/23/63. Ab. sick with scabies in Charlottesville hospital 12/22/63-1/2/64. Ab. on leave through 2/64. Present 3-12/64. 52 horses present 8/1/64. 53 horses present 9/1/64. Signed for 2 jackets, 6 pr. pants, 5 shirts, 5 pr. drawers, 7 pr. shoes & 9 pr. socks 9/30/64. 42 horses present 10/1/64. Paroled Augusta, Ga. 5/20/65. Res. Aiken Co., S. C. 12/19/96. d. 5/13/09. Bur. Hammond Cem., Beech Island, Aiken Co.

WHERRY, WILLIAM L. Pvt., Co. K. b. Chester Dist. circa 1844. Res. York Dist. Enl. Rock Hill 4/15/63 under age 18. WIA (left breast) Upperville 6/21/63. Ab. wounded in Richmond hospital 7/12-9/22/63. Returned to duty 10/9/63. Present through 2/64. Detached Green Pond 4/23-30/64. Present as Teamster 5-6/64. Detached horse recruiting station Lancaster Dist. 7-8/64. d. in Charleston hospital 10/20/64 age 20. Bur. Magnolia Cem., Charleston. Effects: Sundries.

WHITE, ALEXANDER. 4thCpl., Co. G. Enl. Co. A, 2nd Ga. Inf. 6/20/61. Transferred Co. B, 15th Ga. Inf. 5/12/62. Transferred Co. G, 1st S. C. Cav. 1/30/63. Joined Camp Pines, Va. 2/1/63. Ab. detached 3-4/63. Present 5-6/63. Ab. on horse detail 8/31/63. Present 9-12/63. Ab. recruiting station for horses 1/20-2/29/64. Ab. on leave 4/25-30/64. Promoted 4thCpl. Present 5-12/64. NFR. Receiving pension in Ga. 1916. d. Charlton Co., Ga. 6/6/25.

WHITE, HENRY. Colored Servant, Co. H. On roster but not on muster rolls.

WHITE, JAMES. Captain, Co. B. On postwar roster.

WHITE, JAMES D. Pvt., Co. A. b. S. C. 3/29/34. Enl. 11/7/61. Detailed Gen. Ripley 12/24/61. Reenlisted Co. K, 4th S. C. Cav. Grahamville 3/25/62. Detailed by Gen. Pemberton until discharged for disability 12/8/62. NFR. d. Lancaster Co. 8/13/76. Bur. Douglas Presb. Ch., Lancaster.

WHITE, JAMES WILSON. 1stLt., Co. H. b. S. C. 12/26/39. Farmer, age 21, Chester Dist. 1860 census. Enl. Rock Hill 1/9/62 age 21 as 2ndLt. Present through 6/62. 65 horses present 3/31/62. AWOL 7-8/62. Present 9-10/62. Ab. on CM 11-12/62. Present 3-4/63. 48 horses present 2/1/63. 37 horses present 3/1/63. 56 horses present 5/1/63. Promoted 1stLt. 6/10/63. Commanding Company 8/31/63. Signed for 4 caps, 14 jackets, 12 pr. pants & 18 shirts 8/24/63. Signed for 6 jackets, 3 caps & 1 shirt 8/26/63. Signed for 8 pr. boots, 14 pr. socks, 3 blankets, 22 pr. drawers, 12 pr. pants & 3 overcoats 9/8/63. Ab. on horse detail 9-10/63. Commanding Company 11/63-2/64. 20 horses present 12/15/63. Present 3-4/64. 36 horses present 3/27/64. Commanding Company 5-6/64. 16 horses present 5/15/64. Signed for 33 pr.

pants, 35 jackets, 36 shirts, 34 pr. drawers, 17 caps, 6 Army tents, 5 camp kettles, 12 mess pans, 1 spade & 2 axes 6/30/64. Present 7-8/64. 58 horses present 8/21/64 & 9/1/64. Signed for 1 jacket, 7 pr. pants, 6 shirts, 8 pr. drawers, 7 pr. shoes, & 6 pr. socks 9/30/64. 59 horses present 10/1/64. Ab. on leave 10/24-31/64. Detached Pocotaligo 12/10-31/64. Paroled Charlotte, N. C. 5/15/65. Returned to Ft. Mill, S. C. after the war. Merchant & Farmer. Moved to Graham, Alamance Co., N. C. 8/26/77. Manufacturer. Graham, N. C. 1880 census. d. Ft. Mill S. C. 7/12/87.

WHITE, JOHN JAMES. Pvt., Co. H. b. S. C. circa 1846. On roster but not on muster rolls.

WHITE, P. R. Pvt., Co. H. b. S. C. 1827. Enl. by 8/64. Ab. on detail 8/15/64. NFR. d. Charleston 6/30/92. Bur. St. Lawrence Cem., Charleston.

WHITE, ROBERT P. Pvt., Co. H. b. Abbeville Dist. 8/33. Farmer, age 26, Clarendon, Colleton Dist. 1860 census. Enl. Rock Hill 1/9/62 age 27. Present through 4/62. Ab. on sick leave 6/29-30/62. Present sick in camp 7-8/62. Present 9-12/62. Ab. sick with "ulcer caused by vacination " in Farmville hospital 4/11-5/7/63. Present through 8/31/63, however, ab. sick with debilitias in Richmond hospital 7/22-8/17/63. Ab. on horse detail 9-10/63. AWOL 10/23-12/31/63. Present 1-2/64. Ab. on leave 424-52/364. Present sick in camp through 6/64. Present 7-10/64. Present sick in camp 11-12/64. NFR. Farmer, age 38, North Fork, Polk Co., Ark. 1870 census. Farmer, age 66, Council, Washington Co., Idaho 1900 census.

WHITE, T. FRANKLIN. Pvt., Co. H. b. S. C. 1846. Enl. Fredericksburg, Va. 3/6/64. Present through 12/31/64. NFR. Farmer, age 34, Washington, Bradley Co., Ark. 1880 census. d. 9/11/10.

WHITE, WILLIAM COKE. Pvt., Co. K. b. S. C. 8/14/47. Age 13, Charleston 1860 census. Enl. Columbia 5/1/64. Present through 8/64. Ab. sick in Columbia hospital 10/24-31/64. On picket Chatman's Ford 12/29-31/64. NFR. Farmer, age 62, Leggett, Marion Co. 1910 census. d. 8/25/16. Bur. Centenary Meth. Ch. Cem., Marion Co.

WHITE, WILLIAM W. 1stSgt., Co. A. b. S. C. 12/30. Enl. 11/7/61. Detached to Engineers 2/2/62. Reenlisted Co. K, 4th S. C. Cav. 1862 age 32. WIA & captured Hawes Shop 5/28/64. Sent to Old Capitol & Elmira. Exchanged 2/25/65. Paroled 4/65. Farmer, St. John's Berkeley C. 1880 census. d. Charleston 3/9/13.

WHITESIDES, JAMES MEEK. 1stSgt., Co.'s D & K. b. S. C. 1839. Enl. Chester CH 10/9/61 as Pvt. Appointed Cpl. 10/24/61. Present 11/61-3/62. Transferred Co. K 6/25/62. Promoted 2ndSgt. Present 7-8/62. Ab. detached 9-10/62. Promoted 1stSgt. On scout 11-12/62. Present 3-12/63. Ab. sick with chronic rheumatism in Richmond hospital 2/4-3/28/64. Present through 6/64. On duty James Island entrenchments 7-8/64. Ab. on sick leave for 20 days 10/28/64. Present 11-12/64. Paroled Raleigh, N. C. 4/65. Farmer, Broad River, York Co. 1880 census. d. Broad River, York Co. in family history.

WHITESIDES, JAMES H. Pvt., Co. H. b. Spartanburg Dist. 10/31/32. Trader, age 26, Rock Hill, York Dist. 1860 census. Enl. Rock Hill 1/9/62 age 28. On picket Jacksonborough 3-4/62. Present 5-10/62. Present as wagon guard 11-12/62. d. of pneumonia in Lynchburg hospital 3/21/63. Bur. Old City Cem., Lynchburg. Effects: $12.50.

WHORTON, SAMUEL M. Pvt., Co. F. b. S. C. 12/47. Age 12, Charleston 1860 census. Enl. Columbia 3/2/64. Detached Green Pond 4/25-30/64. Present 5-8/64. Detached Dill's Bluff 9/17-12/31/64. NFR. Gold Miner, age 52, Auraria, Lumpkin Co., Ga. 1900 census.

WIDEMAN, C. A. Pvt., Co. A. Enl. by 10/11/61. NFR.

WIDEMAN, JACOB J. Pvt., Co. I. b. Charleston, S. C. 10/27/45. Enl. Columbia 8/12/64. Ab. sick in hospital 10/16-31/64. Present 11-12/64. NFR. Farmer, age 64, Waltertown, Ware Co., Ga. 1910 census. Farmer, age 74, Jamestown, Ware Co., Ga. 1920 census. d. Waycross, Ga. 7/28/22. Bur. Oakland Cem.

WIDEMAN, JAMES A. Pvt., Co. A. b. S. C. circa 1845. Res. Abbeville Dist. 1860 census. Enl. Columbia 1/25/64. AWOL through 4/64. Present 5-12/64. NFR. Farmhand, age 27, Bordeau, Abbeville Co. 1870 census.

WIDEMAN, JAMES WARREN. Pvt., Co. A. b. Long Cane, Abbeville Dist. 9/16/46. Att. Erskine College. Enl. Columbia 4/16/64. Detached Green Pond through 4/30/64. Present 5-12/64. Paroled Hillsboro, N. C. 4/26/65 and again at Augusta, Ga. 5/29/65. Farmhand, Bordeau, Abbeville Co. 1870 census. M. D., Bordeau, Abbeville Co. 1880 census. M. D., Due West, Abbeville Co. 1900 census. d. 4/17/18. Bur. Due West Cem., Abbeville Co., S. C.

WILES, JOHN MARTIN. Pvt., Co. E. b. S. C. 1830. Farmer, age 20, Orangeburg Dist. 1850 census. Enl. Ft. Motte 10/26/61 age 30. Present through 12/61. Ab. on leave 1/62. Ab. sick in hospital 3-4/62. Present 5-8/62. Present sick in camp 9-10/62. Ab. sick with debility & hemorrhoids in Richmond hospital 11/28/62-12/21/62. Transferred Farmville hospital with bronchitis. Transferred Gordonville hospital with "Hemiplygia." Furloughed for 30 days 3/7/63. Ab. sick in Columbia, S. C. hospital 5-10/63. Present 11-12/63. Ab. on sick leave 2/15-29/64. Ab. on horse detail 4/20-5/20/64. Present through 6/64. Ab. sick in hospital 8/12/64. Retired by Medical Board to Invalid Corps 8/30/64. NFR. d. Orangeburg, S. C. 6/68.

WILKES, ELI CORNWELL. Pvt., Co. D. b. near Baton Rouge, Chester Dist. 11/17/35. Medical Student, age 24, Chester Dist. 1860 census. Enl. Chester CH 7/28/62. Present sick in camp 9-10/62. Present 11-12/62. Ab. sick with "Catarrhal Fever or Febris Intermittens" in Farmville hospital 4/25-5/963. Ab. sick in Rockingham Co., Va. 7/25/63. d. of disease Luray, Va. 9/4/63. Bur. Calvary Bapt. Ch. Cem., Chester Co. Brother of John W. & William T. Wilkes.

WILKES, GEORGE WASHINGTON. Pvt., Co. D. b. S. C. 1/9/32. Res. Chester Dist. Enl. Chester CH 9/10/61 age 29. Present 11/61-4/62, detailed as Company Quartermaster. Present 5/62-4/63. d. of disease Winchester, Va. 7/28/63 age 30. Bur. Confederate Cem., Winchester.

WILKES, JESSE C. Pvt., Co. D. b. S. C. circa 1844. Res. Chester Dist. Not on muster rolls. KIA Brandy Station 6/9/63.

WILKES, JOHN WESLEY, JR. 2ndLt., Co. D. b. near Baton Rouge, Chester Dist. 3/4/41. Farmer, Chalksville, Chester Co. 1860 census. Enl. Co. E, 6th S. C. Inf. Chester Dist. 4/11/61 age 20 as 3rdLt. Present through 8/61. Present sick 9/61. Ab. on leave for 30 days 10/30/61. Returned to duty 12/9/61. Present through 12/31/61. Promoted 1stLt. 1/1/62. Company disbanded 4/11/62. Enl. Co. D, 1st S. C. Cav. Chester CH 7/28/62 as Pvt. Ab. on sick leave 7/30/62. WIA 8/29/62. Present 9-10/62. On scout 12/25-31/62. Present 3-6/63. Ab. on horse detail 8/18-31/63. Present 9-10/63. Detailed Brigade Hqtrs. 11/2-12/31/63. Promoted 2ndLt. Present 1-4/64. 27 horses present 3/31/64. Ab. on leave 6/20-30/64, however, signed for 43 pr. pants, 47 jackets, 2 pr. shoes, 1 pr. socks, 54 shirts, 67 pr. drawers & 34 caps James Island 6/30/64. Present 7-8/64. Ab. on sick leave 10/21-31/64. Commanding Company 11-12/64. Paroled Hillsboro, N. C. 4/65. Farmer, County Commissioner, Post Master, Baton Rouge 1888-1908. Magistrate 1888-1928. d. 1/1/28. Bur. Calvary Bapt. Ch. Cem., Chester Co. Brother of Eli C. & William T. Wilkes.

WILKES, WILLIAM THOMAS. 2ndCpl., Co. D. b. near Baton Rouge, Chester Dist. 5/16/39. Att. Trinity College, N. C. (now Duke U.). Student, age 21, Chester Dist. 1860 census. Enl. Chester CH 10/18/61 as Pvt. Ab. sick in hospital 11-12/61. Present 3-10/62. Promoted 2ndCpl. Present 11/62-4/63. KIA Brandy Station 6/9/63. Bur. Calvary Bapt. Ch. Cem., Chester Co. Brother of Eli C. & John W. Wilkes.

WILKES, WASHINGTON. Pvt., Co. E. Not on muster rolls. d. Manchester, Va. hospital in post war account. No dates.

WILKINS, CHARLES D. Pvt., Co. B. b. S. C. circa 1838. Farmer, age 22, Paris, Lamar Co., Texas 1860 census. Enl. Camp Frigg 5/19/62. Present through 10/62. Sold horse 10/1/62. Bought a horse 10/11/62. On picket Ellis Ford 11-12/62. Present as Teamster 3-4/63. Present 5-6/63, dismounted. Ab. on horse detail 8/18-31/63. AWOL 9-10/63. Present 11/63-12/63, however, ab. sick with mumps & "Parotitis" in Richmond hospital 11/29-12/15/63. Present 1-8/64. Detailed Courier for Colonel Campbell 10/31/64. Present 11-12/64. NFR.

WILKINS, JOHN J. Pvt., Co. D. b. S. C. circa 1847. Res. Marshall Co., Miss. 1860 census. Enl. 1865 age 18. Paroled Charlotte, N. C. 8/17/65 & Mobile, Ala. 9/20/65. Age 18, 5'6", ruddy complexion, hazel eyes, dark hair, res. Marshall Co., Miss. Farmer, Marshall Co., Miss. 1880 census. Retired, Abbeville, Lafayette Co., Miss. 1910 census.

WILKINS, WILLLIAM D. Pvt., Co. B. b. Cherokee, S. C. circa 1847. Student, age 13, Fingerville, Spartanburg Co. 1860 census. Enl. Spartanburg 10/18/63. On picket Fredericksburg 11-12/63. Ab. horse recruiting camp 1-2/64. Present 3-8/64. Courier, Dr. Lebby 10/15-31/64. Detached Pocotaligo 12/2/-31/64. NFR. Merchant, age 33, Fannin Co., Texas 1880 census & Honey Grove, Fannin Co., Tex. 1900 & 1910 censuses.

WILLBANKS, WILLIAM SANFORD. Pvt., Co. F. b. S. C. 5/12/31. Res. Pleasant Hilltop, S. C. Enl. Pickens CH 12/4/61 age 30. Present through 4/62. Detached Church Flats 5-8/62. On detached service 9/62-8/31/63, driving QM wagon. Present 9-12/63. Detached driving wagon for Signal Corps 2/15-29/64. Detached Green Pond 4/25-31/64. Present 5-6/64. Detailed to drive wagon for QM 8/25-31/64. Present 9-12/64. NFR. Farmer, age 69, Pleasant Hill, Newton Co., Ark. 1900 census. d. Silgam Springs, Ark. 11/2/08 age 77.

WILLIAMS, ALEXANDER H. Pvt., Co. H. b. S. C. circa 1824. Res. York Dist. Enl. Rock Hill 1/9/62 age 38. Present 3-6/62. Ab. on sick leave 7-8/62. Present 9-10/62. Detailed as wagoner 11-12/62. Present under arrest 3-4/63. Present 5-8/63, however, ab. sick with "kidney afliction" in Richmond hospital 7/28-8/12/63. Ab. in arrest Richmond 9-10/63. Present 11/63-4/64, however, CM'd 1/64. Transferred Co. H, Lucas's 15th Bn. S. C. Arty. 5/13/64. Joined 5/20/64. Present through 12/64. Captured Cheraw, S. C. 3/6/65. Sent to David's Island, N. Y. Released 6/23/65. 5'5 1/2", light complexion, light hair, blue eyes. Farmer, Pendleton, Anderson Co. 1900 census.

WILLIAMS, AMBROSE W. Pvt., Co. A. b. S. C. 9/22/45. Pension application only record of service. Farmer, Highland, Greenville Co. 1900 census and O'Neal, Greenville Co. 1920 census. Receiving pension Helena, Newberry Co. 4/16/23. d. Chick Springs, Greenville Co. 11/6/29. Bur. Pleasant Hill Ch. Cem.

WILLIAMS, ARCHIBALD CAMPBELL. Pvt.., Co. I. b. Walterboro, S. C. 12/19/47. Res. York Dist. Att. Citadel. Enl. Columbia 1864 in postwar account. Served along the S. C. coast and was in the battle of Telafinne. Paroled Augusta, Ga. 5/9/65. Citadel records says served in Co. F, 3rd S. C. Cav. School Teacher, Macon, Ga. postwar. d. there 9/7/12. Bur. Riverside Cem.

WILLIAMS, CHARLES C. Pvt., Co. I. b. S. C. 1843. Age 7, Marlboro, S. C. 1850 census. Enl. Parker's Ferry 4/3/62. AWOL through 4/30/62. NFR. Had enlisted Co. B, 3rd S. C. Cav. Walterboro 3/14/62. Present through 5/63. Ab. on leave 10 days 6/63. Present 7-8/63. Ab. detached for 30 days 10/10/63. Present 11-12/63. Ab. in arrest Savannah, Ga. 1-2/64. Present 3-4/64, (AWOL 7 days). Present 5-6/64 (AWOL 20 days). Present 7-8/64. Present sick in camp 9-10/64. Present 11-12/64, Courier, Cavalry Hdqtrs. NFR.

WILLIAMS, E. PRESTON. Pvt., Co. H. b. York Dist. circa 1810. Tinner.. Enl. Rock Hill 1/9/62 age 52. On picket Routole Bridge 3-4/62. Detailed to drive ambulance 5-6/62. Ab. on sick leave 7-8/62. Ab. driving ambulance 9-12/62. Present sick in camp 3-4/63. Ab. sick in hospital through 8/31/63. Ab. on sick leave 9-10/63, however, ab. sick with debility & as nurse in Richmond hospital 10/5/63. Furloughed for 30 days 10/27/63. On detail Gordonsville 11/5/63-2/64. Transferred Yorkville, S. C. 1/4/64. Detached with Enrolling Officer, York Dist. 3-6/64. Present 7-8/64. Transferred Lafayette S. C. Light Arty. 10/5/64. Present through 10/31/64, age 55. NFR.

WILLIAMS, FRANK. Pvt., Co. G. b. Orangeburg Co. 1846. Enl. age 21 on postwar roster. Living

Abbeville Co. 1902.

WILLIAMS, GEORGE. Chaplain. Not on muster rolls.

WILLIAMS, HUMFARA P. Pvt., Co. C. b. S. C. circa 1841. Enl. Hamburg 8/27/61 age 20. Detached on Wadmalaw Island 11-12/61. Present 3-12/62. On detached service 3-4/63. Present 5/63-10/64. Detached Pocotaligo 12/10-31/64. NFR. Farmer, age 39, Millbrook, Aiken Co. 1900 census.

WILLLIAMS, JAMES FRANKLIN. Pvt., Co. G. b. McCormick, S. C. 2/14/37. Enl. Abbeville CH 3/15/62. On picket Simmons House through 4/62. Present 5-6/62. Ab. sick Charleston hospital 7/30-10/31/62. Discharged Columbia 12/11/62. NFR. Claimed to have reenlisted and present Greensboro, N. C. 4/65 at surrender. Receiving pension McCormick, S. S. 9/25/19.

WILLIAMS, J. ROBERT. Pvt., Co. B. b. Edgefield Dist. 5/20/48. Student, age 13, York Dist. 1860 census. Enl. 1865. Paroled Augusta, Ga. 5/31/65. Farmer. d. Barnwell, S. C. 1/5/16 age 68. Bur. Ulmers, S. C.

WILLIAMS, JASPER A. Pvt., Co. F. b. S. C. 1845. Enl. Columbia 2/18/64. Detached Green Pond 4/25-30/64. Present 5-10/64. Ab. sick 12/25-31/64. NFR. School Teacher, age 35, Montevallo, Shelby Co., Ala. 1880 census. d. Jacksonville, Texas 1/12/38.

WILLIAMS, JOHN SAMUEL. 3rdCpl., Co. G. b. Abbeville Dist. 10/19/39. Enl. Abbeville CH 1/8/62 as Pvt. Present 3-4/62. Present sick in camp 5-6/62. Promoted 4thCpl. Ab. sick in Charleston hospital 7/30/62. Promoted 3rdCpl. 9/15/62. Ab. sick through 10/62. Reduced to Pvt. Present 11/62-4/63. WIA (fractured humeros lower 3rd of arm) Brandy Station 6/9/63. Ab. wounded in Richmond hospital 6/12/63. Furloughed for 60 days. Ab. wounded at home 8/31/63-4/64. Retired to Invalid Corps 5/24/64. NFR. Farmer, Abbeville 1880-1920 censuses. Receiving pension Abbeville 9/8/19. d. Abbeville 12/24/25. Bur. Lebanon Ch. Cem.

WILLIAMS, ROBERT GILLIAM. Pvt., Co. K. b. Newberry Dist. 8/20/42. Enl. Chishomville 12/30/64. On picket Chatman's Ford 12/31/64. Paroled Raleigh, N. C. 4/65. Planter, postwar. d. Greenville, S. C. 7/6/17. Bur. Springwood Cem.

WILLIAMS, SAMUEL. Pvt., Co. G. On post war roster.

WILLIAMS, THOMAS LAFAYETTE. 2ndCpl., Co.'s D & K. Enl. S. C. 5/38. Enl. Co. D Chester CH 9/10/61 age 23 as Pvt. Present 11/61-4/62. Transferred Co. K 6/26/62. Present through 8/62. Ab. on detached service 9/12-10/31/62. Present 11-12/62. Ab. on detached duty 4/8-8/31/63, however, WIA Brandy Station 6/9/63 on post war roster. Present 9-12/63. Ab. detailed horse recruiting station 1/10-2/29/64. Present 3-4/64. Promoted 2ndCpl. Ab. sick in Charleston hospital 5-6/64. Ab. on sick leave 7-8/64. Ab. on duty at Provost Marshal's Office, Charleston 9-10/64. On picket Chatman's Ford 12/31/64. Paroled Raleigh, N. C. 4/65. Flagman on Railroad, age 62, Rock Hill, York Co. 1900 census. d. 1904. Bur. Laureland Cem., Rock Hill, S. C.

WILLIAMS, WILLIAM. Pvt., Co. C. b. S. C. 7/18/43. Enl. Camp Butler 9/10/61 age 18. Present through 6/62. Ab. on sick leave 8/8-21/62. Present 9/62-6/63. Ab. on horse detail 8/15-31/63. Ab. sick at home 9-12/31/63. Present 1-10/64. Detached Pocotaligo 12/10-31/64. NFR. d. S. C. 1/27/09 age 61.

WILLIAMS, WILLIAM E. Pvt., Co. G. b. S. C. 7/36. Age 14, Orangeburg Dist. 1850 census. Enl. Abbeville CH 2/27/62. On duty Simmons House 3-4/62. Ab. sick in Charleston hospital 6/12-30/62. Present 7-8/62. Ab. in arrest 9-10/62. On scout 11-12/62. Present 3-10/63, however, ab. sick with "Varioloid" in Richmond hospital 7/3-8/5/63. Under arrest in guard house 12/26-31/63. Ab. on leave 2/12-29/64. Detached Green Pond 4/17-30/64. Present 5-6/64. On detached service 7-8/64. Ab. on leave 15 days 10/31/64. Detached Pocotaligo 12/10-31/64. NFR. Grocer, age 63, Orangeburg Co. 1900 census. d. Augusta, Ga. 9/15/25.

WILLIAMS, WILLIAM W. Pvt., Co. G. b. S. C. 1848. Age 12, Emanuel Co., Ga. 1860 census. Not on muster rolls. Wife receiving pension Anderson Co. 1901.

WILLIFORD, A. S. Farrier, Co. K. b. S. C. circa 1845. Res. York Dist. Enl. Hamilton's Crossing 2/20/64. Present 3-4/64. Ab. sick in Charleston hospital 6/27-30/64. Present 7-8/64. AWOL 10/15-31/64. Courier, Tar Bluff 12/29-31/64. Paroled Charlotte, N. C. 5/22/65. Married 12/14/65. Baptist Minister, age 35, DeKalb, Kershaw Co. 1880 census.

WILLIFORD, JESSE B. Pvt., Co. H. b. S. C. 10/23/26. On roster but not on muster rolls. d. 12/16/05. Bur. Laurelwood Cem., York Co.

WILLIS, ALFRED D. Pvt., Co. I. b. S. C. circa 1832. Farmer, age 18, Spartanburg Dist. 1850 census. Enl. Parker's Ferry 4/3/62. AWOL through 4/30/62. NFR. Had enlisted Co. C, 5th S. C. Cav. Cheraw 3/12/62. Present through 7/63. WIA Brandy Station 6/8/63 on post war roster. CM 8/13/63 for 3 days AWOL. Present through 1/64. Transferred Co. G, 11th S. C. Inf. 2/1/64. d. in hospital Lake City, Fla. 4/8/64.

WILLIS, ANDREW P. Pvt., Co. B. b. Spartanburg Dist. circa 1844. Age 6, Spartanburg Dist. 1850 census. Enl. Adams Run 9/26/62 age 18. Present through 4/63. WIA Brandy Station 6/9/63. DOW's in Gordonsville hospital 6/25/63 age 23. 5'8", red complexion, sandy hair, blue eyes, Farmer. Bur. Maplewood Cem. Gordonsville. Effects: $4.50.

WILSON, J. F. W. Pvt., Co. F. b. S. C. circa 1837. Farmhand, age 23, Cherokee, Pickens Co. 1860 census. Enl. Pickens CH 12/4/61 age 28. Present through 6/62. Ab. on leave 7-8/62. Present 9/62-6/63. Ab. on detached service 8/18-10/31/63. AWOL 11/1-2/64. Dropped as a deserter 3/1/64, however, transferred Co. A, 20th S. C. Inf. 5/13/64. Present until WIA "fractured tibia at the knee" 6/24/64. In Williamsburg hospital 73/64. DOW's 7/18/64. Left widow. Age 27, Farmer.

WILSON, JAMES K. 2ndLt., Co. D. b. S. C. 1/37. Age 23, Cedar Shoals, Chester Dist. 1860 census. Enl. Chester CH 9/10/61 age 24 as Pvt. Present through 4/62. Ab. on sick leave 6/29-30/62. Present 7-8/62. Ab. sick in Summerville hospital 9/29-4/63. Present through 8/31/63. Ab. recruiting horses 9-10/63. Promoted 4thCpl. Ab. on horse detail 12/22/63-3/64. Present 3-8/64. Promoted Brevet 2ndLt. 8/27/64. Ab. sick Summerville hospital 9/24-10/31/64, however, signed for 1 jacket, 6 pr. pants, 7 pr. drawers, 7 pr. shoes, 9 pr. socks & 1 wall tent, James Island 9/30/64. Present 11-12/64. Paroled Hillsboro, N. C. 4/65. Farmer, age Landford, Chester Co. 1870 census. Farmer, Rossville, Chester Co. 1880 & 1900 censuses. d. 5/30/09. Wife receiving pension Richburg, Chester Co. 10/29/19.

WILSON, JOHN. Sgt., Co. A. b. S. C. circa 1838. Enl. 11/11/61 as Cpl. Promoted Sgt. 1/25/62. Ab. sick 2/62. NFR.

WILSON, JOHN ROBERT F. 4thSgt., Co. G. b. S. C. 2/28/40. Enl. Abbeville CH 1/8/62 as 3rdCpl. On picket Simmons House 3-4/62. Ab. sick in hospital 6/9-30/62. Present 7-10/62. Promoted 2ndCpl. 9/15/62. Present 11-12/62. Promoted 1stCpl. 11/27/62. Present 3-6/63, however, ab. sick with "dysenteria" in Farmville hospital 5/8-6/1/63. Ab. on horse detail 8/31/63. Present 9-10/63. Promoted 4thSgt. Present 11/63-10/64. Detached Pocotaligo 12/10-31/64. NFR. Farmer, Belton, Anderson Co. 1870 census. d. Abbeville 2/26/17. Bur. Melrose Cem., Abbeville, S. C.

WILLIAMS, JOHN SAMUEL. 3rdCpl., Co. G. b. Abbeville Dist. 10/19/39. Enl. Abbeville CH 1/8/62 as Pvt. Present through 6/62. Promoted 4thCpl. & ab. sick isn Charleston hospital 7/30/62. Promoted 3rdCpl. 9/15/62, but ab. sick through 10/62. Reduced to Pvt. Present 11/62-4/63. WIA Brandy Station 6/9/63. Lower third of arm amputated in Richmond hospital 6/12/63. Furloughed for 60 days 6/23/63. Ab. wounded at home until retired to Invalid Corps 5/24/64. NFR. Farmer, Abbeville 1880-1920 censuses. d. Abbeville 12/24/25. Bur. Lebanon Cem.

WILSON, JOHN SIMONTON. Captain, Co. D. b. Chester Dist. 7/4/20. Att. Mt. Zion College 1838 & S. C. College 1842. Lawyer. Served in S. C. Legislature 1856-61. Enl. Chester CH 9/10/61 age 40 as 1stLt. Present through 10/62.144 horses present in squadron 10/3/62. Signed for 10 horses & 14 mules, Columbia 10/4/62. 151 horses present in squadron Lynchburg 10/24/62. Ab. on General CM, Fredericksburg 12/27-31/62. Present 3-4/63, WIA Brandy Station 6/9/63. Present Gettysburg. 77 horses present 5/5/63. Signed for 1 blank book, 3 tent flies, 1 coffin, & 21 lbs. horse shoe nails, Stevensburg 6/10/63. Signed for 2 pr. horse shoes 8/2/63. Ab. sick with "Int. Fever" in Charlottesville hospital 8/15-9/21/63. Transferred to Lynchburg hospital. Present 10/63. 29 horses present 10/23/63. On picket 12/30-31/63. 35 horses present 12/31/63. Present 1-6/64. Captain & 34 men present 3/31/64. Signed for 25 pr. shoes, 6 blankets, 2 fly tents, 1 camp kettle, 1 overcoat, 6 shirts, 1 pr. socks, 8 jackets & 11 pr. drawers 3/31/64. Signed for 10 pr. drawers, 3 pr. shoes, 1 jacket & 3 pr. pants, Columbia 4/11/64. Signed for 12 caps, 16 jackets, 16 pr. pants, 12 shirts, 11 pr. drawers, 5 pr. shoes & 8 pr. socks, Green Pond 4/20/64. Signed for 1 shovel, 6 Army tents, 2 camp kettles & 7 mess pans 6/30/64. 65 horses present 8/1/64. Ab. on leave 8/20-31/64. 55 horses present 9/1/64. Ab. detached guarding POW's Florence 9/18-12/31/64. 56 horses present 10/1/64. Signed for 1 wall tent 11/24/64. NFR. Grocer, Chester 1870 census. Merchant, Chester 1880 census. Probate Judge, Chester Co. 1890-1902. d. 10/3/02. Bur. Evergreen Cem., Chester, S. C.

WILSON, M. J., Pvt., Co. A. Enl. age 19 on postwar roster. Living Abbeville Co. 1902.

WILSON, MARTIN. Servant/Cook, Co. B. (Colored). Served under J. H. Copeland & Joseph Little, 1861-1865. Receiving pension Laurens Co. 5/8/23.

WILSON, MATTHEW WALLACE. Pvt., Co. H. Enl. b. S. C. circa 1830. Rock Hill 1/9/62 age 32. Ab. sick in hospital 3-4/62. d. of disease 5/12/62.

WILSON, ROBERT S. Pvt., Co. H. b. S. C. circa 1831. Enl. Rock Hill 1/9/62 age 31. On picket Rautale Bridge 3-4/62. Present 5-10/62. On picket 11-12/62. Present as wagon guard 3-4/63. Present 5/63-2/64. Detached Green Pond 4/16-30/64. Present 5-10/64. Detached as Carpenter, Stono Bridge 12/7-31/64. Paroled Charlotte, N. C. 5/22/65. Carpenter, age 39, Ft. Mill, York Co. 1870 census.

WILSON, SAMUEL T. Pvt., Co. I. b. S. C. circa 1842. Enl. James Island 6/16/64. Present 7-8/64. Ab. on leave 10/19-31/64. Present 11-12/64. NFR. Farmer, age 28, Sumter, S. C. 1870 census. Farmer, age 38, Easley, Pickens Co. 1880 census.

WILSON, URIAH. Pvt., Co. C. b. S. C. circa 1837. Enl. Wadmalaw Island 3/8/62 age 25. Ab. on sick leave 5-12/62. Ab. on detached service recruiting horses 3-12/63. Present 1-2/64. Detached Green Pond 4/16-30/64. Present 5-12/64. NFR. Alive 1898.

WILSON, URIAH J. Pvt., Co. A. b. Abbeville Dist. circa 1837. Enl. Abbeville CH 8/13/61 age 24. Present through 4/62. Ab. sick 5-6/62. WIA & ab. wounded 7-10/62. Present sick in camp 11-12/62. Discharged for disability 3/16/63. Age 31, 6'2", dark complexion, grey eyes, dark hair, Farmer. Enl. Co. B, 1st S. C. Arty. 12/31/63 as conscript. Present 1-2/64. Ab. sick leave for 30 days 10/19/64. Present 12/64. Paroled Greensboro, N. C. 4/26/65, res. Abbeville Dist.

WILSON, W. H. Pvt., Co. H. b. S. C. circa 1839. Enl. Rock Hill 1/9/62, age 23. Present 3-4/62. Discharged 5/28/62. NFR.

WILSON, W. W. Pvt., Co. H. Res. York Dist. Enl. by 1862. d. of disease 3/15/62.

WILSON, WILLIAM A. R. Pvt., Co. D. b. S. C. 11/28/47. Enl. Chester CH 5/12/62. Present until captured Upperville 5/21/62. Exchanged by 9/62. Ab. sick in Richmond hospital 10/11/62. Present sick 11-12/62. Furloughed for 40 days 12/2/62. Present 3-10/63. Ab. sick with bronchitis in Richmond hospital 11/11-20/63. Ab. on horse detail 12/22/63-2/64. Detached Green Pond 4/21-30/64. Present 5-10/64. Detached Coosawhatchie 12/10-31/64. Paroled Hillsboro, N. C. 4/65. d. 5/9/91. Bur. Evergreen Cem., Chester, S. C.

WILSON, WILLIAM H. Pvt., Co. D. Enl. Rock Hill 1/9/62. On picket Rautole Bridge 3-4/62. Present 5-6/62. Present sick in camp 7-8/62. Ab. under arrest Raleigh, N. C. 9-10/62 & Hillsboro, N. C. 11-12/62. NFR. d. Columbia Co., Fla. 6/22/76.

WIMBERLY, JOEL GEORGE WASHINGTON. Pvt., Co. E. b. S. C. 6/10/38. Res. St. George, Colleton Dist. 1850 census. Enl. Coosawhatchie 12/13/61. Present through 12/31/61. Ab. sick at home 2/10-4/62. Ab. on sick leave 5/18-6/30/62. Present 7-10/62. On scout 11-12/62. Present 3-8/63. Ab. on horse detail for 30 days 10/24-12/31/63. Present 1-2/64. Detached Green Pond 4/16-30/64. Present 5-6/64. Present sick in camp 7-8/64. On picket 9-10/64. Present 11-12/64. Paroled Hillsboro, N. C. 4/65. d. 4/30/85. Bur. Indian Fields Meth. Ch. Cem., Rosinville, S. C.

WIMBERLY, JOSHUA L. Pvt., Co. E. b. S. C. circa 1839. Enl. Coosawahatchie 12/13/61. On picket White Point through 12/31/61. Present 3/62-2/64. Detached Green Pond 4/16-30/64. Present 5-8/64. Detailed to guard POW's Florence 9/18-12/31/64. NFR. Farmer, age 71, St. George, Dorchester Co. 1910 census. d. 9/20/06. Bur. U. D. C. Cem., St. George, Dorchester Co.

WINGO, WILLIAM J. Pvt., Co. B. b. Va. circa 1845. Enl. Columbia 5/20/64. Present through 12/64. NFR. Store Clerk, age 35, Richmond, Va. 1880 census. Farmer, McMinnville, Tenn. 1900 census. Living Pontotoc, Oklahoma 1920.

WISE, ALEXANDER WALKER. Pvt., Co. D. b. Baton Rouge, Chester Dist. 9/13/39. Carpenter, age 21, Baton Rouge, Chester Dist. 1860 census. Enl. Co. E, 6th S. C. Inf. Chester CH 4/11/61. Present through 4/30/62. Company disbanded. Enl. Co. D, 1st S. C. Cav. Chester CH 8/24/62. Present through 8/31/63, however, brought Capt. Jones's body back to Rock Hill, 6/63. Ab. on horse detail 10/23-12/31/63, however, ab. sick with "Feb. Remit." in Charlottesville hospital 8/10-25/63. Ab. sick with "scabies" in Richmond hospital 2/18-4/6/64. Age 25, 5'10", fair complexion, dark hair, blue eyes, Mechanic. Detached Green Pond 4/16-30/64. Present 5-12/64. Paroled Greensboro, N. C. 4/26/65. Farmer, Halsellville Chester Co. 1870-1900 censuses. Farmer & Judge, Chester CH 1920 census. d. 6/11/29. Bur. Calvary Bapt. Ch. Cem., Leeds, Chester Co.

WISE, JAMES H. Pvt., Co. I. b. S. C. circa 1844. Ferryman, Myers Ford, Lexington Dist. before enlistment. Enl. Columbia 4/12/64. Detached Green Pond 4/22-30/64. Present 5-8/64. AWOL 10/21-31/64. Present 11-12/64. NFR. Farmhand, age 26, Lexington CH 1870 census. d. McCormick, S. C. 5/20/22 age 75.

WISE, WADE HAMPTON. Pvt., Co. E. b. S. C. 12/45. Student, age 15, St. Mathews, Orangeburg Dist. 1860 census. Enl. Adams Run 1/6/62. Ab. sick in hospital 3-4/62. Present 5-12/62. Detailed as wagon guard 3-4/63. Present 5-6/63. Ab. sick with debility in Richmond hospital 7/4-14/63. Detailed as Courier, Gen. Stuart's Hqtrs. 9-10/63. Detailed as Courier, Brigade Hqtrs. 12/17/63-2/64, however, dismounted 1/27/64. Ab. on horse detail 3/31-4/30/64. Transferred Captain Bachman's S. C. Battery 5/10/64. Present through 12/64. NFR. Farmer, Amelia, Orangeburg Co. 1880 & 1900 censuses. Possibly the W. H. Wise 12/29/44-9/17/-. Bur. West End Cem., St. Matthews Co.

WISE, WILLIAM. Pvt., Co. E. b. S. C. circa 1825. Farmer, age 35, Orangeburg Dist. 1860 census. Enl. Coosawhatchie 11/14/61. Present through 8/62. Present sick in camp 9-10/62. On scout 11-12/62. Ab. sick in Lynchburg hospital 4/1-10/19/63. Ab. sick in Staunton hospital 11-12/63. Ab. sick Columbia, S. C. hospital 2/5-29/64. Ab. on horse detail 4/13-5/13/64. Transferred Co. D, 20th S. C. Inf. 5/16/64. NFR. Farmer, age 45, Elmore, Ala. 1870 census. Farmer, age 55, Amelia, Orangeburg Co. 1880 census. d. 1900 age 75.

WISE, WILLIAM HATTAN. Pvt., Co. C. b. S. C. circa 1834. Age 26, Barnwell Dist. 1860 census. Enl. Camp Butler 9/10/61 age 23. Detached Wadmalaw Island 11-12/61. Present 5-6/62. Ab. on sick leave 7/31/62. Present 9/62-8/31/63. Ab. on horse detail 10/23-12/31/63. Present 1-2/64. Detached Green Pond 4/16-30/64. Present 5-8/64. Detached to guard POW's Florence 9/17-12/31/64. Transferred 1865 on post war roster. d. S. C. 7/14/10.

WISE, WILLIAM W. Pvt., Co. E. b. S. C. circa 1835. Farmer, age 35, Orangeburg Dist. 1860 census. Enl. Coosawatchie 11/14/61. Present through 8/62. Present sick 9-10/62. On scout 11-12/62. Ab. sick in Lynchburg hospital 4/1-10/19/63. Ab. sick in Staunton hospital 11-12/63. Ab. sick Columbia, S. C. hospital

2/5-29/64. Ab. on horse detail 4/13-5/13/64. Transferred Co. D, 20th S. C. Inf. 5/16/64. NFR. Farmer, age 45, Elmore, Ala. 1870 census. Farmer, age 55, Amelia, Orangeburg Co. 1880 census. d. 1900 age 75.

WITHERS, JOHN B. Pvt., Co. H. b. S. C. circa 1845. Res. York Dist. 1850 census. Enl. Rock Hill 1/9/62 age 17. Ab. on sick leave 3-4/62. Present 5-8/62. Detailed to drive horses from Summerville, S. C. to Staunton, Va. 9-10/62. Ab. on leave 11-12/62. Present 3-8/31/63. Ab. on horse detail 9-10/63. Present 11-12/63. Ab. detached 1/15-2/29/64. Detached Green Pond 4/26-30/64. Present 5-6/64. Ab. on leave 8/20-31/64. Present 9-12/64. Paroled Charlotte, N. C. 5/13/65. Age 25, no occupation, Ft. Mill, York Co. 1870 census. Lumber Manufacturer, age 35, Naylor, Lowndes Co., Ga. 1880 census.

WITHERS, W. R. Pvt., Co. A. b. S. C. circa 1844. On post war roster. Reenlisted Co. K, 4th S. C. Cav. Charleston 11/21/63. Present until arm mashed by a horse & in Richmond hospital 6/2-27/64. Present - until ab. on horse detail 8/11/64 for 40 days. Present 9-10/64. NFR. Cotton Mill Worker, age 56, Youngs, Laurens Co. 1900 census.

WITSELL, EMANUEL. Pvt., Co. A. b. S. C. 1818. M. D. On post war roster. Reenlisted Co. K, 4th S. C. Cav. Grahamville 3/25/62 age 40. Present until detailed with Gen.'s Pemberton & Walker 8/25/62. Transferred To. Co. I, 11th S. C. Inf. 11/18/62. Appointed Asst. Surgeon. Reassigned back to 4th S. C. Cav. as Asst. Surgeon 7/1/63. Present until dropped because of excess officers 4/9/64. NFR. d. d. 1/9/97. Bur. Live Oak Cem., Walterboro.

WITSELL, WALTER HAMILTON. Pvt., Co. A. b. S. C. 1825. Farmer, Colleton Dist. 1860 census. On post war roster. Reenlisted Co. K, 4th S. C. Cav. Grahamville 3/25/62. Present until ab. sick 10/1/62-2/64. Dropped 4/9/64, reduction of Company. NFR. Farmer, Walterboro 1870 census. d. 1872. Bur. Live Oak Cem., Walterboro.

WIX, WILLIAM RILEY. Pvt., Co. D. b. S. C. 11/4/30. Overseer, age 29, Chester Dist. 1860 census. Enl. Chester CH 7/28/62. Ab. on sick leave 8/16-31/62. Present 9/62-6/63. Captured near Martinsburg, Va. 7/19/63. Sent to Camp Chase. Age 32, 6' 4 1/2", dark complexion, gray eyes, black hair, Farmer. Transferred Ft. Delaware 2/29/64. Released 6/210/65. Farmer, Blackstock & Halsellville, Chester Co. 1870-1900 censuses. d. 2/5/06. Bur. Woodward Bapt. Ch. Cem., Chester, S. C.

WOOD, ROBERT A. Pvt., Co. I. b. S. C. 9/43. Enl. Parker's Ferry 4/3/62. Present 5-6/62. Ab. on sick leave 7-8/62. Present 9-10/62. On picket 11-12/62. Present 3-4/63. Detailed as a Teamster 5-6/63. Present through 8/31/63, however, ab. sick in Richmond hospital 8/12-17/63 & 10/12/63. Present 11/63-2/64. Detached Green Pond 4/16-30/64. Present 5-8/64. AWOL 10/31/64. Detached Pocotaligo 12/10-31/64. NFR. Cotton Mill Worker, age 56, Youngs, Laurens Co. 1900 census.

WOOD, THOMAS. Pvt., Co. H. b. Rock Hill, York Dist. 3/10/10. Enl. Rock Hill 1/9/62 age 46. Present 3-4/62. Ab. on sick leave 6/28-7/31/62. Present 8-10/62. Detailed to drive ambulance 11-12/62. Present sick in camp 3-4/63. Present 5-6/63, dismounted. Ab. on horse detail 8/18-31/63. Present 9/63-4/64. Transferred Co. A, Lucas' 15th Bn. S. C. Arty. 5/16/64. Present through 12/64. NFR. Brick Mason age 58, York Co. 1870 census. d. Rock Hill 12/12/88. Bur. Laurelwood Cem., York Co. Wife receiving pension Rock Hill 1903.

WOODHURST, ANDREW JACKSON. Pvt., Co. G. b. S. C. 11/32. Res. Abbeville Dist. 1860 census. Enl. Abbeville CH 1/8/62. On picket Simmons House 3-6/62. Ab. detailed in Surgeon's Dept. 7-8/62. Present 9/62-6/63. Ab. on horse detail 8/31/63, detailed to drive wagon near Rockville, Md. & lost horse 7/63. Paid $500.00. Present 9/63-12/64. NFR. Farmer, age 65, Abbeville Co. 1910 census. d.4/11/16. Bur. Lebanon Presb. Ch. Cem., Abbeville, S. C. Brother of George W. Woodhurst.

WOODHURST, GEORGE WASHINGTON. Pvt., Co. G. b. England 1836. Came to U. S. 1850. Brick Layer, Greenwood, Abbeville Dist. 1860 census. Enl. Abbeville CH 3/15/62. Present through 4/62. Present sick in camp 5-6/62. d. of typhoid fever in College hospital, Columbia, S. C. 7/18/62. Brother of Andrew J. Woodhurst.

WOODRUFF, JOHN C. Pvt., Co. B. b. Spartanburg Dist. 3/35. Farmer, age 24, Poolsville, Spartanburg Dist. 1860 census. Enl. Clinton 8/22/61 age 25. Ab. sick 10/31/61. Present 11/61-6/62. On scout Edisto Island 7-8/62. Present 9-10/62. On picket Ellis Ford 11-12/62. Present 3-4/63. WIA & horse killed Upperville 6/21/63. Ab. wounded (shell wound left side) in Richmond hospital 6/22-10/28/63. Furloughed for 30 days. Paid $175.50 for horse. AWOL 12/20-31/63. Ab. sick in Columbia, S. C. hospital 1/20-2/29/64. Ab. on horse detail 3/31-4/30/64. Present 5-10/64. Detached Pocotaligo 12/2-31/64. NFR. Farmer, Woodruff, Spartanburg Co. 1880-1910 censuses. d. by 1919. Probably buried in Antioch Cem., Woodruff, but no stone. Wife receiving pension Spartanburg Co. 4/10/19. Brother of Phillip P., Samuel P.& William C. Woodruff.

WOODRUFF, PHILLIP P. Pvt., Co. B. b. near Woodruff, Spartanburg Dist. 7/6/40 or 6/16/42. Farmer, age 19, Poolsville, Spartanburg co. 1860 census. Enl. Camp Ripley 1/20/62. Present 3-6/62. Present sick in camp 7-8/62. Present 9-10/62. On picket Ellis Ford 11-12/62. Present 3-6/63. Present Gettysburg. Ab. on horse detail 8/18-31/63. Present 9-10/63. Detailed with Signal Corps 12/18/63-2/64. Ab. on horse detail 4/26-30/64. Present 5-10/64. Detailed as Courier, Secessionville 11-12/64. Discharged Florence 3/65 on pension application. NFR. Farmer, Woodruff, Spartanburg Co. 1870-1900 censuses. Pumper, Railroad Tank, Woodruff, 1910 census. Receiving pension Woodruff 3/25/19. d. Woodruff 10/5/26. Bur. Antioch Cem., Woodruff, S. C., but no stone. Brother of John C., Samuel P. & William C. Woodruff.

WOODRUFF, SAMUEL PINCKNEY. Pvt., Co. B. b. S. C. circa 1835. Res. Spartanburg Dist. 1850 census. Enl. Clinton 8/22/61 age 27. Ab.sick 10/31/61. Present 11/61-6/62. Present sick in camp 7-8/62. Present 9/62-4/63. Detailed as Teamster for Brigade 5-6/63. Ab. on horse detail 8/31-12/31/63. Detached horse recruiting camp 2/23-29/64. Ab. on leave 4/26-30/64. Present 5-8/64. Present sick in camp 9-10/64. On picket West lines, James Island 11-12/64. NFR. Farmer, age 67, Riverton, Campbell Co., Ga. 1900 census and age 78, 1910 census. Brother of John C., Phillip P. & William C. Woodruff.

WOODRUFF, WILLIAM C. Pvt., Co. B. b. S. C. 5/9/45. Age 13, Poolsville, Spartanburg Dist. 1860 census. Enl. Columbia 4/14/64. Present 5-8/64. Present sick in camp 9-10/64. Present 11-12/64. NFR. Farmer, Woodruff, Spartanburg Co. 1870-1910 censuses. d. 6/13/16. Bur. Antioch Cem., Woodruff, no marker.

WOODS, JOHNSON C. Pvt., Co. B. b. S. C. circa 1816. Res. Chester Dist. 1860 census. Enl. Clinton 8/22/61 age 45. Ab. sick 10/31/61. Present 11/61-6/62. Present sick in camp 7-8/62. Present 9-10/62. Ab. sick with "Endocarditis" in Richmond hospital 11/12-12/31/62. Ab. sick with "nebula in eye"in Palmyra, Va. hospital 3/4/63. Transferred Farmville hospital 3/9/63-5/3/63. Furloughed for 30 days. Ab. sick with "Endocarditis" in Richmond hospital 7/23/63. Returned to duty 9/25/63. Present through 12/63. Ab. sick with "Carditis"in Richmond hospital 1/5-2/27/64. Furloughed for 40 days. Present 3-4/64. Transferred to Invalid Corps Columbia, S. C. 6/17/64. NFR. Farmer, age 55, St. Peters, Beaufort Co. 1870 census. Stock Raiser, age 65, Robert, Hampton Co. 1880 census. Alive 1898.

WOODSIDE, JOHN LAWRENCE. Pvt., Co. C. b. Simpsonville, S. C. circa 1844. Farmer, Greenville Dist. 1860 census. Enl. Columbia 7/6/64. Ab. sick in Columbia hospital 8/30/64. On picket 9-10/64. Present 11-12/64. NFR. Post Master Woodville, Greenville Co. 1876.Farmer, Austin, Greenville Co., 1880 census. Alive 1898.

WOODWARD, HANSFORD D. Pvt., Co. C. b. S. C. circa 1834. Res. Barnwell Dist. Enl. Camp Butler 9/10/61 age 27. Present 11/61-6/62. Ab. on sick leave 7/26-31/62. Ab. on sick leave 9-10/62. d. of disease Staunton, Va. 12/3/62. Probably buried in unmarked grave in Thornrose Cem., Staunton.

WOODWARD, JUDSON D., SR. 2ndCpl., Co. C. b. S. C. 1840. Tanner, age 20, Woodward, Barnwell Dist. 1860 census. Enl. Camp Butler 9/10/61 age 20 as Pvt. Detached Wadmalaw Island 11-12/61. Present through 6/62. Ab. on sick leave 8/17-31/62. Promoted 2ndCpl. Present 9-12/62. Ab. sick with chills & fever in Staunton hospital 3/12/63. Ab. sick at home 4/63. Present through 8/31/63. Ab. on horse detail 10/23-12/31/63. Ab. on detached service 1/20-2/29/64. Present 3-12/64. NFR. Farmer, Millbrook, Aiken Co. 1880-1910 censuses. d. Columbia, S. C. 6/18/19, age 79. Bur. Montmorenci Bapt. Ch., Aiken

Co. Brother of Samuel A. Woodward.

WOODWARD, SAMUEL A. Pvt., Co. C. b. Aiken, S. C. 8/8/42. Enl. Camp Johnson 10/18/61 age 19. Present through 6/62. Ab. on sick leave 8/14-31/62. Present 9-12/62. Ab. on sick leave 3/63. Ab. on horse detail 4/1-6/12/63. Present through 8/31/63. Ab. sick with "Remit. Fever" in Richmond hospital 9/23/63. Furloughed for 30 days 9/26/63. Present 11/63-12/64. NFR. Farmer, Millbrook, Aiken Co. 1880-1910 censuses. d. there 12/12/24. Bur. Millbrook Bapt. Ch. Cem., Aiken Co. Brother of Judson D. Woodward.

WORKMAN, JOHN J. Pvt., Co. H. b. S. C. circa 1823. Farmer, age 35, York Dist. 1860 census. Enl. Rock Hill 1/15/62 age 38. Present through 4/62. Ab. on leave 5-6/62. Present 9-12/62. Ab. on horse detail 4/8-30/63. Ab. horse recruiting camp 5-8/63. Present 9-12/63. Ab. on leave 2/12-29/64. Transferred Co. E, 17th S. C. 3/12/64. WIA (middle toe left foot amputated) 6/3/64. Ab. wounded in Williamsburg hospital until furloughed 6/6/64. Ab. wounded in Columbia hospital 7/22-8/31/64. Present 9/64-2/65. NFR. Farmer, age 57, Catawba, York Co. 1880 census.

WORTHY, EDWARD JAMES. Pvt., Co. D. b. Chester CH, S. C. 3/24/38. Age 22, Chester CH 1860 census. Enl. Chester CH 9/10/61 age 21. Present through 6/62. Ab. on sick leave 8/8-31/62. Ab. sick in Richmond hospital 10/12-15/62. Present sick in camp 10/31/62. Ab. sick with debility in Charlottesville hospital 11/26/62-1/1/63. Present 3-8/63. Ab. on horse detail 10/23-12/31/63. Present 1-2/64. Detached Green Pond 4/16-30/64. Present 5-12/64. Paroled Hillsboro, N. C. 4/65. Farmer, Baton Rouge, Chester Co. 1870 & 1880 censuses. Farmer, Tate Co., Miss. 1900 census. d. Roscoe, Nolan Co., Texas 1/19/07.

WORTHY, HENRY. Pvt., Co. D. b. S. C. 1843. Age 7, Chester Dist. 1850 census. Enl. Chester CH 9/10/61 age 19. Present 11-12/61. d. of disease at home in Chester Dist. 4/5/62.

WRAGG, ANDREW MC DOWALL. Pvt., Co. A. b. S. C. 3/21/38. Merchant, Charleston 1860 census. On post war roster. Reenlisted Co. K, 4th S. C. Cav. Grahamville 3/25/62. Present until detailed with Gen. Walker 10/13/62-10/63. Present 11/63-2/64. Present until WIA (right thigh) Hawes Shop 5/28/64. Ab. wounded in Richmond hospital until furloughed 6/4/64 for 60 days. In Charleston hospital 10/31/64. Paroled Greensboro, N. C. 5/2/65. Accountant, Columbia, S. C. 1870 census. Retired, Charleston 1910 census.

WRIGHT, BEDFORD B. Pvt., Co. D. b. S. C. 8/35. Res. Chester Dist. 1850 census. Enl. Chester CH 9/10/61 age 26. Entry cancelled. Enl. Co. H, 7th S. C. Bn. Inf. 6/30/62. Present through 2/63. Ab. on leave 5/28-6/15/63. Present 7-8/63. Ab. sick in Charleston hospital 9/23/63. Furloughed for 20 days. Present sick in camp 11-12/63. Present 1-2/64. WIA (left forearm) 5/16/64. Furloughed for 60 days 5/22/64. Ab. wounded through 8/64. Ab. on leave through 11/64. NFR. Farmer, age 63, Blackstock, Chester Co. 1900 census.

WYLIE, WILLIAM. Pvt., Co. D. b. S. C. 1828. Student, age 23, Chester Dist. 1850 census. Res. Chester Dist. 1860. Enl. Chester CH 9/10/61 age 33. Present 11/61-4/62. Present, driving wagon for corn 5-6/62. Present 7-10/62. Driving wagon for QM Dept. 11-12/62. Present sick in camp 3-4/63. Present 5-6/63. Ab. on horse detail 8/3-31/63. Detailed in rear with horses 9-12/63. Present 1-2/64. Detached Green Pond 4/16-30/64. Present 5-8/64. Detached to Chester CH for boxes 10/24-31/64. Present as Blacksmith for Regiment 11/14-12/31/64. Paroled Hillsboro, N. C. 4/65. NFR. Farmer, age 41, Catawba, York Co. 1870 census & age 52, 1880 census. d. Rock Hill, S. C. 12/3/20.

YANCEY, B. N. 1stLt., Co. B. b. 3/20/36. On roster but not on muster rolls. d. 5/13/03. Bur. Vise Cem., Scott Co., Ark.

YARBOROUGH, EDWARD TENNANT. Sergeant Major. b. S. C. 5/26/42. Student, age 16, Abbeville Dist. 1860 census. Enl. Co. A Abbeville CH 3/19/62 as Pvt. Present through 6/62. Promoted Sergeant Major. Present 7/62-6/21/63. Signed for forage for 7 horses Culpepper Co. 2/28/63 & for 7 horses, Rockingham Co. 3/31/63. Signed for forage for 6 horses in the field 6/30/63. Ab. sick with "Int. Fever" in Richmond hospital 7/27/63. Furloughed for 30 days 8/19/63. Ab. sick in S. C. 8/31/63. Present 9/63-12/64. NFR. Clerk in Warehouse, age 38, Augusta, Richmond Co., Ga. 1880 census. Farmer, age

58, Uniontown, Perry Co., Ala. 1900 census. d. 11/6/02. Bur. Uniontown Cem., Perry Co., Ala.

YARBOROUGH, HENRY PLEASANT "HOOD." Pvt., Co. G. b. Union Co. 12/24/47. Enl. age 22 on postwar roster. Farmer, Cross Keys, Union Co. 1880 census. d. Woodruff, Spartanburg Co. 2/27/13. Bur. Hobbyville Cem.

YATES, ANDREW. Pvt., Co. K. b. Charleston, S. C. 1846. Student, age 14, Ward 1, Charleston, S. C. 1860 census. Enl. Columbia 4/13/64. Courier, Colonel Black 5-12/64. Paroled Raleigh, N. C. 4/65. M. D., age 23, John's Island, Charleston 1870 census. M.D., age 32, Hodges, Abbeville Co. 1880 census. d. Charleston, S. C. 6/15/84.

YATES, ELLIOTT V. Pvt., Co. F. b. Sumter Dist. 1835. Student, age 25, Pickens Dist. 1860 census. Enl. Pickens CH 12/4/61 age 24. Present through 4/62. On picket Camp Means 5-6/62. Ab. on leave 7-8/62. Ab. sick 9-10/62. Present 11/62-6/63. Ab. sick in Richmond hospital 8/1-1/31/64. Returned to duty 2/1/64. Present 3-12/64. NFR. Farmer, age 36, Salubrity, Pickens Co. 1870 census.

YATES, FRANKLIN. Surgeon. Not on muster rolls. NFR.

YATES, JOSEPH ATKINSON. Surgeon. b. Charleston, S. C. 9/4/29. Gd. S. C. Medical College 1860. Appointed Asst. Surgeon, Columbia 11/16/61. Assigned to 24th S. C. Inf. Transferred 1st S. C. Cav. 9/25/62. Present 9-12/62 & 4-6/63. Left with sick & wounded and captured Gettysburg 7/4/63. Sent to Ft. McHenry 7/31/63. Exchanged 8/10/63. Present Hamilton's Crossing 11-12/63. Ab. on leave for 30 days 2/1/64. Present 3-6/64. Promoted Surgeon 7/11/64 to rank from 5/18/64. Present 9-12/64. NFR. M.D., Moultireville, Charleston, S. C. 1880 census. d. Bessemer, Ala. 8/3/88. Bur. Charleston, S. C.

YEADON, WILLIAM J. Pvt., Co. I. b. S. C. 4/24/46. Enl. Co. G, Palmetto Bn. Arty. Summerville, S. C. 1/2/63. Present through 10/63. AWOL 11-12/63. Ab. sick leave 1-2/64. Present 3-4/64. Transferred Co. I, 1st S. C. Cav. 6/16/64. Present 7-8/64. Detached as Nurse, McLeod's hospital, James Island 10/17-12/31/64. Paroled Greensboro, N. C. 5/1/65. Store Clerk, Sumter, S. C. 1870 census. Grocer, Sumter, S. C. 1900 census. Receiving pension Sumter, S. C. 4/15/19. Retired Salesman, Sumter, S. C. 1920 census. d. 2/26/200. Bur. City Cem., no marker.

YOUNG, BENJAMIN NATHANIEL. 1stLt., Co. B. b. S. C. 3/20/36. Farmer, age 24, Spartanburg Dist. 1860 census. Enl. Clinton 8/22/61 age 25 as 3rdCpl. Present through 12/61. Ab. on picket Hallover Bridge 3-4/62. Present 5-2/62. Promoted 2ndLt. 6/18/62. Present 7-10/62. Ab. sick in hospital 11-12/62. Present 3-6/63. Ab. sick with chronic diarrhoea in Richmond hospital 8/31-10/31/63. Present 11/63-2/64. Detached Green Pond 4/26-30/64. Present 5-10/64. Promoted 1stLt. 9/15/64. Present 11-12/64. NFR. Farmer, Boone, Scott Co., Ark. 1870 census. Married Logan Co., Ark. 1878. Farmer, Scott, Brawley Co., Ark. 1900 census. d. Scott, Ark. 5/13/03.

YOUNG, JOHN H. Pvt., Co. A. b. S. C. 10/40. Res. Abbeville Dist. 1850 census. Enl. Abbeville CH 8/13/61 age 21. On duty Bear Island 11-12/61. Present 3/62-8/31/63. Ab. sick in hospital 9-10/63. Ab. as safe guard near camp 11-12/63. Ab. on leave 2/16-29/64. Ab.on sick leave 4/25-30/64. Present sick in camp 5-6/64. Present 7-12/64. NFR. Married 1877. Farmer, Lowndesville, Abbeville Co. 1880 census. Farmer, Corner, Anderson Co. 1900 census.

YOUNG, THOMAS L. Pvt., Co. B. b. S. C. 1/26/46. Age 14, Spartanburg Dist. 1860 census. Enl. Spartanburg 1/26/64. Ab. on leave 4/26-30/64. Present 5-12/64. NFR. Farmer, Brawley, Scott Co., Ark. 1880-1910 censuses. d. in Texas 5/19/25.

YOUNGBLOOD, JAMES T. Pvt., Co. C. b. Houston, Ga. 1848. Age 11, Blount Co., Ala. 1860 census. Enl. Columbia 11/20/64. Detached Pocotaligo 12/10/64. NFR. Farmhand, age 22, Drayton, Dooly Co., Ga. 1870 census. d. in Ga. 1900.

YOUNGBLOOD, JOSEPH C. Pvt., Co. C. b. S. C. circa 1837. Enl. Aiken Dist. age 25 on post war roster. Dry Goods Merchant, age 35, Williston, Barnwell Co. 1870 census. Alive 1898.

ZANER, SAMUEL CHRISTOPHER, JR. 2ndCpl., Co. G. b. S. C. 1842. Enl. Abbeville CH 1/8/62 as Pvt. On picket Jacksonborough 3-4/62. Ab. sick in hospital 6/62. Present 7-12/62. Promoted 4thCpl. 11/27/62. Present 3-10/63. Promoted 3rdCpl. Present 11/63-6/64. Promoted 2ndCpl. Ab. sick in Charleston hospital 7/15-8/31/64. Present 9-12/64. NFR. Married Cleburne Co., Ala. 1878. d. 1902. Bur. New Harmony Bapt. Ch. Cem., Cleburne Co., Ala.

ZEIGLER, JOHN JOSEPH. Pvt., Co. E. b. S. C. circa 1831. Enl. Ft. Motte 10/26/61 age 30. Present through 6/62. Ab. on leave 8/24-31/62. Driving wagon 9-10/62. Ab. detached 11-12/62. Ab. sick with debility in Charlottesville hospital 1/24-25/63. Ab. sick with chronic rheumatism in Richmond hospital 4/26/63. Transferred to Lynchburg hospital 5/11/63. Transferred Danville hospital 6/27/63. Returned to duty 7/12/63. Ab. sick with pneumonia in Richmond hospital 10/12/63. Furloughed for 30 days 11/4/63. Present 1-2/64. Ab. on leave 4/21-30/64. Present 5-10/64. Detailed to work at Stono Bridge 12/3-31/64. WIA (left arm) and in Petersburg hospital 5/27/64. Discharged 7/14/64. Paroled Hillsboro, N. C. 4/65. Farmer, age 51, Amelia, Orangeburg Co. 1880 census.

ZEIGLER, J. M. Pvt., Co. I. Not on muster rolls. d. and buried Two Mile Swamp Bapt. Ch. Cem., Bamberg Co., S. C., no dates. Tombstone only record of service.

BIBLIOGRAPHY

MANUSCRIPTS:

Applications for Home for Confederate Veterans, Columbia, S. C. 1909-1958.

Bamberg County S. C. Confederate Soldiers and Pension Applications.

Bailey A. Barton Muster Roll Book of Pickens Dist., S. C. circa 1858. Oconee County Historical Society, 1990.

Burial Ground of the Confederate Dead, Magnolia Cemetery, Charleston, S. C., n.d.

Confederate Cemetery Burials, University of Virginia, Charlottesville, Va.

Confederate Rolls of Honor, Thomas Cooper Library, U. of South Carolina.

Confederate Roll of Honor, Museum of the Confederacy, Richmond, Va.

Confederate Soldiers Buried in Colorado.

Confederate Veterans Camp of New York, 1903.

Confederate Veterans Magazine, 1897-1932.

Duke U. Library. Turner W. Holley letter 13 June 1863.

Florida Civil War Pension Records, Florida Dept. of Archives and History.

Furman University General Catalog, 1852-1899. Sumter, S. C., 1899.

Georgia State Archives. "Recollections and Reminiscences 1861-1865 through World War I." South Carolina Division, United Daughters of the Confederacy, Vol. 6, 1995.

Georgia Division, UDC, Confederate Reminiscences and Letters. 1861-1865., Vol. VII, Atlanta, 1998.

Horry County. CSA Soldiers Who Died in the War Between the States.

Laurens County Enrollment Book of Confederate Veterans. Laurens District Chapter of South Carolina Genealogy, 1998.

Memory Rolls, South Carolina Dept. of Archives and History, Columbia, S. C.

Muster Rolls Company C, 1st South Carolina Cavalry, 1 November - 31 December, 1863. The South Carolinian Library, Columbia, S. C.

National Park Commission. 1st South Carolina Cavalry.

Oakwood Cemetery, Richmond, Virginia Register of Internments in Confederate Plots, n. d.

Pay Rolls Company C, 1st South Carolina Cavalry, 1 November -31 December, 1863. The South Carolinian Library, Columbia, S. C.

Roll of Dead. South Carolina Troops Confederate State Service. South Carolina State Archives and History, Columbia, S. C. ,1994.

Roster of Confederate Dead, Charleston, S. C.

South Carolina In the Civil War. 1st South Carolina Cavalry.

South Carolina Pension Records. South Carolina Dept. of Archives and History, Columbia, S. C.

South Carolina Physicians in the Civil War.

South Carolina Civil War Pension Applications, 18888-1906. Genealogical Society of Utah.

Spring Farm Letters, Southern Historical Collection, Wilson Special Collections Library, University of North Carolina. ((War Letters of A. M. Kee).

Theodore C. Barker Papers, The South Carolinian Library, Columbia, S. C.

James Wilson White Correspondence, Southern Historical Collection, Wilson Special Collections Library, University of North Carolina.

York County Confederate Survivors Association 1880.

York County in the Civil War.

PRINTED SOURCES:

Baker, Gary R. Cadets in Gray. Palmetto Books, Columbia, S. C. 1989.

Bell, Louise M. Rebels in Grey, Soldiers from Pickens District, S. C. 1861-1865. Joyce's Print Shop, Clemson, S. C., 1984.

Blackford, William W. War Years With Jeb Stuart. Charles Scribners's Sons, New York, 1946.

Brooks, U. R.. Butler and His Cavalry in the War of Secession, 1861-1865. The State Co., Columbia, S. C. 1909.

_____., Stories of the Confederacy. The State Co., Columbia, S. C. 1912.

Busha, Charles H. The Enduring Legion of An Anderson County, South Carolina Confederate Veteran: Unreconstructed Rebel and Folk Hero.

Coffin, Richard M. A History of the Phillips Georgia Legion Cavalry. Mercer U. Press., Macon, Ga. 2011.

Confederate Military History, Volume II, South Carolina. Extended Version. Broadfoot Pub. Co., Wilmington, N. C. 1987.

Cooke, John Esten. Wearing of the Gray. Indiana U. Press, Bloomington, Ind., 1959.

Easterby, J. H. A History of the College of Charleston. Scribner Press, 1935.

Emerson, W. Eric. Sons of Privilege: The Charleston Light Dragoons In the Civil War. University of South Carolina Press, 2005.

Ferguson, Chris. L. Confederate Dead at Hollywood Cemetery. Angle Valley Press, Winchester, Va., 2008.

Green, Edwin L. A History of the University of South Carolina. The State Co., Columbia. S. C., 1916.

Guevarra, Mark B. & Chris S. Price. The Chester County (S. C.) Confederate Compendium. , n. p., 2001.

Hart, Joseph E., Jr. Supplement to Confederate Veterans Enrollment Book of York County, S. C. 1902.

Hartley, Chris J. Stuart's Tar Heels. Butternut & Blue, Baltimore, 1996.

Heller, J. Roderick III and Carelyen A. Helle, editors. The Confederacy Is On Her Way Up the Spout, Letters to South Carolina 1861-1864. U. of Georgia Press, Athens, Ga., 1992.

Hemphill, James C. Men of Mark in South Carolina, Volume 4. Men of Mark Pub. Co., Washington, D. C., 1909.

Henderson, William D. The Road to Bristoe Station. H. E. Howard, Inc., Lynchburg, Va. 1987.

Hodge, Robert A., compiler. A Death Roster of the Confederate General Hospital, at Culpeper, Virginia. Robert A. Hodge, Fredericksburg, Va. ,1977.

_____. Tombstone Locations in the Fredericksburg Confederate Cemetery. Robert A. Hodge, Fredericksburg, Va. 1988.

Ingram, Virginia M. Confederate Veterans of Abbeville County (S. C.) from Enrollment Books Taken About 1902. Abbeville Books, Abbeville, S. C. 1998.

Jamison, Jocelyn P., compiler. They Died at Fort Delaware 1861-1865. Delaware City, Del. Fort Delaware Society, 1997.

Jones, J. Keith, editor. The Boys of Diamond Hill: The Lives and Civil War Letters of the Boyd Family of Abbeville County, South Carolina. McFarland & Co.; Jefferson, N. C., 2011.

Kirkland, Randolph W., Jr. Broken Fortunes. The South Carolina Historical Society, Charleston, S. C., 1995.

_____. Dark Hours: South Carolina Soldiers, Sailors and Citizens Who Were Held in Federal Prisons During the War For Southern Independence. The South Carolina Historical Society, 2002.

Kohn, Augusta. Confederate Rolls of South Carolina. n.p., 1898.

Krick, Robert Edward Lee. Staff Officers in Gray: A Biographical Register of the Staff Officers in the Army of Northern Virginia. U. of North Carolina Press, Chapel Hill, N. C. 2003.

Krick, Robert K. The Gettysburg Death Roster. Bookshop of Morningside Press, Dayton, Ohio, 1981.

_____. Lee's Colonels. A Biographical Register of the Field Officers of the Army of Northern Virginia. Bookshop of Morningside Press., Dayton, Ohio, 1991.

Kurtz, Lucy F. and Ritter, Ben. A Roster of Confederate Soldiers Buried in Stonewall Cemetery, Winchester, Virginia. Farmer & Merchants National Bank, Winchester, Va., 1984.

Markham, Jerald H, The Diuguid Records, 1861-1865 and Biographical Records. Heritage Books, Westminster, Md. 2007.

McClellan, H, B. The Campaigns of Stuart's Cavalry. Blue & Grey Press, Secaucus, N. J. , 1993.

Mesic, Haret B. Cobb;s Legion Cavalry. McFarland & Co., Jefferson, N. C., 2009.

Mitchell, Adele H. The Letters of General J. E. B. Stuart. Stuart-Mosby Historical Society, 1990.

Moore, Thomas Craig, editor. Upcountry South Carolina Goes to War: Letters of the Anderson, Brockman & Moore Families, 1853-1865. U. of. South Carolina Press, 2009.

Longacre, Edward G. The Cavalry At Gettysburg. Associated University Presses, Inc., Cranbury, N. Y., 1986.

McSwain, Eleanor D., Editor. Crumbling Defenses or Reminiscences of John Logan Black, Colonel, C.S.A. Macon, Ga. 1960.

Mosby, John S. Stuart's Cavalry in the Gettysburg Campaign. Moffat, Yard & Co., New York, 1908.

O'Neill, Robert F., Jr. The Cavalry Battles of Aldie, Middleburg and Upperville, June 10-27, 1863. H. E. Howard, Inc. Lynchburg, Va. 1993.

Purifoy, John. Farnsworth's Charge and Death At Gettysburg. Confederate Veterans Magazine, Volume XXXII, pp. 307-309.

Racine, Philip N, Editor. Gentlemen Merchants: A Charleston Family Odyssey 1828-1870. U. Tenn. Press, 2008.

Salley, Alexander S., compiler. South Carolina Troops In Confederate Service. The State Co., Columbia, S. C., 1914.

Smith, C. Foster. Jeremiah Smith and the Confederate War. The Reprint Co., Publishers. Spartanburg, S. C. 1993.

Southern Historical Society Papers, Volume 27. Sick and Wounded Confederate Soldiers in Hagerstown and Williamsport Hospitals. Richmond, Va.

Southern Historical Society Papers, Vol. 32. Page 307. Farmsworth's Charge and Death at Gettysburg, by David Caldwell, Richmond, Va.

Southern Historical Society Papers, Volume 16, pp. 228-230. "Charge of Black's Cavalry at Gettysburg" by P. J. Malone, Richmond, Va.

Sturkey, O. Lee. Hampton Legion Infantry, C.S. A. Broadfoot Pub. Co., Wilmington, N. C., 2008.

Supplement to the Official Records of the Union and Confederate Armies, Part II, Record of Events, Volume 64, South Carolina Units. Broadfoot Pub. Co., Wilmington, N. C. 1994-1999.

Thomas, John P. Historical Sketch of the South Carolina Military Academy. Walker, Evans & Cogswell, Publishers, Charleston, S. C., 1879.

United Daughters of the Confederacy. Partial List of Confederate Soldiers Buried in Churchyards in Columbia, S. C. and Vicinity. n.p.

U. S. War Department. War of the Rebellion: A Compilation of the Official Records of the Union and Confederate Armies. 128 volumes. Washington, D. C. 1880-1901.

Towles, Louis P., Editor. The World Turned Upside Down: The Palmers of South Santee. U. of. South Carolina Press, 1996.

Trout, Robert J. Memoirs of the Stuart Horse Artillery, 2 Volumes. U. of Tennessee Press, Knoxville, Tenn., 2008.

Watkins, Raymond W. Confederate Burials, Thornrose Cemetery, Staunton, Va. LCDA & H. Inc., Meridian, Miss., 1996.

_____. Confederate Burials, Volume III. Blandford Church Cemetery, Petersburg, Va. LCDA & H., INC. Meridian, Miss., 1992.

Wells, Edward L. Hampton and His Cavalry in 1864. B. F. Johnson Publishing Co., Richmond, Va., 1899.

Wert, Jeffrey D. Gettysburg: Day Three. Simon & Schuster, New York, 2001.

Wittenberg, Eric J. The Battle of Brandy Station. The History Press, Charleston, S. C., 2010.

NEWSPAPERS

Abbeville (S. C.) Press.
Camden (S. C.) Confederate.
Edgefield (S. C.) Advertiser.
Yorkville (S. C.) Enquirer.

ONLINE PUBLICATIONS

Ancestry.Com.
  Freepages.
    Company H, 1st South Carolina Cavalry Roster.

C. S. A. of America. Army, South Carolina Cavalry, 1st, Company K. Muster Rolls and Pay Rolls 1863-1864.

Find A Grave.
All South Carolina Cemeteries.
Loudon Park Cemetery, Baltimore, Md.

Fold 3

1st South Carolina Cavalry Records and Muster Rolls.

Grave sites of Confederate Soldiers of South Carolina

Pickens District, South Carolina War Dead

mwycoff.tripod.com.
South Carolina In the Civil War.
South Carolina Cavalry Units.
1st South Carolina Cavalry

COLONEL JOHN LOGAN BLACK

COLONEL JOHN LOGAN BLACK

ROBERT ALEXANDER STEELE
(No. 373)

CPL. ROBERT ALEXANDER STEELE, COMPANY H

PVT. MILES A. EADES, COMPANY F

LT. ISHAM IRVINE FOX, COMPANY I

PVT. JOHN WILLIAM POWER, COMPANY G

WALLACE MILLER (IN 1909)

PVT. WILLIAM WALLACE MILLER, COMPANY C

WALLACE MILLER (OF THE SIXTIES)

PVT. WILLIAM WALLACE MILLER, COMPANY C

JOHN H. PIERCE, SCOUT

PVT. JOHN HENRY PIERCE, COMPANIES C. & K

LT. LEROY WORTH LUSK, COMPANY F

LT. JAMES WILSON WHITE, COMPANY H

PVT. WILLIAM RILEY WIX, COMPANY D

LT. COLONEL JOHN DAVID TWIGGS

PVT. ROBERT JAMES MC BRYDE, COMPANY G

PVT. ALEXANDER BARNETT BIGGER, COMPANY H

PVT. DAVID BENJAMIN ROTHROCK, COMPANY K

ABOUT THE AUTHOR

Robert Jett Driver, Jr. was born in Staunton, Va. He grew up in Greensboro, N. C., Charlottesville, Va. and Elkin, N. C. Driver served as an enlisted Marine during the Korean War. He graduated from Guilford College and was commissioned in the Marine Corps in 1959. Following service with the 4th Marines in Hawaii and a tour at Camp Pendleton, California, he was promoted to Captain. Driver commanded Company A, 1st Battalion, 9th Marines in Vietnam in 1965 and Company E, 2nd Battalion, 9th Marines there in 1966. He was awarded a Bronze Star and Naval Commendation Medals for valor in combat. After a tour at Headquarters, Marine Corps in Washington, D. C., he was promoted to Major. Driver returned to Vietnam as the Operations Officer of 2nd Battalion, 5th Marines and later, 2nd Battalion, 7th Marines, being awarded a Legion of Merit for his bravery and leadership in combat. Following a tour of duty at Quantico, Va. he graduated from the Marine Corps Command & Staff College. Driver commanded the Marine Corps Recruiting Station in New Orleans, covering the state of Louisiana, for four years. During this period, he was promoted to Lieutenant Colonel. He retired in 1979, and received another Legion of Merit for his successful tour on recruiting duty. He and his wife Edna reside in the historic village of Brownsburg, Va.

"Driver is the author of 12 books on Virginia and Maryland Confederate soldiers and "Lexington and Rockbridge County in the Civil War." He also wrote a book on his Marine Corps career, "More Than A Few Good Men," including his two tours in Vietnam."

www.ingramcontent.com/pod-product-compliance
Lightning Source LLC
Chambersburg PA
CBHW081840230426
43669CB00018B/2768